PHANTOMS OF THE SKIES

THE LOST HISTORY OF AVIATION FROM ANTIQUITY TO THE WRIGHT BROTHERS

PHANTOMS OF THE SKIES

THE LOST HISTORY OF AVIATION FROM ANTIQUITY TO THE WRIGHT BROTHERS

BY

J. ALLAN DANELEK

AND

CHUCK DAVIS

First Edition,
First Printing, 2011

Book design and layout by J. Allan Danelek
Book editing by Michele Brown
Cover art and design © by J. Allan Danelek
(Blueprints of the Wright Flyer used on the cover by Herb Kelley, courtesy of Chuck Davis)
All photos, unless otherwise designated, are public domain.
AUP is a registered trademark of Adventures Unlimited Press, Ltd.

Danelek, J. Allan, 1958-
Phantoms of the Skies: The Lost History of Aviation from Antiquity to the Wright Brothers / J. Allan Danelek. — 1st ed.
ISBN 978-1-935487-38-8

Adventures Unlimited Press
303 Main Street
Kempton, Illinois 60946-0074 U.S.A.
www.adventuresunlimitedpress.com

OTHER BOOKS BY J. ALLAN DANELEK

Mystery of Reincarnation: The Evidence & Analysis of Rebirth

The Case for Ghosts: An Objective Look at the Paranormal

Atlantis: Lessons from the Lost Continent

UFOs: The Great Debate: An Objective Look at Extraterrestrials, Government Cover-Ups, and the Prospect of First Contact

2012: Extinction or Utopia? Doomsday Prophecies Explored

The Great Airship of 1897

The Case for Reincarnation

TABLE OF CONTENTS

ACKNOWLEDGMENTS

Largely because time has a way of relentlessly erasing witnesses and key evidence from history, when one sets about producing a history of something that happened more than a century ago, they are forced to depend upon the reliability of others. While it would be nice to have access to original records and possess the resources to travel around the globe in an effort to uncover never-before known facts about each person in this work, such is not possible, forcing us, like most historians, to rely upon other sources for much of our information. Fortunately, in this day and age we have access to reams of information that would have required weeks or even months of tedious work to have uncovered in years past, making acquiring the prerequisite data, if not simple, at least easier than it has been to previous generations of authors. Of course, when one does this, they naturally run the risk that not every source they appeal to may be reliable. In the case of the Internet, this is a special danger, and one we are well aware of. However, if one learns to cross-check carefully and compare information from multiple sources—which we have endeavored to do whenever possible—it is easier to ensure that the information one obtains is reasonably accurate. This doesn't guarantee that everything written in this work is one hundred percent correct, of course, but it's as close as we were able to make it.

During the course of our investigations, several sources especially stood out for their usefulness in ensuring the highest degree of accuracy possible. Of course, Wikipedia is always a useful—if uneven—source that has frequently proven invaluable in terms of obtaining basic information and a generally sound, if sparse, narrative of events. Unfortunately, the site often gets a bad rap due to the fact that anyone can make entries to it, making it an unacceptable source to many. Some even go so far as to equate the information culled from it as little better than material one might get off a blog, but this is an unfair criticism. Wikipedia makes every effort to ensure its sources are accurate and frequently demands that information include citations much as would be demanded for a college thesis or peer-reviewed work. In fact, with literally millions of people accessing Wikipedia each day, it is perhaps the most peer-reviewed site on the planet, making the information within it, in some ways, more accurate than that which might be found in some library texts. As such, we have used it liberally throughout this work with little trepidation.

Among the many other sites we looked to for information, two especially stand out. One is Carroll Gray's Flying Machines website (http://www.flyingmachine.org/) which contains a wealth of information on lesser known aviators throughout history. While every effort has been made to maintain a unique and original narrative throughout this work, Mr. Gray's work can be seen from time-to-time, if not in direct quotations, at least in essence. The other site we found helpful is John Ayre's Century of Flight website (http://www.century-of-flight.net), an extremely well-done and informative site that provided voluminous information on a number of subjects (the article on early flying wings by E.T. Wooldridge being especially useful). The Wright Brothers Aeroplane Company Museum also maintains a nice site (http://www.wright-brothers.org) that contains considerable amounts of useful information, particularly in regards to the alleged first flights of Gustave Whitehead. Special thanks are also due to Paula Mercado and Justo Miranda for their excellent work on the Saqqara Glider and, especially, for Paula's very fine illustration (which can be seen in chapter one). Additionally, we must tender a special note of thanks to our publisher, David Childress, for his contribution of the drawings of vimana aircraft found in chapter one, as well as for his general encouragement in moving forward with this project. His willingness to not only agree to publish our book but permit us the freedom to lay it out ourselves—something that rarely happens outside of self-publishing—demonstrates a degree of confidence uncommon among publishers today, and is most appreciated.

As part of this process of designing and laying out the book, we were required to seek the services of our own editor. We were lucky enough to find a talented individual willing to take on the job in Michele Brown, who did a remarkable job of editing this work. Unfortunately, as it was necessary to make some minor revisions after she had completed her task, it is likely a few typos or minor errors may have since crept back into the manuscript, for which we take full responsibility.

Finally, we would be remiss not to include a special word of thanks to our families and, especially, our wives, Carol and Claudia, who, during the course of putting this work together, had to endure many lonely evening wondering whatever became of their spouses as well as put up with endless "shop talk" over dinner. Their long-suffering and eternal patience is a gift that is greatly appreciated and never taken for granted.

INTRODUCTION

It has been traditional to teach school children that Christopher Columbus discovered America, that Charles Lindbergh was the first man to fly across the Atlantic, and that the Wright brothers' invented the airplane. Time has since taught us that the first Europeans to touch foot on the continent were the Vikings, while Lindbergh was the 85th man to cross the Atlantic by air[1]. Would it be any more remarkable to learn, then, that the Wright brothers brief flight in December of 1903 may not have truly been the first successful flight of a manned, heavier-than-air powered machine?

This is not to diminish their success, but recognition that progress in any field of endeavor is seldom a solo affair. Almost every great inventor built upon the work of others, which is as true for the airplane as it is for the computer, the microwave oven, and the flat screen TV. As such, it's not unreasonable to maintain that without incorporating the work of others—most of whom remain unheralded and even largely unknown to this day—it is difficult to imagine the Wright brothers enjoying the success they did, at least at such an early date. Of course, the airplane was an idea whose time had come, making it inevitable that a successful flying machine would have been built and flown at some point. However, it's difficult to imagine how it might have happened when it did without the foundations having been laid by others years and, in some cases, even decades earlier.

But who are these early pioneers of aviation who made this possible? In this work, we will examine that question by tracing the largely forgotten history of aviation from antiquity through the Wright brothers and beyond, all in the quest to tell the history of those who first dared to dream of ways of challenging the age-old belief that man should remain content to stay earthbound as God designed him to be. It is the story of the first men—and even a few women—who dared to believe that humans could—and, indeed, should—challenge gravity, sometimes even to the point of paying for such brashness with their very lives.

While most of the people mentioned in these pages will be known to aviation enthusiasts and historians, many remain largely unknown to the average reader. It is the purpose of this book to rescue those names from the obscurity time and distance has banished them to and bring them out of the shadows of the better known giants of aviation in an effort to recognize their substantial contributions to humanity's quest to fly. Some of these people were real scientists and engineers who worked from sound scientific principles, pursuing their dreams with grim determination and clever innovation, while others were mere tinkerers, dreamers, or even idealistic eccentrics who played with the idea of flight without any real idea of what they were doing. Some spent small fortunes in a frequently futile quest to

---

1. The first successful crossing of the Atlantic by air was made by RAF pilots Captain John Alcock and Lieutenant Arthur Whitten Brown in a Vickers Vimy bomber flown from Newfoundland to Ireland in June of 1919. Several other fixed-wing aircraft also made the crossing, along with two airships—the British R-34 and the German-built zeppelin LZ-126 (later to be renamed the U.S.S. Los Angeles)—before Lindbergh's attempt. Lindbergh's flight was, however, the first solo crossing.

fly like a bird, and a few even paid the ultimate price with their very lives. While some of them enjoyed a modest degree of success, most of them failed in their efforts, but all of them need to be applauded for even their failures contributed to the evolution of flight by demonstrating to others who followed what didn't work, thereby eliminating them as options and bringing man one step closer to success.

Much of the fascination with this story, however, comes in watching how each attempt to soar became but one more stepping stone on the path to making air travel the convenience we so largely take for granted today. In doing so, we can come to understand not only how the foundation for even greater accomplishments were laid, but learn to appreciate how the early aviators managed to shrink our planet to a size where today anyone can step onto an aircraft and be in another city or even another country—trips which used to take days or even weeks a century ago—in a matter of hours. The fact that we can do this in safety and comfort is also a direct by-product of the fact that they were willing to put themselves in danger in an effort to understand the principles of flight, thereby making air travel the gift it is for all of us today.

But we can have some fun looking at the process as well, which is the reason for the patent plates section at the back of the book. It is always interesting to journey into the past to see how our ancestors looked at the world of aviation and technology in their day, which is what these plates allow us to do. Most of these ideas never saw the light of day, of course, but each is imbued with the sort of imagination that exemplified an age when many believed technology was the savior of mankind and that anything was possible with a bit of determination and hard work. To many of the authors of these patents, flight had an almost magical quality to it which gave them permission to let their imaginations flow freely with the most bizarre notions. Most of what they came up with was woefully impractical and some of it even downright dangerous, but it's still both interesting and, at times, even amusing to look at their ideas and shake our heads in wonderment.

Finally, we write this book as a tribute to the early aviators throughout history in the hopes that others will be inspired by their accomplishments. It is also designed to remind us that just as the aeronautical pioneers of the past found ways to overcome the tremendous technological hurdles they faced in their efforts to make their dreams a reality, we too can find ways to overcome the technological obstacles that stand in our way in our quest to reach the stars. The ability to fly must have seemed as unlikely to men who lived two hundred years ago as interstellar flight does to us today; in accomplishing the feat, however, they pointed the way for the rocketeers and engineers of the future to find the means of one day making our way to other worlds. For this, the aviators of the past who flew those rickety machines and fragile balloons deserve our undying gratitude and respect, for without them we should never know what it is to soar above the clouds and, as the famous poem says*, *"reach out our hand and touch the face of God."*

Jeff Danelek and Chuck Davis
July, 2010

---

*From High Flight, a poem written by John Gillespie Magee, Jr. (1922-1941), a Royal Canadian Air Force pilot killed in World War II.

Chapter One: Aviation in Antiquity

1

At some point in the distant past, someone likely fashioned a handful of feathers around a rock or other projectile, took it to a precipice, and threw it into the air in hopes that it would soar like the very bird from which the feathers had been taken. Most likely they were disappointed by the results, but it ignited in them the belief—however ridiculous and unlikely it must have seemed at the time—that flight was possible, and humans have worked from that premise ever since.

It's unlikely there was any one man who first considered the possibility that one might fly like a bird. Most likely the quest to fly has always been a part of mankind's basic urge to push the boundaries of his abilities, as is implied by the ancient Greek myth of Daedalus and Icarus, who used candle wax to affix feathers to their arms and took flight. Though their epic ended in tragedy when Icarus ignored his father's advice and flew too close to the sun, thereby causing the wax to melt and leaving him suddenly featherless, the mythology itself fairly shouts that even in antiquity the quest to find a way to soar like an eagle was very much a part of the human mind-set, even if well outside man's ability to realize a means of doing so at the time.

Of course, the quest to fly was not realized purely though ancient mythology. It's a well established fact of history that humans have been experimenting with aviation for over two thousand years. Perhaps the earliest reliable example of making something fly comes from the fifth century BCE, when a Chinese man named Lu Ban supposedly constructed a "wooden bird" that may have been a proto-kite, demonstrating that even in the earliest years of recorded human history, man was already finding ways to circumvent Earth's gravity. In the third century BCE, the Chinese also were making small, paper balloons inset with candles that they called Kongming Lanterns, which they sent aloft in an attempt to frighten enemy troops into thinking they were seeing divine forces floating above their adversaries. Apparently, such devices were successful in inducing terrified soldiers to flee a battlefield in the belief they were being opposed by an angry spirit, making them a remarkably simple and effective way of unnerving one's enemy (and perhaps initiating the first use of psychological warfare in history).

But frail kites and tiny balloons weren't the only flying "machine" being contemplated thousands of years ago, nor was China the only country in which they were being attempted. In 400 BCE the Greek philosopher and astronomer Archytas supposedly designed and built a bird-shaped, steam-powered mechanical bird that, at least according to the Roman historian Aulus Gellius, actually flew some 200 meters before coming to a landing. If true—and who are we to challenge the word of ancient Roman historians?—the little device was the first powered, unmanned and heavier-than-air flight in history—and one that predated the earliest attempts at powered flight by at least 2,300 years!

It's impossible to gauge how much ancient man may have understood about the physics of aerodynamics, of course, though it is evident that the quest to fly is as old as civilization itself and, possibly, as old as mankind. Most scholars, however, assume that even if the ancients could build things like kites and tiny paper balloons—or even tiny mechanical birds for that matter—it was unlikely they understood the processes that made them work. This, however, is assumed. For all we know, they might have had a very good understanding of the forces that permitted certain objects to go aloft while others remained firmly earthbound, but since much of the evidence of these early attempts at harnessing flight have been lost to history—the accounts being handed down to us purely through the writings or, in some cases, oral traditions, of others—we can never know if they were merely playing on the periphery of flight or were working towards a practical application of what they knew. In either case, without blueprints, drawings, or working mechanisms we might examine, the earliest evolution of aviation remains as much a mystery to us today as it has always been.

At least, that was the situation until a little over a century ago. It was right around the turn of the century when things began to change in regards to our appreciation of just how much the ancients may have understood about flight and forced some to reconsider aviation's earliest manifestations. And it all began in a pharaoh's tomb half a world away.

The Saqqara Bird

The year was 1898 when British archeologists, working on the tomb of Pa-di-Imen near Saqqara[2], Egypt, found a tiny wooden carving of what appeared to be a falcon made from sycamore. Dated at over two thousand years old, at the time it was assumed to have been just another of scores of similar bird carvings found over the decades at various locations throughout Egypt, and so it was promptly catalogued, thrown into a box of similar artifacts, and tucked away in a dusty corner of the Cairo Museum where it lay largely forgotten and undisturbed for the next seventy years. It likely would still be there today had it not been "rediscovered" in 1972 by renowned Egyptologist Dr. Khalil Messiha (1924-1998) when he was going through boxes of artifacts in the Cairo Museum and came across the unusual object. An expert in such carvings, what he found especially intriguing about this particular artifact was that the "falcon" had an unusual vertically oriented tail feather rather than the traditional horizontal tail, completely in contrast to every other bird carving he had ever previously examined. He also noted that, in addition to the vertical tailplane, he could find no trace of any decorative feathers painted on the body as were common to similar carvings. There were eyes painted on its head, he noted, and two faint reddish lines surrounding the belly under the grooves, but he felt the lack of feather decoration and the absence of legs—also common on

The Saqqara bird as it sits at the Cairo Museum today. Modern scholars believe it is a simple carving of a falcon, but what bird has vertical tail feathers?

[2] Saqqara lies just west of the ancient Egyptian capital of Memphis, which itself is some twelve miles south of Egypt's modern capital of Cairo.

other carvings found in the tombs—indicated that the model may have been more than just a simple carving of a bird as first assumed. In fact, in that it lacked the standard characteristics of a common falcon carving and seemed to possess all the necessary elements—accurately angled wings for lift and a vertical rudder for control—the good doctor came to the controversial conclusion that the "Saqqara Bird"—as it came to be later known—was nothing less than an aerodynamically flyable glider, thereby implying that the basic principles of flight may have been known by the ancient Egyptians over two thousand years before the Wright brothers ever flew.

Naturally, the implication that the Egyptians may have understood the basic dynamics of aeronautics was received with considerable skepticism by Egyptologists the world over, but Dr. Messiha stuck to his hypothesis. In fact, he even went so far as to speculate it could well have been a scale model for larger, flyable examples potentially capable of lifting a man aloft. He even wondered if there were not full-scale examples still buried beneath the constantly shifting sands waiting to be uncovered, which, if there were, would force historians to rewrite everything they knew about the history of aeronautics and Egyptologists to include the possibility of manned flight as an element of ancient Egyptian knowledge.

Of course, such a claim was instantly discounted by most Egyptologists at the time (just as it remains so to this day), because of the assumption that the Egyptians, despite understanding enough about aerodynamics to have fashioned working boomerangs, couldn't possibly have understood the basic principles of flight. Additionally, it was pointed out that without an elevator—necessary to prevent the glider from pitching over into a nose dive during flight—the artifact wasn't flyable as found either, thereby compounding the mystery. As a result, most scholars concluded that the object was probably a child's toy of some kind, or perhaps a wind vane—the vertical rudder working to turn the object as the wind direction shifted.

Dr. Messiha was not easily dissuaded from his hypothesis, however, and was determined to dem-

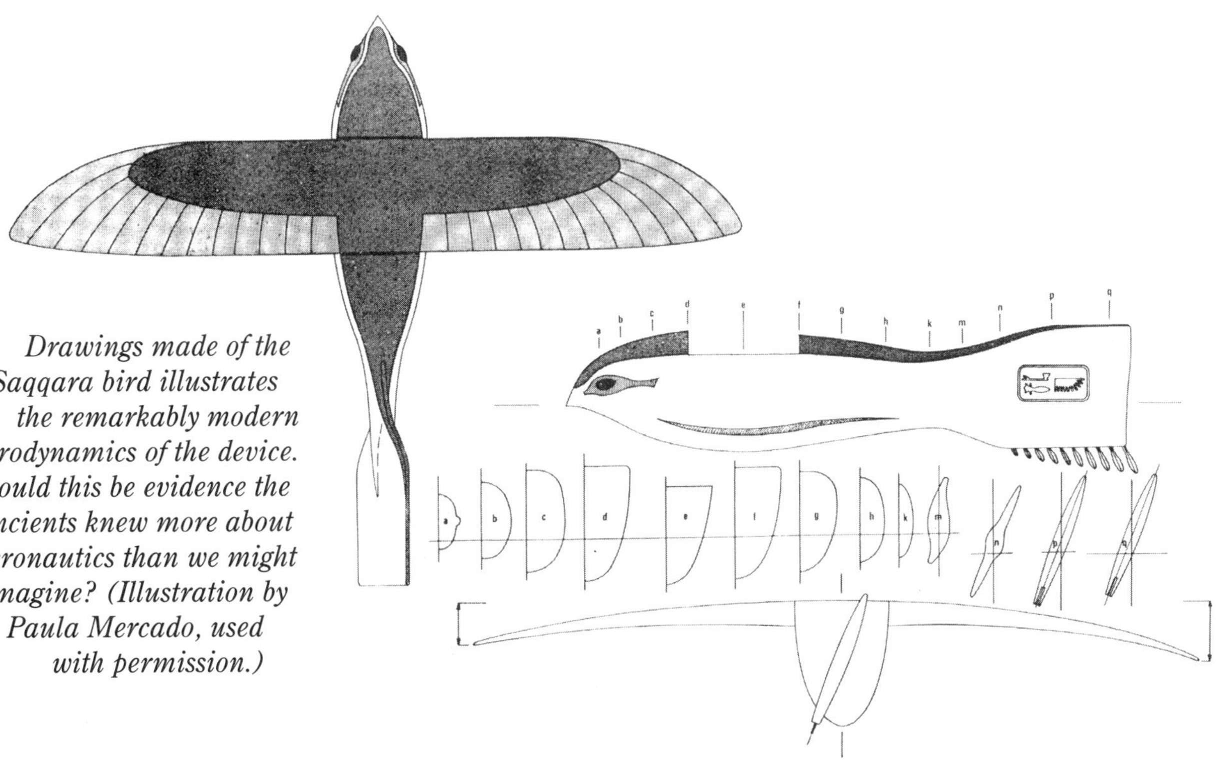

Drawings made of the Saqqara bird illustrates the remarkably modern aerodynamics of the device. Could this be evidence the ancients knew more about aeronautics than we might imagine? (Illustration by Paula Mercado, used with permission.)

onstrate that the Egyptians may have understood the basic principles of aviation much more thoroughly than anyone was willing to give them credit for. Noticing that the lower part of the tail fin was broken off, suggesting that it may have in fact once had a stabilizer attached to the bottom of the "rudder" as on a modern aircraft[3], Dr. Messiha fashioned a larger replica of the artifact out of balsa wood and, attaching a tailplane at the point he suggested one might have existed before it was broken off, he found that it could easily sail in the air for a few yards when thrown by hand. Further, and while a few skeptics went to some pains to demonstrate that as a glider the object lacked genuine aerodynamic stability, flight simulator tests done in a 2009 History Channel program entitled *Ancient Aliens* demonstrated that the glider is indeed quite flyable once an elevator is attached to the rudder, seemingly exonerating Dr. Messiha's hypothesis and potentially changing our understanding of the ancient Egyptian's capabilities in regards to aeronautical understanding. Despite this, however, the idea that the object was a flyable glider remains the minority viewpoint among archeologists today and probably will for some time—at least, until a more complete example is unearthed or, as Dr. Messiha speculated, a full-sized version is pulled from the sands of the vast Sahara.

Of course, it's not difficult to understand why archeologists are so hesitant to accept the Saqqara bird as a glider. It is, after all, a small thing that, even if capable of gliding for a few feet under ideal conditions, remains very reminiscent of a simple carving of a falcon. However, it's level of sophistication should give many cause for wonder. Vertical rudders and elevators, as well as aerodynamically practical chambered wings as seen on the model, would not be incorporated into aircraft designs until late in the nineteenth century, making the craftsman behind the model a true visionary. Additionally, when one considers the question more objectively, the idea that the Saqqara Bird is a glider isn't as remarkable as it may seem. After all, the Egyptians had been using boomerangs for hunting and sport for a millennia,[4] demonstrating that they had a decent understanding of basic aerodynamics. While Dr. Messiha's contention that it could be a model of a larger manned craft seems a bit of a stretch, the possibility that it may have been a glider is not as remarkable an admission as might be imagined, especially when one considers that these same Egyptians were the same people who had constructed some of the most complex and architecturally sophisticated structures the world had ever seen. Further, considering how little we really know about the ancient past, it seems presumptuous to assume it's impossible that the ancient Egyptians may have possessed considerably more aeronautical knowledge than many may be comfortable granting them. The implications of such a possibility would be staggering, of course, but then the past is often full of such surprises.

So what is the final verdict on the Saqqara Bird? While it's exciting to contemplate what it would mean if the ancient Egyptians—or any ancient civilization for that matter—really had grasped the fundamentals of lift versus weight and thrust versus drag, even if on a primordial level, unless and until similar, more complete examples—or a full-size airplane—are unearthed from the dry sands of the desert, it must remain a mystery. In the end it may turn out to have been merely a child's toy or a wind vane—precisely as many insist it is. However, it's also conceivable that even it that's all it turns out to be, even then that simple toy may have been the by-product of years or even decades of experimentation with flight. In other words, it may well have been a toy, but a toy based off some very sound aerodynamic principles, which will continue to entice us into considering the possibility that the distant past may have included, at least on a simplistic level, the very real prospect of—at a minimum, unmanned flight.

[3.] The top of the rudder also appears to have broken off, suggesting an elevator could have been mounted there as well. Either location would have work in terms of stability.

[4.] King Tutankhamen, the famous boy Pharaoh who died over 3,000 years ago, owned a large collection of boomerangs of both the straight flying and returning variety.

Two of the pre-Columbian "airplane" trinkets crafted by the Zenu people of Columbia more than 1,000 years ago. Notice the very modern-looking delta wings, rudder and elevators.

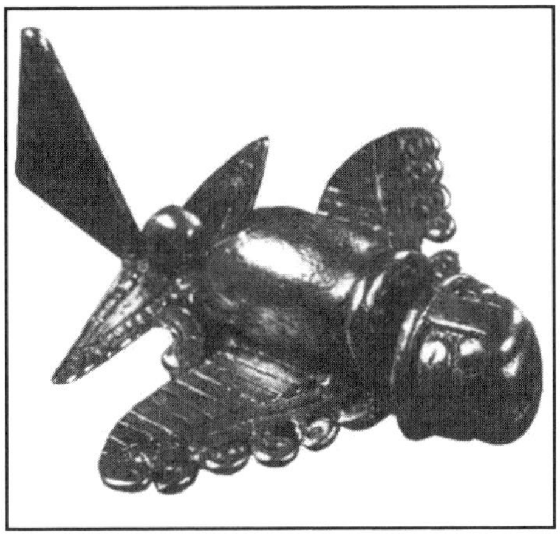

The Delta Wing "Fighters" of Columbia

While the Saqqara Bird is an intriguing artifact in its own right, an even more controversial discovery has lead some ancient astronaut theorists to suggest that some ancient peoples may have known far more about aviation than we might begin to imagine. In fact, if they are correct, they may have known as much about it as we do today!

It started in 1954 when the government of Colombia sent part of its collection of ancient gold artifacts—believed to have been crafted by the Sinu (or Zenu) peoples of Columbia, (a pre-Incan culture that existed from approximately 200 BCE to 1600 CE)—on a tour of various museums in the United States. Superbly crafted by some of the finest sculptors and goldsmiths of pre-Columbian South America, among the artifacts were a series of gold trinkets thought to be highly decorative representations of birds, fish, and insects. Several of these trinkets, however, were considered unusual in that they appeared to possess unusually modern aerodynamic elements such as wings, elevators, and rudders—features which, of course, would be entirely foreign to the Sinus or any pre-Columbian civilization a thousand years ago. In any case, castings of the artifacts—each about two inches long and just over an inch wide—were made prior to the collection being returned to Colombia, one of which ended up in the hands of biologist-zoologist Ivan T. Sanderson (1911-1973), a Scottish naturalist and writer known for his ground-breaking work in the field of cryptozoology and his lifelong interest in the paranormal. It was Sanderson who, after a thorough examination of the objects, concluded that the artifacts did not represent any known winged animals. In fact, he even went so far as to suggest that the objects were models of high-speed aircraft, even

Is the remarkable similarity in shape between this one thousand year old gold trinket and this 1960's-era F-102 delta wing fighter nothing more than mere coincidence?

Many suggest the Columbian "jet" have been nothing more than a highly stylized image of a flying fish, which is not as far-fetched as it seems. Notice that many species of flying fish have a second set of horizontal fins very similar to those seen on the trinket, as well as a vertical tail. Additionally, from 1900 to the 1930s, flying fish were studied as possible models to develop airplanes.

though they were crafted ten centuries before the first modern jet ever flew!

Obviously, such a claim was considered impossible. No civilization on the planet at the time could have such knowledge or have come in contact with anything like a modern supersonic aircraft, resulting in a wholesale denunciation of Sanderson's suggestion as so much pseudoscience. However, when one takes an objective look at the objects, it is difficult not to come away with the feeling that they appear more mechanical than biological. For example, the front wings are delta-shaped and rigidly straight edged, which is very different from what one sees in nature. Even more astonishing, however, is the triangular shaped rudder and what appears to be elevators at the back of each artifact, which are entirely consistent with what one would expect to see on a modern airplane. In fact, the delta-shaped wings and perpendicular rudder suggests nothing less than a supersonic fighter of the type seen first during the 1950s. Adding to the mystery, an insignia appears on the left face of the rudder on one of the objects that has been identified as the Aramaic or early Hebrew letter beth or B (suggesting that the original plane did not come from Colombia, but was the product of a very early people inhabiting the Middle East who knew the secret of flying).[5] Even more perplexing is that when a team of researchers constructed models of the craft—first powered with a propeller but afterwards with a jet engine—both proved to fly well.

The problem, of course, is if the artifacts are representations of jet aircraft, where did the ancients—either early Colombians or Middle Easterners—get their inspiration? Or is it all a case, as many contend, of pure coincidence that a fanciful model of an insect or a fish managed to fly when scaled up to size and given a proper power source?

Orthodoxy would argue for the latter, though it should still give one pause to wonder. Clearly, if the Incan "airplane" is a representation of an insect, it is curious at how un-insect-like it is. For example, the wings on the artifacts are attached at the base of the "body", whereas the wings of all flying insects are attached at the top of their torso. Additionally, no known insect has a vertical tail fin or the sort of appendages seen on the tiny trinket, making it extremely unlikely that either is a representation of an insect.

As far as the artifact possibly being a stylized representation of a fish, however, there does seem to be some support for such a possibility—especially in regards to one specific type of fish known to the natives of the region for a millennia. In effect, are the delta-wing "fighters" of Colombia highly stylized representations of a flying fish?

There are some similarities between the trinkets and the common flying fish, which does, after all, have a vertical tail fin and triangular-shaped wings mounted on the side of its body. One species in particular—the *genus Exocoetidae* being the most noteworthy—even have a second, smaller set of wings

[5] This contention is disputed, however, with many claiming the resemblance between the ornate symbol seen on the artifact and a letter of the Hebrew alphabet to be purely coincidental.

immediately behind the first set, precisely as is seen on the artifacts. Further, the addition of what appears to be ornate eyes on the front of the "nose" suggest the craftsman of the object may have had a flying fish in mind after all. There are a few discrepancies with this theory as well, however; for example, the flying fish has a distinctly forked-shaped tail fin whereas the Colombian artifacts show a large, triangular-shaped tail fin. Additionally, the "rudder/fin" on the artifacts are not symmetrical as they are on a fish, but are asymmetrical—flat on the bottom and extending upwards only. Why the craftsman of the object went to such pains to make the rest of the image anatomically accurate but choose to give it a very different—and aerodynamically workable—rudder-like tail remains as much a mystery as the artifacts themselves. Could these have been evidence of some early technology, lost to us but known to at least some people in ancient times, or is the past merely playing tricks on us?

The Mysterious Flying Machines of Ancient India

While the Saqqara Bird and the Sinu artifacts are interesting hints that there may be more as-yet-to-be discovered mysteries about the past than we are aware of, there is one mystery that makes both pale in comparison when it comes to the question of just how much man may have understood about flight in the ancient past. The evidence in this case are not carved pieces of wood or gold trinkets, but a narrative contained within one of the most ancient and revered texts in the world that some believe contain descriptions, as difficult as it is to imagine, of what could possibly be ancient aircraft and even spacecraft that flew thousands—and in some cases, even tens of thousands—of years ago!

The ancient texts in question are known as the Hindu Vedas, which are a prodigious collection of poems thought to be among the oldest writings

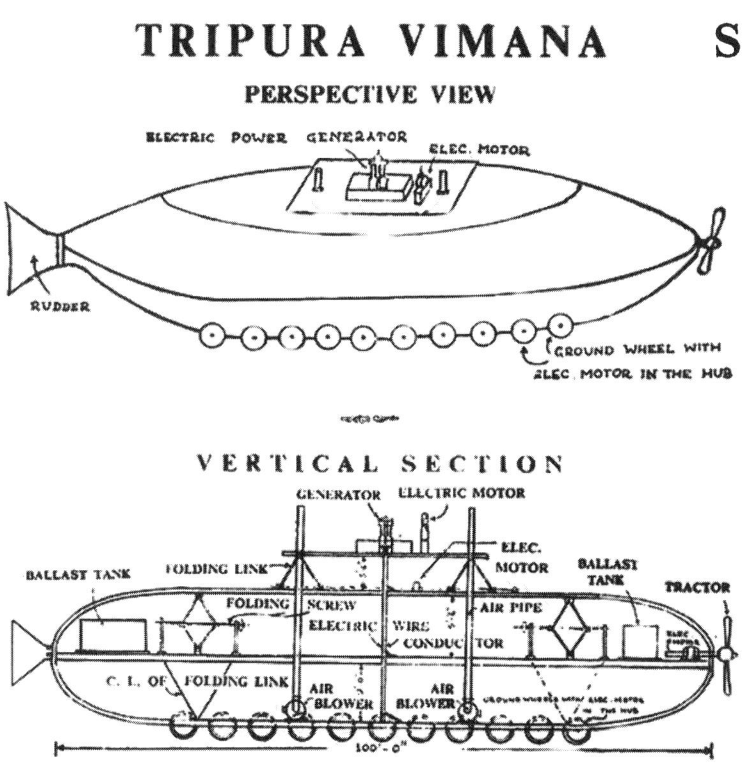

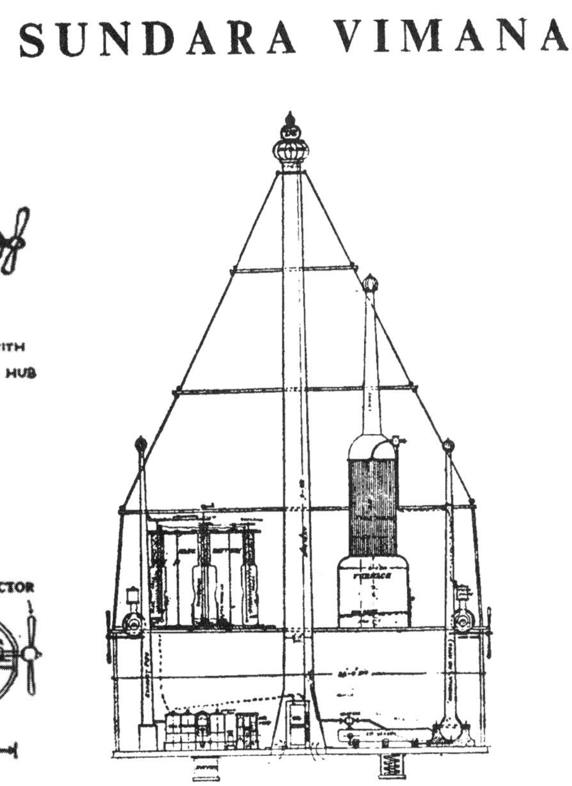

Modern interpretations of two of the many different types of mysterious "vimana" flying machines that are described in some detail by the ancient Indian Vedas.

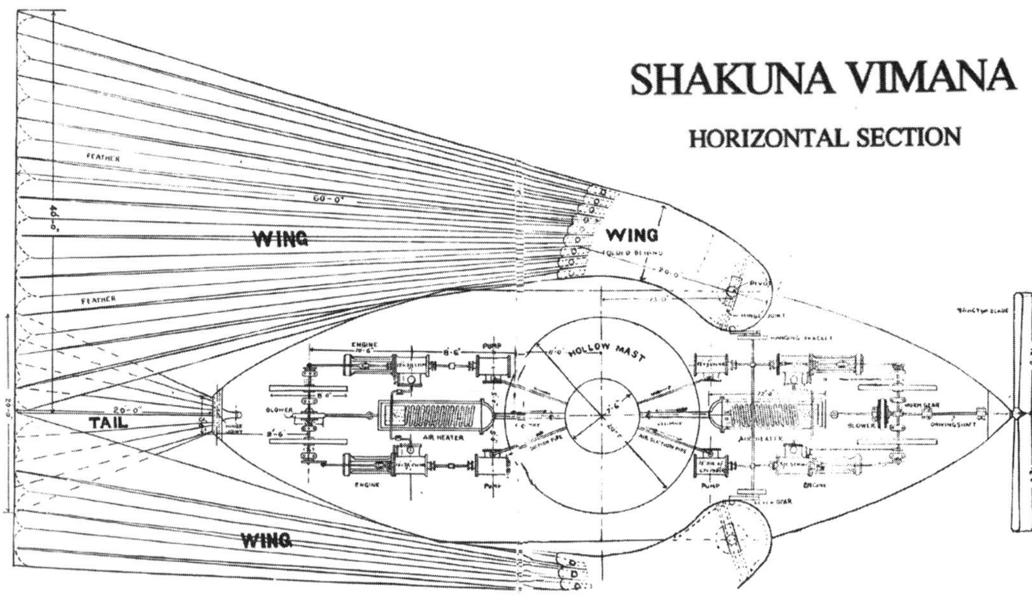

An illustration of a Vimana drawn by T.K. Ellappa of Bangalore, India in 1923. Notice how remarkably modern the craft appears.

known to exist on Earth. Composed in Vedic Sanskrit, the texts constitute the oldest layer of Sanskrit literature and the oldest scriptures of Hinduism, some of which may be as old as 3500 years old. Additionally, they may have even been transcribed from earlier texts or oral traditions that predate them by thousands of years, making the stories contained in them among the oldest known to man to have survived to modern times.

It is within these books that the ancients refer to something known as "vimanas", which were apparently types of flying ships of different configurations and capabilities supposedly seen in the skies over ancient India thousands of years ago. Generally dismissed by modern science as poetic and highly imaginary references to the various types of mythical air cars the gods moved about in (literally, flying palaces or chariots), there are some who maintain that, due to the complex nature of the descriptions and the apparent sophistication of how the vehicles operated, they may have been actual aircraft that flew during a time before recorded history. While it's beyond the scope of this work to examine each of these "evidences" in any detail, they are worth taking a brief look at if for novelties' sake if nothing else.

Perhaps what's most intriguing about these ancient references is that there are so many different kinds of craft. Some are fairly small and appear to be bound by the atmosphere while some are huge—several stories in height—and are capable of traveling to the moon and planets of our solar system. One is even a double-decker, circular machine with portholes and a dome that might be used to describe a "flying saucer" today.[6]

Supposedly, these craft flew with the "speed of the wind" and gave forth a "melodious sound" while another Indian source, the Samar, describes vimanas as "iron machines, well-knit and smooth, with a charge of mercury that shot out of the back in the form of a roaring flame." One ancient text, the Sutradhara, even includes 230 stanzas dealing with the construction, take-off, and operation of these ships, while other texts—equally as old and detailed—are practically owner's manuals that include information on how the craft were steered, what precautions needed to be taken for long flights, how to protect the airships from storms and lightening and even how to switch the drive to "solar energy" from a free energy source—a power plant that sounds suspiciously like zero-point energy. Still another text, the Vaimanika

6. One suggestion is that these were extraterrestrial craft being flown by "sky people"—effectively making them among the earliest alien visitations in history—but this is no less fantastic than is the claim they are evidence of a very advanced but now defunct human culture that flourished thousands of years ago

Sastra, even has eight chapters that include diagrams that describe three types of aircraft and discusses the materials each are made from.

Additionally, the texts talk about the ancient civilization that flew these craft as being known as the "Rama Empire", a civilization made up of many large, sophisticated cities that supposedly developed at least fifteen thousand years ago on the Indian sub-continent. Interestingly, Rama warred against a parallel civilization that existed in the mid-Atlantic called the "Asvins"—the apparent conflict between the two even possibly being the catalyst for both their destruction. (Some believe the "Asvins" to be a reference to Plato's mythical Atlantis, which supposedly existed about that same time, but this may well be simply a case of merging competing mythologies.)

So what are we to make of these ancient texts? Certainly, the claim that ancient Indians knew about what could only be described as futuristic technologies (even from our perspective in the twenty-first century) is nothing short of astonishing, yet there can be no doubt that most of these texts are authentic, and there are literally hundreds of them. Skeptics maintain they are simply metaphorical in nature, but if that's the case, how could ancient Indians come up with such exotic ideas from within the context of a pre-industrial, largely technology-vacant world? Even mythology is based on known factors, so where did the ancients get their ideas regarding the sorts of energy sources the "god's" ships used, what they were made of, and how they operated?[7]

For now the question of just what precisely the ancient Indians were describing must remain unknown and, most likely, unknowable, but it is interesting to wonder just how certain we can be that we really know what happened in the distant past. Could there be room in the prehistoric past for things like vimanas or are they nothing more than the fantastical ramblings of an ancient imagination?

Conclusions

There is a tendency among scholars to dismiss the notion that ancient man might have grasped the basic fundamentals of flight and possibly even experimented with it, but we suspect that is hubris talking. Intelligence and ingenuity are not modern inventions, but are a part of even the most ancient civilizations and so have to be factored in when considering what could or could not have happened so long ago. While little golden trinkets and anecdotal stories of fantastic aircraft battling in the skies over India don't prove that the ancients built flyable aircraft, they do suggest that humans have been looking skyward from the earliest inception of civilization, making our assertions of what they did and didn't know subject to both subjective interpretation and modification as new evidences are uncovered.

The fact is that we don't really know what was known to the ancients. Too much has been lost over the centuries for us to affirm with any degree of certainty what the Egyptians or Sumerians or Babylonians—or even for that matter the Greeks and Romans—knew about flight. When one considers the vast wealth of knowledge that may have been lost when the library at Alexandria repeatedly burned down over the centuries, can we really be so sure that among the many things written of on the thousands of scrolls contained within its walls there was nothing of flight? It seems impossible to imagine that a thousand years of scholarship could have nothing to say of the greatest—and possibly earliest—of man's dreams. As such, science would be well-advised to withhold judgment until every last ancient scroll and petroglyph has been uncovered and deciphered before deciding what we really know about our own past.

7. A study by aeronautical and mechanical engineering students at the Indian Institute of Science in Bangalore in 1974 ("Flights of fancy? (Part X of XII)". The Week. 2001-06-24. http://www.hvk.org/articles/0601/100.html) concluded that the aircraft described in the text were "poor concoctions" and that the author showed a complete lack of understanding of aeronautics. However, even bad science fiction must have some basis in real technology upon which to build.

CHAPTER TWO: THE EARLIEST ATTEMPTS AT FLIGHT 2

When it comes to finding examples of humanities' earliest attempts at mimicking the birds, it's disappointing to see how little there is to be found in the history books. Undoubtedly much of this can be ascribed to the difficulty of recording important events of the past as well as the tremendous amount of material that has been lost over the centuries. Then again, it may simply be the case that attempts to fly were rarely attempted for any number of reasons. However, that's not to imply that the historical record is completely vacant. In fact, there are a few accounts from antiquity that do make claims of men having flown, or describe examples of flying machines, that are worth examining. Whether any of these early attempts at flight have any bearing in the truth will always be open to debate, of course, but they do provide us with intriguing glimpses into the past and provide us with clues that the history of manned flight may go back much further than most people are aware. In this chapter we will look at these stories and in so doing, consider the possibility that the story of aviation is richer and more curious than we have been led to believe.

The First Manned Flight...in 875 CE?

It's a fair bet that when the average person considers the great aviation pioneers of the past, one name that does not frequently come up is that of Abbas Ibn Firnas, the ninth century Spanish astronomer, musician, and engineer. However, this man may well have been the first considered by historians to have flown, and it is a remarkable story to tell.

Abbas Ibn Firnas (810-887 CE), was born in Izn-Rand Onda (near present day Ronda, Spain) and lived all his life in the Emirate of Cordoba. Something of an inventor in addition to his many other skills, it was while he was serving in the court of the Caliph 'Abd al-Rahman II that he was inspired to attempt flight by watching a man named Armen Firman[8] jump from the minaret of the Grand Mosque in Cordoba, in an effort to test what could only be described as a "flying suit". Not surprisingly, the attempt ended in failure, though fortunately the suit (basically a loose cloak stiffened with wooden struts that he intended to glide upon like a bird) acted like a parachute, breaking the man's fall and permitting him to walk away from the event with only minor injuries. The attempt itself, however, was enough to get the brilliant Firnas's fertile imagination working on coming up with his own flying contraption—though it would be many years before he would produce a comparable device as that the mysterious Firman used.

Working on his own version of a flying suit intermittently for much of the rest of his life, it wasn't until somewhere around 875 CE that Ibn Firnas, now an old man, was ready to try it out. Unfortunately, we have very little idea of what he finally came up with,

[8] There is some confusion as to whether Armen Firman is simply the Latinized version of Abbas Ibn Firnas or another person entirely. Muslim scholars generally consider the two to be separate individuals while Western sources believe the unsuccessful tower jump may have been an earlier attempt by Firnas to fly.

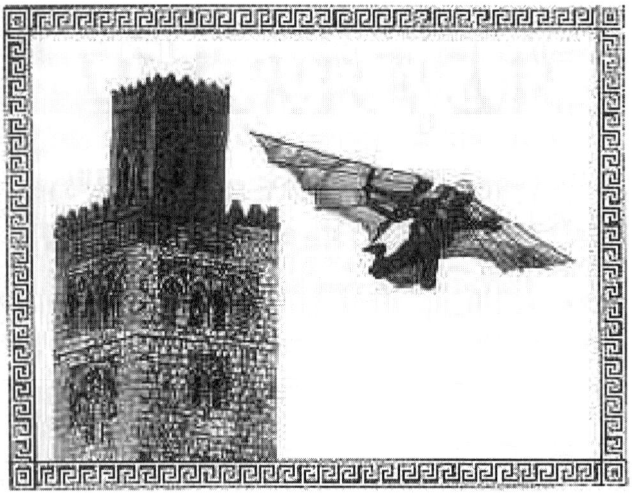

Highly stylized illustration of a man named Armen Firman's attempt to fly from the Grand Mosque in Cordoba, Spain. The experiment could have easily ended in tragedy had not the device acted like a parachute and broken the man's fall.

This purely speculative, modern drawing of Abbas ibn Firnas in flight shows him attached to a surprisingly contemporary-looking glider. While there is no firm proof that Firnas actually flew, the evidence that he did is compelling.

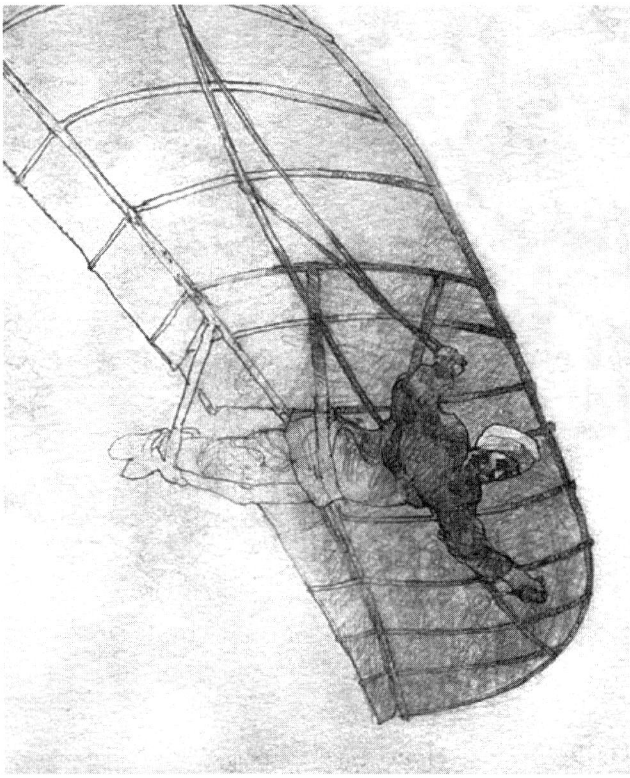

though legend has it he built some sort of wing-type device made of silk and eagles' feathers probably not much different from the gliders designed by Otto Lilienthal in the late nineteenth century, though this is pure conjecture. Not surprisingly, Firnas didn't provide drawings or descriptions of what the device looked like or precisely how it worked, forcing us to speculate about what design he might have come up with or—more correctly—what sort of design might have actually flown using the technologies and materials available in the ninth century.

In any case, it is believed that the daring inventor took the contraption to the top of a nearby mountain and, in front of a considerable crowd who appreciated a good spectacle from time to time, he jumped. Whether he launched himself from a cliff face or simply ran down a steep embankment in much the same way modern hang-glider pilots do is uncertain, but what is known—at least according to legend—is that much to the delight and utter astonishment of the crowds watching from below, he flew. In fact, it is said he supposedly managed to remain airborne for a solid ten minutes and even at one point reached an altitude of several hundred feet before returning to the ground. Unfortunately, however, Firnas had neglected to build a tail plane into his design, giving him no means by which he might slow his descent—an oversight which resulted in a hard landing that injured him badly enough that he would live out the final years of his life in continual pain.

But did the flight happen? There's no way of knowing, though it is generally considered an established fact among Muslim scholars and historians today. Certainly, the man appeared to be intellectually capable of such a feat; according to Islamic historians, Firnas was studied in chemistry, physics, music, and astronomy. He also appeared to possess the engineering skills to pull it off as well, as attested to by the fact that he had a number of important inventions on his résumé—from a clock that ran on water to corrective lenses. It was even said he built a room in his home in which spectators could witness stars, clouds, thunder, and lightning—all of which were produced by mechanisms located in his basement laboratory. As such, if any man alive at the time was capable of figuring out a way to fly, Firnas was that man.

Unfortunately, we will never know if the story of Abbas Ibn Firnas and his remarkable flying machine are true. Like so much of history, it must be taken largely by faith. However, if it is true, then the man can be credited for being the first to achieve sustained flight, accomplishing it over a millennia before either the Wright brothers or Otto Lilienthal (see chapter seven) tried their luck at gliding in the waning years of the nineteenth century. In fact, if he actually remained aloft for ten minutes as was reported by some sources, he also established a time and distance record that was to stand for the next thousand years, which would have been quite an accomplishment in any era, much less at the height of the Dark Ages.

A wonderful rendering of Eilmer of Malmesbury's daring—and, some might say, foolhardy—attempt to fly.

The Flying Monk

Abbas Ibn Firnas was not the only man who had a penchant for attaching a crude set of wings to himself and jumping from great heights in the hopes of soaring like a bird. There was another who appears in the history books who also may have successfully—if briefly—made a sustained flight using a glider. That man? An eleventh-century English Benedictine monk by the name of Eilmer of Malmesbury—but who is perhaps better known to history as simply "the flying monk."

According to the writings of the respected medieval historian William of Malmesbury, somewhere around the turn of the millennium (most scholars maintain around 1010 CE), Eilmer fixed wings to his hands and feet and launched himself from the top of a tower at Malmesbury Abbey. It is perhaps best to let the noted English historian describe what happened next:

> *"He was a man learned for those times, of ripe old age, and in his early youth had hazarded a deed of remarkable boldness. He had by some means, I scarcely know what, fastened wings to his hands and feet so that, mistaking fable for truth, he might fly like Daedalus, and, collecting the breeze upon the summit of a tower, flew for more than a furlong [201 meters]. But agitated by the violence of the wind and the swirling of air, as well as by the awareness of his rash attempt, he fell, broke both his legs and was lame ever after."*[9]

If true, it appears that Eilmer's glider may have suffered from the same design flaw that Abbas Ibn Furnas' machine possessed—a lack of tail surface to

9. William of Malmesbury, Gesta regum Anglorum / The history of the English kings, ed. and trans. R. A. B. Mynors, R. M. Thomson, and M. Winterbottom, 2 vols., Oxford Medieval Texts (1998–9)

slow its descent. In any case, the disastrous crash wasn't enough to keep the determined monk from modifying his design and preparing to make a second flight. It was only through the timely actions of the abbot of Malmesbury Abbey, who forbade the young monk from trying it again, that probably saved Eilmer from quite possibly flying with the angels and effectively brought to an end one of the more bizarre episodes in church history.

Whether the flight actually took place must remain, like so much of history, remains purely speculative. However, the story of Eilmer's flight has been retold many times through the centuries by scholars, encyclopedists, and proponents of man-powered flight, suggesting that there is at least a kernel of truth to the story. Additionally, according to modern researchers, given the specifics of the flight (elevation of the tower, general topography, probable wind and weather conditions, etc.), such a flight would have been feasible, making it possible that Eilmer of Malmesbury may indeed have been the second man in history to fly—although, like Firnas, the attempt cost him much in terms of pain and permanent injury.

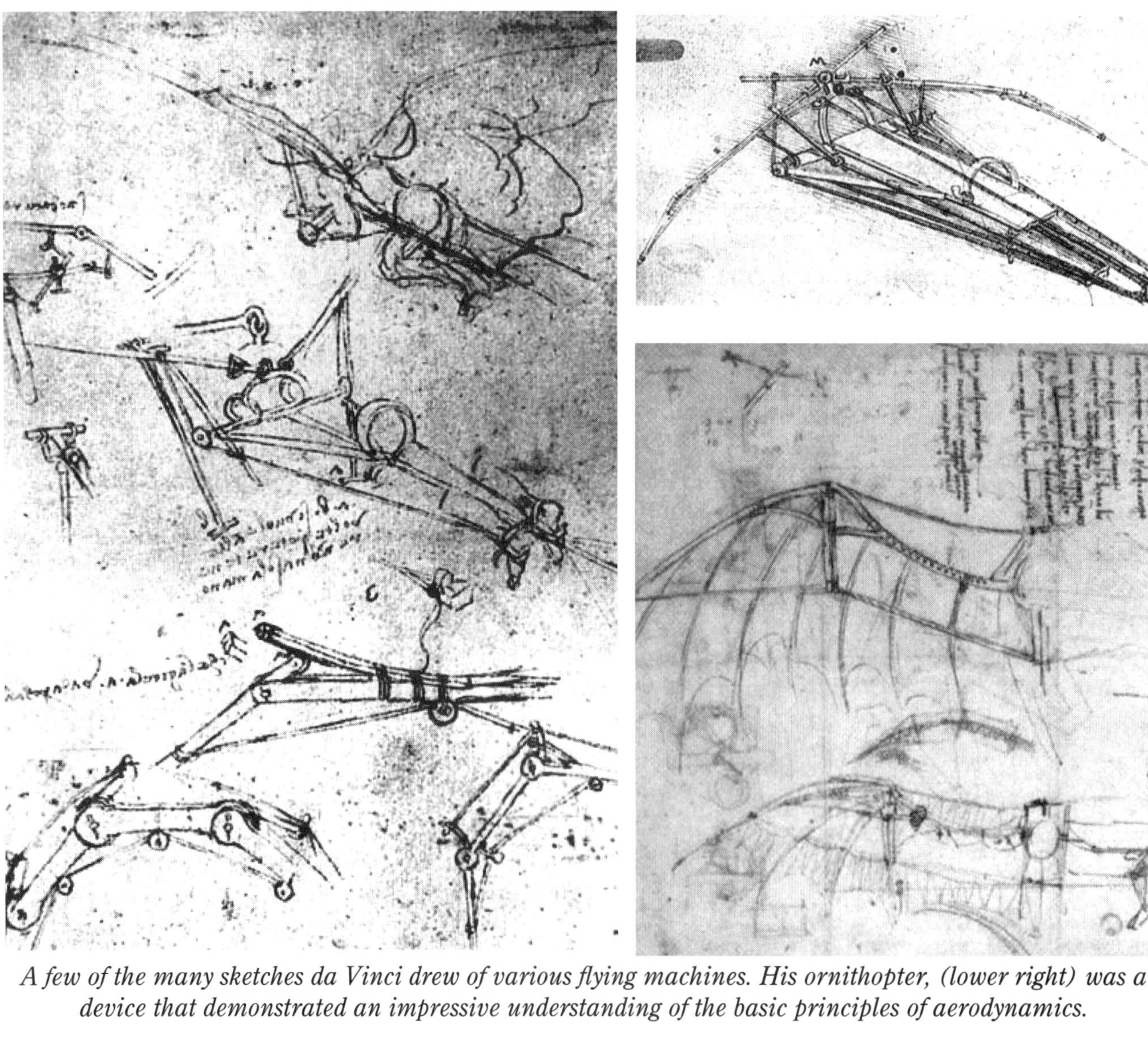

A few of the many sketches da Vinci drew of various flying machines. His ornithopter, (lower right) was a device that demonstrated an impressive understanding of the basic principles of aerodynamics.

Leonardo da Vinci and his Flights of Fancy

While the names of Abbas Ibn Firnas and Eilmer of Malmesbury remain largely unknown to most people, there is one man who also played around with the idea of flight whose name is among the most famous in history. That man is the legendary Florentine scholar and scientist, Leonardo da Vinci (1452-1519)—a man who, despite the fact that he never built a full-sized flying apparatus or attempted to fly personally, is still considered by many to be one of the earliest aeronautical engineers of the modern era.

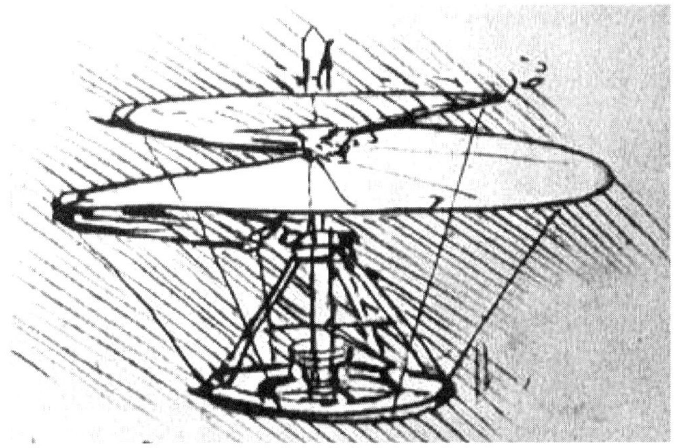

Da Vinci's sketch of the first "helicopter". Below: Modern model of the machine proved it to be an innovatiove but ultimately impractical device.

Though better known for his war machines (the tank and the scythe-wielding chariot being two of his more famous ideas), he also came up with a number of flying machine concepts, many of which he recorded in meticulous detail in his journals. While none of them appear to be airworthy by modern standards, some of his ideas showed real merit and, with a bit more work, may have even proved flyable. Perhaps the most promising idea da Vinci came up with was something known as the ornithopter, which was basically a bat-like hang glider with retractable wings designed to be flapped precisely as a bird's wing. This flapping motion was to be achieved through a series of ropes and pulleys that permitted the pilot to change the inclination of the wing's outer edges[10] while attempting to keep the whole thing balanced by constantly shifting weight back and forth. While such a device would have been difficult to fly even by a modern, experienced hang-glider pilot, there is contemporary evidence that, if it had been built, it would likely have flown.[11] However, there is no evidence that da Vinci ever actually built the device, though it does remain evidence of one of the first flying devices ever recorded.

Another interesting idea da Vinci came up with was a corkscrew-type helicopter which, while certainly innovative, was technically unfeasible because it used a sail-like helical screw instead of rotor blades to provide lift. There is some evidence the idea wasn't his alone, however, but may have been inspired by a small toy of the era; a corkscrew-shaped contraption whose rotor was sent spinning by pulling sharply on a cord wrapped around the shaft. Da Vinci was likely aware of this toy and may have been considering ways of turning it into a full-size flying machine, but probably realized it would never be able to lift its own weight—much less the weight of a pilot—with

10. This is perhaps the first known example of a control technique known as "wing warping," which was later used with great success by the Wright brothers.

11. A prototype was built in the 1990's, based on da Vinci's drawings and using materials that would have been available to him in the sixteenth century, that actually flew for a short distance. However, his sketchy design was interpreted using modern knowledge of aerodynamic principles, likely making it more flyable than it would have been had da Vinci built and flown the device in his day.

the technologies available to him, and so it never went beyond the sketch stage. (Da Vinci's sketches of the strange craft may have been useful in another way, however: it is said that the great twentieth-century aircraft designer and inventor of the modern helicopter, Igor Sikorsky, was shown this drawing by his mother as a little boy and found within it inspiration to take up the study of helicopters, eventually making a name for himself by doing so. As such, the great Italian master's influence was to find its way even into the modern age of aeronautics.)

Probably da Vinci's most bizarre creation in regards to aeronautics, however, was a bowl-shaped contraption inside which a man stood and operated a set of flapping wings through a series of hand and foot operated treadles, pulleys, and drums. It's hard to imagine that a man as gifted as da Vinci wouldn't realize that such an apparatus could scarcely have been expected to fly, making us wonder if it wasn't merely a whimsical bit of fun by the great master, who may have been simply letting his imagination run wild. On the other hand, he may have considered it a serious proposal though, again, there is no evidence he ever acted upon it, which in retrospect is probably a good thing.

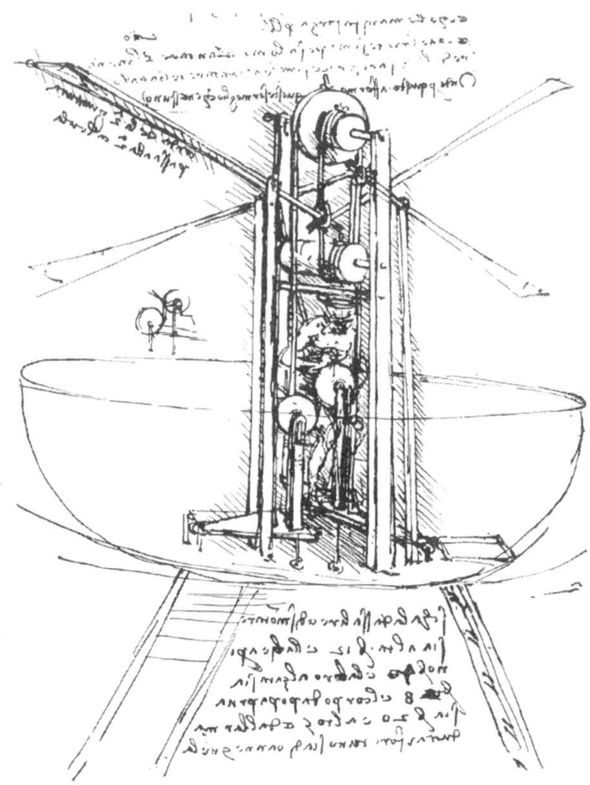

Da Vinci's paddle powered bowl may have been more of a product of the man's noteworthy imagination than an actual attempt to design a real flying machine.

It's unfortunate that da Vinci didn't put more thought and energy into his flying machines. It would have been interesting to see what a mind like his might have come up with had he pursued the study of aeronautics more seriously. Unfortunately, he was a man of varied interests, and as is the case with such geniuses, he apparently lacked the ability to devote himself to any one idea at the exclusion of all others. This made him in some ways an impractical genius—that is, a man who could conceive of many remarkable things but who lacked the ability to bring any one of them into existence.

What da Vinci did demonstrate with his drawings, however, was that Renaissance man was beginning to consider the principles of flight in a more scientific manner. Even in his corkscrew-shaped helicopter he demonstrated that man was capable of conceiving a means of achieving flight well outside the orthodox concepts of the era. This made his many sketches and notes the first primordial manifestations that flight was possible, thereby both laying the foundation for the serious study of aeronautics and providing inspiration for later aviators.

The Flying Boat of Francesco Lana de Terzi

I think it's safe to say that when one considers who the early pioneers of aviation might have been, few have a seventeenth century Jesuit priest in mind, but then again, the field of aviation is chocked full of such surprises (as the eleventh century English Benedictine monk, Eilmer of Malmesbury—previously mentioned—proved when he ascended to the top of a church tower and hurled himself into space). Regardless of stereotypes, however, it was a Jesuit priest by the name of Francesco Lana de Terzi

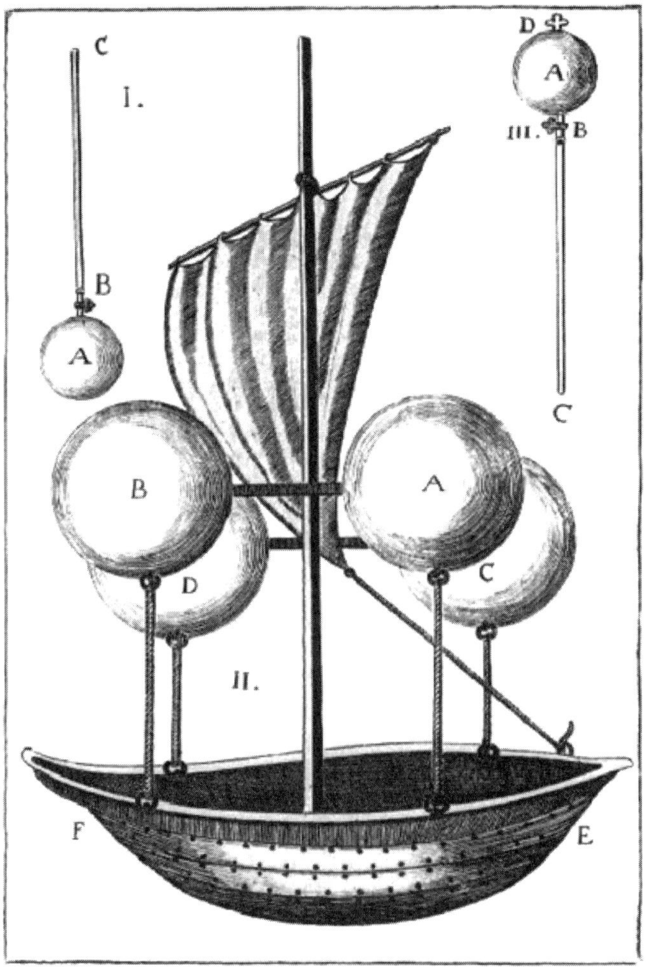

Francesco Lana de Terzi's flying boat, circa 1670. The ship was to be held aloft by four copper vacuum spheres, which, while a very futuristic concept at the time, would have proven entirely incapable of lifting the craft.

(1631-1687) who many consider to be the Father of Aeronautics—a title he acquired despite the fact that he never actually built anything or ever attempted to fly a machine himself throughout the course of his lifetime. So how did he acquire such a lofty title? He acquired it by being the first to turn aeronautics into a field of science by establishing "a theory of aerial navigation verified by mathematical accuracy", thereby laying the foundation from which later aviators would work.

A professor of physics and mathematics at Brescia, Italy, de Terzi became known for his aeronautical ideas when he published a book that included a chapter entitled *saggio di alcune invenzioni nuove premesso all'arte maestra* ("To test some promised new inventions of the master artist"). In this book, de Terzi described something he called a "flying ship" that was to ascend into the air via a series of vacuum spheres.

While his airship design might be considered quaint to modern sensibilities, it was quite revolutionary for the time. Basically a single-masted boat, what made it a flying machine was the attachment of four smaller masts that each held a single 24-foot diameter, 400-pound hollow copper sphere, which were to be pumped to vacuum conditions to ascend and then evacuated when one wished to descend. The idea was that the air-free sphere, in being lighter than the surrounding air, would presumably provide enough lift that six passengers could ride serenely over the Italian countryside with minimal effort. Unfortunately, no one possessed the ability to manufacture copper sheets thin enough to do the job, and even if they had, once the air was pumped from the spheres, they would have been flattened because of the pressure differential between them and the surrounding air.

While an impractical solution to the question of flight, of course, it was an interesting concept for its time and demonstrated that man was already thinking of different ways he might circumvent the pull of gravity. In any case, de Terzi had serious doubts about his machine, though it wasn't in regards to its airworthiness. He was afraid of how it might be used were it to be successful—especially in its potential military applications. As though he were capable of foreseeing the great zeppelin raids of World War I, in a somewhat prophetic statement, he wrote: "God will never allow that such a machine be built...because everybody realizes that no city would be safe from raids... iron weights, fireballs and bombs could be hurdled from a great height".

Perhaps this warning was de Terzi's greatest gift to humanity, which would ultimately fail to heed his words in later centuries, much to its own peril.

Bartolomeu de Gusmão's Balloons

De Terzi was not the only "man of the cloth" interested in aviation during this period. Shortly after learning of de Terzi's experiments with his flying boat, a young Portuguese priest named Bartolomeu de Gusmão (1685-1724) was inspired to present a petition to King John V of Portugal in 1709 in which he outlined, in some detail, his own idea for an airship.

His "airship" was quite different from de Terzi's concept, however; instead of proposing achieving lift through vacuum-filled copper spheres, Gusmão proposed spreading a huge sail over a boat-like craft that had tubes through which bellows would fill the sail with air. The vessel was also to be propelled by magnets encased in two hollow metal balls—though it's not entirely clear how the magnets were to pull the vessel through the air. Needless to say, the unconventional design would never have gotten off the ground, but that didn't quell the king's—or young Bartolomeu's—enthusiasm for aviation, and, with the King's spiritual and financial support, he spent the next few years undertaking a serious study of the field of aeronautics (which was quite an undertaking, especially considering how little was known of the field at the time).

It wasn't long before the young priest made a revolutionary discovery for the time: that heated air is lighter than the surrounding air and, when trapped within a lightweight, sealed envelope, can be induced to make an object rise into the air (a fact known to the ancient Chinese centuries earlier only to be "rediscovered" by Gusmão almost two thousand years later). This compelled Gusmão to produce a number of balloon-like designs, some of which were not all that different from what was to be seen flying over the skies of France less than a century later.

Not content to merely sketch out his ideas, however, Gusmão set about testing his discovery by building a small paper balloon, which he intended to send aloft before the royal court. Unfortunately, the balloon caught fire and had to be extinguished before it could rise. A few days later, however, the apparently optimistic and no less enthusiastic Gusmão was back, this time with a larger balloon. This one, however, proved successful when it rose nearly to the ceiling of the palace before the king's servants, fearing that the thing might start a fire if it made it to the rafters, brought it down in pieces. Wisely deciding to move his experiment outdoors, three days later a third flight was made before the king and his empress, this time with spectacular success; the balloon—which by some accounts was built in the shape of a bird with a tail and wings—rose nearly a kilometer (nearly 3,200 feet)[12] into the air before exhausting its flame and slowly falling into the Terreiro do Paço square in Lisbon. The king was so impressed with the demonstration that he granted the rights to any and all future flying ships to Gusmão, with death being the penalty for those who dared to intervene or copy his ideas. (Apparently, this was when patent and copyright laws had teeth in them.) The king also rewarded the inventor by appointing him to a professorship and making him a member of the prestigious Academia Real de Historia, which was quite an accomplishment for the twenty-five year old priest.

The Passarola—Gusmão's rather fanciful design for a "flying ship", circa 1709.

12. While this estimate of the balloon's altitude is probably exaggerated, it is not impossible it could have reached such a height.

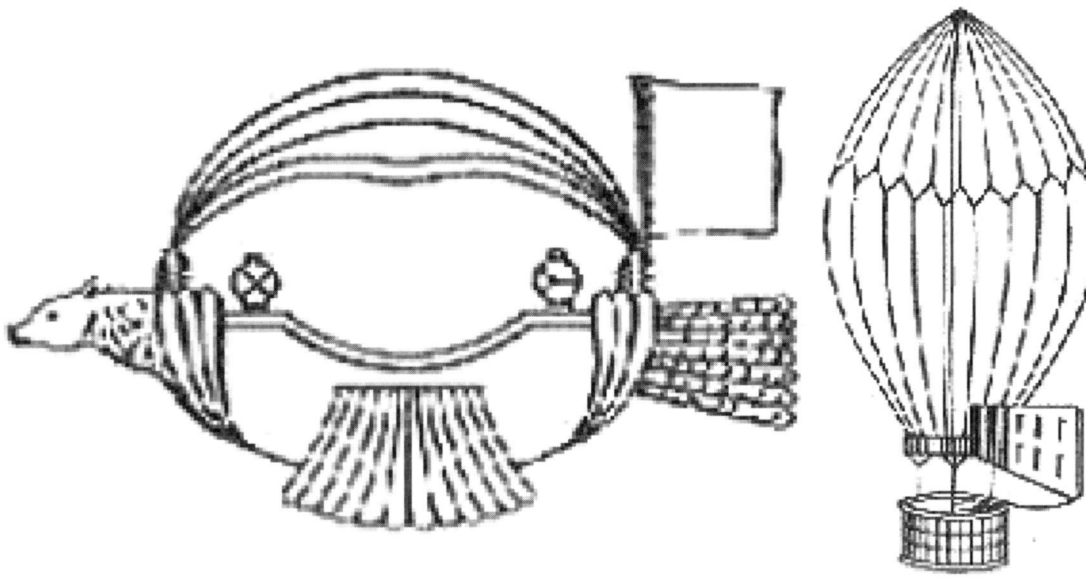

Speculative drawings of some of Gusmão's earliest balloon designs. The device on the right is very reminiscent of an early Montgolfier brothers balloon.

Buoyed by his success and now with the king's blessing, young Bartolomeu immediately busied himself with further experiments on his balloons and even started working on an airship design—a triangular gas-filled pyramid sent aloft by a flame in its center. Unfortunately, at some point he fell afoul of the religious authorities, who apparently considered his machines evidence of wizardry. It's unclear precisely what happened at that point; some sources claim he was brought before the Inquisition to explain his infernal devices, though others maintain he was brought before the group on an unrelated matter. In either case, deciding the atmosphere in Lisbon was not conducive to further airship research, he burned most of his papers and fled to Toledo, Spain to continue work on his pyramid-shaped balloon. Unfortunately while there he contracted a fever and died, never having a chance to see if his revolutionary airship might have flown. The unfortunate thing is that there was a good chance his balloon might have worked and, if it were big enough, that it might even have carried the first man aloft, beating the Montgolfier brothers (see next chapter) into the air by over half a century. As it is, we can only wonder how far the gifted priest from Lisbon might have gone had the political/religious atmosphere in Portugal been more open to his radical new inventions.

Swedenborg's "Reasonable Suggestion"

It was around this period that another man came on the scene with ideas about flight that were considered quite advanced at the time. Like da Vinci, this man is also considered one of the greatest minds of his era, and also like the great Florentine scholar, he too was also known for wide-ranging interests. That man was one of Sweden's greatest scientists, philosophers, and theologians, the Christian mystic Emmanuel Swedenborg (1688-1772).

Though better known for his later and somewhat controversial theological concepts, in his earlier years Swedenborg dabbled with, among other things, aeronautics. To that end, in 1714, while still a young man, he sketched out what is widely recognized as the first published description of a practical flying machine, which he later went on to describe in a periodical entitled *"Machine to Fly in the Air."*

Though fairly crude by modern standards, Swedenborg's design, known as the *Daedalion*, was actually quite advanced for the time. Essentially a glider that sported an elliptical wing with the long axis pointing in the direction of flight, the machine even had a cockpit, centered in the middle of the wood-framed wing. The forward-looking Swedenborg even equipped the *Daedalion* with a four-legged landing gear and a balance weight slung beneath the wing—two ideas that would be rediscovered a couple of cen-

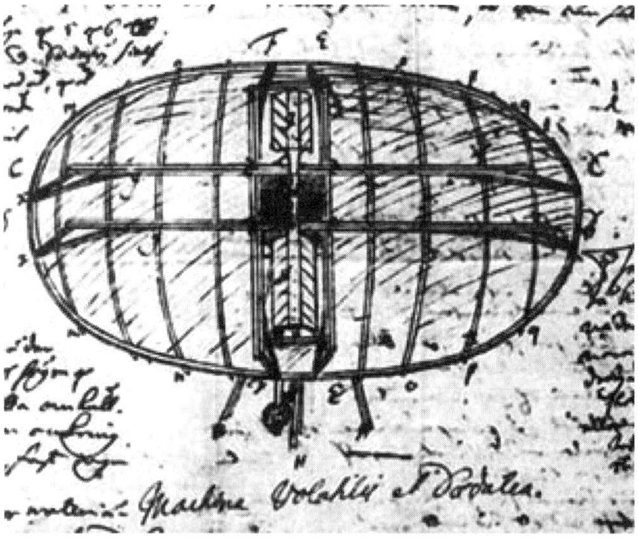

Sketch of a potentially viable flying machine, the Daedalion, described in one of Emmanuel Swedenborg's notebooks. It was, in effect, a hang glider with circular flapping wings; Swedenborg's craft even possessed a landing gear—a very futuristic feature at the time.

turies later. Power was to have been provided by a pair of vertically flapping paddles or oars, making Swedenborg's machine essentially a hybrid ornithopter.

Swedenborg had no interest in actually building his machine, of course, but he didn't consider that a problem. His motive in designing the *Daedalion* was merely to suggest a starting point for later serious aeronautical research. In fact, he was so confident that the problem of manned flight would soon be solved, he wrote in his journal:

> *It seems easier to talk of such a machine than to put it into actuality, for it requires greater force and less weight than exists in a human body. The science of mechanics might perhaps suggest a means, namely, a strong spiral spring. If these advantages and requisites are observed, perhaps in time to come someone might know how better to utilize our sketch and cause some addition to be made so as to accomplish that which we can only suggest. Yet there are sufficient proofs and examples from nature that such flights can take place without danger, although when the first trials are made, you may have to pay for the experience, and not mind an arm or leg.*

Curiously, Swedenborg's flying machine remained largely forgotten until the nineteenth century when a visiting American researcher discovered it in one of his notebooks. At first it was considered merely a curiosity, but in 1910 the Royal Aeronautical Society studied the sketch in more detail and declared it "the first reasonable suggestion to build a heavier-than-air machine." Later the *Daedalion* was analyzed by the Smithsonian Institute, which compared its features favorably with those of modern aircraft, and even built a model of Swedenborg's creation which it displayed for many years.

Marin Aguilera's Flying Machine

One more name belongs in the pantheon of aviation pioneers from antiquity, though it is one that few people outside of his native Spain know about. What makes this man a candidate for inclusion is that, unlike many of his contemporaries, he didn't just draw pictures of flying machines but, like Ibn Firnas and Eilmer of Malmesbury, the man may have actually built and successfully flown a heavier-than-air machine—and he did it all a century before either Lilienthal or the Wright brothers.

Diego Marín Aguilera (1757–1799) was a Spanish inventor who was famous for having devised a number of labor-saving devices, any one of which were impressive in their own right and should have earned him considerable respect in his native country. However, it was the flying machine he supposedly built in 1793 that he was best remembered for, despite the fact that it proved to be a personal disaster to him that nearly cost him both his life and his livelihood.

Of course, Marin didn't come up with the idea of flying like a bird spontaneously. In fact, he had spent most of his life studying the way eagles flew and had spent six years designing a device intended to duplicate their feat, but on a larger scale. Convinced he could fly if he built a machine that mimicked the way a bird's wings and tail worked in maintaining lift, balance, and control, he had the local blacksmith construct a pair of wrought iron "joints" devised to move like a fan and on the night of May 15, 1793, accompanied by the blacksmith and one of his sisters, Marín took his glider to the highest part of the castle of Coruña del Conde and, by the light of the full moon, launched himself into space. Remarkably, instead of plunging to his death as both his sister and the blacksmith probably expected, he supposedly glided some distance (by some accounts, nearly one thousand feet) before crash landing next to the Arandilla River when one of the metal joints broke. Fortunately Marin was not seriously injured in the crash landing, though he was angry at the blacksmith for failing to attach the joint properly.

Instead of being hailed as a hero for his daring feat, however, the intrepid aeronaut found himself shunned by most of the residents of the town, many of whom considered him a lunatic. Obviously unable to appreciate the magnitude of his accomplishment or realize the value of manned flight, it is said the townspeople even found and burned his flapping-wing creation, much to Marin's dismay. Disgraced and all but an outcast within his own town, he never attempted to fly again and spent the last years of his life depressed and despondent—a sad fate for a man of such intellect and bravery. One can only imagine how history might have been affected had the people of his era only understood the significance of what he accomplished and embraced it. If they had, aviation historians today may be speaking of Spain as having been the cradle of modern aviation and of Marin with a reverence usually reserved for the likes of Orville and Wilbur Wright.

Fortunately, history left Marin a better legacy. Today he is considered the "Father of Aviation" in Spain, with the Spanish Air Force even dedicating a monument to him next to the castle where he tried his hand at flying like an eagle. While it is not known with any degree of certainty if he actually flew, it is noteworthy to recognize what the American Institute of Aeronautics and Astronautics wrote about him: "It is impossible to determine how much truth there is to the story of Marín, but it seems that he did achieve some gliding flight, surviving after structural failure and a crash landing. Marín, who had no formal scientific education, was endowed with a special technical ingenuity and is a good example of the ageless human aspiration toward flight." Obviously, the people of his era owe Marin a much deserved and long overdue apology.

An old Spanish Air Force Lockheed T-33 jet trainer stands as a tribute to the unheralded Spanish Father of Aviation, Diego Marin Aguilera, next to Castle Coruña del Conde, the site of Marin's alleged 1793 flight.

Conclusions

Since there is no physical evidence, detailed drawings, or living witnesses to attest to their efforts, it's impossible to determine if men like Ibn Furnas, Eilmer of Malmesbury, or Diego Marin actually flew, making it easy for some to consign these stories of early flight to the realm of fiction. However, this is premature at best and downright cynical at worse. While such feats may appear most improbable from our lofty vantage point in the twenty-first century, it's equally difficult to imagine each of these stories to be purely products of the imagination or ancient mythology without at least some basis in fact.

And if there is any truth to any of these stories, we need to be impressed with these accomplishments. In fact, when one considers how these men achieved so much without the benefit of prior experience, a good foundational understanding of the basic physics of flight, or the technology available to us today, their accomplishments could be considered in many ways even more impressive than our own. As such, we need to acknowledge these early pioneers of flight—both those who merely conceived of flight and those who actually made hands-on attempts at accomplishing it—for their ingenuity, their courage, and their vision, for without them to inspire future generations of aviators, the extraordinary progress of the nineteenth and twentieth centuries may not have been possible.

Chapter Three: The Great Eighteenth Century Balloon War

For some mysterious reason, the years between the American and French revolutions saw a flurry of aeronautical activity, especially in France, and specifically having to do with balloons. Exactly why this incredible era of innovation came about is uncertain. Some claim it was a result of the Age of Enlightenment then sweeping across Europe which created a newfound atmosphere which not only made creativity possible but even encouraged it. More likely, however, it was a by-product of two things: the introduction of rubber and other light-weight materials that made constructing an impermeable envelope possible; and the discovery of hydrogen in 1766, the combination of which made lighter-than-air flight inevitable. How these elements came together during the last two decades of the eighteenth century to kick off a period of incredible aeronautical progress—one marked by innovation and ingenuity, along with extraordinary bravery on the part of many—is a story that is well worth the retelling.

And to think, it all started, just as it seems do with all aeronautical advances, with the work of a pair of brothers.

The Montgolfier Brothers

When speaking of the era, no names stand out as prominently as do those of the famous Montgolfier brothers, Joseph-Michel (1740-1810) and Jacques-Etienne (1745-1799), two men who did more to initiate an era of unprecedented aeronautical innovation than any men in modern history. Their story is especially remarkable in light of the fact that neither of them were scientists or engineers. Instead, they were a pair of highly successful and wealthy paper manufacturers from Annonay, France. However, as is so often the case in history, it was they fate chose to play a most unlikely role in the development of aviation that was to resonate well into the twentieth century and beyond.

In retrospect, it may not be all that remarkable that two paper merchants from France would be the ones to introduce the world to hot-air ballooning. First, they worked on a daily basis with the very materials required to construct a practical balloon which—along with having considerable financial resources—made them one of the few people in the world at the time with the means to undertake such a project. Even more, the brothers were a perfect mix of inventiveness and practicality which, when combined, gave them the discipline and ingenuity to make them the ideal candidates to design and build the first successful modern hot-air balloons the world had ever known.

The Montgolfier brothers, the fathers of modern ballooning.

The brothers didn't simply decide one day to test their hands at lighter-than-air travel, of course. As is frequently true of many great inventors, their interests came about through a series of random events—in this case, as a result of Joseph's observation that laundry drying over an open flame occasionally formed pockets and billowed upwards from the heat. This piqued his curiosity as to whether such a phenomenon might be able to lift more than laundry, so he built a box-like chamber out of very thin wood and, covering the sides and top with lightweight taffeta, lit some paper under it to see what would happen. Somewhat surprised to see the contraption quickly lift off its stand and collide with the ceiling, the excited Joseph immediately recruited his brother to help him build an even larger cube to see if he could replicate the feat. Quickly constructing a second, larger box, together they watched in amazement as the heated cube quickly rose and floated away, and together they immediately began considering how such a phenomenon could be harnessed in a practical way.

What induced them to take the next step and build an even larger balloon is not entirely clear, but after performing more experiments on ever larger paper envelopes, they decided to build a much larger balloon with the intention of demonstrating it to the public. Abandoning their pyramid and cube shaped envelopes in favor of a spherical envelope, after several months of tedious and careful work, they produced the world's first hot-air balloon.

The craft was quite an engineering marvel for the time. Made of sackcloth and composed of three thin layers of paper to make it airtight, it was held together by 1,800 buttons and a reinforced by a fish net of cord covering the outside of the envelope. While this gave it a somewhat ornate appearance, the brothers remained uncertain as to whether such a massive device could actually fly. Realizing there was only one way to find out—and with their rather substantial reputations on the line in case of failure—on June 4, 1783 they lit the stove at the base of their creation and, in front of a group of dignitaries, watched as the monstrosity majestically rose into the sky to an estimated altitude of nearly a mile and floated off, much to the astonishment and delight of the crowd.

As a result of this rather modest attempt at flight, Paris was immediately smitten with what was later referred to as "balloon fever," making the brothers instant celebrities and putting considerable pressure on them to build another balloon. They were intent on doing more than merely replicating their earlier success, however; they wanted to see if it were possible to build a balloon large and robust enough to carry a man into the air—a feat that up to then would have been considered the stuff of fairy tales. Bringing their full attention and talents to bear on this new challenge, within a few months they unveiled their newest creation to an anxious public and the era of manned flight was set to dawn.

Uncertain what possible physical affects ascending hundreds or even thousands of feet into the sky would have on a human being, however, before sending a man aloft in the unproven device, they decided to send some non-human aviators up first to see if they would suffer any ill effects. Selecting a sheep, a duck, and a rooster, on the morning of September 19, 1783, the balloon—since bestowed with the romantic sobriquet *Aerostat Réveillon*—was sent aloft into a perfect blue sky with its three occupants bleating, quacking, and crowing in unison. Not only did the balloon fly perfectly—even reaching an altitude of 1,500 feet and floating a distance of nearly two miles before being brought down safely (all before the watchful eyes of King Louis XVI and Queen Marie Antoinette at the royal palace at Versailles)—but it was quickly determined that the three unwilling aeronauts were none the worse for the venture. Declaring the flight a great success, they immediately set about building an even bigger balloon for their next human-manned ascent, and soon they unveiled their latest masterpiece to an anxious public. This one, at 75 feet tall and with a diameter of nearly 50 feet, was close in size to a modern hot-air balloon and, as such, the largest man-made aerial device ever seen.

Also, and probably in recognition that they were making history, they even had their balloon garishly decorated with golden flourishes, signs of the zodiac, and other embellishments, making it as much a work of art as a work of aeronautics. It even had the sun with Louis XVI's face in the center—evidence that the brothers understood the value of keeping in the king's good graces. Such a paint scheme all but guaranteed that even if the balloon failed to fly, it would be a memorable sight to behold.

Finally, on the morning of October 15, 1783, the brothers and their spectacular balloon were ready. Even though Joseph was the chief designer, it was his brother Etienne who had the honor of being the first to ascend in the craft, which he did to great applause. The balloon remained tethered to the ground throughout the entire flight, however, denying Etienne the honor of being the first man to fly in a free-floating balloon. (That honor would go to another Frenchman, Jean-François Pilâtre de Rozier, who would make the ascent a few weeks later in one of their balloons—see next entry.) Why the brothers chose to keep the balloon tethered is unclear. Perhaps it was because they knew that, once the balloon left the safety of the ground, it would be at the fickle mercy of the wind, and since neither of them knew how to fly a balloon, they wisely decided to play it safe. On the other hand, they may simply have not wanted to let it float off and so be forced to build a new one. In either case, the cautious flight didn't seem to hurt their reputation, as their accomplishment was heralded throughout France as a first, and honor and praise were heaped upon the brothers by their grateful fellow citizens.

The Montgolfier brother's first manned balloon launch, October 15, 1783.

Curiously, it seems that after this triumph—and especially once de Rozier made his series of flights—they largely lost interest in ballooning. While Joseph was to make only one more flight in January of 1784—and one that nearly ended in disaster when the balloon was forced to make a crash landing just fifteen minutes after taking off—after that the brothers largely put their balloon experiments behind them and moved on to other pursuits. Why they did this remains a source of some speculation. Some conclude that after Joseph's close call, they decided it was too dangerous a pursuit and gave it up. Others, however, speculate they may have given ballooning up in frustration once they were unable to find a way to make the balloons steerable. But perhaps the simplest explanation is that once they accomplished their goal of building and flying a manned balloon, they were content to move on to more traditional pursuits. Whatever their rationale, the brothers returned to the family paper mill in Annonay, where they put their minds together producing other devices of more practical value until Etienne's death in 1799 and Joseph's eleven years later.

So what conclusions can be made regarding the Montgolfier brothers brief excursion into ballooning? Certainly, in retrospect, it could be argued that they really hadn't invented anything new, for the ancient Chinese and, quite possibly, the Portuguese priest Bartolomeu de Gusmão (see chapter two), had both discovered that heated air had the capacity to make an airtight envelope float. What they did do, however, was successfully send a man aloft for the first time in recorded history and introduce flight to the industrialized nations of Europe in a spectacular way, thereby setting the stage for the next great leap in aviation. Once they flew their massive balloons over Paris, it could truly be said that man was never going to be content to remain earthbound again, making their contribution to the world of aviation among the most important of all the earliest efforts to fly.

The Flights of Pilâtre de Rozier

While the flights of the Montgolfier brothers are much better known to aviation buffs, it was another Frenchman who was to actually initiate the age of flight by being the first man—along with the Marquis d'Arlandes—to go aloft in a free flying balloon. Unfortunately, he would also be known as the first man to give his life for his passion, though he would be far from the last. That man's name was Jean-François Pilâtre de Rozier (1754-1785).

A chemistry and physics teacher from Metz, de Rozier was apparently so inspired when he witnessed the first public demonstration of the Montgolfier brothers' animal-laden balloon flight in June of 1783, that he approached King Louis XVI to ask if he might not be the first man to make a manned ascent. At first the king was hesitant to send a member of the French aristocracy airborne in an effort to prove a largely untested technology and proposed sending two condemned prisoners aloft in case something went wrong, but de Rozier was able to persuade him that the honor should go to men of some stature instead of two mere criminals. Once the Marquis d'Arlandes volunteered to accompany de Rozier on the ascent—and perhaps aware that the flight of a noblemen would be good press—the king finally relented and, after several tethered test flights proved successful, the men were ready to attempt an untethered flight. Secure in the basket of one of the Montgolfier brothers' balloons (the brothers, while not inclined to fly their own creations, were quite willing to let others fly them and to that end built a number of balloons between 1783 and 1784), shortly after

Highly idealized etching of the first untethered balloon flight in history made by de Rozier and the Marquis d'Arlandes on November 21, 1783.

2:00 P.M. on the afternoon of November 21, 1783, the duo took off from the garden of the Château de Muette and for the next twenty-five minutes flew serenely over Paris before setting down safely some miles away. Accomplishing the feat in the presence of the king and thousands of enthralled Parisians, the flight made de Rozier and d'Arlandes overnight celebrities and afforded them the honor of being the first men to ascend in a free-flying balloon.

Buoyed by this success, de Rozier chose to make ballooning his passion and, working closely with the Montgolfiers over the next few months, he made several more untethered flights, all of them quite successful. His most memorable flight, however, occurred in June of 1784, when de Rozier, accompanied by fellow balloonist Joseph Proust, made a special flight in a modified Montgolfier balloon before the kings of Sweden and France and the French king's wife, Marie Antoinette. The massive balloon, appropriately christened *La Marie-Antoinette* after the queen, travelled over forty miles in just forty-five minutes and reached an altitude of 10,000 feet before cold and turbulence forced it to land near the Chantilly forest, making it one of the most observed balloon flights in history up to that time.

De Rozier quickly demonstrated himself to be a man who lived for a challenge, and he was soon determined to attempt an even more daring flight. What he had in mind, however, especially given the still primitive technology available to them, was particularly dangerous: a crossing of the English Channel. Such a flight, coming so soon after the balloon's invention and with much of the technology still unproven, would be truly a remarkable accomplishment, but de Rozier believed it could be done. Knowing that a hot-air balloon would not have the necessary range to make a flight of such a distance, however, de Rozier set about building a hybrid balloon that would operate off both hot air and hydrogen. Before he could perfect his design, however, another Frenchman, Jean-Pierre Blanchard, beat him across the channel in a fully hydrogen-inflated balloon, rendering any further attempts at making the crossing redundant. However, this was not enough to deter de Rozier from his ambitions. As Blanchard had made the crossing from the English side of the channel, he would simply make the crossing from the French side. Working feverishly to perfect his balloon and despite the fact that his hybrid system remained largely untested, by June of 1785, he decided he was ready to attempt the crossing and set out for the coast. It was to prove to be a fateful decision.

Accompanied by his companion, Pierre Romain, early on the morning of June 15, 1785, the two men set off from Boulogne-sur-Mer on their historic attempt and at first things appeared to be going well as they started drifting beneath a flawless blue sky towards the coast of England, less than twenty miles away. Moments later, however, it became apparent the flight was in trouble when the wind abruptly shifted and started pushing the balloon back over dry land. With no means of countering nature, it was immediately apparent that the attempted crossing would have to wait for more favorable conditions and de Rozier began looking for a place to set his balloon down. Just as he began to descend, however, tragedy struck when the envelope suddenly and inexplicably deflated and, while hundreds of onlookers watched in horror, the craft plunged to the ground from an altitude of nearly 1,500 feet, killing both men on impact.

Dramatic illustration of the deaths of de Rozier and Romain after their balloon suddenly deflated while attempting to cross the English Channel, plunging them 1,500 feet to their deaths.

The crash not only ended the life of one of France's most promising aviators, but it threw the country into a state of morning (and may have even played a role in persuading others—especially the Montgolfier brothers—from attempting any more flights in the future). It also leaves one to wonder what the man—whose ambition, ingenuity, and personal bravery were beyond question—might have accomplished had he lived. Unfortunately, the world would never know. De Rozier did, however, leave one lasting legacy: the modern hybrid gas and hot-air balloon is named the Rozier balloon after his pioneering design.

Jacque Charles; Father of the Hydrogen Balloon

The problem with the hot-air balloon was that it could not carry enough fuel to keep its burner lit more than a few minutes, giving it a very short range and reducing it to little more than a novelty. Additionally, since they were sent aloft by heated air generated from an open flame—a most dangerous proposition considering the fragile and highly combustible nature of balloons at the time—balloons were considered too dangerous to be of practical value. What was needed was a lifting agent that wasn't dependent upon an open flame and that would work in all weather. It wasn't long before the answer was found when English scientist Henry Cavendish discovered hydrogen in 1766, giving the world the one element it needed to make ballooning not only relatively safe[13], but give balloons nearly unlimited range.

Villagers in the town of Gonesse, convinced that Professor Charles' hydrogen balloon was an emissary of the devil, are portrayed here attacking the diabolical device with axes and pitchforks.

It was the French inventor, scientist, and mathematician Jacques Alexandre César Charles (1746-1823) who was the first to take full advantage of this new element and use it in a balloon, however. Working in conjunction with two French engineers of some renown, Anne-Jean and Nicolas-Louis Robert (see next section), he designed and built the first hydrogen-filled balloon in the brother's workshop in Paris. Though small when compared to the Montgolfier brothers' monstrous balloons, the rubberized silk balloon more than proved to live up to expectations. Launched in August 1783 from the Champ de Mars in Paris with six thousand paying customers watching from below, the crowd stood mesmerized as the tiny, unmanned balloon easily ascended to a height of some 3,000 feet and floated serenely away before finally landing in the village of Gonesse—some fifteen miles to the north of the city. Unfortunately, the frightened villagers of Gonesse had no idea they were being visited by the first hydrogen-filled balloon in history and dutifully set upon the thing, tearing it to shreds with knives, axes, and pitchforks.

Eighteenth-century illustration of the first flight of a hydrogen balloon made by Jacques Charles and Nicolas-Louis Robert on December 1, 1783.

[13] Of course, hydrogen is highly combustible itself, but with a degree of care it is reasonably safe to use, especially in an unpowered balloon and before the era of electrical powerlines.

Undeterred by the local populaces decidedly hostile reaction to their new device and convinced it could safely carry a man aloft, Charles and the Robert brothers set about building a much larger balloon ten times the size of their earlier effort. In the first of many technological innovations that were to be developed as a result of ballooning, the men developed and fitted a hydrogen release valve to the envelope, which would make it possible to descend (thereby giving the pilot some degree of vertical control). Curious as to whether their new balloon would live up to expectations, on December 1, 1783, Charles and Nicolas-Louis Robert ascended from the Jardin des Tuileries in Paris and, reaching a height of nearly 1,800 feet, flew an astonishing twenty-five miles in just over two hours. In doing so, they not only easily broke the altitude and distance records set by de Rozier just weeks earlier, but proved that hydrogen was a superior lifting agent.

But the day almost ended in tragedy when, within minutes of landing, the professor, perhaps in a fit of over exuberance, decided he wanted to make a second flight. This time, however, he would make the flight alone because the balloon had lost some of its hydrogen, making it less buoyant. Unfortunately, he overestimated how much gas had been lost and the balloon, now carrying only half the weight it had earlier, ascended more rapidly than he had anticipated. Quickly reaching the dizzying altitude of some 10,000 feet, Charles frantically began valving off much of the hydrogen in an effort to descend. He made it back to terra firma safely, but the descent had been so rapid he suffered acute pain in his ears, apparently persuading him to give up flying his unruly devices personally. As a result—and possibly also because he had accomplished what he intended to do—Jacques Charles never flew again but was content to spend the rest of his life teaching physics at the Conservatoire des Arts at Métiers. His legacy as the first to use hydrogen successfully in a balloon was enough, however, to forever enshrine his name in aviation history.

The Robert Brothers

While Professor Charles was content to remain earthbound the rest of his life, his partners in the balloon-building business, Anne-Jean Robert (1758-1820) and Nicolas-Louis Robert (1760-1820), were not. Their interests were not confined to merely ascending in a balloon, however. They, instead, were intent on creating a balloon that could be steered—a prospect they were to spend a number of years pursuing.

Working in cooperation with Professor Charles (who apparently was willing to keep his hand in the business even if no longer an active participant), they decided to build an elongated, steerable craft based loosely of the design of another Frenchman, Jean Baptiste Meusnier (1754 to 1793), who had presented a design for an early dirigible to the Academy of Sciences that the brothers felt showed great promise. Of course, Meusnier's proposed airship was nearly 200 feet long and was to be powered by three propellers driven by 80 men from inside a boat-shaped basket suspended from the canopy—obviously far too large to be constructed with the technologies available at the time—but the brothers felt the basic idea sound. Following his general elliptical-shaped design, they even incorporated an internal ballonette to help the craft

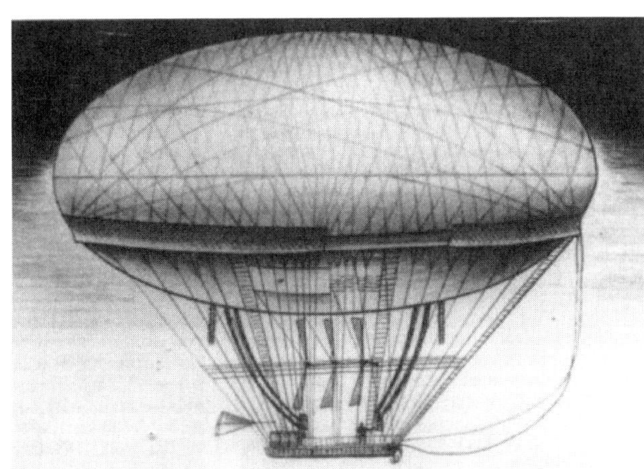

Meusnier's design for a steerable airship served as the inspiration for the Robert brothers' craft, which, though setting a distance record, proved incapable of being controlled in flight.

maintain its shape. Instead of eighty men turning three propellers to steer the craft, however, they settled on several sets of oars to provide propulsion and directional control.

Flying from Saint-Cloud to Meudon on July 15, 1784 with two men, M. Collin-Hullin and Louis Philippe II, the duke of Chartres, onboard, the flight proved to be something of a disappointment. Not only did the oars prove useless in steering the craft, but the flight nearly ended in disaster when the lack of a gas release valve induced the duke to slash the ballonette with his infantry sabre in an attempt to prevent the balloon from rupturing when they reached 12,000 feet, resulting in a barely controlled crash landing that left the balloon badly damaged but its passengers fortunately uninjured.

Undeterred by their spectacularly unsuccessful first test flight, however, the brothers quickly repaired the damage, and together with M. Collin-Hullin (the duke apparently having chosen not to try his luck a second time), the three went aloft again in September, 1784, this time managing to cover the one hundred forty miles from Paris to Beuvry in just under seven hours. In so doing, they made the first flight in aviation history of over one hundred kilometers and set a distance mark that was to stand for many years. Of course, they never were able to control the balloon's direction of flight, but the flight was still considered an overall success in that the craft flew as advertised—though without directional control it was little more than a misshapen elliptical balloon rather than a true airship.

Curiously, and like a number of other balloon pioneers of the era, after this success the brothers apparently lost interest in any further aeronautical endeavors and moved on to other pursuits. This is unfortunate, for it appears that with their elongated rather than spherical configuration, they were onto something, which, with a bit more work, might have eventually evolved into the first true dirigible and rewritten history in the process.

Jean-Pierre Blanchard and the First Crossing of the English Channel

After the success of Jacque Charles and the Robert brothers, hydrogen balloons became the preferred lifting agent for all balloons of the era and a part of all major ballooning accomplishments over the next few years. Among the most notable—and perhaps most dangerous at the time—was the attempt to cross the English Channel which, while only a dozen miles across at its narrowest, could be an unpredictable body of water prone to sudden shifts in wind direction, poor visibility, and unexpected rain squalls. Regardless of the danger such a feat promised, however, there were several balloon enthusiasts intent on being the first to make the crossing, the best known being the aforementioned de Rozier. However, it would be another Frenchman—little known up to that time—who would be the first to actually accomplish it.

Jean-Pierre Blanchard (1753-1809) was a French inventor who was so impressed by the flights of the Montgolfier brothers and Jacques Charles that he decided to build his own balloon. Instead of using hot air to provide lift, however, he decided to use the recently discovered element hydrogen instead.

His first experience flying the balloon did not start out well, however, when a drunken spectator slashed at the balloon's mooring ropes and oars with his sword after being refused a place on board. After subduing the inebriated man and determining that the damage was superficial, Blanchard decided to press on. Flying the balloon a considerable distance before gently landing in a nearby field, the flight not only clearly demonstrated the superior lifting qualities of hydrogen, but made him something of an overnight celebrity in the process.

Invigorated by his success and realizing that hydrogen made flights of great distance and duration possible, he decided to attempt a crossing of the channel itself. After determining that the air currents would be more favorable if he flew from England to France, he left France and carried out a number of short test flights over the English countryside as he continued to refine his hydrogen balloon design. Finally, on January 7, 1785 he—along with his American companion, Dr. John Jeffries of Boston—decided the conditions were perfect for a crossing and they took off from Dover Castle and set out over the channel. Fortunately, the balloon performed perfectly and in just over two-and-a-half hours the duo landed safely in Guînes, France, a feat which, considering the primitive level of technology available at the time and the still dangerous nature of ballooning in general, made the accomplishment every bit as remarkable as Lindbergh's solo crossing of the Atlantic in 1927.

But that wasn't the end of Blanchard's endeavors. Awarded a substantial pension by King Louis XVI for his accomplishment, soon the tireless aviator was touring Europe in his balloon, making flights on an almost weekly basis throughout the continent. He even travelled to America, where, in January of 1793, he made the first balloon flight in North America. What was especially interesting about this flight is that it was witnessed by then President George Washington and no less than four future presidents (Adams, Jefferson, Madison, and Monroe).

Another interest of Blanchard's was the parachute. Up to then, parachutes were not particularly practical because of their large size, but Blanchard came up with a much lighter and more compact version made out of folded silk. His new parachute proved to not only be a great improvement over previous configurations, but it quite possibly saved his life when he had to use one to jump when one of his balloons ruptured during a flight. The fact that the chute worked as advertised not only proved to be a boon to ballooning safety, but also proved to be a popular attraction to visitors, who were willing to pay good money to watch Blanchard jump from his balloons. This, of course, afforded him the steady income he needed to continue his various aeronautical pursuits and guaranteed he would remain flying as

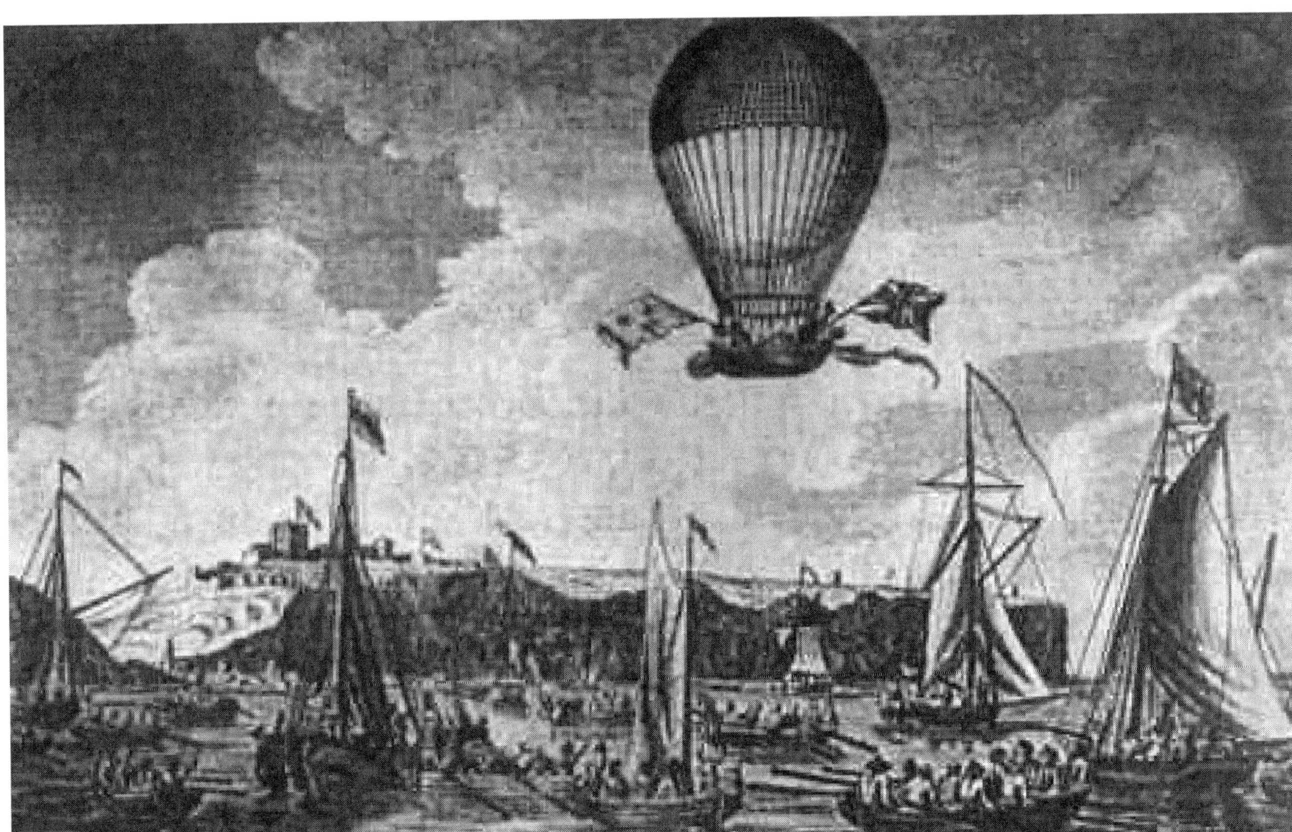

Jean-Pierre Blanchard and John Jeffries crossing the English Channel by air, January, 1785.

long as the crowds wanted to be thrilled by watching a man jump from a balloon and, of course, as long as the parachutes continued to work as designed.

Unfortunately, not all of his ideas worked as well. One curious aspect of Blanchard's early balloons was his insistence—much like the Robert brothers—that a balloon could be steered through the air with oars much like a boat. As others quickly learned, however—and Blanchard eventually came to realize himself—was that such a device was futile. While oars and paddle wheels were ineffective as a means of propulsion, curiously no one thought of trying a propeller (which, interesting, had been incorporated into Meusnier's elongated design that the Robert's brothers had used as a template for their own "steerable" balloon). Although no power plant existed at the time capable of turning such a device at a sufficient speed, the fact that it was not at least attempted seems inexplicable and remains one of the great imponderables surrounding early balloon designs.

In any case, Blanchard continued to wow crowds with his balloon antics for almost twenty years, making quite a name for himself and becoming the preeminent balloonist in all of Europe. In 1804 he married and even somehow managed to convince his wife, Marie Madeleine-Sophie Armant (better known as "Madame Sophie") to join him in the skies and together they made numerous ascents together as the first husband and wife balloonist team.

In 1809, however, it all came to an abrupt end when Blanchard suffered a heart attack during one of his flights and fell from his balloon, seriously injuring himself. Several weeks later he died from his injuries (though it's not certain whether he died from injuries sustained during the fall or from a heart attack), thus ending the life of one of the great aviators—and showmen—of the era and bringing an end to the exploits of Jean-Pierre and Madame Sophie Blanchard.

Actually, that's not quite the end of the story. While it might be imagined that Blanchard's death would have ended Madame Sophie's aerial exhibitions as well, it seems that during their high flying exploits, Madame Sophie had become so enamored with the "ballooning bug" that she decided to pick up where her husband left off and continued to support herself with ballooning demonstrations for the next ten years. In fact, the novelty of being a woman balloonist made her in some ways even more famous than her husband and she was called upon to entertain such notables as Napoleon Bonaparte and King Louis XVIII. It could even be argued that for a time, she was one of the most famous women in Europe renowned not only for her aerial exploits but for making her own death-defying parachute jumps in front of huge crowds.

Highly romanticized portrait of the indomitable Sophie Blanchard. Note the ludicrously undersized but suitably ornate gondola.

But it wasn't all glory for Madame Sophie. She had to endure numerous hard landings, and lost consciousness on several occasions when her balloon rose too high into the air. She also endured frostbite and even almost drowned once when her balloon crashed in a marsh. It was the inherent danger of her chosen profession, in fact, that was to eventually cost her her life when in July of 1819 her balloon caught fire and crashed while she was putting on an exhibition at the Tivoli Gardens in Paris. Apparently, it was common for her to attach fireworks to the gondola of the balloon and fire them off once she reach altitude, putting on a dazzling display. On this day, however, it appears that some of the fireworks had been knocked out of position by a tree as she ascended and, despite the cries of some in the crowd trying to warn her, she lit the fuses. Inevitably, the pyrotechnics ignited the hydrogen in the envelope, bringing the balloon down in flames.

Actually, the intrepid aviatrix almost survived the mishap. Demonstrating superior piloting skills, she somehow managed to check the speed of the balloon's fall by jettisoning ballast, making it appear that she might just suffer another hard landing. Unfortunately, however, the balloon's basket smashed into a roof of a tall building and Madame Sophie, tangled in the basket's ropes and unable to free herself, was jettisoned onto the street below, where she died of a broken neck. This tragedy made her famous as not only being the first aviatrix in history, but the first woman to die in an aeronautical accident.[14]

Conclusions

It's apparent that the later decades of the eighteenth century were an important time in the history of aviation, for that was when human beings took their first tentative steps into the world of flight. Plus, with the discovery of hydrogen, the foundation had been laid for the ever more sophisticated and successful airships that would emerge in the next century, culminating in the massive behemoths of Graf von Zeppelin a century later and his technological marvels of the skies.

It was only the lack of a proper power-plant—and with it, the means of controlling the direction of a balloon through the sky—that prevented the earliest aviators from making the balloon more than just a curiosity. However, as the nineteenth century dawned, new technologies were beginning to emerge—most importantly among them being the steam engine—that promised to one day make controlled flight possible. It would take many decades for the technology to be developed to the point that would make the airship truly practical, but it portended that the remarkable era of aviation that was to play such a major role in the twentieth century was on the horizon, after which the world would never be quite the same again.

14. It should be noted that Sophie Blanchard, while the first woman to be known for her exploits in the air, was not the first woman to fly in a balloon. That honor goes to Madame Thible, who first flew on June 4, 1784. (See chapter eighteen.)

Chapter Four: Sir George Cayley: The "Father of Aviation" 4

During the late eighteenth century, aeronautics was seen more as a form of entertainment than a serious field of study. One problem was that while the balloons flown by the Montgolfier brothers, Jean-Pierre Blanchard, and Jacques Charles were technological wonders of their era, the principles by which they flew were simple and readily understood, giving the science community little to contemplate in regards to flight. Ballooning was very straight forward; inflate a sealed envelope with hot air or hydrogen, and it ascended. But while this allowed man to leave the ground, it was not truly flying in the truest sense of the word. As far as controlled flight—that is, flight the way the birds did it—was concerned, science was no closer to understanding the forces that governed flight than they were a century earlier, and it was this lack of understanding that remained the greatest single obstacle to man ever joining the birds in the sky.

By the dawn of the nineteenth century, however, all of that was to change when an English scientist took it upon himself to undertake the first truly scientific study of the aerodynamic forces that governed flight, thereby laying the groundwork for scores of later aviators to follow. That man's name was Sir George Cayley, whom many today consider to be the "Father of Aviation."

Sir George Cayley (1773-1857) was a man who seemed uniquely born into the role fate was to have him play. Born in Scarborough, England, in 1773 to an important Yorkshire family (his father was Sir Thomas Cayley, the fifth Baron of Brompton), young George grew up in a privileged background, which made it possible for him to acquire a good education—an opportunity few of his countrymen were so privileged to enjoy at the time—thereby providing him the basic intellectual discipline needed to pursue any of the sciences. But even more important was that he was blessed with a strong mother who was, according to most biographers, a well-educated and artistic woman who encouraged young George in many of his interests and took it upon herself to make sure that he had every opportunity to develop his talents. One way she did this is by allowing him considerably more freedom to explore various interests than was common for higher class children in the eighteenth century. For example, noticing that George was developing a keen interest in engineering, she permitted him to form a relationship with the village watchmaker despite the fact that such relationships—a boy from the aristocracy making friends with trades-people—were contrary to the social norms of the times. She also encouraged him to keep a journal of his observations, which she read on a regular basis to determine his mental development and note his wide range of interests.

While young George could have pursued any area of study he chose (his range of interests during his adolescence ran the gamut from nature and botany to engineering and science) it was aviation that held a special fascination for him (which may have been a result of reading of the exploits of de Rozier and Blanchard, which were taking place at the time). However, it wasn't lighter-than-air flight that young Cayley was most interested in; he wanted to understand how heavier-than-air objects could fly through the air without the benefit of a gaseous lifting agent. As a result, he spent hours watching birds in flight and carefully observing the way they flew. He was especially intrigued by the way they gained lift by both flapping their wings and simply soaring on air currents, forcing him to wonder if the "magic" of flight wasn't in some way associated with the shape of their wings. These early studies were to prove instrumental in helping him later to come to understand the most basic of all aeronautical principals: that air flowing over a curved lifting surface provided lift—an important observation that would prove vital to later scientists and inventors.

It wasn't until 1792, however, shortly after Cayley graduated from London's Hackney College, that he began exploring the nature of flight in earnest. He started small by building himself a simple helicopter rotor from feathers and other bits and pieces and studying it to see if he could understand the principles that made it fly. Through this—and as a result of his experience working in a number of fields of science and technology—he hit upon the basic forces that govern flight, which he decided to illustrate by etching the first modern airplane—complete with a fuselage and wings—onto a small silver disk in 1799. What he came up, while crude and undeveloped, was in effect a "template" which later aviators would use to construct the first heavier-than-air machines of the modern era.

What was especially interesting about the design of Cayley's plane was how far ahead of its time it was. In fact, it compared favorably to aircraft designed a century later, possessing a fixed main wing, a fuselage, and even a cruciform tail unit with surfaces for vertical and horizontal controls. In addition, Cayley incorporated such things as a cockpit for the pilot and a rudimentary means of propulsion that consisted of revolving vanes, a precursor to the propeller. On the opposite side of the coin he etched a simple diagram outlining the three basic elements of flight—lift, drag, and thrust—that would be the key to heavier-than-air flight. In doing so, and a full century before the Wright brothers would fly their first glider, Cayley established the basic configuration of the modern airplane and articulated the principles that made flight possible, making him truly a visionary of the highest magnitude.

Not content merely with hypothesizing his ideas, however, Cayley next went on to construct a series of models designed to test his theories. Building a number of devices designed to test various configurations of air foils, through these devices Cayley came to appreciate the fact that, while a curved airfoil would provide lift, it was thrust that was the key component in achieving that lift. In fact, he very

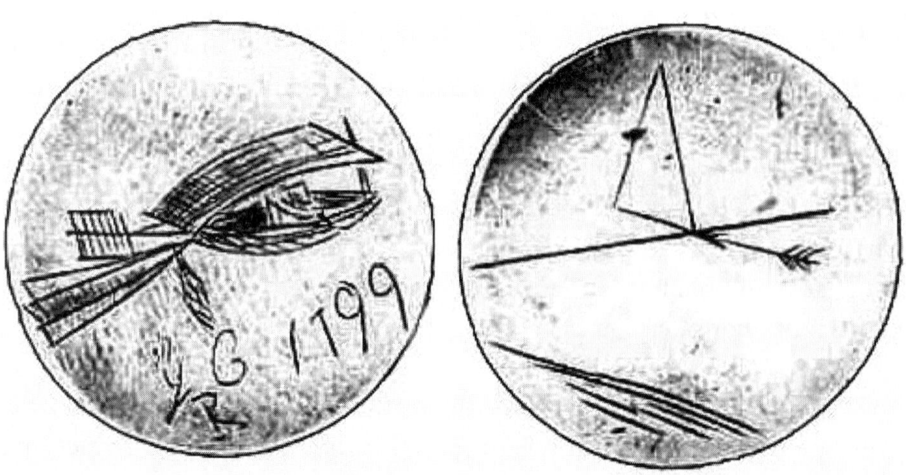

Cayley's 1799 etching of the first true airplane, while crude from our modern perspective, possessed all the basic elements of a modern aircraft, thereby creating the template by which future aviators would one day conquer gravity.

quickly came to understand that, without sufficient forward thrust, no airfoil could achieve enough lift to carry its own weight. Unfortunately, Cayley was handicapped by having only one proven power plant available to him with which to work, the steam engine, and even that was still under development at the time. He quickly saw, however, that even if he could build a functional steam engine, it would prove entirely too heavy and underpowered to do the job. To produce the amount of power needed to lift both a man and his machine off the ground would require building an engine so large it would be unable to lift its own weight. This problem became known as the power-to-weight ratio, which was to prove one of the greatest obstacles to achieving heavier-than-air flight throughout the balance of the century.

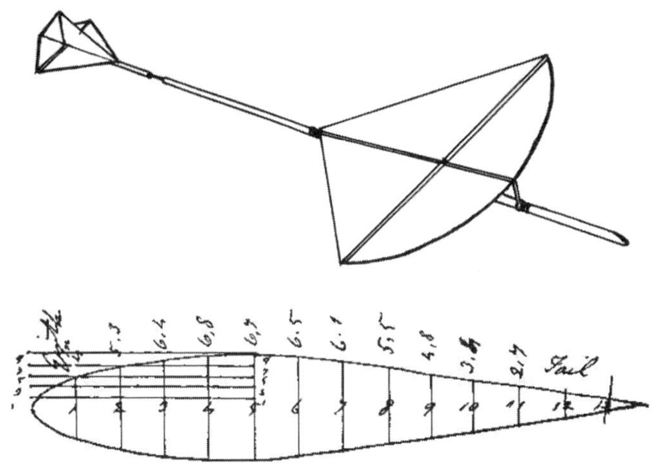

Cayley's study of small gliders (top) led him to understand the basics of airfoil design, which he was able to describe with remarkable accuracy just a few years later.

Undeterred, Cayley began to experiment with a number of other potential engine ideas over the next few years. One that seemed to hold promise was a combustion engine fueled by gunpowder, while another was an engine that ran off compressed air, but none of these ultimately proved to be capable of providing the degree of thrust required. This inability to acquire an engine capable of putting a manned aircraft into the sky was to dog him the remainder of his life, and remain the one issue he would never be able to solve.

Even if he couldn't produce a practical engine, he was able to use what he had learned of aerodynamics up to that point by experimenting with gliders. Building a series of progressively larger gliders over the next few years, Cayley soon learned exactly how the forces of lift and drag operated. In fact, to measure how drag worked on objects at different speeds and angles of attack, he built a "whirling-arm apparatus" (a forerunner of the wind tunnel), which he had developed as a result of his earlier experiments in ballistics and air resistance. Additionally, by observing that birds soared long distances by simply twisting their arched-wing surfaces, he deduced that fixed-wing machines would fly if the wings were cambered, compelling him to undertake a scientific study of airfoils. Working with a small glider—basically a kite on top of a pole about five feet long, with a fixed wing and a cruciform tail that attached to the

One of Cayley's early gliders. Note the curvature of the wings, indicating that he was already beginning to grasp the basic dynamics of air foils, as well as the necessity of a rudder/elevator tail assembly for balance and control.

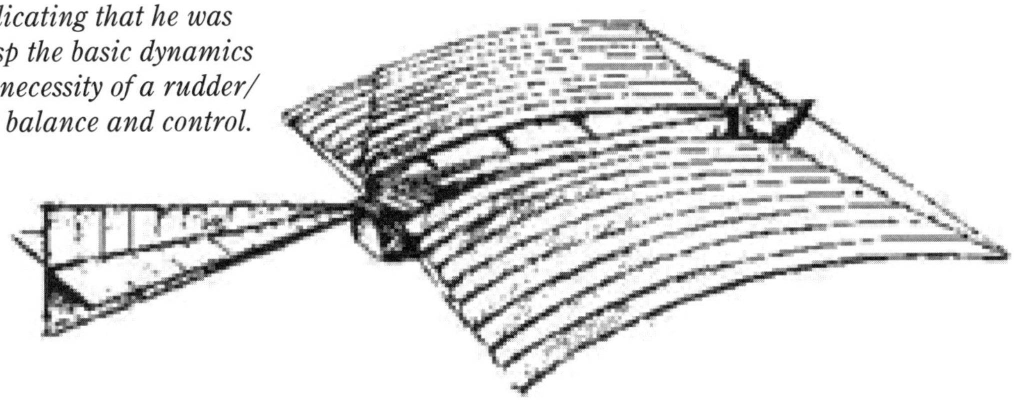

fuselage by universal joints with movable ballast to control the center of gravity—Cayley was able to investigate the improved lifting capacities of cambered wings. Through this, he came to realize that an area of low pressure is formed above the wing—a valuable breakthrough that was to have extensive repercussions in the future. He also demonstrated the use of inclined, rigid wings to provide lift and roll stability, and how the use of a rudder provided longitudinal stability.

Eventually Cayley felt he had learned enough from his small model gliders and he decided to build a full-sized glider, which he completed during the summer of 1809. With a total wing area of approximately 172 square feet (18.5 square meters) many consider this creation to have been the first real airplane in history, making him the first to accomplish what it would take others literally decades later to achieve.

Tests on this craft apparently proved so successful that he wrote a paper, *On Aerial Navigation,* in 1810 that articulated his discoveries in some detail. Thus, at the age of thirty-seven, George Cayley had provided all the ingredients necessary to make heavier-than-air flight possible. Remarkably, this paper, which painstakingly laid out the very basis for the science of aerodynamics, would remain largely unknown for decades to aviation enthusiasts of the period, who would not discover his writings until fairly late in the century. Unfortunately, by the time his work was to come to light and began enjoying widespread distribution, much effort on the part of aviators who had little scientific basis upon which to create their designs was wasted.

Curiously, shortly after his paper appeared, it seems Cayley all but abandoned his study of aeronautics. Why he did this is uncertain, but it is likely he felt he could go no further until he had a proper engine and so chose to pursue other interests in the interim, which he did with similar degrees of success for the next few decades while he and the world waited for technology to catch up to human ingenuity.

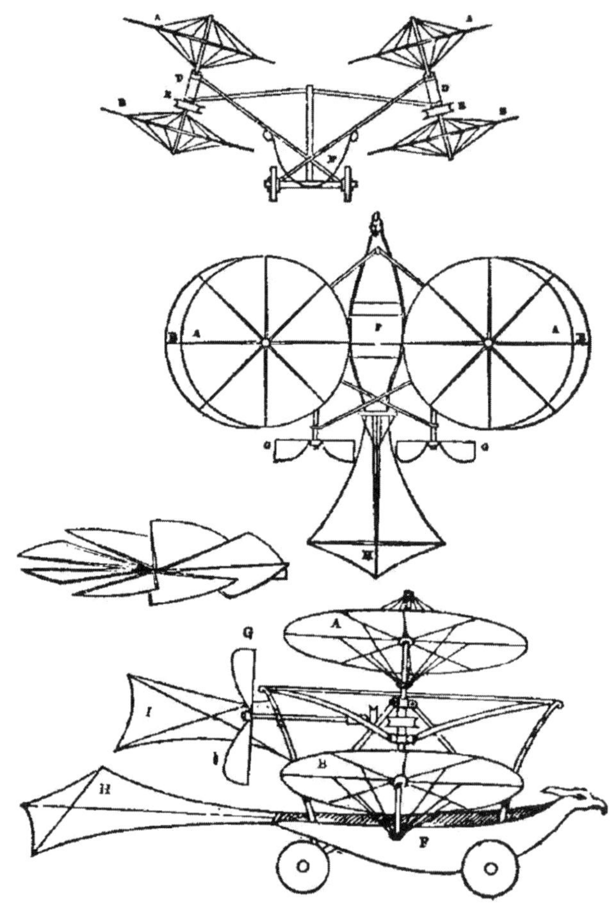

Cayley's somewhat fanciful "aerial carriage"—drawn in 1843—while a interesting concept, proved to be unworkable and was quickly abandoned.

Sketch made by Cayley of his 1849 triplane glider, which was supposedly successfully flown with a ten-year-old boy at the controls, earning it the nickname, the "boy carrier."

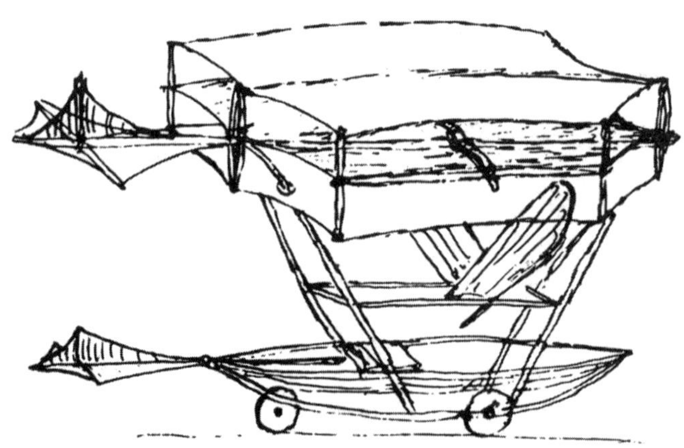

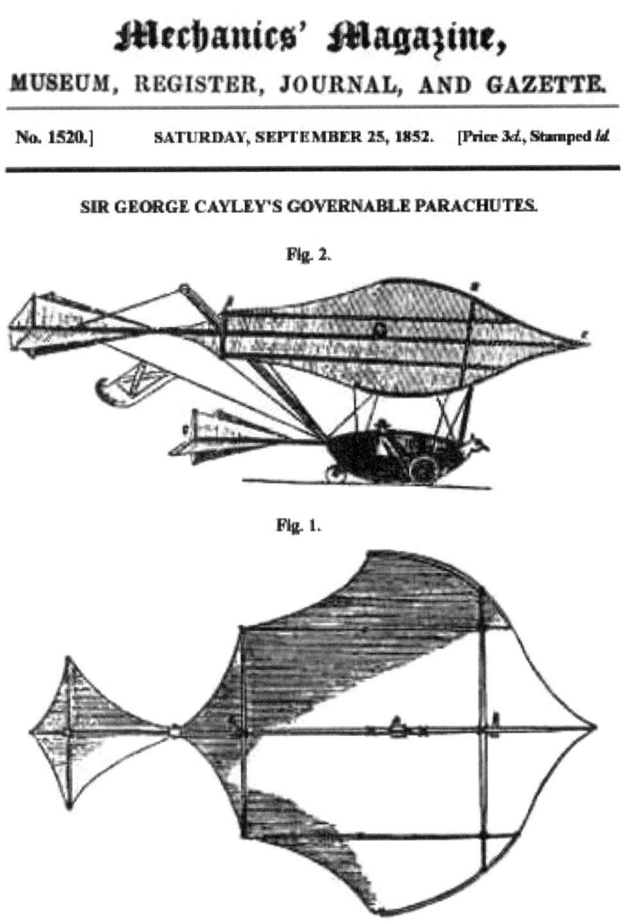
Mechanics' Magazine,
MUSEUM, REGISTER, JOURNAL, AND GAZETTE.
No. 1520.] SATURDAY, SEPTEMBER 25, 1852. [Price 3d., Stamped 4d.
SIR GEORGE CAYLEY'S GOVERNABLE PARACHUTES.
Fig. 2.
Fig. 1.

Cayley's 1852 manned glider—interestingly labeled a "governable parachute" by Mechanics Magazine—may have actually flown a short distance.

A modern replica of one of Cayley's manned gliders being successfully flown. (Notice the towline, which forces us to ask whether Cayley may not have towed his gliders in the same way.)

Later Aeronautical Efforts

It would be thirty years before Cayley—now in his seventies—would turn his attention once again to aeronautics, but it was this phase of his work that has received the most attention.

Cayley's initial efforts, however, appeared more speculative than a serious attempt at engineering. An example of this can be seen in a series of drawings he did in 1843 that described what he called an "aerial carriage," which was basically a wheeled boat hull that was to be lifted by counter-rotating rotors and propelled by pusher propellers like an auto-gyro. It never got beyond the concept stage, of course, most likely because of the lack of a proper power plant necessary to lift his futuristic craft off the ground. Likely realizing the impossibility of building such a craft himself, Cayley eventually abandoned his "aerial carriage" idea entirely and returned to working with gliders.

Largely picking up where he had left off forty years earlier, Cayley began building a series of progressively larger craft, some of which were large enough to carry a man. At this point, however, determining exactly what he built and when he flew it becomes more difficult. By most accounts, it appears he built a three-winged glider in 1849 (the three wings stacked on top of each other like a triple-decker sandwich) which he supposedly tested by allowing the ten-year-old son of his coachman (some accounts maintain it may have been his own grandson) to pilot it. (The story may well have been more anecdotal than factual, though the fact that his glider was later sometimes referred to as the "boy carrier" suggests there could be some element of truth to it. It's difficult to imagine Cayley endangering the life of a child in an effort to test one of his gliders, however, though it is possible he was unaware of the potential danger involved. The craft, after all, would likely have flown only a short distance and then only a few feet above the ground, making it no more dangerous than sledding down a snow-packed hill—a common child's pastime of the era.)

Whether the 'boy carrier" actually flew or not, it appears that Cayley did build at least one glider that definitely was capable of carrying a man and that he may well have seen it successfully flown.

The device in question appears to have been quite advanced for the era, and very unlike anything we would imagine a manned glider should look like. Abandoning his earlier triple wing configuration in favor of a monoplane design, the resulting aircraft was a flimsy affair with a diamond-shaped wing made from a linen sheet shaped by cane and held together with string. Below the wing was a boat-like fuselage, containing a seat for the pilot and a three-wheel undercarriage—a feature that wasn't to be seen in aircraft until the turn of the twentieth century. It also sported a pair of tail planes, one at the back of the wing and one behind the cockpit. (Aware that weight was a major consideration, Cayley even had the rims of the tires tied to the axle with string, thereby inadvertently inventing the world's first spoked bicycle wheel in the process.)

By all accounts, Cayley tested the machine in 1853. At the age of eighty, he was too frail to fly it himself, but he apparently managed to persuade his coachman to get into the thing and try his hand at being the world's first test pilot.[15] Apparently unwilling to refuse his employer's requests, the man and the machine were both pushed down a steep hill at Brompton Dale and the glider, its fabric wing billowing in the wind, supposedly lifted gently into the air. Almost immediately, however, the glider dipped to one side but the coachman, apparently a man capable of learning quickly, used the tiller connected to the tail surfaces to level the machine out. Now running straight and true, legend has it that the device flew some nine hundred feet before the apparently shaken coachman coasted the thing to a hard landing. (While it may be more anecdotal than factual, there is a often repeated story that the coachman, upon making his way back to terra firma, climbed out of the glider and blurted to Cayley, "Please, Sir George, I wish to give notice! I was hired to drive, not to fly!") While determining precisely what happened that day may never be finally settled, it does appear that Cayley's glider may well have achieved lift off and flown for a short distance, In that it may have carried a frightened coachman onboard, Cayley's craft may have been, then, the first heavier-than-air machine to send a human aloft in "modern" times (the earlier legendary solo flights of Abbas Ibn Firnas and others notwithstanding).

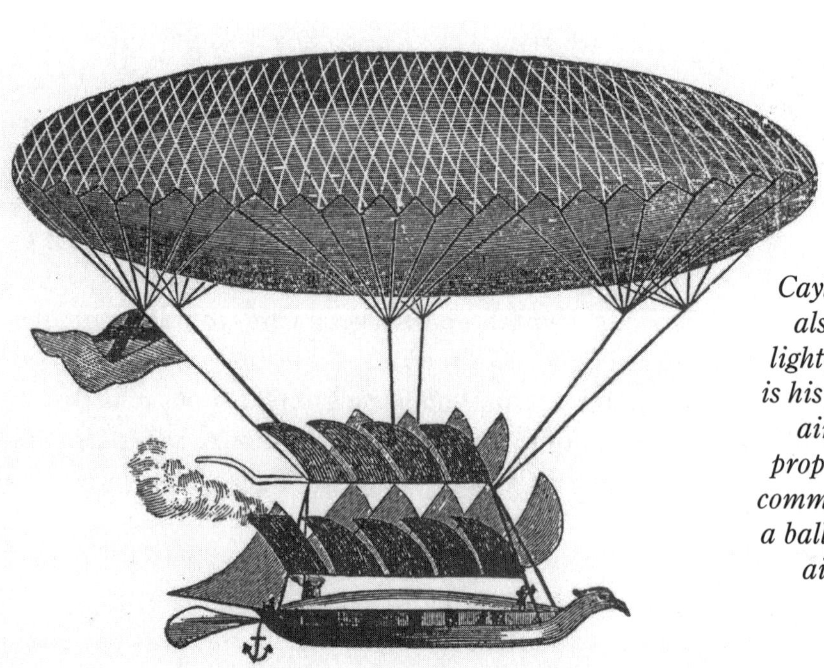

Cayley didn't design only gliders. He also played around with ideas for lighter-than-air aircraft as well. Here is his 1816 design for a steam-powered airship. Note the use of oars as a propelling element, demonstratting a common misconception of the time that a balloon could be "rowed" through the air much like a boat on a pond.

[15] There is another story that the first to pilot his glider was an employee of Cayley's named John Appleby, while still another names Cayley's own grandson, George John Cayley, as the pilot, though none of these stories have ever been confirmed.

What the event signified is difficult to assess today. In that it's unclear whether the craft was truly controllable (though the fact that the pilot supposedly was able to straighten out the plane by making use of a control surface would seem to argue otherwise) this brings up the question of whether it even constituted genuine flight at all. Certainly, considering the primitive level of technology available at the time and the fact that Cayley was working in largely uncharted waters in terms of aeronautics, it has to be considered—if not an astonishing success—certainly a remarkable attempt at achieving manned, controlled flight.

What Cayley might have achieved had he access to an adequate power-plant can only be surmised. It's certainty possible that had he found a sufficiently light and powerful motor to attach to his glider, the history of aviation might well have started a half century earlier than it did. Possibly it was best said by author Richard Hallion in his book *Taking Flight: Inventing the Aerial Age from Antiquity Through the First World War*, when he wrote: "Had Cayley not lacked a proper engine to power his concepts, Brompton Dale might today be regarded as Kitty Hawk is now."[16]

Conclusions

Cayley died in 1857 at the age of eighty-four, but his legacy lives on to this day. In being the first to identify the four forces of flight—thrust, lift, drag, and weight—he is rightfully regarded as the intellectual forefather of every fixed-wing aircraft flying today. Additionally, not only did he build and fly the first manned gliders, but he also was the first to describe in some detail the theoretical problems of flight and suggested using multiple wings—biplanes or triplanes—to produce maximum lift with minimum structural weight. But perhaps his greatest achievement came in recognizing that curved surfaces (cambered airfoils) provided greater lift than flat surfaces, and that flight is possible with wings that are held entirely rigid, thereby dispensing with the then popular notion that propulsion should generate both lift and forward motion at the same time, as birds were able to do. Cayley was also the first to subscribe to the idea that the propulsion system should generate thrust but that the wings should be a separate element of the aeroplane, shaped so as to create lift. Further, he was the first to articulate such forward-looking concepts as ratio of wing area to lift, determining a wing's center of pressure, the importance of streamlined shapes, and the recognition that a tail assembly was essential to stability and control. He even demonstrated the need for bracing the basic biplane structure for strength and extolled the need for a wheeled undercarriage (a point the Wright brothers failed to recognize until several years after their first flights). Finally, and perhaps most important of all—at least in regards to practical, controllable flight—was his recognition that sustained flight would not be possible until a lightweight engine was developed to provide adequate thrust and lift, a development which would not take place until the first flight of the *Wright Flyer* in 1903.

It's not surprising, then, that whenever one reads about the history of aviation, Sir George Cayley's name frequently crops up. Unfortunately, many of the aviation pioneers who followed either weren't aware of his work or disregarded it, preferring to start from scratch, which may be one reason there were so many failed attempts at heavier-than-air flight through the remainder of the century, and why it took nearly a century for the first man to fly. But those aviators who looked at his studies and took them to heart achieved a degree of success many of their contemporaries could only dream of. As such, it can truly be said that without Cayley's work laying the foundation, the later successes of Chanute, Lilienthal, and the Wright brothers would not have been possible, making him truly deserving of the title the "Father of Aviation."

16. Oxford University Press, 2003.

CHAPTER FIVE: AVIATION PIONEERS THROUGH 1880 5

While George Cayley was working on understanding the principles that governed flight, there were a handful of others who were also experimenting with manned flight. In fact, even before Cayley's first largely successful glider experiments at Bromley Dale in 1853, a few of them were already experimenting with manned flight and, in a few cases, even powered flight. One pair of inventors, in fact, were even making plans to establish the world's first airline—a full half a century before anyone had even heard of the Wright brothers!

Unfortunately, confirming some of these early efforts is nearly impossible, leaving many claims unverified. There are a few that have a better pedigree than others, however. Among the earliest and potentially most credible of these came from Norway in 1825, where a watchmaker named Hans Andreas Navrestad is rumored to have dragged his homemade glider—who had apparently built the thing after studying the shape and nature of butterfly wings—to the top of a mountain overlooking Lake Navrestad and launched himself into space. What's most interesting about this report is not only the assertion that he successfully flew, but that he glided an impressive distance of nearly four hundred meters before safely landing, a claim which, if true, would have made him the first man—at least in modern times (Ibn Firnas and Eilmer of Malmsbury being much earlier)—to have flown. Of course, without the ability to verify the event, it must remain firmly fixed in the realm of local legend, but considering the sometimes bizarre world of aeronautics, it isn't out of the question that it may well have happened.

But there were two other men from this period—contemporaries and fellow countrymen of Cayley, in fact—who were not only working on the problem of flight at the same time Cayley was, but did their work under the careful scrutiny of the public eye, thereby removing them from the realm of mythology. And, further, there is evidence that they were moving in the right direction.

Henson and Stringfellow's Aerial Transit Company

John Stringfellow (1799-1883), a maker of bobbins for the lace industry and a carriage builder by trade, and William S. Henson (1812-1888), a lace maker from Chard, England, were inventors not only by trade, but by temperament. Henson appears to have been the more prolific of the two, having been awarded a number of patents for everything from a lightweight steam engine to the T-handled safety razor, though Stringfellow was also an engineer and inventor of some note. While the specific particulars of how their partnership came to be remains fuzzy, it is evident that at some point the two men met—probably as a result of their shared employment in the lace industry—and soon struck up a friendship that centered around their mutual interest in aeronautics. Eventually combining their considerable engineering and business expertise, at some point the two men decided to embark on a project that would consume the better part of a decade of their lives but would, if they were successful, make them among the most famous men in the world.

What they had in mind was truly revolutionary. Convinced that the problems of overcoming gravity was on the verge of being solved and being entrepreneurs of the first order, they decided to build an airplane that would become the nucleus of what was to be the world's first "aerial transit company" (in other words, an airline). Additionally, the airplane they had in mind was to be equally as revolutionary. Their *Henson Aerial Steam Carriage*—as they called it—was to be large enough to carry a dozen passengers in relative ease and comfort for considerable distances. Basically an oversized monoplane built over an enclosed carriage, with a single horizontal rudder reminiscent of a bird's tail and two pusher propellers, what made their machine truly remarkable was its gargantuan size. Henson, the primary designer of the craft, called for it have a wingspan of no less than one hundred fifty feet—as large as many jet airliners today! To power the monster through the air, he proposed equipping it with one of Stringfellow's patented steam engines, which would have made the craft, in addition to be extremely large, extremely heavy as well.

However, the machine also incorporated some fairly advanced features as well. For example, it incorporated a remarkably modern wing configuration consisting of three lateral spars serving as the primary load-bearing structures (a fairly modern concept) and airfoil-shaped ribs evenly spaced along the spar to give the wing its cambered profile. It also possessed an entirely enclosed and aerodynamically streamlined cockpit mounted to a triangular bicycle landing gear—both features that were not to appear in early aircraft for over half a century.

Convinced they had a winning formula—and design—the men set about constructing small scale models of their craft (which they had since rechristened the *Aeriel*) for a series of test flights while they worked on acquiring the substantial funds required to build a full-sized prototype. Unfortunately, they quickly learned that their design proved incapable of flight. Imagining that the problem lay in the size of their models, over the next two years they constructed a number of ever larger models (some with wingspans of up to twenty feet) but, regardless of their determination, they could not get any of their airplane to take flight. The problem appears to have

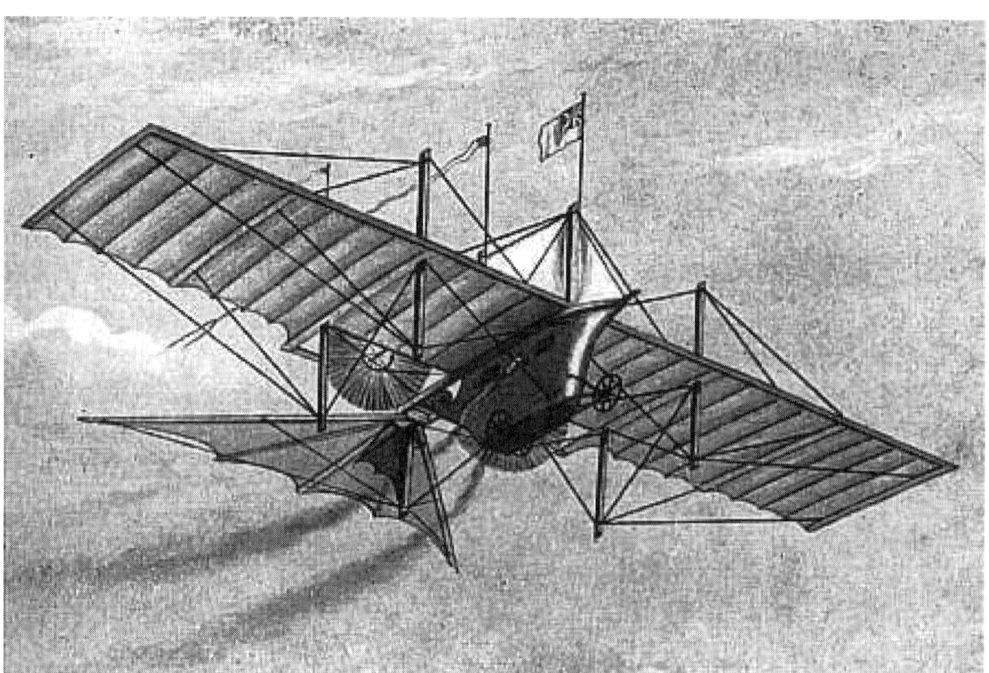

Stringfellow and Henson's design for an Aerial Steam Carriage, circa 1842, complete with a British naval ensign flying from the foremast. Though difficult to perceive here, the machine was much larger than it appears in this drawing; had it actually been built, it would have been nearly as large as a modern airliner.

Stringfellow's successful 1848 steam-powered model.

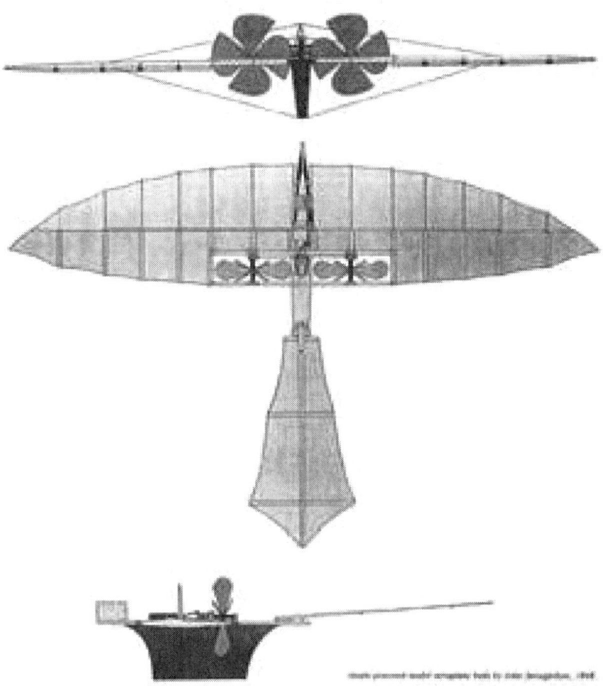

Stringfellow's very successful 1868 steam-powered triplane model (currently on display in the Early Flight Gallery of the National Air & Space Museum in Washington, D. C.)

been Henson's belief that an aeroplane, like a horse-drawn carriage, should fly largely in a straight line, meaning that he was mostly interested only in countering pitch and yaw motions. While his ideas were in line with Cayley's, which only allowed for pitch and yaw control—2-axis control—without the ability to control an aircraft's inherent tendency to roll as well, it was doomed to fail. While it's extremely unlikely that, even with proper funding and an adequate power plant, the craft would have flown, had Henson given serious thought to 3-axis control, it is possible he might have achieved at least a small degree of controlled flight.

In any case, after countless failed efforts and with all attempts at securing additional funds having failed, Henson was ready to walk away from the entire project. Disillusioned, in 1849 he emigrated to America, where he spent the rest of his life working as a machinist and civil engineer, never again to try his hand at anything even remotely aviation-oriented. His partner, Stringfellow, however, was more persistent and continued to experiment with various aircraft configurations for the next several years, using his small steam engines for power. It wasn't until he abandoned Henson's original design, however, and began working on those of his own, that he began having some success with his small scale models. In fact, in 1848 one of them, a small monoplane design with a ten-foot wingspan, flew for a respectable distance of thirty feet which, considering the limited aeronautical engineering knowledge available at the time, had to be considered a significant accomplishment.

Buoyed by his success and now joined by his son, Frederick (who was to build a number of powered glider models of his own design years later), over the next twenty years the two collaborated on a number of flying machines, the most successful being a steam-powered triplane design that made a number of short distance flights and was even successfully flown at London's Crystal Palace in the summer of 1868.

In demonstrating a degree of engineering acumen was well ahead of its time, Stringfellow may well have been very close to building a manned version of his steam-powered airplanes, but age and illness finally caught up to him and he never had the chance to take his ideas further. This was a great pity, for it is conceivable, given his talent and determination, that with more time and a more powerful engine, he might well have succeeded in conquering the challenge of flight where many of his contemporaries failed. Unfortunately, history is often much more about timing than talent, leaving the world to ponder what might have been.

Jean-Marie Le Bris: the Sailor of the Air

Another valiant attempt at achieving flight was attempted by the French sea captain, Jean-Marie Le Bris (1817-1872) who, despite having no formal engineering training, built a machine that came remarkably close to actually flying. Basing his design on the proportions and configuration of an albatross—a bird he had spent years studying—and after several months of feverish work, he finally produced what could be considered one of the most bird-like airplane designs ever seen. With a bullet-shaped hull and 50 foot-wide wings that looked conspicuously like those of the albatross, (he even named it, not surprisingly, *L'Albatros artificiel*—"The artificial Albatross"), along with a triangular tail that also looked suspiciously bird-like, the craft was truly an engineering marvel. Outfitted with hand-operated levers to change the angle of incidence of the wings and with foot-operated pedals designed to alter the relative position of the tail, Le Bris' simple but effective flight controls made it years ahead of its time. The craft had no landing gear, however, forcing it to be placed upon a carriage and towed behind a horse-drawn cart until it lifted off, at which point the make-shift landing gear would fall away and the aircraft would—hopefully—fly off unimpeded by such protuberances.[17]

Completing the glider in December of 1856, Le Bris decided it was time to unveil his creation to the general public and, taking it to the beach at Sainte-Anne-la-Palud, he pulled himself into its cockpit and ordered the cart's driver to head the horse into the wind. What happened next is still debated to this day, but if reports are to be believed, it must surely go down in the annals of history as one of the most unusual events in aviation history. According to eyewitness reports, just as the craft appeared about to rise, its restraining rope became snagged, and while the carriage driver was trying to unsnag it, he somehow became entangled in it. As such, when the craft suddenly broke free, it rose into the air with the carriage driver still attached to the tethering line. Soaring to a height of about three hundred feet, the craft supposedly sailed along for a distance of some six hundred feet with the cart driver dangling precariously beneath the machine before it mercifully and gently landed in the sand. Fortunately, the glider was only slightly damaged and the cart driver walked away shaken but unhurt, but it was a close call for all involved and no doubt the most excitement the beach at Sainte-Anne-la-Palud had seen in years.

While there is no way to know exactly how high or how far Le Bris flew his machine (the distances probably being rough estimates) or whether there was an unwilling cart driver hanging from a rope beneath at the time, it seems entirely credible that he did manage to catch enough of an updraft to get his

Le Bris' flying machine may not have been the first plane to fly, but in 1868 it did manage to become the first heavier-than-air craft ever to be photographed.

17. This was precisely the method German aircraft designer Willi Messerschmitt used to launch his famous World War II rocket-powered fighter, the ME-163; it had a detachable dolly that was shed upon takeoff, forcing the plane to land on skids—often with unfortunate results.

machine to glide for a few seconds. If so, that would make Le Bris, along with the hapless cart driver, among the first men to achieve successful, extended, unpowered flight. (If true, it also demonstrated his craft achieved enough lift to not only lift itself, but the weight of two grown men as well.)

But the intrepid aviator wasn't quite done yet. After repairing the damage to his glider, Le Bris decided to make a second attempt, this time flying from a large mast he had built over the floor of a nearby stone quarry. With his repaired *Artificial Albatross* suspended on a movable yard arm some one hundred feet above the quarry's floor, the fearless aviator had the glider turned into the wind and, after adjusting the incidence of the wings to a positive angle, slipped the restraint and began a long glide towards the ground below. Unfortunately, no sooner had Le Bris begun to drop when errant winds rising over the lip of the quarry caught his machine and sent him out of control. Over-correcting, Le Bris put the craft into a steep dive and made a hard landing on the quarry floor, utterly destroying the plane and breaking one of his legs in the process.

Le Bris proved to be a difficult man to discourage with something as trivial as a destroyed aircraft and a fractured limb, however, and he immediately set about raising the funds required to build a second glider. Fortunately, by this time the French navy had begun taking serious interest in the military potential of aeroplanes and gave Le Bris the funding he required to carry on his research. Now with the resources required to build another airplane, he set about building his second glider, which, for some reason, took him nearly a full decade to complete. An almost identical duplicate of his first glider—with the exception that it was somewhat lighter and carried a weight meant to provide some additional stability—by early 1868 he was ready to try again. This time, perhaps reconsidering the wisdom of launching his craft into mid-air from a crane suspended over a rock quarry, he decided to launch his craft from a stationary cart. Testing his machine near Brest, France, while he made several attempts with his new and improved *Artificial Albatross,* the best he could achieve was a brief hop that reached an altitude of 35 feet and glided about 70 feet. (However, he was able to achieve some more significant glides, one of nearly 600 feet, with only ballast aboard.) Landing without the aid of a landing gear, however, quickly took its toll on the craft and after a few more unmanned attempts, the machine was so badly damaged it never flew again, bringing an end to Le Bris' aviation experiments just as they seemed to be bearing fruit.

Le Bris was never able to acquire the resources to build another *Albatross* and he died just five years later, but his contribution to the world of aviation cannot be underestimated. The man was a pioneer and an innovative designer who drew on the familiar aspects of naval architecture to build his craft while his ability to translate natural forms into their mechanical equivalents rather than just attempting to replicate nature—as most before him had done—was a remarkable achievement in its own right. Despite personal injury and an occasional lack of funds, Le Bris persevered and even found a degree of success that has been rarely acknowledged.

Jan Wnek, The Bird Man of Krakow

Up until the mid-1860s the study of aeronautics appeared to be centered almost exclusively in England and France, making western Europe the undisputed center of the aviation universe at the time. There were others outside this region, however, who were also experimenting with unpowered, manned flight, most of whom remained almost entirely unknown to the rest of the world. Perhaps the one man who enjoyed the most success—and possibly least fame—was an illiterate and self-taught Polish peasant, sculptor, and carpenter named Jan Wnek (1828-1869), who not only built but repeatedly flew a controllable glider numerous times during a three-year period between 1866 and 1869 in the skies over Austro-Hungary—making him the first aviation pioneers to herald from central Europe.

Though lacking formal education, Wnek

proved early on to not only have a keen mind, but an instinctive talent for mechanics. While he spent the first years of his adult life using that talent to make improvements on farm machines in the area, he eventually found his interests turning towards the study of birds (which he could sculpt with great accuracy) and how they flew. Like many aviation inventors of the era, he noted how they made use of rising air currents to facilitate climbing and, after making a detailed study of a duck's wing, made a number of small, hand-thrown flying models to test.

Eventually satisfied that he'd learned as much as he could from his models, in 1866 he began work on a glider capable of carrying a man. Covering a skeleton of ash-wood ribs with a varnish-coated linen, after several months of work he completed a glider he named simply *"Loty"* (Flyer), which he first flew in June of that year—apparently successfully—from a small hill near his home. After making several flights to better learn how to control his glider, Wnek finally felt confident enough in his abilities that he was prepared to attempt a take-off from a considerably higher vantage point. Realizing that the local church was the ideal location, he asked for and received permission from the village priest to build a special ramp on top of its one hundred foot high tower from which to launch himself into space. Of course, he knew that launching from such a height would be fatal should his glider prove uncooperative, but the fledgling aviator was nothing if not optimistic.

Finally ready, and with huge crowds in attendance from throughout the area, Wnek ascended the tower and, after carefully strapping himself into his still-experimental glider and checking to ensure he had the most favorable winds possible, he gingerly flung himself off the ramp. Instead of falling to his death as many in the crowd assumed he would, however, he caught a fortuitous updraft and made a flawless flight of nearly a mile before coming to a perfect landing, much to the astonishment and delight of the crowds below.

After that he became a local favorite and was frequently called upon to perform his flying leap at numerous carnivals and festivals over the next three years, which he did with great dash and daring. Sadly, it was while performing the stunt in May of 1869 that Wnek's luck finally ran out. It's not clear exactly what happened, but for some reason he suffered a hard landing that injured him so badly that he died a few weeks later, leaving a wife and three children behind. (There was some speculation the "accident" may have been an act of sabotage by his assistant, though this has never been proven.)

Unfortunately, Wnek left no known written records or drawings of his glider, but if the flights he was reported to have made from the church tower are true—and there is good evidence to suggest they are—the illiterate peasant from Krakow

A speculative model of Jan Wnek's glider based upon eyewitness descriptions. If accurate, Wnek's gliders were remarkably similar to those that Otto Lilienthal would build forty years later.

not only beat the great Otto Lilienthal by a good twenty-five years, but may have even bested the great German aviator in terms of distance and duration. This makes it all the more astonishing that his flights left no lasting legacy upon aviation, but this is probably due to his having the misfortune of being born outside of western Europe, where his feats would have enjoyed considerable press coverage. Also, the fact that his flights appear to have been performed more for entertainment than for the sake of science may also have kept him from acquiring serious attention by scientists and aviators of the era. In any case, it remains a mystery to this day how such an impressive array of accomplishments could remain so little known and barely acknowledged to this day.

Felix Du Temple and his Aluminum Monoplane

While most people assume it was the Wright brothers who made the first manned, powered flight in history, there is some evidence that the honor may actually belong to a French naval officer by the name of Felix Du Temple de la Croix (1823-1890) who, according to legend, sometime in the year 1874, actually flew a short distance under power in an aluminum-framed monoplane.[18]

It all began some fifteen years earlier. Du Temple had begun tinkering with the idea of flight in the mid-1850s and even patented a design for an aerial machine in 1857. While his design proved impractical, it did feature a number of futuristic innovations such as a retractable, wheeled landing gear, a tractor propeller, and even an internal engine, making it a truly forward-looking craft for its time.

Unfortunately, it weighed more than a ton, making it far too heavy to be flown by the 6-horsepower steam engine Du Temple proposed. After constructing and testing numerous bird-shaped models and even a nearly full-sized version with his brother, he finally came to this conclusion himself and set aside his aviation ambitions for a time while he waited for the technology to catch up to his dreams.

He wasn't entirely idle, however. During the interim he began tinkering with various engine designs, the result being the creation and patenting of a very compact, high-speed circulation steam engine of his own design. Now in possession of a power plant potentially capable of lifting an aeroplane into

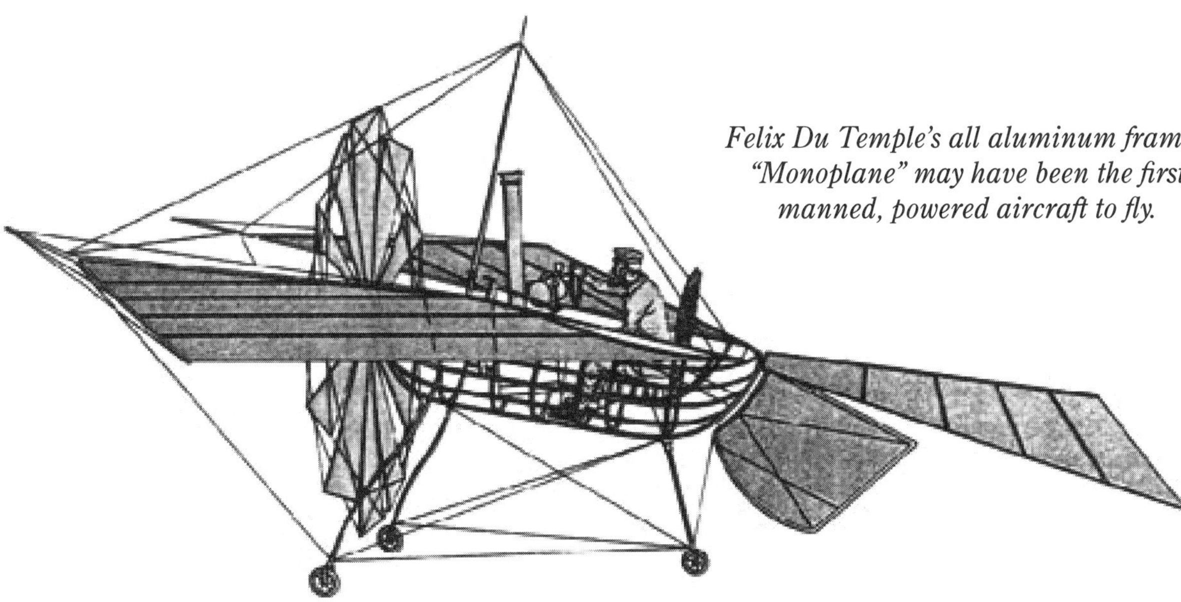

Felix Du Temple's all aluminum framed "Monoplane" may have been the first manned, powered aircraft to fly.

[18] Though precisely what constitutes "powered flight" is still debated. It is generally believed to mean a manned, airborne ascent from level ground under internal power that demonstrates 3-axis control. However, by this criteria, most of the pioneers do not qualify (including the Wright brothers, who used an inclined track to launch their first plane). For the purpose of this work, "flight" can be more loosely defined as the successful ascent of a manned aircraft under power for a measurable distance.

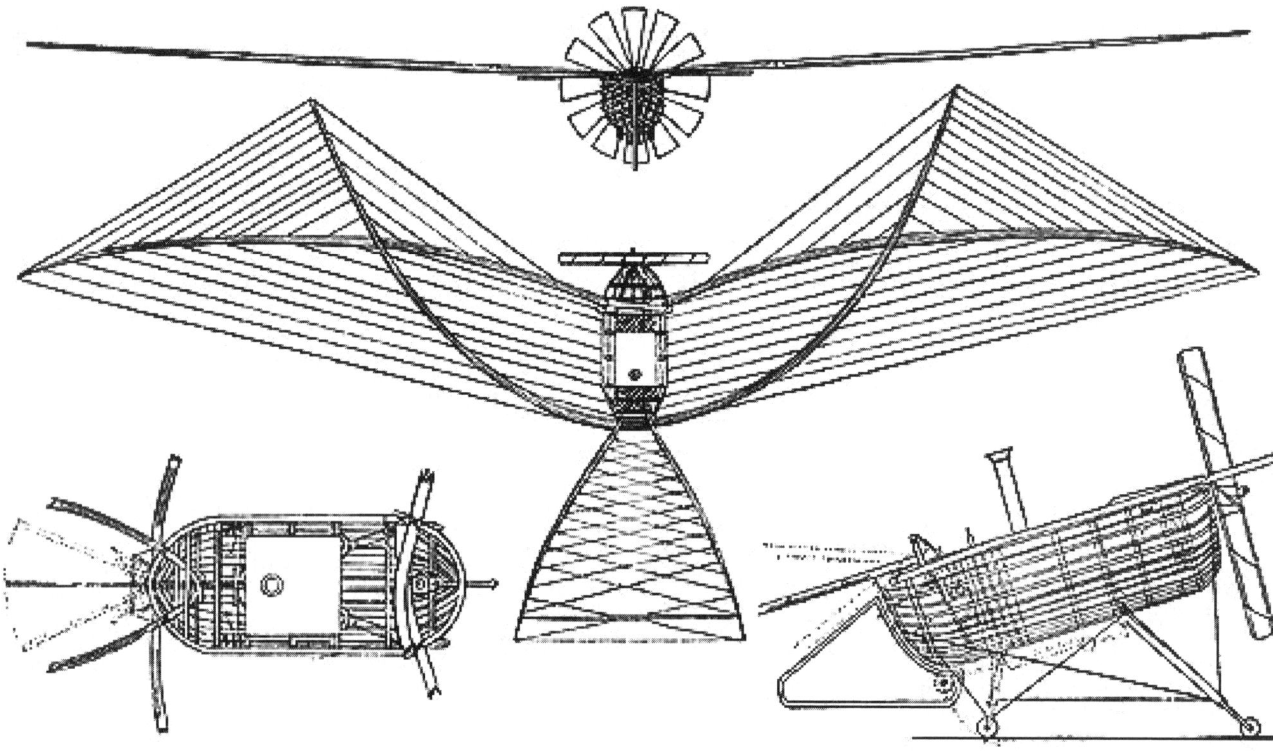

While it is still debated whether Du Temple's 1857 aeroplane actually flew, it is undeniable that it incorporated some remarkably forward-looking features, among them a retractable landing gear and a tractor propeller.

the air, with the aid of his brother, Louis, Du Temple built a small, powered monoplane and attached his revolutionary engine to it. The key to making it airworthy, he knew, was in keeping it as light as possible, which he managed to do by making much of the craft out of a relatively expensive and extremely light new metal known as aluminum—an element which had only been identified in 1808. In this way he was able to keep the entire craft's empty weight to under two hundred pounds—a truly remarkable achievement for any era.

While at first glance it would not appear his plane should have been capable of flight (its wings were actually canted forward—a truly unusual configuration that was as unconventional back then as it would be today) after making several trial runs, it is generally agreed that Du Temple may have gotten the machine airborne, at least briefly. According to witnesses, the craft achieved lift-off under its own power after sliding down a ski-jump ramp and glided for a short time before returning safely to the ground, making it the first successful powered flight in history. If true—and it is generally conceded by historians that the flight did in fact take place—the craft made it into the air almost thirty years before the Wright brothers, an achievement the French naval officer and inventor had every right to be proud of.

Curiously, and like so many aviation pioneers of the era, after this flight Du Temple apparently lost interest in further refining his aeroplane or in making more flights, and never flew again. Instead, he chose to refine, patent, and market his revolutionary little engine instead, which turned out to be a good idea for not only did his motors make it onto numerous French naval craft throughout the century, but they eventually made him a very wealthy man as well.

The Brilliant Alphonse Penaud

Another man who probably did more for the science of aeronautics than anyone else during the mid-nineteenth century and yet remains among its least known contributors today was French aviation pioneer Alphonse Penaud (1850–1880), a man who proved to be one of the truly gifted if deeply troubled minds of his era.

What makes Penaud unusual is that, unlike the others mentioned in this chapter, he never built or attempted to fly a full-scale aircraft. Instead, his claim to fame came in respect to his ability to understand of the basic dynamics of aeronautics in ways that was years ahead of most of his contemporaries. So advanced were his ideas, in fact, that he was to become a source of inspiration to future generations of aircraft inventors—including the Wright brothers themselves.

Penaud's fututistic little Planophere, a rubber-band powered model capable of flying distances of over 100 feet. Peanaud was a master at working out basic flight and control issues on small scale models—a practice few later inventors tried, despite the benefits that doing so would have afforded.

Penaud took an unusual path into the study of aviation. The son of a Navy admiral, he would most likely have ended up a naval officer like his father, but because of health problems he was unable to attend the French Naval Academy (the École Navale) at Brest. Fortunately, he had an interest in and an aptitude for the science of aeronautics as well, and began pursuing that field of study instead—something that likely would have been impossible had he embarked on a naval career.

In any case, he soon demonstrated himself to be so adept at the field that he was allowed to join the prestigious Société aéronautique de France when barely twenty years old. Even more impressive, by the time he was just twenty-six, he had become that society's vice-president and even participated in the publication of its monthly journal *L'Aéronaute*.

Penaud's clever little rubber band-powered helicopter not only flew well, but it was to become a source of inspiration for a number of later aviators—among them two brothers from Dayton, Ohio named Wilbur and Orville Wright.

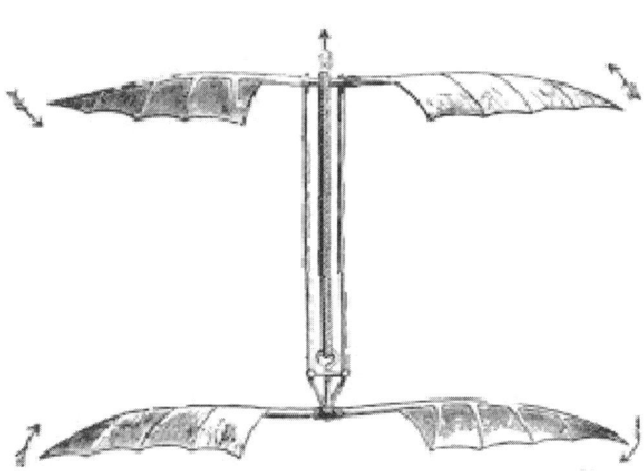

When he wasn't writing about or studying the physics of flight, Penaud kept busy by constructing toy gliders, helicopters, and ornithopters that demonstrated extraordinary flight capabilities while being powered with nothing more than rubber-band driven motors. It was through these small models, however, that he was able to develop many theories concerning airfoils and the problems of stability and

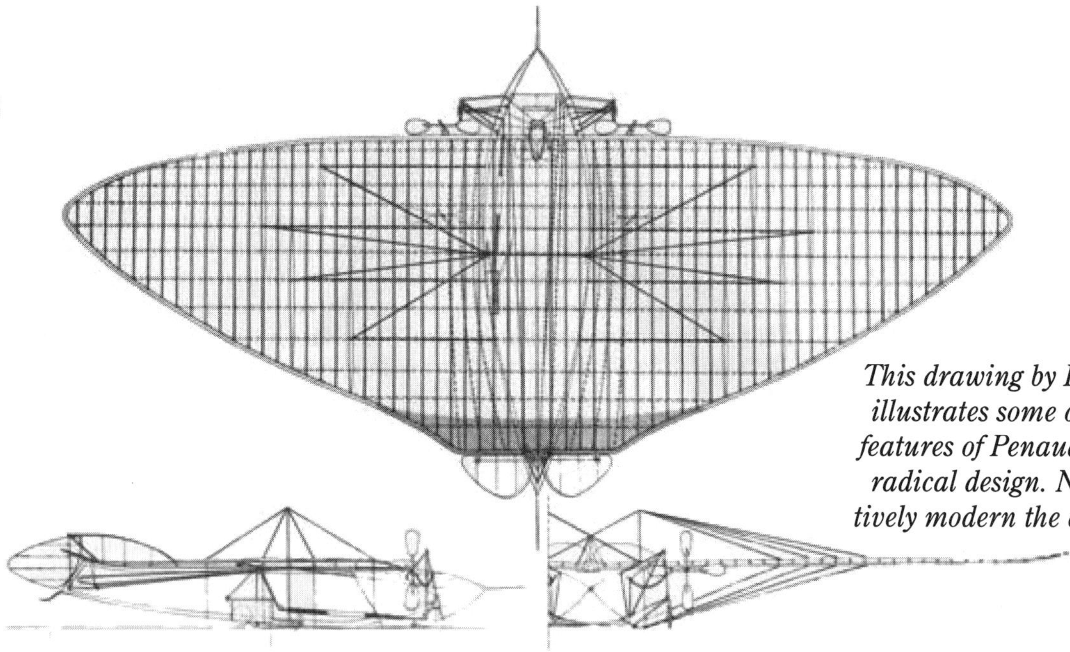

This drawing by Paul K. Guillow illustrates some of the advanced features of Penaud and Gauchot's radical design. Notice how relatively modern the aircraft appears.

control in flight, which quickly made him one of the top experts in the field.

In 1871, Penaud demonstrated one of his best known and most capable model airplanes, the *Planophore*—a device that perfectly demonstrated that achieving sustained, controlled flight was possible by adding a rudder/elevator assembly for lateral and longitudinal stability some distance behind the wings. Flying his *Planophore* a distance of one hundred thirty feet to prove his theory, the simple little toy not only flew especially well, but looked remarkably modern. In fact, the configuration of the tail was eventually to become known as the "Penaud tail" and was to become the template for most twentieth-century aircraft.

Penaud soon grew bored with experimenting with small runner-band powered toys and in 1876 he and mechanic Paul Gauchot combined several years of intensive research and experimentation with aerodynamics, materials, and engines into designing what they believed would be an airplane capable of carrying a human being. What they came up with could only be called revolutionary by any standard. Something of a "flying wing", the amphibious monoplane—with its retractable four-wheel undercarriage and rudder and elevators that could be moved by means of a single control column in a glass-canopy enclosed cockpit—was years ahead of its time. Even its twin, wing-mounted propellers, which were connected by crankshafts to the engine mounted within the fuselage where the flight crew could have direct access for maintenance and repairs—another modern concept—were made entirely of metal and even had variable pitch (an idea that wasn't to see the light of day again until just before the Second World War.) The list of innovations even extended to an instrument panel which included an angle-of-attack indicator, an airspeed indicators, and even aneroid altimeters—something one might expect to see in the mid-twentieth century rather than in the nineteenth—thereby demonstrating once-and-for-all Penaud's aeronautical and engineering brilliance.

Whether the airplane was capable of flight or not is purely conjecture. Certainly, it would have been difficult to find a power plant capable of producing enough power to make it airborne; however, had a powerful enough engine been available, there is no reason it couldn't have flown. Unfortunately, the world would never find out as Penaud was unable to raise the necessary funds to build a flyable example of his remarkable machine. Even worse, his revolutionary concepts were met with skepticism and even ridicule by officials and many of his fellow aeronautical enthusiasts, which deeply depressed him to the point that, in October of 1880, he committed suicide, bringing to an end one of the most gifted minds of the era and leaving many of his ideas to be "rediscovered" years later by others.

Victor Tatin's Powered Model

Another man who significantly influenced aeronautics though he only worked with small, powered models, was Victor Tatin (1843-1913), a French inventor who may have created the first true airplane in 1879.

Like his contemporary, Penaud, Tatin started working with small rubber-band powered models at first, trying out a number of different shapes and sizes (and even experimenting with ornithopter—craft with wings that flapped like a birds) before finally settling on a fixed-wing design. Recognizing that using twisted strands of rubber could not provide a suitably stable and predictable power source, he decided to construct a craft designed to be driven by a reservoir of compressed air. Naming it, appropriately enough, the *Aéroplane*, the sturdy little craft—with a wing span of over six feet and weighing just under four pounds—had two counter-rotating propellers that powered the craft to a speed of 18 miles per hour (a considerable speed at the time, especially in a small model airplane). When it was functioning correctly, he could make it fly for several seconds at a time, making it the first powered model aeroplane in history. Though only a model, the little craft demonstrated all the ingredients necessary to achieve flight: enough power to provide lift and enough lift to overcome its own weight, making it a true airplane in every sense of the word and one that would be studied by later designers with great interest.

The *Aéroplane* wasn't Tatin's last aeronautical triumph however. In 1890, working in collaboration with French physician, Charles Richet (1850-1935), he built a considerably larger steam-driven model that also proved to be successful. Finally in 1911, near the end of his life, he built something called the *L'Aéro-Torpille* (literally, the flying torpedo), a monoplane with upwardly arched wings and a streamlined monocoque fuselage. Powered by a 50 horsepower Gnome rotary engine, the plane flew successfully—if only briefly—in 1912, one year before Tatin's death, thereby bringing one final accomplishment to a man who had enjoyed so many throughout the course of his long life.

Victor Tatin's original compressed air model on display in the French Air and Space Museum (Musée de l'Air et de l'Espace) near Le Bourget Airport, north of Paris.

CHAPTER SIX: THE ADVENT OF THE AIRSHIP 6

While the development of heavier-than-air flight during the middle of the nineteenth century was the more interesting aspect of man's quest to fly, it wasn't the only research that was going on. Ballooning remained as popular in the mid-nineteenth century as it had in the eighteenth century. In terms of advancements in lighter-than-air technology, however, little had changed since the Montgolfier brothers had launched their first balloons in 1783. New and more resilient materials had been developed and other small innovations had been introduced since then, of course, but for the most part lighter-than-air development seemed to be stuck at an evolutionary dead end.

The problem remained, as it had from the beginning, the inability to make balloons go where their pilot wished them to go. While a good deal of vertical control was to be had, balloons continued to be subject to the vagaries of the wind, often with unfortunate consequences. In the middle of the nineteenth century, however, things promised to change when engines capable of providing the sort of power required to propel a balloon through the air started to become available. Unfortunately, they were still years away from fulfilling their complete potential, but they were a sign that things were beginning to change, This ignited a flurry of activity in regards to airship development, with a number of inventors working hard to produce the first steerable balloon—or airship, as such machines were known—but it would be a little known Parisian engineer by the name of Henri Giffard who would be the first to actually succeed, and in so doing kick-start an era of airship development that was to last well into the succeeding century.

The Little Big Airship of Henri Giffard

Baptiste Henri Jacques Giffard (1825-1882) was a French railway engineer known mainly for his expertise with steam generation and boiler technology. However, he also had an interest in aviation and to that end, in 1844 he acquired his ballooning license at the remarkably young age of nineteen. That might have remained the extent of his aeronautical interests, too, had he not been called upon to help a fellow Frenchman, Pierre Jullien of Villejuif, demonstrate a streamlined model for an airship he named *Le Précurseur*[19] at the 1850 Paris Hippodrome. Fascinated by the futuristic shape of Jullien's revolutionary airship, it wasn't long before Giffard decided to construct his own fully flyable airship, based in large part on Jullien's design.

Apparently a man of some means, Giffard spent the next two years of his life pursuing his dream, which he finally completed in 1852. Unlike most balloons of the era—and in keeping with the general shape of Jullien's model—his was not spherical like a standard balloon, but cigar-shaped, with conical ends and a sail-like triangular vertical rudder mounted aft. To power the machine, he constructed a small steam engine that weighed just 350 pounds (including the boiler and the coke to fire it), which

19. Jullien's airship model, in mounting a rudder, elevator, and gondola, possessed all the elements of a modern airship, making it decades ahead of its time.

Jullien of Villejuif's highly streamlined airship Le Précurseur, a small model of which was displayed at the Paris Hippodrome in 1850, proved to be the source of inspiration for Giffard's airship, which would incorporate many of Jullien's ideas.

drove a large, ten-foot diameter, three-bladed propeller mounted at the aft end of the gondola. (The engine was also unique in that it was fitted with a downward pointing funnel and featured a new system designed to prevent sparks from rising upward and igniting the gas bag by mixing the combustion gases, thereby neutralizing any sparks.) Although it produced a mere three horsepower, he believed that would be sufficient to demonstrate that controlled flight was possible. At 144-feet in length, it was also the largest balloon ever built up to that time, and so, not surprisingly, it caused quite a sensation when it was unveiled to the public for the first time.

Anxious to test his new machine, on September 24, 1852, Giffard, flying from the Hippodrome in Paris, cast off the tethers that held the ship in place and the craft rose majestically into the calm autumn air. Stoking the machine's tiny boiler, soon its massive propeller was spinning at maximum velocity and the ship slowly made its way almost fourteen miles to the nearby town of Trappes where, despite the fact it could barely manage a tepid six miles per hour, it was observed to make a number of slow turns and circles. In doing so, Giffard demonstrated for the first time that a powered airship could be steered. Of course, he flew his craft under ideal conditions on a dead calm day; had there been even the slightest breeze, it's likely the underpowered craft would have proven unsteerable. Nonetheless, he did prove that the direction of flight of a lighter-than-air craft could be controlled, even if only for short distances under ideal circumstances, marking a milestone in the development of the airship.

Unfortunately, the enthusiastic Mr. Giffard got caught up in his own success and, instead of further refining his airship and perhaps learning as much as he could about controlling it in flight, he decided to dismantle it and start work on an even larger version, which he was to spend the next three years working on. This one, however, was to have a larger and more powerful engine capable of pushing the craft along at a more useful ten miles per hour, but to carry the additional weight, he needed to increase the length of the envelope by nearly one hundred feet, making the airship nearly as large as a modern day Goodyear blimp. However, while this provided enough lift to get the larger engine off the ground, Giffard had neglected to widen the diameter of the envelope, resulting in the ship being unable to maintain its cigar shape. This nearly proved to be a fatal mistake when during its first and only flight, the envelope sagged dramatically and began to pitch violently, resulting in the gas bag becoming detached from the netting and splitting. Releasing much of the 113,000 cubic feet of hydrogen contained within it, the ship began descending rapidly and it looked as though Giffard was going to go down with the ship, but he was able to set it down successfully and escape before the limp gas bag fell onto the hot boiler and ignited, completely incinerating the ship.

Despite the disastrous flight, Giffard was undeterred in his efforts to make his dreams a reality and he soon set about planning an even larger airship. He turned out to be quite the dreamer, however, for his new airship was to be nearly two thousand feet in length and with a maximum diameter of just under one hundred feet (twice the size of the famous German dirigible *Hindenburg*). Of course, such a craft was far beyond the capacity of the nascent technology of the age to build and so it never got beyond the planning stage, but no one could ever accuse Henri of not thinking big.

Solomon Andrews' Remarkable Airship

It's always interesting to see how some people easily find their way into the history books while others remain almost entirely forgotten even though their contributions may be, in many ways, every bit as significant as those who do make it. Such is the case, it would seem, with Doctor Solomon Andrews (1806-1872), a prolific inventor and medical doctor from Perth Amboy, New Jersey, who came as close to aviation immortality as any man ever has, and yet somehow managed to remain virtually unknown to this day.

Certainly Andrews seemed to be the sort of man capable of great things. A man of considerable talent and a seemingly inexhaustible passion for innovation, during the course of his lifetime Andrews accumulated a number of patents on everything from an improved sewing machine to the first practical kitchen range. (He even designed a padlock which has been used by the U.S. Post Office since 1842, demonstrating that he definitely knew how to make things that would last.) With such a far-ranging mind, then, it's not remarkable that at some point in his life he started playing around with airships. Unlike most airship inventors of the era, however, Andrews took a novel approach to the idea of airship propulsion. While others were looking at steam

Henri Giffard's 1852 airship, though tiny and slow, was the first steerable airship in history, which portended the future of airship development well into the twentieth-century..

Solomon Andrews' first airship in 1863. The craft was revolutionary in that it had no engine, directional control being accomplished by the movement of the pilot along the gondola and basic gravity. (Drawing courtesy Popular Science Magazine, January 1932 edition.)

as a propellant, he decided he could get his airship to fly simply through what he called "gravitation"; effectively, his craft would glide through the air by making use of the specific gravity between the balloon and the surrounding atmosphere through a system of inclined planes designed to steer the craft. With this technique, the airship would cycle between positive and negative buoyancy, resulting in an airflow across the body of the craft and attached airfoils. As such, to ascend, one would simply drop ballast from the bow (or have the pilot move aft) until the craft reached an angle of approximately 10 to 15 degrees, forcing it to rise. Then, by valving off some of the hydrogen and shifting weight forward, the craft would descend, again providing forward momentum. In this way the craft would effectively tack against the wind much like a sailing ship would when running against a headwind, providing an energy-free means of travel. It was a curious idea that, if successful, would eliminate the need for a power plant—with all the resultant weight and fuel such entailed—providing humanity with a cheap and simple means of conquering the skies.

How far Andrews got with this idea is unclear. It appears that as early as 1842 he built and tested a small scale model (though still of substantial size). It's not clear whether these tests went anywhere, however, but they must have been enough to convince Andrews that he was onto something, for he continued to work on his "gravitation-powered" airship right up to the start of the American Civil War.

Though the war forced him to curtail his motorless airship experiments for a time while he volunteered his services as a medical doctor for the Union army, he never gave up trying to find a way to bring his idea to life. After observing how inadequate the fixed balloons of the Union army were in terms of providing useful reconnaissance of enemy positions, he hit upon the idea of presenting his airship idea to the government as a legitimate military device.

While he waited for a response from the government—which seemed to be dragging its feet on his proposal—Andrews set about building a full size example of his airship—which he called the *Aereon*—in an effort to prove the practicality of his novel approach to manned flight. What resulted from his ambitious efforts was a most usual-looking craft made up of three 80-foot-long cigar-shaped balloons attached to a rudder and gondola, which Andrews was ready to test for the first time in the summer of 1863.

One might imagine that an unpowered airship would not do well in terms of maneuverability or directional control, but in fact it seemed to have actually worked. Flying the *Aereon* over Perth Amboy, New Jersey in June of 1863, the craft seemed to perform as advertised, with Andrews making it turn and circle merely by moving ballast and himself forward and backwards along the length of the narrow gondola. After completing some modifications, he apparently flew it again in August of that year, during which time, at least according to an article that ran in the *New York Herald* on September 8, 1863, Dr. Andrews' craft was observed to climb in a tight upward spiral until it was almost lost in the clouds. If true, this made Andrews' unusual machine the first dirigible capable of being flown against the wind—and for considerable distances at that.

It's not clear, however, exactly how much actual directional control Andrews had over the craft or whether he was merely riding the prevailing currents much like a kite. It's also not certain how much control he might have maintained over the craft against a strong breeze, or how much straight-line control he would have been able to maintain over an extended period. It certainly appears unlikely that the craft could have been made to travel along a fixed course for any period of time, as it would have been forced to fly an exhausting zigzag path while it climbed and descended much like a roller coaster, making its practicality open to debate. It was still an interesting concept that showed promise, however, and perhaps that's all the intrepid Mr. Andrews wanted to demonstrate through his novel design.

Convinced he had largely perfected his craft, Andrews grew increasingly excited about using his unusual airship for the Union cause, even going so far as to write to Abraham Lincoln offering the *Aereon* for use in the war as an observational vehicle. It appears that his concept of "gravitational flight" was not taken seriously by the government or the military, however, despite the fact that he had repeatedly demonstrated that the concept worked. Through sheer persistence, however, he did finally manage to arrange a demonstration before the Smithsonian Institution early in 1864, only to have his demonstration repeatedly postponed. Unfortunately, by the time officials in Washington appeared to be finally ready to give his airship a look, the war ended and Andrews and his dirigible were quickly forgotten.

However, Andrews wasn't one to let a good idea go to waste. Deciding his *Aereon* had civilian applications as well, he organized the Aerial Navigation Company in the hopes of establishing a regular airline between New York and Philadelphia—a fantastic proposal at the time and one that predated the advent of the world's first airline, DELAG (see chapter twelve) by over fifty years.

Unfortunately, Andrews' second craft, the *Aereon #2*, which worked along the same principle as his earlier craft, proved far less successful. Considerably larger than his first craft and more lemon-shaped than cigar-shaped, he quickly discovered he had less control over its direction of movement than he had with his previous airship. In spite of that, Andrews still made two controlled flights in the thing, one over New York City on May 25, 1866 and a second just a few days later on June 5, both of which were enthusiastically covered by the local press and both of which carried passengers (the first airship passengers in history, it appears). It was even claimed by one paper, the *New York World*, that the ship obtained a height of some 2,000 feet and drifted lazily over the city for some time before landing safely on Long Island. Obviously enthused that it appeared the age of air transport was at hand, the paper wrote: "Navigation of the air is a fixed fact. The problem of the century has been solved."[20]

What should have been heralded as one of the great triumphs of the ages, however, ended abruptly—not by technical difficulties or unforeseen catastrophe, but by economic collapse. It seems that just

20. New York World, May 28, 1866.

as Andrews was beginning to find the capital to fund his airship service, many banks collapsed, leaving his fledgling airship company/airline bankrupt and ending both his flights and his dreams for a New York to Philadelphia air service. It also leaves us to wonder what might have happened had Andrews found the capital to continue to refine his design; would the dirigible have become a common sight in the sky through the last decades of the nineteenth century and the history of aviation been dramatically rewritten, or would his airship have flown off the pages of history as the exhausting and impractical (imagine having to have scores of passengers continually move from one end of the airship to the other to keep it in motion) form of travel it really was? Of course, the world will never know.

The Human-Powered Airship of Dupuy de Lome

It's curious how much impact the brief Franco-Prussian War of 1870 had on aeronautics. When Paris was surrounded by Prussian forces in mid-September of 1870, the French came up with some innovative ideas to keep their forces inside the city in communication with the rest of the country. Perhaps the most unique strategy that came out of it was the idea of using hot-air balloons to fly over the Prussian lines to deliver letters and communiqués—and frequently people—outside the city. (It was an idea that seemed especially appropriate, considering the city's rich aeronautical history.) In any case, among those who participated in this scheme was a man named Henri Dupuy de Lôme who, while not personally piloting any balloons over Paris himself, did create and build one that, were it not for the end of the war, may have resulted in the first human-powered airship in history.

Stanislas Charles Henri Dupuy de Lôme (1816–1885) was considered by many to have been the greatest French naval architect of the nineteenth century and, arguably, the greatest in French history. Mostly known for his work on steam-driven ironclads, battleships, and submarines, what's less known about him is that he also put his engineering prowess to work trying to create a controllable balloon designed to circumvent the siege of his beloved Paris.

The problem de Lôme faced, as it was for all balloon enthusiasts of the era, was the balloon's dependence upon the whim of the weather and

Dupuy de Lôme's innovative hand-cranked airship might have become a reality had not the Franco-Prussian War ended before it was ready for its first flight. Note the large two-bladed propeller that was to propel the craft at a leisurely 8 knots, along with the quaint boat anchor hanging from its bow.

the direction of the prevailing winds. De Lôme, however, believed it might be possible to develop a practical, navigable balloon that could fly to specific destinations without having to depend upon favorable winds to do so. The airship he came up with was large for the time, with a length of one hundred twenty feet and a nearly fifty foot diameter at its maximum. It also had an elongated basket that could carry up to eight people, most of whom would be busy turning a hand-driven shaft that would drive the craft on a predetermined course at—in theory—a respectable 8 knots. Christened the *Dupuy de Lôme* after himself (the man obviously being no stranger to self promotion), France capitulated just a few days before it was ready for its maiden voyage, thereby depriving the world of seeing whether it would have flown or not. As it is, it was quickly dismantled a few weeks after the armistice was signed and forgotten (except, apparently, by Mr. de Lome and his biographers).

Whether it would have worked is anyone's guess, of course, but it demonstrated that even in 1870 man was already imaging ways lighter-than-air craft might be used in military operations—a prospect that was to become a reality less than a half century later when France and Prussia (later united into the country of Germany) would clash again in the skies of Europe—and this time airships would play a much larger role.

Paul Haenlein's Hydrogen-Powered Airship

Another man to have some success with a steerable airship was the German engineer Paul Haenlein (1835-1905), who was to build and fly one of the first successful semi-rigid dirigibles[21] in history.

A graduate of the prestigious Technische Hochschule in Karlsruhe, Haenlein was a pattern maker and civil engineer who had worked for several companies in Germany before turning his attention to airships. What he eventually came up with was revolutionary—not in its size or configuration—but in its propulsion system. What Haenlein proposed using to power his craft was not steam (Giffard) or gravity (Andrews) or even human power (de Lôme), but a small Lenoir-type internal combustion engine that actually ran off of the hydrogen contained in the airship itself!

With his system, the gas was actually pumped from the envelope of the balloon—which was kept fully inflated by pumping in compensating air to the bags inside the main envelope—directly into the engine itself to serve as the fuel for combustion. This would deliver about 6 horsepower to a 15-foot diameter propeller, causing it to spin at a respectable 40 revolutions per minute. The problem was that the revolutionary engine consumed about 250 cubic feet of gas per hour which, in continually consuming its own lifting agent, greatly reduced its range. The engine also had the disadvantage of requiring

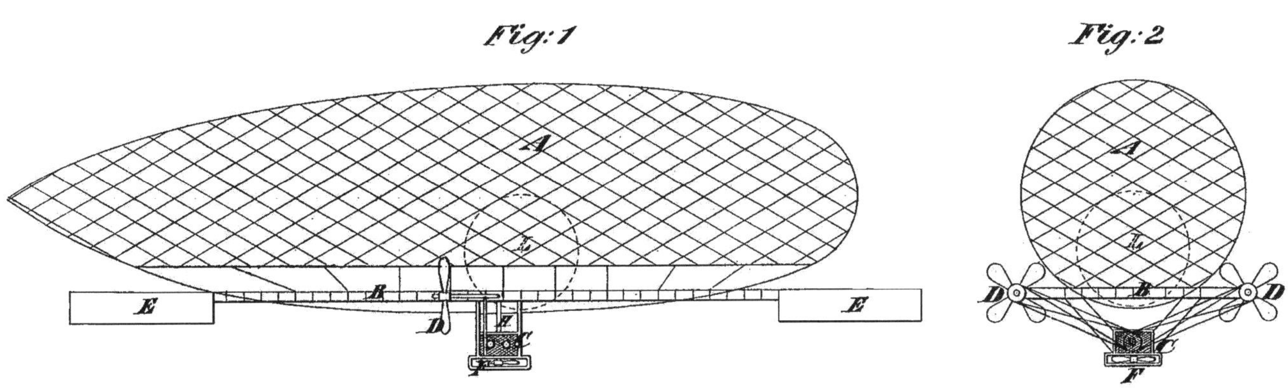

Patent drawing from Paul Haenlein's 1870 hydrogen-powered airship design.

21. A semi-rigid dirigible is differentiated by the fact that the nonrigid envelope is attached to a bow-to-stern keel that gives it some degree of rigidity, as opposed to a "blimp", which has no rigid structure to help maintain the envelope's shape, or a fully rigid airship, which has a complete internal skeleton within the body of the envelope.

a large storage capacity. It did, however, prove to be an elegant solution to the problem of fuel storage and was the first instance of the use of an internal combustion engine in connection with aeronautical experiments.

Interestingly, Haenlein actually went on to build and test his novel airship. On December 13, 1872, he flew his airship near Brunn, Germany, where he was able to get the 164 foot long, 30 foot diameter craft to reach a sustained speed of 15 knots, which was not only quite an accomplishment for the time but was nearly three times faster than what Giffard's earlier airship had been able to achieve. The trial was made with the dirigible tied down, however, and while Haenlein intended to perform an untethered flight next, he ran out of money at the worst possible moment and was forced to call the whole thing off.

It's difficult to imagine that the funds couldn't have been found to see what the man's ship might have accomplished, but even what little was shown demonstrated the craft to have been a major advance over all previous designs. In using its own lifting agent as an energy source, however, it was doomed to be a short-range novelty at best, so perhaps it was better the idea died an early death.

The Airship of Albert and Gaston Tissandier

Other than the Montgolfier brothers, there are few siblings as famous in France for their work with lighter-than-air aviation than the famous Tissandier brothers, each of whom were to be credited with a number of firsts in a century known for many aeronautical firsts.

The Tissandiers were from a family of well-educated architects, artists, and scientists. Albert (1839-1906) was an architect of some renown as well as a proficient illustrator while his brother Gaston (1843-1899) was a well-known chemist and meteorologist with a fervent fascination for balloons and airships.[22] Together they founded and edited the scientific journal *La Nature*, a publication that continues to this day under the name *La Recherche*.

It was Gaston who first took up the cause of ballooning when he made an ascent in one at Calais in 1868. He even had the distinction of being spirited out of a besieged Paris during the Franco-Prussian War of 1870 by balloon, but his most adventurous flight aloft took place in April 1875 when, together with fellow aviation enthusiasts Joseph Croce-Spinelli and Theodore Sivel, he ascended in a hot air balloon to an unheard of altitude of nearly 28,000 feet,

Albert Tissandier's hauntingly beautiful pencil-sketch of the Zenith floating beneath a haloed moon at night, made from the gondola of the craft at 3,600 feet, March, 1875.

22. He is also the father of Paul Tissandier (1881-1944) who became a well-known aviator in his own right.

The Tissandier's elegant 1883 airship proved to be as underpowered as Giffard's earlier machine had been. Without access to a more powerful engine, the brothers apparently gave up their lighter-than-air pursuits and moved on to more lucrative ventures.

setting a new world altitude record in the process. Tragically, the flight almost proved to be Gaston's last, for without portable oxygen, the men quickly succumbed to the lack of air and both Spinelli and Sivel died. Remarkably, Gaston survived, though he was rendered deaf by the experience—an affliction he would struggle with the rest of his life.

Albert was no less adventurous and every bit as committed to aviation as his brother, having first flown in a hot-air balloon between Melun and Paris in 1868—a flight which was made during a snowstorm. Later he served as a pilot on one of the first balloons to fly in and out of Paris during the Franco-Prussian War, carrying letters and dispatches—a feat for which he was awarded the Medaille Militaire for bravery. Like his brother, he also escaped from the city in the gondola of a balloon during the height of the siege.

It was their flights together, however, for which they are best remembered, the most famous being one they took in 1875 onboard a hydrogen-filled balloon called the *Zenith*[23]. Flying from Paris to Arcachon near Bordeaux—a record-setting distance of nearly four hundred miles—in two days, during the flight Albert drew pictures of the countryside below and even sketched a drawing of the balloon flying over a shimmering field of clouds at night, demonstrating his considerable talents as an illustrator.

Eventually the brothers became interested in the prospect of controlled flight and decided to experiment with the concept using small electric motors, which they demonstrated at an exposition by attaching one to a large model of a dirigible. After determining that it may be possible to control an airship in this manner, they set about building a full-sized ship that they named *Aerostat* to test out their ideas. Unfortunately, the electric motors of the era were not powerful enough to power a full-sized dirigible (the motor they used could only produce 1.5 horsepower). It appears that at this point, perhaps because of the failure of their *Aerostat* to perform satisfactorily, they gave up any further studies of airships and, like many pioneers who also walked away just at the point when success seemed inevitable, they moved on to other pursuits—much to the detriment of the aeronautical community.

23. It was in this same balloon that Gaston took his nearly fatal flight a month later.

It's always difficult to determine what the history of aviation might have looked like had the brothers continued their experiments, but it is likely, had they pursued their experiments with more determination and made a concerted effort to obtain a more powerful engine, that they might well have become one of the premier names in lighter-than-air development. As it is, the world was left with only a tantalizing taste of what they might have accomplished—along with, of course, Albert Tissandier's wonderful illustrations.

Renard and Krebs's Successful *La France*

Up to this point, successfully controlling an airship had met with mostly disappointing results. Giffard, Haenlein, and the Tissandier brothers all had workable designs, but none of them had designed a consistently controllable airship, largely due to the lack of a light and powerful enough motor to move a craft against the wind. With each attempt and subsequent failure, however, the aviators of the day got ever closer to realizing their dream of a genuinely steerable airship. This trial-and-error approach finally paid off just a year after the Tissandiers' attempt when another pair of Frenchmen, Charles Renard (1847-1905) and Arthur C. Krebs (1850-1935), built and flew the first truly practical airship in 1884.

Officers in the French Army Corps of Engineers, Renard and Krebs were well suited for the task fate put before them. Renard had started to work on airship designs at the French armies' aeronautical department shortly after the Franco-Prussian War of 1870, while Krebs was an inventor of considerable engineering expertise. With the help of a grant from the French government, over the course of six years together they quietly constructed a 170-foot long airship that would finally demonstrate that airships could be controlled in powered flight.

Named *La France,* the craft proved to be a vast improvement over earlier attempts. Like the Tissandier brothers' airship, it also had an electric, battery-powered motor, but while the Tissandiers' mo-

The hugely successful La France, which made no fewer than seven flights between 1884 and 1885.

tor could produce only an anemic 1.5 horsepower, theirs produced 8.5 horsepower, making it capable of pushing the ship through a calm sky at an impressive thirteen miles per hour—an extraordinary accomplishment at the time. The *La France* also had a number of other fairly sophisticated design features, including a 23-foot diameter, four-bladed wooden tractor propeller that could be inclined upwards when landing to avoid damaging the blades. It also had a futuristic rudder and elevator assembly, a sliding weight to compensate for any shift in the center of gravity, and even ballonettes—small air-filled bladders located inside the envelope of an airship that are deflated or inflated as necessary to maintain the envelope's external shape during ascent or descent (a feature still used on today's blimps to perform exactly the same purpose.)

The first flight of the *La France* took place on August 9, 1884 and, though the craft flew only five miles, Renard and Krebs managed to maintain complete control of the ship throughout the flight, even returning to their starting point and landing successfully, thereby demonstrating that controlled flight was possible if an airship had a sufficiently powerful motor. The *La France* proved to be more than a single-flight wonder as well, making no less than seven flights between 1884 and 1885, each of them under control and in five of which Renard and Krebs managed to return to their starting point, thereby establishing that the first flight had been no fluke. In many ways their achievement also signaled that the age of the controllable airship had at last arrived, though it would be a few decades before the fullest extent of that fact would be realized.

However, Renard and Krebs's airship, while impressive, was fragile and, like all electric-powered craft of the era, had too limited a range to be practical. It did, however, have one unanticipated side-benefit: It served as a source of inspiration to science fiction writer Jules Verne, who wrote of "the striking experiments of Captain Krebs and Captain Renard" in his 1886 novel *Robur the Conqueror*.

Reminiscent of photos of dirigibles taken in the 1930s, this photo of the La France in its hangar shortly after one of its successful flights looks remarkably modern from this perspective. Note the enormous size of the propeller blades.

Conclusions

Renard and Krebs' *La France* demonstrated that the world was on the brink of a new era in aviation and that truly practical, passenger-carrying craft were on the horizon. It would take another two decades for all the elements to come together, but the important foundation upon which lighter-than-air aviation in the twentieth century was to be built upon had been laid. However, as the century began drawing to a close and the advances in technology were becoming more marked, it was the aeroplane that was to rise to the forefront of aviation development, forcing the airship to take a back seat until the turn of the century, at which time it would come to dominate the skies once again.

CHAPTER SEVEN: THE AGE OF THE "GLIDERNAUTS" 7

While the last decades of the nineteenth century were a frustrating time for proponents of powered flight due to the lack of sufficiently powerful motors, the era was known for being a time of fairly rapid advances in the area of manned gliders, the development of which would prove vital in laying the foundation for later powered, manned flight. As such, the era between 1883 and the late 1890s belonged to the "glidernauts"—men like John Montgomery, Percy Pilcher, and Otto Lilienthal. These would be the men who would push the limits of human ingenuity and determination, though their efforts would come at a high price, as most of the aviators examined in this chapter were to lose their lives in the quest to fly like a bird, making this period in aviation history among the most dangerous. However, it would prove to be among the most important as well, for without the bravery of these men the story of twentieth century aviation might never have been told. So who were these men who gave so much?

John J. Montgomery

Up to this point, most of the work in aviation was confined almost entirely to England and France, but by the 1880s the United States was also starting to become a player as it became increasingly scientifically and technologically advanced as well, and perhaps the earliest and most promising of the new American inventors of the era was a physician and aviation enthusiast by the name of John Montgomery.

Doctor John Joseph Montgomery (1858-1911) had developed an interest in aviation during boyhood, but it wasn't until he had graduated from St. Ignatius College in San Francisco, California, that he found the time and means to begin experimenting with gliders. Returning to his home near Otay Mesa, with the help of his sister, Jane, he set up a lab and, after studying how birds of various sizes flew in an effort to ascertain how their wings warped while in flight, he began constructing the first of what was to eventually be many gliders.

His first effort was not an auspicious one, however. Like many early inventors at the time, he took the way birds flew far too literally and built a glider with flappable wings. This, not surprisingly, proved to be woefully unsatisfactory, induced Montgomery to return to the proverbial and, in this case, literal drawing board. Carefully rethinking his initial premise in regards to how birds achieved lift, he abandoned the flapping wings concept and instead drew up a design for a fixed-wing monoplane glider with curved or cambered wings. This proved to be a far more efficient means of achieving lift, but he also found that with his new design he lacked the ability to control his machine laterally. He solved this problem on his next glider, however, by adding hinged surfaces at the rear of the wings to maintain lateral balance, thus anticipating ailerons—a mandatory feature on aircraft today—by literally decades.

Testing on the new design proved to be more promising, as his ailerons appeared to provide a degree of lateral control. Surprisingly, however, instead

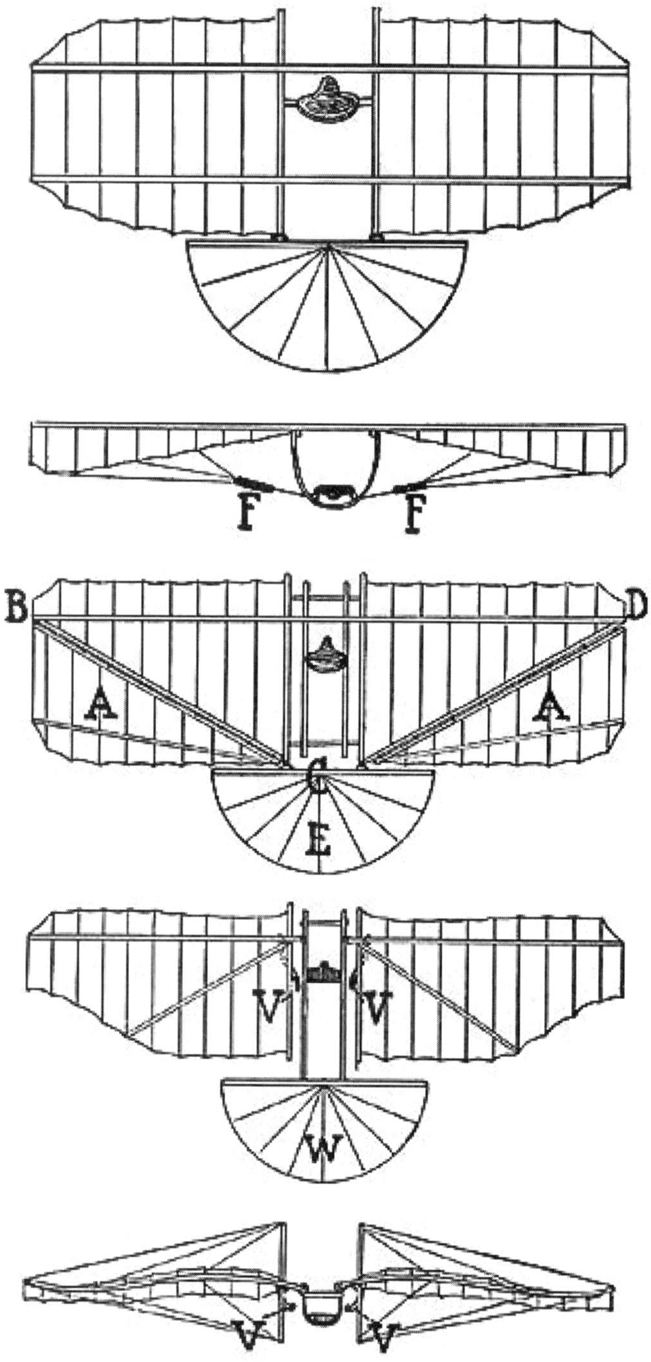

The evolution of an idea: at top is Montgomery's first fixed-wing glider, which lacked lateral stability. His second glider (middle figures) addressed this problem by incorporating early ailerons (letter A). Montgomery's third design (at bottom) eliminated the ailerons and used a curved wing design to "rock" the craft from side-to-side, thereby providing a degree of both lateral and pitch control.

of developing his aileron idea further, he abandoned them on his third glider, deciding that rocking the wings to maintain lateral balance as well as to provide pitch control was a superior solution. It's unclear how successful this proved to be for he was not careful about recording many of his flights, nor were they done in public. They must have been successful enough, however, to convince him to proceed with even more ambitious gliders later in life.

While Montgomery's glider experiments were to eventually prove useful in the development of manned airplanes, perhaps the man's greatest contribution to aviation came in the field of aerodynamics. Like George Cayley nearly half a century earlier, Montgomery too was quite visionary in regards to studying how various airfoil configurations created different degrees of lift. However, unlike Cayley—and in quite a stroke of genius—he tested his airfoils using water rather than simply observing how they performed in the air. Building a large, flat tank over which water was pumped onto airfoils of different configurations, this technique gave him a much clearer visual understanding of how the forces of lift operated and, subsequently, made it easier to identify which configurations provided the maximum lift possible. These studies were to prove crucial to later inventors as well, who frequently referred to his work in their efforts to create the most ideal lifting surface possible. In fact, without them, progress would have been delayed by years, making the name John Montgomery as important to those early pioneers of aviation as that of Cayley and Lilienthal.

Despite Montgomery's important contributions to the field of aeronautics, however, he remains largely unknown to many people today, his name being superseded by those who were to follow in his footsteps. This was partly because he was a private person who preferred to work in private and produced few writings on his efforts. However, a larger reason may have been because while others were making the transition from gliders to manned aeroplanes, Montgomery continued to work exclusively on gliders, leaving him in the shadows of many of his contemporaries.

Right and right center: one of Montgomery's last gliders. Note its substantial size and sophistication, especially when compared to his earlier designs.

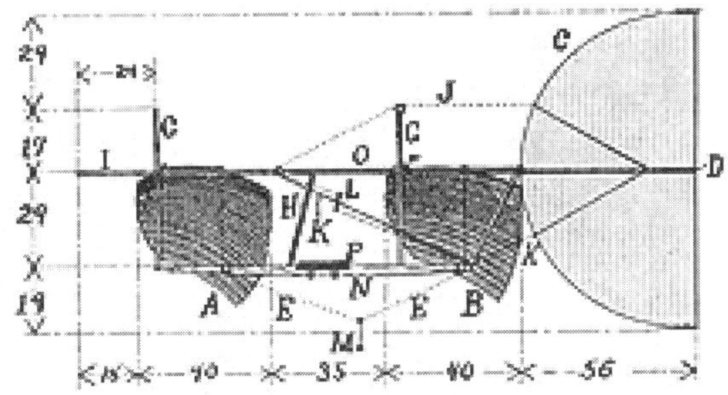

Montgomery's last glider, the "Evergreen", being flown in 1911. It would be in this machine that Montgomery would ultimately lose his life.

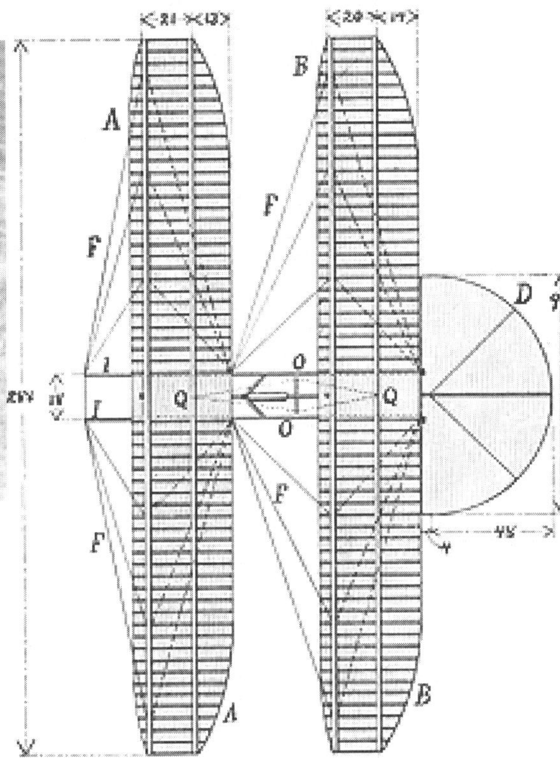

Montgomery and one of his tandem-wing gliders—circa 1905. It's unfortunate the man didn't choose to attach a motor to his machine as it would likely have easily flown if he had.

Montgomery's gliders did, however, continue to grow in sophistication and size until by 1904 he was building and flying large, tandem-wing machines that would be considered advanced even by modern standards. By then, of course, aviation was moving towards powered flight, rendering his large gliders passe. They also proved to be increasingly dangerous to fly as well, as his hired test pilot, Daniel Maloney, found out when a structural failure caused one of Montgomery's experimental gliders to crash, killing him instantly. Ironically, in October, 1911 the fifty-three-year-old Montgomery would learn first-hand just how dangerous even unpowered gliders could be when he crashed while at the controls of yet another of his gliders, a monoplane he christened the *"Evergreen"*, bringing to an end the life of one of America's most prolific aviation pioneers.

Lawrence Hargrave

Like John Montgomery, there was another man whose contribution to heavier-than-air flight was profound but whose name today remains largely unknown people today despite being one of the finest engineers, explorers, astronomers, inventors, and aeronautical pioneers in Australia. The man's name? Lawrence Hargrave.

Born in Greenwich, England, but an émigré to Australia, Lawrence Hargrave (1850 –1915) had been interested in aviation from an early age. Living in Sydney, Australia, however—some twelve thousand miles from Europe where most of the early advances in aviation were being made—combined with his heavy workload as an assistant astronomical observer at Sydney Observatory, made pursuing his hobby difficult. That changed in 1883, however, when his father passed away and he inherited enough money to finally afford him the time and resources to devote the rest of his life to the study of flying machines, which he was to do right up to the time of his death in 1915.

Unlike most of the early pioneers of aviation, Hargrave never built an aircraft himself.[24] His contributions to the field of aeronautics came through his experiments with kites and his work in creating lighter, more powerful engines for aircraft. In regards to the latter, in 1889 Hargrave designed the first rotary combustion engine, a revolutionary concept that was to be used on many aircraft right up through World War I. (What made it so useful that it was a very efficient solution to the problems of power output, weight, and reliability, while its elevated torque and consequent gyroscopic effect gave

Hargrave working with his box kites at Stanwell Park in 1894.

[24] That's not to say that Hargrave never flew, however. In November of 1894, Hargrave was holding onto a train of four of his box kites when he was lifted off the ground by a sudden gust of wind and flew several dozen feet before gently landing in the sand. Fortunately uninjured, the incident demonstrated the powerful lifting capacity of his design, making his box kitre configuration ideal as a "scientific" kite capable of lifting large loads.

the airplane exceptional agility.) Unfortunately, because of the lack of sufficiently light-weight metals with which to construct his new engine, combined with the lack of quality machining capabilities at the time, he was unable to produce an effective example of his own and so would never receive the proper credit for the idea due him. As such, it would be for others—in this case the French Seguin brothers in 1908—to "reinvent" his rotary design, which would go on to become the preferred means of propulsion for fighter planes in World War I. (Rotary engines became obsolete by 1920, however, largely due to their inability to achieve higher altitudes.)

It was Hargrave's work on kites, however, for which he is best known. Working on a variety of different configurations, in 1893 he designed and built the first box kite, whose cellular construction—two cells arranged in tandem—made it the most stable and best lifting kite ever designed. So successful was his kite, in fact, that it was adapted for use by the U.S. Weather Bureau in the 1890s because of its inherent stability and tremendous lifting capacity (and is also the reason why Hargrave is known today as the Father of the Scientific Kite).

But even more profoundly, what the man eventually came to learn from his kites was to be ultimately applied by others to their glider and aeroplane designs, using the Hargrave box kite as a starting point. (His designs even inspired the noted American inventor Alexander Graham Bell—with whom Hargrave maintained a lifelong correspondence—to begin experimenting with his own large kites.) In fact, within a few years European biplane aviation designs—which tended to concentrate on "built-in stability" versus the Wrights' focus on "flyability"—would be dominated by Hargrave's box kite formula[25]. Further, in June of 1909, Hargrave designed a single cell box kite with reverse curve supporting surfaces like a shallow "S" (like the tilde character on the keyboard) that provided both lateral and longitudinal stability in an aircraft, eliminating the need for a tail plane. This is one of Hargrave's lesser-known designs, and the technique is still used in hang gliders today.

Unlike many inventors of the era, Hargrave did not believe in patenting his ideas or any of his inventions, insisting instead that all discoveries should be shared as a means of furthering progress. This made him unique among inventors of his age—or any age for that matter—and did much to hasten advances in aviation during this period. Unfortunately, as a result of this generous attitude, much of his work, like that of many other pioneers of the era, was not sufficiently appreciated during his lifetime. Today he is remembered, however, as an excellent experimenter and for his ability to remain optimistic and persistent even in the face of repeated failure. Modest and unassuming, he was only interested in adding to the sum of human knowledge—a goal he managed to accomplish magnificently during his long and fruitful lifetime.

Otto Lilienthal, the Glider King

Perhaps no name in aviation is as famous or revered today as is that of the German aviator Otto Lilienthal (1848-1896), a man who was to become so well known for his many glider flights that he would be eventually endowed with the title, the Glider King (Der Schirm König in German).

Lilienthal's interest in flight began as a boy when he and his brother, Gustave, made a set of strap-on wings and tried to fly, though that would prove to be the extent of his aerial pursuits for the next thirty years. In the interim, he graduated from the prestigious Royal Technical Academy in Berlin with a degree in design engineering and, after a brief stint fighting for the Prussian army in the Franco-Prussian War, he turned his attentions to running his own boiler and steam engine manufacturing company, which took up most of his time and made him a wealthy man in the process. Through all this, however, he still found time to study aeronautics and even wrote a famous book on the subject, *Birdflight as the Basis of Aviation,* in 1889.

25. A good example of this can be seen in Brazilian aviator Santos Dumont's successful aircraft, the 14-bis aeroplane, which flew in 1906. (See chapter fifteen.) It was built almost entire from Hargrave "cells"—a series of large box kites affixed together.

It wasn't until the early 1890s, however, that Lilienthal finally found the time to pursue his dreams of flight in a more direct, hands-on way. Likely aware of the glider experiments of John Montgomery and others, he began building and flying his own gliders until, by the mid-1890's, he was considered one of the finest and most prolific glider pilots in Europe. In fact, between 1891 and 1896 Lilienthal would make upwards of 2,000 flights on his home-made gliders—a record even many hang glider pilots today would find impressive.

One of Lilienthal's first gliders circa 1891. While advanced for the time, it proved to be a primitive affair when compared to his later efforts.

One unique characteristic about Lilienthal was that he understood the importance of maintaining public interest in aviation, which is why he allowed himself to be photographed so frequently and is why locating photos of the man in flight is relatively easy to do. (He also proved to be something of a showman as well; on a number of occasions he could be seen shouting instructions to photographers on the ground from his glider telling them where to position themselves to get the best shot.) Being the most frequently photographed aviator of the era, however,

One of Lilienthal's gliders around 1895. Notice how closely the design mimicked the features of a bird.

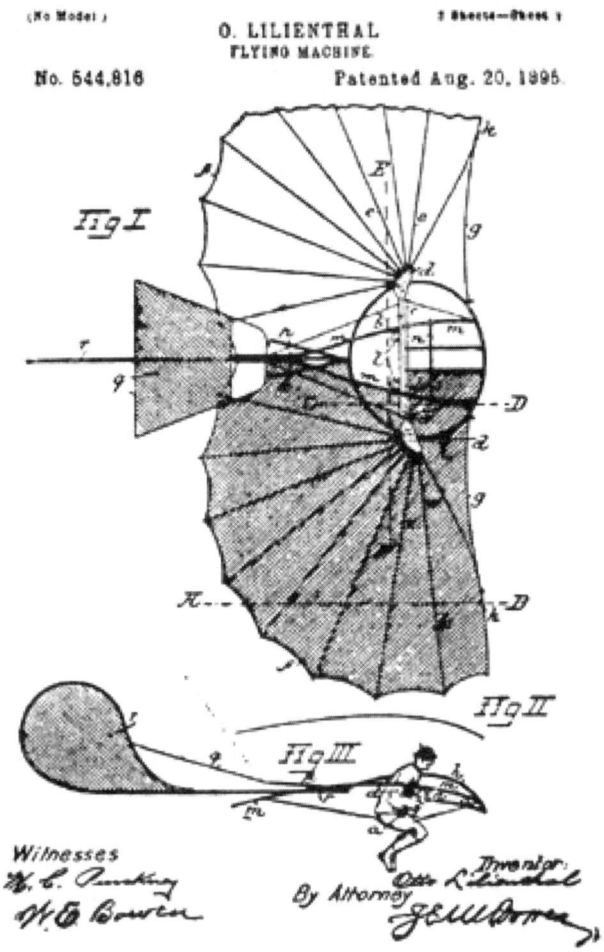

Lilienthal's patent drawings from 1895.

Lilienthal flying in a bi-wing glider just days before his fatal crash in August of 1896.

did much to induce a great deal of enthusiasm among the public for flying and influenced scientific debate as to whether manned flying machines might not eventually be proven to be practical. He also regularly detailed his experiences to the press, which, by the mid 1890s had made him one of the most famous men of his day, not only in Germany, but throughout Europe and, increasingly, around the world. He also maintained friendships with a number of other important aviators of the time such as Samuel Langley, Percy Pilcher, and Octave Chanute, many of whom visited him frequently to study his highly successful glider designs and discuss various solutions to the many engineering challenges of the day.

Like many people who are extremely adept at something, Lilienthal often made it look like flying was relatively simple and safe. However, the fact was that his gliders were often difficult to control, making the ability to make flying them look easy all the more impressive. The biggest problem with his designs proved to be in their basic configuration. Lilienthal controlled his craft by shifting his body to change the center of gravity—much like hang glider pilots do today—but his gliders had a tendency to nose-dive, largely as a result of him mounting the glider around his waist instead of hanging suspended from the bottom as one would in a modern hang glider. This meant that he was only able to shift his legs and lower body to maintain control, greatly reducing the amount of weight shift he could achieve. Lilienthal tried to work on the problem and managed to improve stability with some success, largely through making a bi-wing glider which halved the wing span for a given wing area as well as added a hinged tail-plane that he could move upwards to make the flare at the end of a flight easier, but he never solved the problem entirely.

Sadly, in August, 1896, Lilienthal's luck finally ran out when his glider stalled and crashed from a height of over fifty feet, breaking his back. The man died the following day, his last words being "Kleine Opfer müssen gebracht werden!" ("Small sacrifices must be made!") and was buried at Lankwitz Cemetery in Berlin. With his death, the aviation world lost one of its most articulate and popular proponents. It's also likely that, had he lived, he might well have moved into powered gliders and, with that, may have beaten the Wright brothers by years. (Certainly, he had the technical acumen to handle such a challenge and required only the power plant to make the dream a reality.) Unfortunately, the world would never get the chance to find out what this gifted man might have accomplished if not for making one flight too many.

Percy Pilcher

While Otto Lilienthal is a well known name to the aviation world, another man whose career closely mimicked that of the Glider King but who remains largely unknown outside of his native England was Percy Pilcher (1866-1899), a British inventor and pioneer aviator who was, until his death, his country's foremost experimenter in unpowered flight throughout the later years of the nineteenth century.

Pilcher, probably inspired by the work of the Lilienthal, began pursuing his interest in flight shortly after completing a seven-year stint with the Royal Navy. Learning as much as he could about gliders while he worked for several years as an apprentice shipbuilding engineer and assistant lecturer in Glasgow, Scotland, by the time he was thirty years old, he decided he had learned enough that he decided to begin designing and building his own. The first of these efforts, a little glider he christened the *Bat*, flew well and taught him the basics of how to fly, but soon he was ready to move on to larger projects. To that end, he managed to contact the by-then world famous Lilienthal himself to discuss his gliding experiments, and as a result of their correspondence, he went on to build two more gliders based largely off Lilienthal's machines, which he named the *Beetle* and the *Gull.*

Pilcher, however, was not pleased with the performance of his new designs. (As they were based on the German aviator's gliders, he was probably encountering the same sort of stability issues Lilienthal was dealing with.) Believing his original glider design was inherently more stable, he started work on a fourth glider based off the *Bat*—which he christened the *Hawk.* Something of a cross between a modern ultra light and a hang glider, the *Hawk* was quite modern for its time though a bit unconventional by modern standards. The biggest difference between his craft and most gliders of the era was that the *Hawk* featured a lightweight wheeled landing gear. He also launched his

Pilcher's first simple but remarkably successful glider, the Bat, in 1895.

Pilcher's favorite glider, the Hawk, showing its unique landing gear.

craft in a different manner than other glider pilots did: Instead of simply running down hills or jumping off embankments and then gliding in whatever direction the wind was blowing, Pilcher developed a reliable method of towing his glider aloft. In his new and improved glider, then, Pilcher was to even set a distance record of 820 feet in 1897, making it his most successful glider.

In what likely would have been the direction his late mentor and friend Lilienthal would have gone had he lived, by 1898 Pilcher began giving serious thought to powering the *Hawk* and even went so far as to receive a patent for a powered version of his craft. Before he could turn his glider into an aeroplane, however, he changed his mind about simply adding a motor to his glider and decided instead to build a powered aircraft from the wheels up, which he felt would provide the necessary increased lift required to counter the added weight of a motor. However, he soon ran into the same weight-to-power ratio paradox that had bedeviled so many of his colleagues at the time: Increasing the size of his craft to accommodate a motor meant increasing its overall wing surface (and with it, its weight), which would necessitate the need for an even larger—and heavier—engine to make it airborne, thereby requiring larger wings yet. At first stuck on how to solve the problem, thanks to correspondence with another pioneer, Octave Chanute, he soon had the answer: Chanute advised Pilcher to stack smaller, lighter wings one atop one another in an arrangement we know today as the biplane or triplane configuration, which would allow the wings to generate much more lift without a corresponding increase in weight. Now with the solution in hand, by early fall of 1899 Pilcher completed work on a triplane glider with a 4 horsepower engine, with which he hoped to be the first man to achieve powered, controlled flight.

Anxious to test his new design—and, if successful, perhaps claim the title of being the first man to fly—Pilcher had intended to make his maiden flight on September 30, 1899, but during powered tests a few days earlier, the engine's crankshaft broke, leaving his plane unflyable until repairs could be made. However, as word had spread of the attempt and crowds had gathered to watch (along with a number of guests and potential sponsors he had invited beforehand—many of whom had come a considerable distance to watch him fly), Pilcher didn't want to send them away empty-handed, and so he decided to put on a show with his trusty *Hawk* instead, despite the fact that the weather that day was far from conducive to flying. After waiting for several hours for a break between storms, by early afternoon he decided the weather was good enough to fly and shortly after 4:00 P.M. he went aloft in his glider. Though he had made scores of successful flights in his *Hawk* before, on this day the unthinkable happened: While making a gentle right-hand turn, he craft suddenly went into a nose dive and hit the ground hard, knocking Pilcher out and inflicting multiple internal injuries. Sadly, he died two days later without regaining consciousness, joining his colleague and friend, Otto Lilienthal—whose career Pilcher aspired to emulate—in death, both men having given their lives in the pursuit of their dreams.

Unfortunately, with his death his powered triplane died with him, leaving the world to wonder what might have been. Certainly, there is good evidence that his powered glider might well have achieved sustained flight—or at least a flight comparable to what the Wright brothers accomplished three years later at Kitty Hawk[26]. As such, his untimely death may well have retarded the advance of manned flight by years and made it possible for the brothers from Dayton to be the first to be afforded the honor of achieving manned, powered flight.

Octave Chanute

While the work of a number of aviation pioneers undoubtedly influenced the work of the Wright brothers, probably no one man was more instrumental in their eventual success than was the brilliant and innovative Octave Chanute (1832-1910)—a man whose efforts are nearly as well known today as the Wright brothers' were then.

Born in Paris, France, in 1832, Chanute became a naturalized American citizen when he moved to the United States as a young man. A self-taught civil engineer, he began his career in railroad construction at the Hudson River Railroad in Ossining, New York, where he eventually invented a method for pressure-treating wood with creosote that is still being used today. Chanute also went on to design the first railroad bridge over the Missouri River and the Union stockyards in Chicago, Illinois, and Kansas City, Missouri, before turning his considerable energies—the man was already in his late fifties—towards furthering the new science of aviation.

Obviously too old to fly himself, he worked in partnership with younger experimenters, most prominently among them Augustus M. Herring, a civil and mechanical engineer, and William Avery, both of whom piloted the flights and assisted with his designs. At first the men started modestly, testing small Lilienthal-type gliders, but Chanute eventually decided their dependence upon weight shifting to achieve control was inferior to the concept

Left: Augustus Herring, one of Chanute's main pilots and designers. Below: one of Chanute's early gliders, derived from the designs of Otto Lilienthal (and being flown here by Herring in1896).

26. In 2003, a research effort carried out at the School of Aeronautics at Cranfield University and commissioned by the BBC2 television series "Horizon", demonstrated that Pilcher's design was more or less workable, making it likely that, had he been able to develop his engine, he would have succeeded in being the first to fly a heavier-than-air powered aircraft with some degree of control, and he would probably have done it years before the Wright brothers.

Herring's bi-wing glider being tested at Dune Park on the shores of Lake Michigan in 1896. Herring's design anticipated the biplane configuration of most aircraft through the first three decades of the twentieth century, making it an important evolution in aerodynamics. Though the design is Herring's, Chanute's engineering acumen can be seen in the precise truss bracing in the wings.

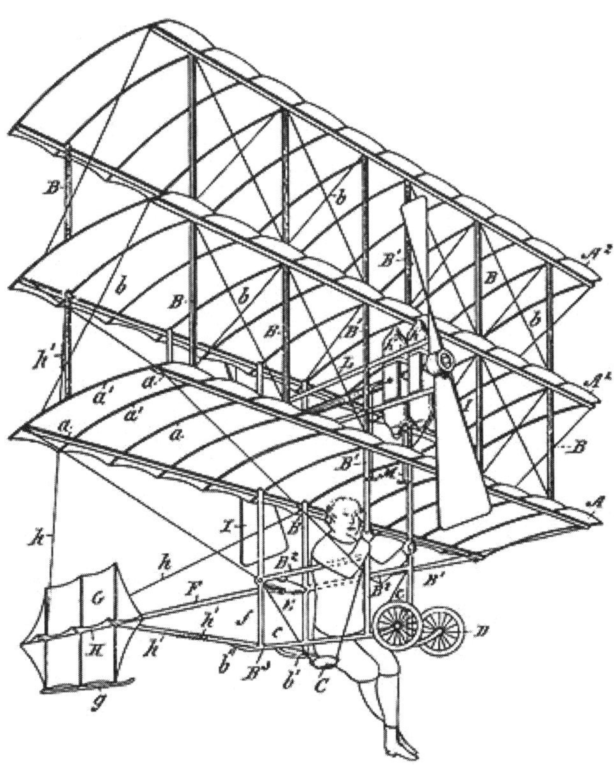

By 1896 Chanute was getting close to designing a powered glider. Though the version above was never built, he did build an unpowered version that proved very successful.

of wing pivoting and, in 1895, he began designing and building gliders of his own design. During this period, his primary goal was to find a method of improving flight control and to that end he built a number of gliders that had either moveable wings or tail control surfaces. Chanute theorized that a glider's wings should pivot on a central frame, rotating slightly if a gust of wind struck the craft. Later he switched to rigid wings, held in place by trusses similar to those he used in bridge construction.

In 1896, Chanute commissioned five gliders to be constructed in Avery's shop. The most sophisticated and successful of these was called the *Katydid,* a craft which featured multiple wings that could be moved about on the fuselage to facilitate experimentation. Flown on the shores of Lake Michigan, the craft made some two hundred successful glides.

Another craft that turned out to be quite revolutionary was the *Chanute-Herring Biplane,* a small but robust glider that used Pratt trussing for strength. (The strut-wire-braced wing structure used in this plane is still used in biplane aircraft and was the model the Wright brothers used in constructing their gliders and first airplane.)

By 1897, Chanute appeared to tire of building gliders and, at the age of sixty-five, essentially dissolved his partnership with Herring and Avery and went into semi-retirement. However, even if he was no longer building airplanes, he remained active in the aviation community as an adviser to, among others, the Wright brothers, with whom he maintained a decade-long correspondence. (In fact, it isn't a stretch to maintain that he was behind many of their best ideas, particularly in regards to strengthening the Wright's machines basic airframe and other vital elements of their design.) Chanute also helped publicize their work and even visited their camp near Kitty Hawk several times, providing constant encouragement and advice on how to improve their designs.

In this regard, it could be said that in many ways, Chanute's greatest contribution to aviation came in the form of advice, support, and encouragement, all of which he gave freely to other aviators of the era. To this end, in addition to his correspondence with the Wright brothers, he exchanged letters with many aviation pioneers, giving them advice and even letting them examine his own designs in great detail. In fact, it was from Chanute that Percy Pilcher got his idea for a powered triplane, but he was killed in a glider crash before he had a chance to fly it.

Unfortunately, Chanute's expectation that others should be as free about sharing their knowledge as he was often led to friction with other aviators of the era, most notably the Wright brothers, who, in marked contrast to Chanute, usually worked to a large degree in secret and were constantly on the lookout for infringement on their patents. It's not entirely clear what was the reason behind their estrangement, however. Some maintain it was their disagreement over how much the brothers should sell their ideas for while others maintain it was a result of Chanute's publicly stated belief that the Wrights' patent, premised on wing-warping, could not be enforced. In any case, their disputes created a schism between the men that was to impair their friendship right up to the time of Chanute's death in 1910. (Wilbur Wright's willingness to deliver the eulogy at Chanute's funeral, however, may have been a sign that the men were able to overcome their differences and acknowledge each other for the true pioneers they were.)

Conclusions

It's impossible to understate how much these men contributed to the evolution of flight. Without their daring and ingenuity, the aviation successes of the twentieth century would have been impossible to imagine. It was men like Montgomery, Lilienthal, Pilcher, and Chanute who worked out the basic physics of flight—often giving their lives in the process—thereby making it possible for their successors to understand what it was that made an airplane fly. While it is possible others could have taken their place, it's difficult to imagine what aviation would have looked like without their contributions, not just in terms of engineering and the science of flight, but in terms of igniting the public imagination and encouraging others to pursue aviation themselves, the fruits of which continued to be seen decades after they were gone. In that, then, is their greatest legacy, for which modern aviation owes them much.

Chapter Eight: The First Unmanned, Powered Flights

While it was the "glidernauts" who received most of the attention during the 1890s, there were others who were working more behind-the-scenes on another aspect of aeronautics; whereas men like Lilienthal, Pilcher, and Chanute were working on the issues of foil aerodynamics, others were working on powering unmanned aircraft through the air—an element of the process that was every bit as important to the future of manned flight. While these experiments—which were noteworthy for their comparative technological sophistication—often proved to be less than successful, they were important steps in laying the foundation for later manned flight, without which the modern airplane might never have gotten off the ground.

Hiram Maxim

One of the first to make a serious effort to build a full-size (and some would say, oversized) aeroplane was the British inventor Sir Hiram Stevens Maxim (1840–1916), which might be a surprise to many people as Maxim was much better known for designing and building weapons and other machinery of war. (In fact, he would be was famous as the man who invented the machine gun, for which he would eventually receive a knighthood from Queen Victoria and so assume the noble title of "Sir".) Therefore, and despite the fact that the aeroplane would one day become a weapon of war, in the 1890s it was still a novelty and, as such, something that would seem outside his area of either interest or expertise. Nonetheless, this American émigré to England maintained a life-long interest in aviation, possibly as a result of his father, Isaac—an inventor and gifted engineer in his own right—who had experimented with aircraft (in this case, early helicopters) as early as the 1860s. Young Hiram, perhaps taken up with his father's enthusiasm for the project, joined him in his efforts to make a flyable machine, and though they failed to do so (most likely because, like many aviators of the period, they could not find a suitably powerful power plant) it apparently instilled in him an interest in aeronautics that would stay with him the rest of his life.

It would not be until 1894, however, until Maxim, now in his fifties and perhaps looking for a new challenge, would finally turn his full attention to building a "flying machine." When he did, however, he would create one of the most memorable—and massive—examples of an aeroplane the world had ever seen.

Whereas most aviation pioneers of the era were working on comparatively small machines, Maxim was a man who liked to think big. The result was a machine that would be considered a behemoth even by modern standards. (With a total length of 145 feet and a wingspan of 105 feet, the machine was approximately the size of a modern 737 jet.) Additionally, it weighed in at a hefty 3.5 tons, making it by far the heaviest flying machine constructed up to that point. Even more unique was that it was entirely steam-powered, with two 360 horsepower engines (easily the most powerful aviation steam engines ever built), all designed to turn two monstrous propellers—each nearly forty feet in diameter.

Maxim, unlike most of the aviation pioneers of the era, was unusual in another way as well: He had absolutely no interest in flying. He was more an idea man—someone who had the engineering acumen and financial resources to build such a massive machine, but who was content to let others actually risk their lives flying the thing. He was also a cautious man who preferred to thoroughly test his designs before blindly starting them up and seeing if they would fly. (This "take-it-slow" approach had a practical rationale as well, for his machine was so expensive he could ill-afford to have damaged or destroyed, mandating he take a more meticulous approach to developing new and—to him—exotic technologies.) As such, he decided to first test its aerodynamic capabilities by running it down an 1800-foot rail, much in the manner of a roller coaster. To keep the machine from inadvertently becoming airborne, however, the ramp was outfitted with outriggers designed to keep the plane from rising (making it the first attempt in history to specifically *prevent* an aircraft from taking off).

In any case, this arrangement seemed to work well, and the machine made a number of successful test runs at increasingly higher speeds during trial runs in the summer of 1894. It worked so well, in fact, that it convinced Maxim that it could handle a full-power test run and so, on July 31, 1894, he had his mechanics "up the ante" by firing up both boilers to full power and sending it on its way. This time, however, the results were both unexpected and very nearly disastrous when the craft, flying down the rails at an estimated 42 mph (a considerable speed at the time), produced so much lift it broke the outriggers and pulled itself free of the track for a few seconds. Though it never got more than a couple of feet off the ground, it did fly some 200 feet before crashing back into the ramp, badly damaging its undercarriage and very nearly turning Sir Maxim's machine into kindling.

While this inadvertent launch could be seen as a positive in that the massive airplane did achieve, if

Maxim's monster plane on its launch ramp—1894. With a wingspan of 105 feet, it was the largest aerial machine ever built up to that time.

No one could ever accuse Maxim of not thinking big. Compare the size of the propellers on his craft to the crew standing in front of them and you get some idea of the scale at which the man worked.

only briefly, free flight, for some reason Maxim stopped further testing and seemed content to abandoned the project, just as he seemed on the verge of success. This proved to be extremely unfortunate, for not only was the man onto something with his steam-powered craft, but his decision to put an end to it deprived the populace of the opportunity to see his great machine wing its way through the sky like some medieval dragon.

While Maxim's monster machine might be considered a failure, it did demonstrate that it was possible to use steam engines to successfully propel an airplane. Though it's unlikely the machine would have been able to sustain flight for any significant distance or be controllable in the air, that it was capable of inducing lift is indisputable. Had Maxim the funds and perseverance to repair and try again, his massive steam-powered airplane might well have flown to become the first—and easily the largest—non-airship ever to fly (a record it probably would have held until the advent of the massive bombers of World War I). Maxim didn't give up on aeronautics completely, however. He later went on to test a number of smaller machines powered by gasoline engines during the 1900s, but by then there was little to be gained by flying unmanned, powered airplanes and his designs went nowhere.

Not all of Maxim's aviation-related projects proved to be failures, however. In 1904 he built an amusement ride for the Earl's Court exhibition in London in which cars spun from a large twirling frame, simulating flight. He intended to use primitive airfoils and wings to allow riders to control their flight as they swung around the arc of the circle, but the idea ran afoul of local authorities and they declared the ride unsafe. They were willing to permit the cars to be fixed in place as they swung around the arc, however, but Maxim declared that then the ride would be nothing more than "a glorified merry-go-round" and he quickly lost interest in the project. Despite Maxim's low opinion of the ride, however, it proved to be wildly popular with the public and it quickly realized a hefty profit, allowing him to soon recoup his losses for his failed aeroplane.[27]

Samuel Langley

Perhaps no man came closer to accomplishing manned flight before the Wright brothers than a contemporary of Hiram Maxim, the American inventor, astronomer, and physicist, Samuel Langley.

Samuel Pierpont Langley (1834–1906) had been interested in flight, by his own admission, all his life, but didn't find the time to begin experimenting with aeronautics in earnest until he was well into his fifties. Precisely the opposite of Maxim, however, he preferred to start small and work his way up to larger designs and so he started by experimenting with rubber-band powered models and gliders, some of which flew up to one hundred feet. He also preferred to test his ideas in a laboratory setting first and so, like George Cayley a century earlier (see chapter four), he also built a rotating arm (that functioned much like a wind tunnel) to test his designs before incorporating them into his designs.

[27] One of these original machines, the Blackpool ride, still operates to this day as part of what is now the Pleasure Beach amusement park, making it one of the oldest operating amusement ride in Europe.

One of Langley's machines mounted atop one of the houseboats he had specifically built to launch his aerodromes – 1897.

By 1891, however, Langley began thinking on a much larger scale and began experimenting with full-sized, tandem-winged, steam-powered aircraft he called *"Aerodromes"* (roughly translated in Greek as "air runner"). Unfortunately, the first of his designs proved to be disappointing, largely because the steam engines he was using were, as usual, too underpowered. But Langley was nothing if not a patient and persistent man and, working hard over the next two years to improve upon the basic engines he had on hand, he finally found a way of heating the steam inside them in a way that allowed them to produce considerably more power than a comparable sized engine of the time was capable of (a noteworthy achievement in its own right). Now with an engine that appeared to be both light enough and powerful enough to work on an aircraft, he immediately incorporated into his latest aerodrome—*Aerodrome No. 4*—and by the summer of 1893 was ready to begin the first test flights.

The next difficulty Langley faced, however, was a result of the way he chose to test his designs. Instead of launching his planes from the ground, he decided to launch them from special-built houseboats anchored on the Potomac River which, while giving him plenty of open space free of obstructions, made it more difficult to recover the craft once they were launched. The problem is that his aerodromes lacked flotation devices, meaning that each time they landed they had to be recovered and extensively rebuilt before their next flight. His designs also had a problem with excessive flexing of the wings, resulting in failure after failure. Langley's doggedness, however, finally paid off when in October of 1894, *Aerodrome No. 4* made a short hop of 130 feet under its own power, which, while modest by most measures, was considered spectacularly successful at the time.

This encouraged Langley to continue to refine his basic designs over the next two years, and in May of 1896, he watched excitedly as his *Aerodrome No. 5* made flights of 3,300 and 2,300 feet, both at a maximum altitude of some 80 to 100 feet and at a speed nearly 25 miles an hour—a distance ten times

Paintings of Langley's most successful unmanned aircraft, the Aerodrome No. 6, which managed to stay aloft a full minute and fly over 4,000 feet—a remarkable achievement in 1897.

Langley standing with his chief engineer and pilot, the fearless Charles Manly, who was to survive both ill-fated attempts to get Langley's Aerodrome "A" into the air.

greater than any previous experiment with a heavier-than-air flying machine. Even more spectacularly, in November of 1896 *Aerodrome No. 6*—considered to be his most successful design of all—flew 4,200 feet and stayed aloft for a full minute—an astonishing accomplishment at the time.

Curiously, just as Langley was experiencing his greatest success, he put an end to his aviation experiments and returned to his work as an astronomer, which is where he probably would have remained the rest of his life—fading into aeronautical obscurity as a result—had not war broken out with Spain in 1898. Prevailed upon by President William McKinley to construct an Aerodrome capable of carrying a human operator, Langley readily agreed to do his patriotic duty. Even though the war was over by the time he was able to get started (the conflict lasted just four months), the government was still interested in the project even after that, and in December of that year the U.S. Army gave him $50,000 (the equivalent of around $1.2 million dollars in today's money) to cover the cost of constructing a new, manned Aerodrome, and work began in earnest.

Building a craft capable of carrying a human pilot significant distances proved to be far more difficult than either Langley or the U.S. government had imagined. As a result, it took him nearly four years before he was ready to test his largest and last aerodrome—the *Aerodrome "A"*—with a human pilot onboard.

The craft would not be flown by the then sixty-nine-year-old Langley, of course, but by a man named Charles Manly (1876–1927), an engineer whom Langley hired as his test pilot. In Manly, Langley easily got his money's worth, for the man proved himself to be not only a most capable pilot, but a superb engineer as well. In fact, it would be his design for an internal combustion engine for the new aerodrome—one that was four times more powerful than the one the Wright brothers used on their plane at Kitty Hawk—that would prove to be the project's main—and some might say, *only*—real contribution to aviation.

What Langley didn't foresee, however, was the problem his fame would bring. By the turn of the century Langley was such a well-known figure that everything he did was extensively covered by the press. Unfortunately, the same press that made him increasingly famous could prove to be most unforgiving when it came to failure, a fact that would come back to haunt him and nearly destroy his reputation when he was finally ready to test his new aerodrome.

Unfortunately, the first attempt, carried out in October of 1903 and observed by literally hundreds of spectators, proved to be an unmitigated disaster when the wing of the plane clipped part of the catapult and sank into the river "like a handful of mortar," according to one reporter. Undaunted,

Langley had the craft repaired and readied for its next attempt, which again was carried out before an even larger audience on December 8, 1903 (just a few days before the Wright brothers' successful flight). Like the first attempt, this one proved to be an even more spectacular failure when the craft broke up just as it left the catapult and again plunged into the river, leaving little more than twisted wreckage in its wake. (Fortunately, the intrepid Mr. Manly was recovered unhurt from the river both times, demonstrating his great faith in his employer and in their design.) The resultant criticism the man endured as a result of these fiascoes stung so badly that he gave up on building a manned aerodrome and never built another aircraft again, which was most unfortunate, for the stately Mr. Langley deserved better for his efforts. Sadly, he died just three years later, still upset with the ridicule his good-faith efforts had incurred and deeply hurt that the failures of his aerodrome appeared to have diminished his many other achievements throughout his life.

Grainy photo showing Langley's Aerodrome "A" coming apart as it leaves the launch ramp, December 8, 1903. Fortunately, the pilot was unscathed but the crash had the misfortune of being witnessed by a number of dignitaries and the press, who proved to be derisive of Mr. Langley's efforts.

The question that remains to this day, however, is would Langley's plane have flown had it not met with a pair of unfortunate accidents? While it can't be known with certainty, most likely the plane would have flown as stably as the models had, but, considering that it had no effective way of controlling the direction of flight would have made it difficult to fly. Additionally, the fact that it lacked any kind of landing gear would also have made it extremely dangerous to land. In that respect, then, it was probably for the best that *Aerodrome "A"* failed to fly, for had it made it into the air for any distance, there was a very good chance it would have cost the pilot his life.

In the end, the failure of Langley's experiments must lie with the man himself. It appears that despite evidence that the basic structure of his aerodrome was too fragile to endure the strain of a high-speed take-off (from a fixed and extremely short platform at that), he failed to appreciate just how much of a problem that was. Additionally, in failing to incorporate control surfaces and some kind of landing gear in his design, he made the plane largely uncontrollable and unlandable and, as such, impractical. The best that could be said for his efforts is that he approached the problem from a scientific and engineering standpoint and was meticulous about documenting and recording everything he did as befits the scientist he was. Had he thought further past the problem of simply how to get the machine into the air, however, he might have taken these issues into account and, with his gifted intellect, may have found a way to produce a genuinely viable airplane.

As a result of these oversights, however, he failed and, worse, failed in full view of the public, which can be a fate worse than death for a man as talented as Samuel Pierpont Langley.

It would not be long before someone would come along an attempt to help Mr. Langley regain his reputation, however. In this case, that man was one of the best known and most successful aviators of the first decades of the twentieth century, Glenn Curtiss (see chapter seventeen) who, just a few years after Langley's death in 1906, attempted to demonstrate that Langley's plane was practical after all by building and flying a replica of the *Aerodrome "A"* with some success. (Of course, the craft was heavily modified, making it a less than fair comparison, but the basic design was still Langley's.) In an even greater irony, Curtiss' model was later exhibited in the Smithsonian with a label claiming that it was "the first man-carrying aeroplane in the history of the world capable of sustained free flight." The claim—entirely erroneous but accepted by many as true because it bore the apparent credibility of the Smithsonian—created quite a controversy and triggered a decades-long feud with the surviving Wright brother, Orville. Eventually the dispute was settled when the Wright estate agreed to donate the *Wright Flyer II*—the earliest of the surviving Wright planes to have flown—in exchange for the museum dropping the claim concerning Curtiss' rebuilt Langley machine and acknowledging the Wright brothers as having been the first to fly a manned aircraft (a policy that remains in effect today). Samuel Langley was still to be honored later, however, by having a number of air and sea craft (including the very first American aircraft carrier, the *U.S.S. Langley*), an air field, and even an award named in his honor, thereby achieving the degree of recognition for his efforts he so richly deserved.

Conclusions

It's not clear exactly how much these early pilotless experiments contributed to the evolution of flight. Certainly they illustrated the difficulties inherent in getting a heavier-than-air craft into the air, especially considering the still relatively primitive power plants available at the time, which was later factored into later manned attempts at achieving flight. It would have been interesting to see what might have happened had either Maxim or Langley continued to refine their aircraft, though in the end it's unclear just how far they could have gotten with their machines even if they had made the effort. The problem was that their designs were simply too large to be controllable, which was an unfortunate consequence of the age's adage that "bigger was better"—which, in these cases, proved to be the exact opposite. In any case, it would be for others to take the next step in heavier-than-air flight by merging the lessons learned from Lilienthal's, Pilcher's, and Montgomery's gliders and Langley's and Maxim's unmanned powered machines and try, before the century was out, to get a man to fly in a powered aircraft.

Chapter Nine: The First Powered, Manned Attempts 9

Until the internal combustion engine and the diesel motor—both of which were to be introduced during the later years of the nineteenth century—could be refined to a degree that would make them a practical means of propelling a craft through the air, aviation pioneers of the era had to be content with experimenting with steam engines to power their craft. Thanks in a large part to the work of Maxim and Langley, however, the prospect of steam-driven heavier-than-air flight was not out of the question (as Langley's successful *Aerodrome No. 6* aptly demonstrated in 1897). In such an atmosphere, then, it wasn't remarkable that a few hearty souls would fashion a steam-engine small enough to fit on an early aeroplane and try their hand at manned flight. While the evidence that they were successful remains scant and debatable, considering the obstacles they faced and the danger inherent in the attempt to power a heavier-than-air machine into the air, their efforts are worthy of examination, if only as a tribute to the extraordinary daring they showed in their quest to be the first men to fly in an airplane.

Aleksander Mozhaysky

A running joke from the darkest days of the Cold War was that the Russians were actually the first to have invented everything, claiming to have introduced the world to everything from the light bulb and the telegraph to the steam engine and the flush toilet (before, of course, seeing their inventiveness coopted by the West). Obviously, most of these assertions were little more than Stalinist-era propaganda, but there was one claim that was made—and that some continue to make today—that may have a kernel of truth to it, and that is the prospect that the Russians may have been the first to fly.

Of course, at first thought such a claim appears absurd and can be easily dismissed as just one more example of ultra-nationalist hyperbole, but there is some evidence that a Russian naval officer by the name of Aleksander Mozhaysky may have briefly powered an untethered, steam-powered machine into the air and, even more remarkable, he may have done it a full twenty-one years before the Wright brothers made their first historic flight!

That there was a man named Aleksander Mozhaysky (1825-1890) is not a myth. According to the historical record, he was, indeed, an Imperial Russian Naval officer who had spent much of his life in service to the czar of Russia. What's also true is that sometime around 1875, this same naval officer, convinced he knew what it took to make an airplane fly by having studied birds in flight from childhood, applied for and received a rather substantial amount of funding from the Ministry of the Military to pursue his research. At this point, however, things begin to get a little more confusing.

What appears to have happened is that the czar's government, growing increasingly impatient with both his excruciatingly slow pace of progress and his repeated requests for more funding, cut off support after several years of unfulfilled promises. Fortunately, it appears that funding was restored shortly afterwards when Mozhaysky—forced to depend upon his own meager financial resources for the time being—was awarded a patent for his design of a large steam-powered monoplane in 1881. After that, and now with the resources required to marry his new steam engine design with an airframe, within a year he completed his revolutionary craft and revealed it to an anxious public.

According to eyewitness testimony and what drawings from the time survived, what Mozhaysky constructed was truly a behemoth machine—especially for the time. With a wingspan of 75 feet and a length of very nearly the same, the plane was propelled by no less than three four-bladed propellers, one mounted on the back of the fuselage and two more inset in slots in the surfaces of the wings, which were driven by two steam engines, one which produced 10 horsepower and was attached to the single pusher propeller on the fuselage and a second, 20 horsepower steam engine, designed to turn the wing-mounted propellers. A large and fairly modern looking horizontal tail surface and a large vertical rudder assembly built around a boat-shaped hull completed the beast, making it truly an astonishing—and likely impressive—sight to behold.

The problem with the design, however, lay in the width (or chord) of the wing, which was an extraordinary 46 feet from leading edge to trailing edge. This, combined with the fact that the wings were completely flat and contained no cambering (curving or airfoil shape), made them extremely limited in the amount of lift they could generate despite having such a large surface area. Apparently unconcerned with such aeronautical trivialities, Mozhaysky remained confident about his design and certain he was on the cusp of being the first man to fly.

A few months later, he was ready to give his machine its first (and what would turn out to be its last) test run. Building a wooden ramp near Krasnoye Selo (a small village south of Saint Petersburg)

Soviet-era illustration of Mozhaysky's monoplane being prepared for launch. Note the massive width of the wings, which probably made them act more like immense sails than airfoils.

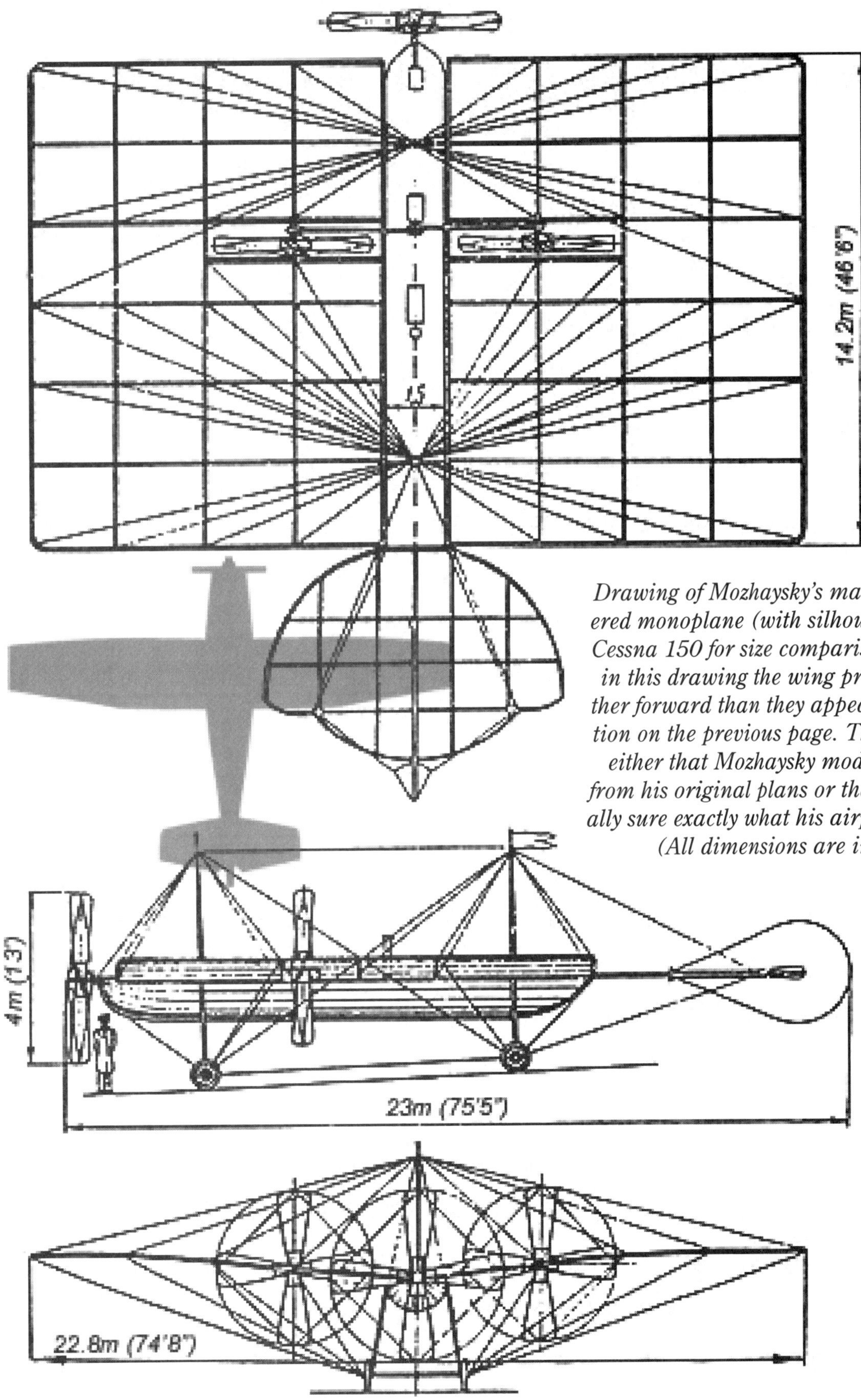

Drawing of Mozhaysky's massive steam-powered monoplane (with silhouette of a modern Cessna 150 for size comparison). Notice that in this drawing the wing propellers are farther forward than they appear in the illustration on the previous page. This demonstrates either that Mozhaysky modified his design from his original plans or that no one was really sure exactly what his airplane looked like. (All dimensions are in meters.)

from which to launch the craft, in July of 1882[28], Mozhaysky[29] powered up the craft's steam engines and let it slide down the ramp. What happened next is still a source of debate today.

Rolling down the massive wooden ramp under full power, apparently the machine actually managed to stagger into the air and fly a short distance (estimates vary between 60 and 100 feet) before crashing back onto its ramp and seriously damaging the wing. As a result of the damage—and perhaps aided by Mozhaysky possibly having second thoughts about entrusting his life to the ungainly machine—it never flew again if, indeed, it ever really flew at all.

Naturally, the event created a controversy which continues in some quarters to this day. While most historians generally concur that Du Temple was the first to have actually flown (see chapter five), some are willing to concede that Mozhaysky's machine may have been the second, though its ascent into the air may have been more attributable to a timely gust of wind than true aerodynamic lift. (The assumption is that a headwind may have caught the underside of the immense wings and flung the craft skyward for a moment until its own immense weight brought quickly back down to earth.) This, however, would not really be considered "flight" except in the loosest definition of the term. Nevertheless, Soviet propagandists later exaggerated the achievement and to this day many Russians remain convinced that Mozhaysky beat the Wright brothers into the air by over two decades.

This bit of revisionist history, however, doesn't detract from the man's accomplishment, especially considering the limits of the technology available to him at the time. At a minimum, Mozhaysky's machine was the first to possess all key components of a modern airplane: fuselage, wing, propulsion, rudder/elevator tail assembly and landing gear, making it a genuine airplane in every sense of the word. Not a successful airplane, perhaps, but an airplane nonetheless.

Clément Ader

Whether Mozhaysky's monster monoplane ever made it into the air or not remains a source of debate even to this day, there was another aviation pioneer from the era whose creations were better known (if no more airworthy) and may have been marginally more successful. In any case, he is certainly worthy of further examination, if only for what doing so

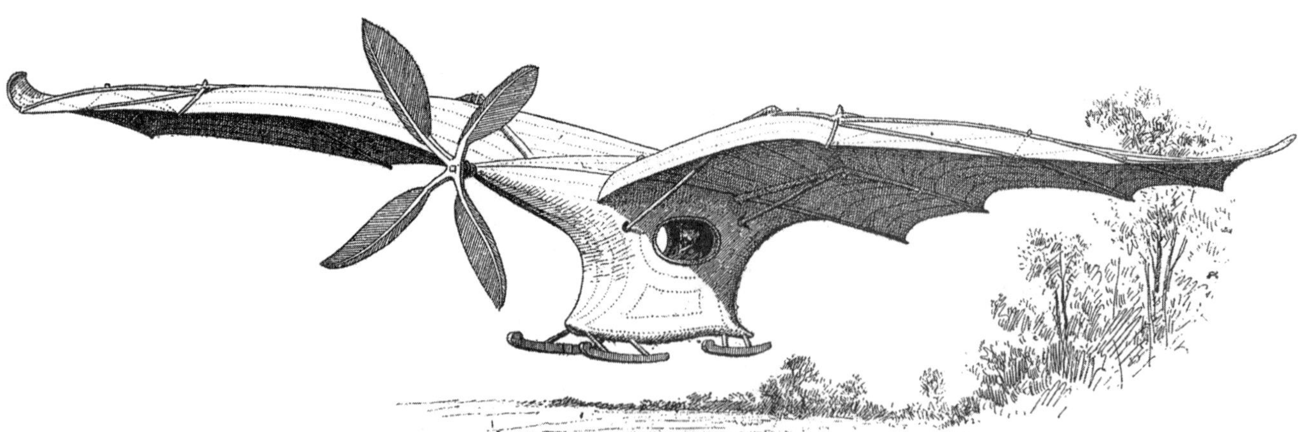

L'OISEAU DE M. ADER.

Was Clement Ader's 1890 test flight in the Éole the first powered flight in history?

[28] Due to a lack of press coverage of the event and the loss—and, some claim, suppression—of many official records afterwards, there is a good deal of confusion concerning precisely when Mozhaysky attempted to fly his plane. Some sources say 1885 while others say 1882. The earlier date is most frequently mentioned, however.

[29] Some accounts claim that the aircraft was piloted not by Mozhaysky himself, but by a volunteer identified only as "I.N. Golubev". Since Mozhaysky was a heavy-set man close to sixty years of age at the time, this is quite possible.

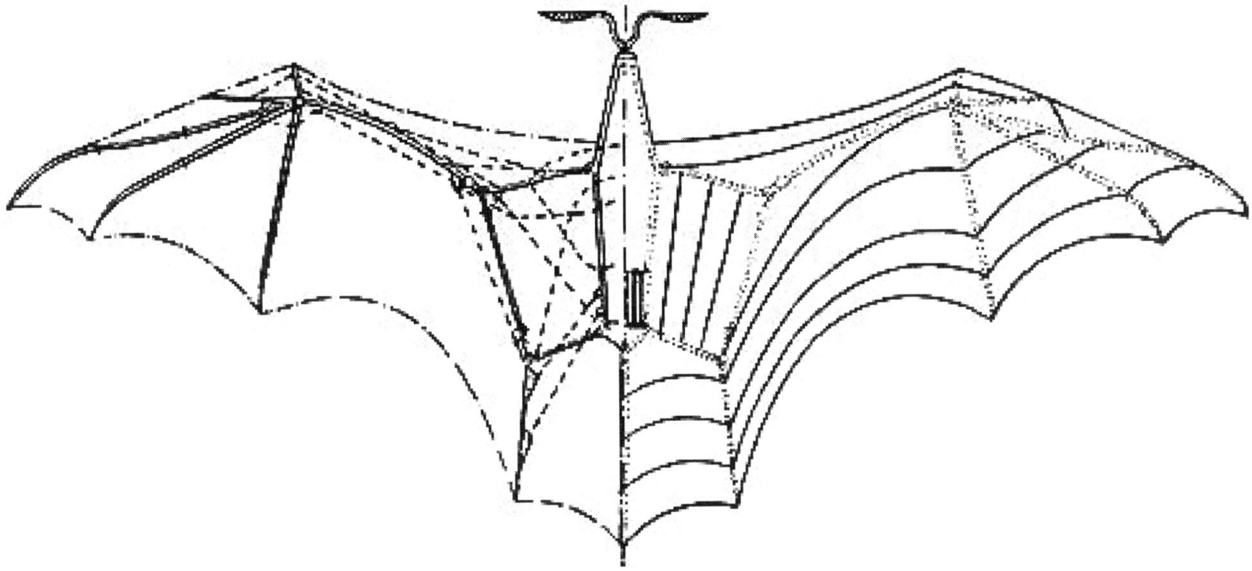

Line drawings of the Éole. Note the almost whimsical propeller, which was designed to fold aft for easy storage.

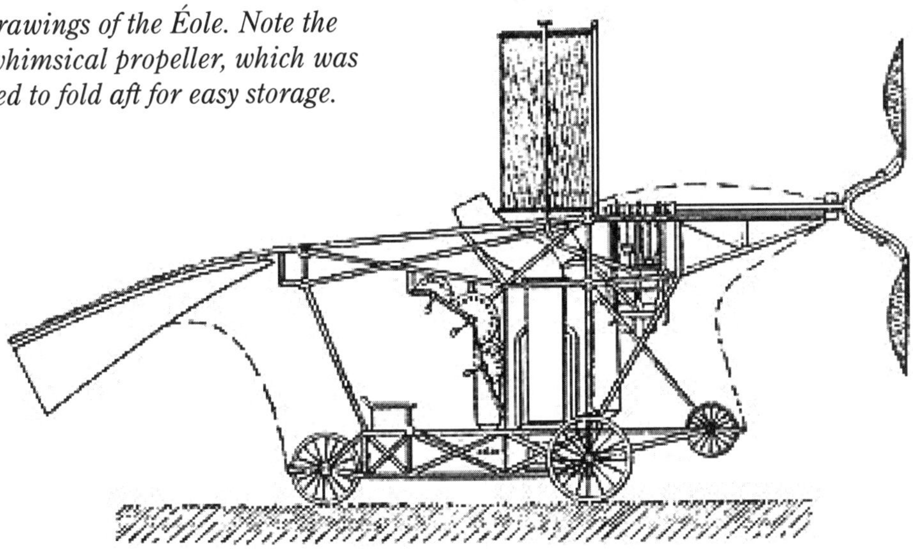

might tell us about the thought processes of the era.

Clément Ader (1841–1925) was a French inventor and engineer born in Muret, France in 1841. Though best known today for his work in aviation, aeronautics was not his primary passion but an interest he played with at various points in his life. However, when he did turn his considerable energies to it full-time, he proved remarkably capable.

Ader's first brush with aviation came about as a result of the Franco-Prussian War of 1871-72. During that brief but important conflict between France and Prussia (the future Germany), Paris became surrounded, resulting in all manner of unusual efforts to find ways in and out of the city. One of the more interesting was by flying balloons over the Prussian defenses, which were powerless to stop them—a technique which was successfully employed throughout the duration of the siege. Ader, it turns out, was one of the men who possessed the ingenuity and daring to build one of these balloons and—at his own expense—for several months he ran passengers and messages in and out of the capital, often at great danger to himself. It's not clear if that was when he got "hooked" on aviation, but it was shortly after that when he began experimenting with an ornithopter (an early type of helicopter)—a full-sized example of which—complete with a 26-foot wing span—he actually built and tried to get off the ground in 1872.

Of course, like every other inventor of the period who had toyed with vertical flying machines, he too discovered that engines powerful enough to lift such a machine did not exist and so he eventually gave up on the idea and went on to pursue other interests.

It would be almost fifteen years before Ader would return to the study of aeronautics, but when he finally did, he would expend considerable time, effort, and money on his newfound passion. The problem for Ader was that, unlike most of his contemporaries who were working on their own aeroplane ideas, he showed little interest in either model building or in exploring the concepts of lift by flying manned gliders as intermediate steps towards flying a powered machine.[30] This likely cost him much in terms of expense and frustration, though in preferring to build full-scale prototypes from the beginning, it sped up the process, thereby putting him in a position of potentially being the first man to accomplish the feat. The only question that remains to this day is whether he actually pulled it off.

All that is known with certainty is that Ader constructed three aircraft between 1890 and 1897, all of them based—fairly faithfully, it appears— on the anatomy of a bat. In fact, all three of his machines were so true to the characteristics of the nocturnal mammal that they were even equipped with folding wings that, when recessed, lay against the fuselage in almost the identical way that a bat's wings do while at rest. Ader felt that such a feature was essential to make the craft easily storable, which he imagined an important feature for any practicable aircraft to possess.

The first of his planes—completed in 1890—was called the *Éole*. The craft had a four-bladed propeller that was run by a lightweight 4 cylinder steam engine capable of developing 20 horsepower. While at first glance it does not appear to be a machine capable of flight, Ader supposedly managed to get the craft airborne in October of 1890 from a field near Paris, making a short flight of around 160-feet (all of it just a few inches off the ground). If true, being that it left

The Avion III on display at the Conservatory of Arts and Industry in Paris around 1900.

30. Some speculate that, had he taken these intermediate steps, he could have become a veritable French Lilienthal.

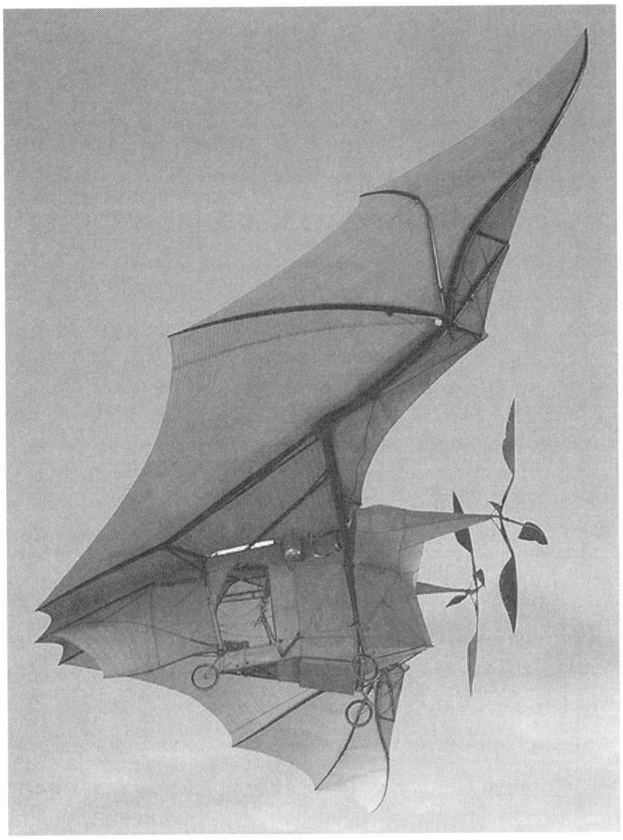

Model of the Avion III clearly illustrates Ader's commitment to mimicking the fluid lines of the bat. Note the almost fanciful propeller blades Ader devised for his aircraft.

the ground under its own power and carried a person through the air for a short distance, that would make the *Éole*'s "flight" the first manned and powered one in history, thereby making Clement Ader the first human to fly in a fixed-wing, airplane.[31]

However, there are some problems with this scenario. First there is no evidence that Ader had any directional control during the flight. As such, if he did achieve lift, it may have been more due to a fortuitous headwind than a result of any thrust delivered by the machine's underpowered engine, thereby reducing his flight to that of a powered glide rather than anything approaching "mechanically powered" flight. (In order to be considered mechanical flight, the craft would need to achieve lift as a result of the forward thrust provided by a mechanical device and not because of natural air currents alone.) Finally, there is scant evidence that he ever attempted to fly the *Éole* again, which seems strange if she really was flyable and serves as evidence—though admittedly circumstantial—that he must have found the plane unflyable. (Ader did later claim to have flown the *Éole* again in September 1891, this time for a distance of 328 feet, but most historians discount this story as there were apparently no witnesses to the event and Ader had a reputation for exaggeration.) As such, his claim to have flown in 1890 remains unsubstantiated.

Regardless of whether the *Éole* flew or not, Ader must have felt he was on the right track for almost immediately he started work on a second, larger craft he called the *Avion II* (also referred to as the *Zephyr* or *Éole II*). While of a similar configuration to his earlier effort, the *Avion II* was to have a larger and more powerful steam motor (30 horsepower versus the 20 horsepower on his earlier creation), though in being forty pounds heavier, such an increase in power would probably have been largely cancelled out by the additional weight. In any case, he never had a chance to find out, for before he could finish the plane[32], he was approached by a French civil engineer named Charles-Louis de Freycinet. Impressed when he saw the *Éole* on exhibit in the Paris Pavilion in October 1891, he persuaded Ader to abandon his newest project in favor of building a larger version that would be capable of carrying up to two passengers or a 165-pound bomb load in addition to a pilot. (In effect, de Freycinet, working on behalf of the French military, wanted Ader to develop the first bomber in history—this even before the first aircraft had ever successfully flown.) Ader immediately agreed to the proposal and was awarded a contract by the French government to construct his most famous aeroplane of all, the *Avion III*.

31. This would be in contrast to Le Bris' plane, which was towed behind a horse, or du Temple's or Mozhaysky's attempts, which were probably the result of a timely gust of wind.

32. Actually, according to some biographers, Ader later claimed that he did finish the plane and even flew it a distance of 200 meters—nearly 700 feet—in August of 1892, but like his earlier claims about the *Eole*, this story is also disputed.

Completed in 1897 after nearly five years of work, the new plane (also called the *Aquilon* or *Éole III*) was similar in design to the *Éole* but considerably larger. Maintaining his trademark bat-like configuration, his new design boasted a 52-foot wingspan (comparable to that of a modern business jet), and was equipped with two four-bladed tractor propellers instead of the single propeller used on his earlier designs. Powered by two 20 horsepower alcohol-burning steam engines, the new plane had twice the power of his earlier effort, making it one of the most powerful aeroplanes of its day (though still woefully underpowered).

In any case, after five years of construction and at considerable cost, the French war department was anxious to see if Ader's plane would fly. To satisfy his patrons, on October 14, 1897, he had the plane rolled out of its hangar near Satory (a military base near Paris still in use today) and readied for its one and only flight. With Ader behind the wheel, after taxiing the plane a number of times, he finally attempted to take off knowing that, if he succeeded in getting the plane in the air, he would be the first man to fly—and this time he would have witnesses present to attest to the fact.

What happened next is a source of some contention. Some witnesses contend that the craft rolled a short distance, took off, and flew more than 300 yards before coming down hard, damaging its undercarriage but otherwise landing intact. Others, however, maintain that it made a few short hops into the air before crashing, but didn't believe it actually achieved flight. In any event, whatever it achieved was not enough to impress the commission sent by the French war department and within days they withdrew funding, bringing an untimely end to Ader's heavier-than-air experiments.

Today the question of whether Ader was the first to fly is still debated, with most researchers coming to the conclusion that, while the *Avion III* probably did briefly leave the ground and rise a few feet in the air, it was not flying in the truest sense of the word. Certainly, if it had flown even a few feet off the ground for a distance of nearly a thousand feet—as some have claimed—that would have constituted an astonishing success and should have been enough to justify further interest and funding by the French government. The fact that the military pulled the plug on the craft is the best evidence that despite Ader's valiant efforts, his aeroplane was simply not flyable.

However, that's not to take anything away from Ader's design. While the craft may appear clumsy to our modern sensibilities, it was extremely advanced for its time and, with a bit of refinement, it might have actually flown—at least a short distance.[33] It was largely because of Ader's stubborn refusal to do preliminary testing on smaller models in an effort to work out the aerodynamics issues and power-to-weight ratio problems that kept it earthbound and cost him the chance to be the first man to fly. Had he taken such necessary if tedious steps, he might have beat the Wright brothers by half a decade and made the Frenchman among the most famous names in aviation history. As it is, and as was often the case with inventors of that era, his impatience may have been his undoing.

Ader never built another aeroplane, either due to a lack of funds or interest. Despite that, however, he remained an active proponent of the development of aviation the rest of his life. In 1909, he even wrote and published a book entitled *L'Aviation Militaire* (Military Aviation) in which he expressed his vision of aerial warfare of the future and even described his concept of an aircraft carrier—which he perceived as having a flat flight deck, an "island" superstructure, deck elevators, and a hangar deck exactly as modern carriers do today. (In fact, this basic configuration was incorporated in the earliest carriers and remained the norm for the entire century.) Even his folding-wings idea was ultimately incorporated into naval aircraft design in an effort to improve storage capacity and so increase total aircraft complement, demonstrating that the man was, if not a successful aviator, certainly a gifted visionary.

[33] Fortunately, his beloved *Avion III* survives to this day and can be seen on display at the Musée des Arts et Métiers in Paris, having undergone extensive restoration in the 1980s.

The succeeding years were not so kind to Clement Ader, who ended up dying in relative obscurity at the age of eighty-four. Despite his engineering acumen (he designed and built each of the steam engines used in his planes) and his passion for aviation, for some reason Ader was one of those early pioneers who ended up being largely overlooked by historians. Of late, though, the man and his machines have enjoyed something of a renaissance among aviation enthusiasts[34], who have come to acknowledge that, despite the man's eccentricities and tendency to exaggerate his achievements, he remains admired for his ingenuity and daring.

Conclusions

In looking over this era in history, what strikes one is the realization of just how close a man came to accomplishing the goal of achieving powered, manned flight. Naturally, this leaves us with a host of questions, none of which can ever be answered. For instance, what if Lilienthal had not died in 1896 but had taken his proven glider designs to the next step and began attaching a small motor to his craft? What if Percy Pilcher's triplane had been ready for flight that day in 1899 instead of being grounded by a broken crankshaft? And what if Clément Ader hadn't run out of funds, forcing him to abandon his powered flying machines just as he was on the verge of success?

The world will never know what might have been but for the whims of fate and financial fortune. The only thing that can be known for certain is that, by the end of the nineteenth century, man was on the threshold of success, having completed all the preliminaries necessary to make the first manned, powered flight possible.

34. In honor of the man, Airbus named one of its aircraft assembly sites in Toulouse after him.

CHAPTER TEN: AIRSHIPS COME OF AGE 10

Up until the mid 1880's, airship designers had little choice in regards to how they might power their ever larger balloons through the skies. Steam engines remained the only proven and readily available power plant, but though it had been tried on airships, it always proved to be too heavy and underpowered to be practical. (Plus it required considerable amounts of fuel to keep its boilers fed, further compounding the weight problem, while the heat from the boilers remained a particularly serious problem around highly volatile hydrogen.) Electric motors were also available and had the advantages of being considerably lighter than a steam engine and not requiring a steady fuel source, (plus heat was much less of an issue) but the technology was still in its infancy and largely experimental. Even though Renard and Krebs had considerable success with their revolutionary dirigible *La France* (see chapter six) using electric motors, the vessel was barely able to fly at a walking speed and lacked the power to counter the effects of even a mild breeze. Plus it had an extremely limited range with even the oversized batteries used on the *La France* providing barely enough power to keep the ship flying for more than an hour. What airship developers needed (as did those working with heavier-than-air flight as well) was a lightweight power plant with a better weight-to-power ratio than a steam engine but with the range that a fuel-fed power plant like steam provided.

As if the aviation gods were listening, in 1885 the German inventor, Gottlieb Daimler, introduced the first gasoline-powered four-stroke internal combustion engine which, though badly under-powered and notoriously unreliable, promised to solve the problem. What it especially offered was the one thing that was crucial if airships or airplanes were ever to become truly practical: it drastically reduced the crucial weight-to-power ratio. (In effect, the internal combustion engine produced more power for its weight when compared to a steam engine of the same size, making it a far superior power plant.) Even better, shortly afterwards another German inventor, Rudolf Diesel, patented a compression engine similar to Daimler's device, but one that ignited the fuel without a spark. Known as the diesel engine, this device promised an even better weight-to-power ratio than the internal combustion engine as well as better reliability, thereby dramatically improving the chances of making airship travel practical.

Of course, both power plants were still in their infancy during the last decade of the nineteenth century (especially in the case of the diesel engines, which had only been introduced in 1893) and difficult to obtain, but there were men who saw in them the future of aviation and so were willing to take a chance on them—especially in the case of Daimler's combustion engine. Perhaps one of the earliest of these experimenters was a German airship enthusiast named Friedrich Wölfert, who was to not only be the first man to make use of Daimler's still largely untested engines on his own dirigible, but would ultimately be the first to discover just how dangerous using gasoline-powered engines alongside highly flammable hydrogen could be.

Friedrich Hermann Wölfert

By all accounts, Friedrich Hermann Wölfert (1850– 1897) seemed an unlikely aviator. A German publisher and author with a degree in philosophy, Wölfert became fascinated with airships after meeting with airship enthusiast Georg Baumgarten—who had patented and experimented with his own airship designs in the 1870s. Apparently, Wölfert was so taken with Baumgarten's airships that he offered both financial support and assistance in designing and building them, and very quickly the two men were working side-by-side on the construction on a number small airships, several of which flew with varying degrees of success.

When Baumgarten died in 1884, however, Wölfert was forced to either carry on the work alone or put airships behind him and return to his publishing business. Apparently deciding his heart lay more with airships than with publishing, he sold his business and devoted his full energies to his great machines, the result being that Wölfert would become synonymous with airship construction over the next decade and one of the best known aviators in Germany at the time. (It would also result in the construction of no fewer than seven airships over the next few years, the most famous being a tiny, human-powered airship he called the *Deutschland.*)

Eventually Wölfert's efforts caught the eye of Gottlieb Daimler, who had just recently patented his new petrol-powered engine and saw in Wölfert's airships a potential platform to try out his new invention. (Actually, Daimler, being a first-class entrepreneur, wanted to try out his engines on as many different machines as possible to prove their many applications, making Wölfert's airship the perfect addition to his collection.) Inviting the man to his home in Cannstatt, the two soon struck up both a friendship and a partnership in which Wölfert would mount Daimler's engines to his airship and demonstrate its capabilities while Daimler would provide him with the best engines available (in this case, Daimler's small but powerful *Standuhr*—German for "grandfather clock"—engines). This joining of technologies proved to be an immediate success when in August of 1888 Daimler's little engine managed to push Wölfert's craft through the air at a speed of almost 16 mph, which was considered a considerable velocity at the time.

Over the next few years the two men worked closely together showcasing both Daimler's new engine and Wölfert's acumen as an airship designer and pilot. In fact, the men's partnership was to last

One of many of Wölfert's gasoline-powered airships being readied for launch. Notice how simple he kept his designs, which made them extremely mobile and easy to maintain and operate.

for nearly a decade and result in dozens of airships and many flights, all of which did much to prove Daimler's new engine to the public and gain recognition for both men. Eventually deciding to exhibit their powered balloon in a more public forum, in 1896 Wölfert flew his airship at a large commercial exposition in Berlin where his ship and Daimler's engine proved to be a huge success, with one flight reaching an altitude of 6,300 feet—a record at the time for dirigibles.[35] (This accomplishment even led the German military to take an interest in the craft, which promised considerable financial support if further tests proved successful.)

Even while Wölfert continued to enjoy success, though, disaster was in the offing. The fact that he had made so many uneventful flights over the years may have lulled him into a false sense of security and perhaps even made him a little careless. As a result, he may have pushed the limit of what his airships and Daimler's engines were capable of doing, which would prove a dangerous combination that was to end in disaster for the former publisher and philosopher from Liepzig. It happened the following summer of 1897 when Wölfert, buoyed by his many successes and with promises of bigger and better things to come, built his largest airship yet, which he named the *Deutschland* in honor of his first successful airship. Eager to show off his new craft before a crowd of dignitaries at Tempelhof airfield near Berlin, on June 12, 1897, Wölfert, accompanied by his chief mechanic, took the ship up to a height of nearly a thousand feet and began circling the field. It's uncertain what happened next, but it appears that for some reason the craft suddenly caught fire and crashed to the ground, killing both Wölfert and his mechanic instantly. While it will probably never be ascertained precisely what it was exactly that caused the *Deutschland* to catch fire, the fact that its 2600 cubic feet of hydrogen was separated from the open-flamed exhaust of its eight horsepower Daimler motor by mere inches made such a mishap inevitable. Was the hot exhaust improperly vented on his new airship or did Wölfert merely get a little careless? We will never know, of course, but in either case the crash brought a tragic end to a very promising career that appeared to be on the threshold of achieving great things.

Certainly, Wölfert seemed on the verge of moving towards building ever larger, and even possibly passenger carrying airships, three years before von Zeppelin's first airship was even built. Had he lived, it's entirely possible the age of the airship would have started a decade earlier than it did, and with a German philosopher turned airship inventor named Wölfert at the helm rather than a former Prussian Army General by the name of von Zeppelin.

The Aluminum Airship of David Schwarz

Fortunately, the fatal crash of the *Deutschland* did not end airship research or development in Germany. Others were working on the airship at the same time, and a few of them appeared to be as much on the threshold of success as Wölfert had been the moment he met his tragic fate. One man who seemed to be especially close to success was a relatively unknown Croatian timber merchant named David Schwarz, who was to introduce the world to a most unconventional and, in some ways, very futuristic machine: the first airship constructed almost entirely out of—and even completely sheathed in—aluminum!

One would not imagine a wealthy wood merchant from Zagreb would take much of an interest in aeronautics, but David Schwarz (1852-1897) was not your typical merchant. Though lacking a formal technical education, he somehow managed to teach himself a great deal about the mechanics of constructing an airship, which over a period of several years he put to use by building a truly unique, never-before-seen vessel that many consider today to have been years ahead of its time. What made it so different was that unlike all previous airships, which had a nonrigid gas envelope that depended upon the gas inside it to maintain its cigar shape, his airship's

[35] Spherical, uncontrolled hot-air and hydrogen balloons had flown much higher than that by this time, of course, but controllable dirigibles rarely flew more than a few hundred feet above the ground.

entire outer casing was riveted to a skeleton of thirty-four aluminum latticed frames of different shapes and sizes. This made it fully rigid, meaning that, unlike other airships of the era, it kept its cylindrical shape whether it was inflated or not. Even more unusual was the fact that the entire envelope was clad in very thin sheets of lightweight aluminum (an expensive metal at the time), making it the first all-metal airship in existence. Even more forward looking, the envelope was divided into twelve compartments, each of which could be individually accessed for ease of maintenance, putting it years ahead of its contemporaries and laying the groundwork for the great airships of the twentieth century. For propulsion, Schwarz mounted a 16 horsepower Daimler engine (very similar to the type Wölfert used on the *Deutschland*, only twice as powerful) to the gondola to turn the airship's four propellers—one on each side of the envelope, another directly behind the aluminum gondola, and a fourth directly below the gondola connected to a long shaft. (The main propeller was even axially movable, enabling the airship to fly in any direction desired—an astonishing futuristic concept at the time.) With a length of just over 123 feet and a diameter of 26 feet (giving it a lifting capacity of just over four tons) Schwarz's craft was theoretically capable of carrying a crew of one and almost three hundred pounds of ballast at a speed of 15 miles per hour, which, if it performed as advertised, would make it a truly remarkable feat of engineering.

A rare three-quarter view of Schwarz's revolutionary airship. Photo probabably taken at Tempelhof airport in Berlin sometime in 1897.

So apparently promising was the new craft that the German government began showing considerable interest in Schwarz's airship—so much so, in fact, that Emperor Wilhelm II planned to attend the craft's maiden flight scheduled for September 27, 1896. Unfortunately for history, the emperor had to

Schwarz's airship was unusual in another way as well: though cylindrical in shape, it had a long, pointed nose and a squat, truncated stern, which gave it the appearance of a stubby pencil (or, in the opinion of military observers, an artillery shell).

cancel at the last moment, forcing Schwarz to have to wait to unveil his masterpiece. A few weeks later, however, after deciding he could wait no longer to test his creation, he had the great ship pulled from its hangar by a troop of soldiers and launched. Though it remained tethered and hovered just a few meters above the ground, it was enough to give Schwarz a chance to start and test the engines, which ran flawlessly, making it appear that a new era in aviation was about to be born.

Unfortunately, Schwarz would not live long enough to see his creation actually fly. Burdened by a weak heart, he died suddenly in Vienna while enroute to a more ambitious test flight scheduled for the following January. Without the designer, further interest in the craft evaporated, leaving David Schwarz' dream to die with him. Though his widow, Melanie—who believed in her husband's work enough to continue to push his ideas even after his death—tried to carry on, his aluminum airship idea never caught on.

Part of the problem was that she faced considerable opposition from some in the German military, who were apparently annoyed by the fact that the airship was not of German origin and preferred to work with German nationals. (It was even implied that some in the German army took an active hand in trying to sabotage Melanie Schwarz's efforts whenever possible, though this has not been proven.) The other problem was that, even when she did manage to arrange to have a large-scale test flight performed, the craft ran into problems. Flying over Tempelhof field on November 3, 1897, the airship initially performed as expected until it reached an altitude of 1500 feet, at which point one of the belts that drove its propellers broke. The pilot, a qualified military balloonist but a man unaccustomed to flying a steerable airship, apparently panicked and valved off all the hydrogen, resulting in a steep descent and disastrous crash landing which, while leaving the pilot unharmed, left the ship damaged beyond repair and ended Melanie's quest to see her late husband's dream realized. The victim of his own poor planning, had he possessed the foresight to have tested his airship earlier and in the presence of the general public, he would have been hailed as a pioneer of lighter-than-air flight and his designs quickly adapted by a new generation of inventors. As it was, he instead remains largely forgotten, the victim of a bad heart and bad timing.

The shattered remnants of David Schwarz's unique aluminum-skinned airship lays irreparably damaged near Tempelhof airport, November 3, 1897. It's destruction signalled the end of both a most unusual machine and a very promising future.

The Curious World of Alberto Santos-Dumont

While Wölfert and Schwarz were experimenting with their airships in Germany, another man happened along who would not only exceed their efforts, but was to become one of the most famous—and some would say, colorful—characters in aviation history. Even more remarkable, even though he lived in Paris much of his life, he was not a Frenchman or, for that matter, even a European. He was, in fact, the son of a wealthy Brazilian coffee grower and exporter whose enthusiasm for aviation would make him a household name not only in France, but around the world. The man? The indomitable Alberto Santos-Dumont, the first man who would not only capture the public imagination in ways it had never been done before, and even make it look fun in the process.

Alberto Santos-Dumont (1873-1932) was a man known for taking on the impossible and making it look easy. Coming to France as a teenager, he quickly set about challenging himself to try just about anything that involved danger or controversy.[36] Being that aviation was becoming all the rage, he naturally decided to get involved in that craze as well and to that end, and though just twenty-five years old, he began working on the first of many tiny airships he was to become famous for.

Working largely alone or with only a small staff of workmen, Santos-Dumont was to build eleven different manned dirigibles between 1898 and 1906, many of which proved to be quite successful. One of them, named simply *Airship No. 6*, even won the prestigious Deutsch de la Meurthe prize in October of 1901—a feat considered as remarkable in its day as Lindbergh's solo crossing of the Atlantic Ocean would be considered twenty years later. (Seated precariously in his tiny, rickety bamboo gondola beneath a one hundred foot long silk envelope, Santos-Dumont flew the hardy but unwieldy gasoline-powered craft a total distance of seven miles from the Parc Saint Cloud in Paris to the Eiffel Tower and back in less than thirty minutes, managing to maintain an average ground speed of around 14 mph throughout the entire route—an impressive airspeed at the time.)

But what made Santos-Dumont truly controversial was his tendency to use his airships as his own private taxi service. With air traffic control restrictions still decades in the future, he would glide along Paris boulevards at rooftop level only to land in front of a fashionable outdoor cafe for lunch. Naturally, this made him something of a character among Parisians, who apparently maintained a high degree of tolerance for such behavior.

This is not to diminish Santos-Dumont's skills as an accomplished aviator as well as a first class engineer, however, who proved that steerable airships were capable of being built and flown far

Santos-Dumont's famous Airship No.6 rounding the Eiffel Tower on its way to winning the Deutsch de la Meurthe prize in October of 1901.

36. Santos-Dumont was one of the first men to own an automobile in France, which he used to race around the city of Paris on various adventures.

The colorful Brazilian aviator at the controls of one of his small but remarkably successful airships.

more easily and cheaply than most people thought possible. Though his craft were tiny compared to David Schwarz's aluminum airship and positively microscopic when seen next to one of von Zeppelin's massive machines, his simple little craft successfully proved that all that was required to build such a vessel was a little money, some "outside-the-box" thinking, and the determination to see it through. In fact, it could be argued that had he started earlier and worked on a much grander scale, he might well have beaten the famous German count in laying the foundation for lighter-than-air travel and in doing so dramatically changed the face of air travel in the early twentieth century.

Unfortunately we'll never know what this Brazilian "Father of Aviation" might have accomplished in regards to lighter-than-air development as he lost interest in airships around 1906 when heavier-than-air aviation caught his attention, causing him to abandon all further airship development in pursuit of being the first European to fly an airplane (a goal he very likely achieved as well; see chapter sixteen).

Whatever can be said about the man, the history of aviation certainly wouldn't have been as colorful without the always flamboyant Santos-Dumont about, and for that alone history owes him a debt of gratitude.

Conclusions

It could be maintained that while lighter-than-air aviation was in some ways considerably more advanced than fixed wing aircraft development, especially during the last years of the nineteenth century, it seemed that the dirigible, for all its potential, appeared to be languishing in regards to being a genuinely practical mode of travel. They were ponderous-

The Brazilian aviator built his tiny airships primarily to allow him to travel to important events around Paris. He was often seen landing in one of the city's parks to have a quick lunch before taking to the sky again on his way to his next engagement.

ly slow, immensely expensive to build and operate, capable of lifting only relatively small payloads and, due to the flammability of hydrogen, inherently dangerous to fly. What ships that were built during the waning years of the century may have been interesting and, in the case of Santos-Dumont's tiny airships, even quaint and a source of great amusement, but they lacked a means of making them practical. As such, they seemed to be an evolutionary dead end—the dinosaurs of their age fated to slip into oblivion unless a way could be found to make them useful to the average citizen.

However, all of that was soon to change. The experiments of Wölfert and the refinements made by Santos-Dumont and others had all—either by design or by accident—laid the foundation for the emergence of a fleet of truly massive airships that were to dominate the skies of the early decades of the twentieth century. And it would all be due to the efforts of a retired Prussian General with the unusual name of Graf von Zeppelin, who would soon see to it that the world of aviation would never be quite the same again.

CHAPTER ELEVEN: THE GREAT AIRSHIPS OF COUNT VON ZEPPELIN 11

When most people refer to a "zeppelin" today, they're usually thinking of some kind of oversized balloon designed primarily to cover sporting events or, if they are of a historical bent, perhaps of a massive World War One airship dropping bombs on London. Few have any idea, however, that a word that has since evolved into a noun actually started out as the proper name of a man who became so synonymous with the early development of airships that we still refer to him today whenever we call an airship a "zeppelin", demonstrating his status as one of the most prominent figures in the history of aviation in the twentieth century.

Ferdinand Adolf Heinrich August Graf von Zeppelin (1838–1917) would not, at least at first thought, have seemed the sort of person who had either the background or the inclination to build airships. A general in the Prussian army with a distinguished military career that spanned over thirty-five years, he was not a formally trained engineer (though he did receive some engineering training as a young officer), nor did he possess the considerable financial resources that would be required to undertake the construction of anything as substantial as an airship. In fact, he didn't even begin turning his attention towards aviation until after his retirement in 1890, putting him already well into his fifties by the time he began considering the idea of building an airship. However, this Prussian army officer was also a man of rare determination and tremendous practicality, which perhaps made him the best possible candidate to initiate the age of the airship after all.

It's not known precisely where or when the Count first developed his interests in lighter-than-air craft. Likely it came when he had the opportunity to make a brief tethered balloon flight as an observer for the Union army during the American Civil War;[37] in any case, it appears to have become an interest that stayed with him for the rest of his life and one he decided to pursue upon his retirement from the army.

Unlike most airship inventors of the period who tended to build small, even delicate, airships and work their way up, von Zeppelin intended to start big—with an airship hundreds of feet in length and containing millions of cubic feet of gas. He knew such a massive machine would be extremely costly and complex to build, of course, and being a meticulous man by nature, he decided to do his homework first. To that end, he hired a qualified engineer to do a study of the possible materials available to build an airship and to assess the engines of the time for both fuel efficiency and power-to-weight ratio. Even more atypical of inventors of the era, upon learning that there were no engines then in existence capable of pushing the massive vessel he had in mind through

[37] Von Zeppelin, then a young lieutenant in the Prussian army, had been sent to Washington in 1863 as a neutral observer on the Union side. It was a practice at the time for nations to send representatives from their armed forces to observe other nation's conflicts, likely as a means of gauging the capabilities of potential future adversaries as well as to keep abreast of technological innovations.

the air with sufficient velocity, he was enough of a pragmatist to put his plans on hold for the time being and wait for the technology to catch up to his ambitions.

That's not to say he gave up working on his idea entirely, however. In the interim, he contented himself with other necessary details of airship design by testing a series of different-shaped propellers and striving to obtain higher purity hydrogen gas from suppliers. He also continued to refine his blueprints and acquire all the necessary patents for his great airship, even if the means to actually build it remained elusive.

The greatest problem he faced, however, was in dealing with the sheer size of his airship. All airships at the time were small, nonrigid types that depended upon the pressure of the gas in the envelope to maintain their elongated cigar shape. Von Zeppelin knew his ship would be too large for that and devised a plan to make the envelope out of a rigid aluminum framework covered in a fabric envelope that would maintain its cylindrical shape whether inflated or not. He also decided to divide the envelope into separate internal gas cells, each of which would be free to expand and contract with changes in altitude and temperature. This would have the benefit of both eliminating the need for ballonettes as well as reducing the chance of a hard landing should the envelope be pierced or torn, as only a single gas cell would deflate rather than the entire envelope. He also devised the idea for a modular frame design to allow for the addition of more gas cells, as well as having all the controls, engines, and gondola rigidly attached underneath. In effect, by 1892 he had already envisioned all the elements necessary for a modern airship and, working with his chief designer, engineer Theodore Kober, he was soon ready to start testing and refining his designs.

Raising the funds required to begin building the great ship, however, proved as large an obstacle to making his dream a reality as were underpowered engines and the massive scale of his project. Fortunately, by 1896 the count's plans were beginning

Von Zeppelin's first airship, the LZ-1, floating over Lake Constance—1900. At over four hundred feet in length, it was, at the time, the largest airborne machine ever built.

to be noticed by German engineering groups and other wealthy patrons, many of whom were willing to invest in his ideas. This made it possible for him to form a joint stock company to raise the required funds to begin work on his first airship, and soon he had the necessary funds to begin construction on his first airship, the *LZ* (for *Luftschiff Zeppelin*) *1*, which, at 420 feet from bow to stern, would be the largest man-made aerial vessel ever built.

Being so large, he decided it would simplify matters greatly if he built it on a lake. Building a massive floating assembly hall anchored on Lake Constance near Friedrichshafen, construction of the massive vessel began in the summer of 1898. However, such constructing such a massive ship proved to be a slow and tedious affair and it wouldn't be completed until the summer of 1900—nearly two full years later. However, by early summer of 1900 the great ship was ready for its maiden flight and on the morning of July 2, 1900, the *LZ-1* was towed from its floating hangar and, with a small crowd of onlookers on-hand to watch the count's magnificent machine take to the sky, it was launched.

Unfortunately, the flight proved to be less than impressive. Though the great ship rose majestically into the air, just minutes after takeoff it developed mechanical problems and had to land, much to the disappointment of the crowd. Further, once back inside its hangar, an apparatus used to suspend it broke loose, causing minor damage that required several weeks to repair.

The count was not easily dissuaded, however, and, after having the necessary repairs completed, he was ready to try again. As such, in October of 1900, he once again had the *LZ-1* towed from its hangar and launched, but this time it performed considerably better, managing to stay aloft for some time and even flying several circles over the lake. A second flight a few weeks later proved even more successful when the ship managed to beat the three meters per second speed record set by Renard and Kreb's successful *La France* in 1885.

It would take six years for the count to find the resources to build his second airship, the LZ-2, but it was considered a big improvement over the earlier LZ-1. Unfortunately, it still lacked the large control surfaces needed to make it truly controllable in more than a slight breeze. It would also be the first of a long line of the von Zeppelin's behemoth vessels to be wrecked during testing.

The LZ-2 after being badly damaged during a storm. The incident demonstrated that, despite the airship's great size, it was far more fragile than it appeared, and was especially susceptable to being damaged by high winds. Its thousands of square meters of highly flammable hydrogen also portended disaster for the great ships.

However, the count very quickly recognized that his ship was still ponderously slow and, possessing undersized control surfaces, extremely difficult to steer, making it commercially unviable. Potential investors seemed to recognize this flaw as well, and the Count quickly found his resources dwindling to the point where he had to reluctantly liquidate his company. What he needed to salvage his efforts was to build a larger and more maneuverable ship that would be commercially useful. However, without sufficient funds, the prospect of doing so appeared unlikely, leaving the future of lighter-than-air aviation—at least in Germany—uncertain.

Fortunately, the man proved to be as persuasive a salesman as he was an airship designer and he soon managed to raise new revenue through private donations and a state lottery (as well as by mortgaging his wife's estate), making it possible for him to begin to build a larger and more powerful airship, the *LZ-2*. Unfortunately, both finding the funds and building the craft proved to be even slower processes than building the *LZ-1* had been, and it would be nearly six years before the new ship was completed. Even worse, the venture nearly threatened to fail entirely when, during an early test flight, both engines on the *LZ-2* failed, forcing it to land in the Allgäu mountains to await repairs. Before the engines could be repaired, however, a violent storm came along and pounded the craft mercilessly, tearing huge gashes in her skin and forcing the workers to deflate her to prevent the ship from being torn to pieces. After it was all over, it was evident the *LZ-2* had been damaged beyond repair, placing both the count's financial fortunes and the future of lighter-than-air aviation itself in serious doubt.

Normally, such a catastrophe would spell the end of such an expensive venture, but in this case something curious happened: the incident actually increased public enthusiasm for airships and donations starting pouring in, which allowed von Zeppelin to begin construction of a third airship—the *LZ-3*. Additionally, by incorporating usable parts from the wrecked *LZ-2*, he was able to build the new ship in record time, and within a few months the count's new airship was ready for her first flight.

With the *LZ-3*, the count's bad luck seemed to evaporate. The ship not only flew well, but made

Tail view of the LZ-3. In marked contrast to its predecessors, the ship's control surfaces have been moved to the tail, making the craft far more manueverable than the count's earlier ships.

The monsterous LZ-4 being pulled out of its floating hangar on Lake Constance. Nearly identical to the highly successful LZ-3 in size and layout, it is distinguished by the addition of a massive tail rudder and an additional dorsal fin. Von Zeppelin's ships still hadn't developed the "X" shape configuration common to airships of the twentieth century, but they were moving in that direction.

numerous successful flights over the next two years, covering an astonishing three thousand miles without a serious mishap. So successful was the ship, in fact, that the German military, which had shown an interest in airships for nearly a decade, purchased the ship from the count directly with the intentions of using it as a training ship for future zeppelin pilots. This not only gave him the funding to build bigger and better zeppelins, but gave him the support of the German government as well, and suddenly the future looked bright for von Zeppelin and his massive vessels. It seemed the long-awaited age of the airship had come at last.

However, fate has a way of intervening at the most inopportune moments. Commissioned by the German government to build another zeppelin specifically for the military, von Zeppelin quickly got to work on the massive *LZ-4* which, at over 450 feet in length, was the largest of his vessels to date. Even though the German Army promised to buy the craft once it was completed, it wanted assurances the ship could remain airborne for at least twenty-four hours straight before it would finalize the purchase, to which von Zeppelin, supremely confident that his great ships were up to the task, readily agreed.

Count von Zeppelin (middle of three men in derbys) standing in front of his most successful airship, the LZ-3, circa 1908. The ship proved so successful, in fact, that the German military purchased the vessel for use as a training ship, and operated it until its retirement in 1913.

Launched in August of 1908, the *LZ-4* appeared at first glance to be everything it was promised to be. Fast and maneuverable, the ship easily took to the air and at first appeared destined to easily break the twenty-four hour endurance limit set by the military. Before it could do so, however, fate intervened when the ship was forced to set down for some minor engine repairs near Stuttgart, Germany. Precisely as had happened with the LZ-2 two years earlier, as luck would have it a storm blew in and pummeled the ship with high winds and heavy rains. Soon the winds became so strong they threatened to tear the machine loose of its moorings and, despite the best efforts of the crew and the military personnel that had been sent to maintain order around the great ship, it finally broke free and crashed into a grove of trees, which tore into several of her hydrogen cells, filling the air with deadly, flammable hydrogen. No one is entirely certain what happened next, but somehow a spark ignited the gas and the entire ship was soon engulfed in flame. While the crew managed to escape with only minor injuries, within minutes the *LZ-4* was nothing more than a smoldering tangle of twisted aluminum. By the time the count returned to see with his own eyes what had become of his latest masterpiece, he was greeted with only smoking ruins which left him not only destitute but now with serious doubt as to whether airships would ever be truly practical. It must have seemed to the count as though that the age of the airship had ended even before it had begun.

But that's when something even more remarkable happened. While the destruction of the *LZ-4* should have been the end of airship development in Germany, precisely the opposite occurred. The crowd that had gathered to watch the ship burn suddenly felt a sense of nationalism at the sight and spontaneously took up a collection of donations, yielding an impressive total of six million marks. The great airship, it seems, had become a symbol of German pride and soon even more donations began pouring in from across the country until the count found he had enough to not only rebuild, but even establish a second airship construction company, the Luftschiffbau Zeppelin Company. After that, zeppelin production went into high gear, and though there would be other crashes and setbacks over the next few years (seven of the twenty-seven zeppelins built by the Luftschiffbau Zeppelin Company before 1914 would be destroyed in accidents—mostly on the ground—though all without injury or fatality), the zeppelin was here to stay, all because of a determined retired Prussian general and his dreams of lighter-than-air flight.

Von Zeppelin's Legacy

There is no doubt that the story of von Zeppelin and his magnificent airships is one of the greatest success stories in aviation history. So successful were the count's majestic ships, in fact, that they were to serve as the backbone of the world's first airline, DELAG (for Deutsche Luftschiffahrts-Aktiengesellschaft, an acronym for "German Airship Travel Corporation"), which, by 1914, would transport over 37,000 people and make over 1600 flights, all without a single accident or fatality—a feat that's been unmatched by any air carrier since.

The story of the great zeppelins is a bittersweet one as well, however, for within a decade of the count's great accomplishments, his airships became synonymous with death and destruction as the mighty ships would be turned into instruments of war. In fact, it was their service as heavy bombers during the World War I that really spelled the beginning of their demise, as images of massive zeppelins dropping bombs on the streets of London quickly

Left: The gutted wreckage of the LZ-4 at Echterdingen, August, 1908. The crash nearly ended von Zeppelin's dreams and could have brought an end to airship development had not the German people rallied to the count's side. Top and bottom right: before and after views of the rudder/tail assembly after the crash.

One of von Zeppelin's hugely successful pre-World War I passenger-carrying airships, the Schwaben. Until 1914, the German airline DELAG carried thousands of passengers without a single injury or fatality—a safety record unparalleled to this day.

destroyed their romantic mystique. Additionally, images of them lighting up the skies as their gas bags were ignited by phosphorus-tipped machine gun tracers from English and French fighters made them appear more vulnerable than they really were. Finally, when the last of the great German-built zeppelin airships, the *Hindenburg,* exploded over Lakehurst, New Jersey, in 1937, the zeppelin became synonymous with technological failure as well, dooming them forever as serious contenders in the passenger air industry. Such a legacy is undeserved, of course, especially when one considers the airship's remarkable safety record (those who died on the *Hindenburg* being the only civilians ever to lose their lives on a German-built and operated zeppelin), yet the stigma remains to this day, which is perhaps the greatest tragedy of all.

In many ways, perhaps it was fortunate that the count did not live to see the demise of his creation. He died in 1917 at the age of seventy-nine, two years before the Treaty of Versailles was signed that forbade Germany from possessing airships and forced her to ignominiously dismantle or surrender what few ships managed to survive the war. No doubt, had he lived to see it, it would have destroyed him emotionally, as it did for many who had built and manned the great ships for so many years.

Curiously, the count's zeppelins were never designed for war, yet it was they who paid so much of the price for losing that conflict, despite the fact that they were only minor players in the affair. It might be argued that they were dismantled not because they posed any great military threat, but because they were such obvious symbols of German power and pride that their destruction served better as a symbol of allied victory and hubris than a military necessity. In retrospect, had the allies allowed the Germans to continue to build and refine their great ships—provided they used them for purely peaceful purposes—and allowed Germany to purchase non-flammable helium to fly them, all the world would have benefitted. As it was, in punishing Germany, the victors only managed to punish themselves by denying the world of the immense benefits the great ships promised had they only been allowed to continue to evolve alongside aircraft.

The count's mighty airships were not perfect vehicles, of course. They were ungainly things that by 1914 had grown to lengths of over five hundred feet and required enormous amounts of ground personnel to control them. They were also expensive to build and operate, and though their massive motors could push them through the air at speeds in excess of 80 mph, they were always in danger of being superseded by the much faster and more cost-effective airplane. However, during their time they were engineering marvels whose only fault lay in their great size and the volatility of the gas that held them aloft. It's fair to say that had the Germans access to non-flammable helium as their American counterparts had,[38] and considering the expertise with which they were able to build and operate their great ships, zeppelins might still be flying today.

Conclusions

It's always been interesting to me to watch how lighter-than-air and heavier-than-air technologies evolved side-by-side over the last few decades of the nineteenth century and into the early years of the twentieth. They were not competitors in the truest sense of the word, nor really even rivals, but parallel technologies that moved forward with considerable speed as the means to power a glider or a maneuver a balloon through the air became increasingly available. But what's most important to remember is that while the airplane remains the better known technology to most people, it was the airship that was king of the skies—at least for the first couple of decades of the twentieth century—as they transported literally thousands of passengers in comfort and safety years before the first rickety passenger airplanes ever struggled into the air. Had helium been as available in the 1920s and 30s, I suspect the air would still be filled with massive airships in the same way the seas are alive with cruise ships and cargo vessels. Had the German zeppelins been allowed to evolve—without the spectacular Hindenburg disaster or the even greater disaster of World War II to contend with—airships capable of carrying five hundred passengers at speeds in excess of 200 mph might well be commonplace in the skies today and, though they could never compete with the speed and convenience of the airplane, they would take a considerable share of the flying public's dollars in exchange for the comfort and simple elegance they offer. It was the great airship engineer, pilot and employee of von Zeppelin, Hugo Eckener, who probably summed up the difference best when he commented that "You travel in an airplane, but you voyage in an airship."

Whether the great airships will ever come back remains to be seen. There are those today who are playing with the notion, and several enterprising companies (most notably and, not surprisingly, several in Germany) have built some sizable, passenger

Artist's rendering of the SF-50 SkyFreighter airship currently under development by the Millennium Airship Company of Spokane, Washington. Is the age of the airship really over or will the twenty-first century see its rebirth? (Used with permission.)

[38] Curiously, even though the U.S. Navy used helium in its dirigibles during the 1920s and 30s, it lost three of its four airships in accidents with heavy loss of life, making it evident that the flammability of hydrogen had never really been the biggest problem airships had to deal with.

carrying ships, perhaps to serve as prototypes for even larger ships to come. However, for the most part it seems that their era has come and gone. With the advent of the massive French airbuses, the airships can't even use their vast payload to justify their immense costs, rendering them economically impractical—at least on a large scale. I suspect we are all the poorer for this, for none of us will ever have the chance to stand and gawk—as many in the first third of the twentieth century had—as one of the great ships floats serenely overhead. It would have been impossible to watch twenty-two million cubic feet of hydrogen gas hovering overhead like some great airborne whale without being impressed at the sight; unfortunately, I suspect it will be something none of us will ever have a chance to experience, and in that is the saddest legacy of all.

12 CHAPTER TWELVE: THE WRIGHT BROTHERS ACCOMPLISH THE IMPOSSIBLE

At the risk of moving beyond the basic premise of this book—that being to catalogue the pre-twentieth century history of aviation—it quickly became apparent that ending it there leaves the story only half told. It is almost as though the climax to an epic movie has been left out, making it necessary to move past the nineteenth century and examine some of the earliest attempts at heavier-than-air manned flight during the first decade of the twentieth century and the men who made it possible. It is only in understanding what happened during this transitional period in aviation's evolution that we can begin to give the successes of the new century some context, and to do that, it will be necessary to briefly examine the remarkable accomplishments of the brothers from Dayton, Ohio.

Of course, it's not necessary to do a detailed recounting of the Wrights' early efforts. Numerous books—many of which are quite exhaustive—have been written on the Wright brothers, making any contribution we might add to the narrative redundant at best, but then revisiting the Wright brothers accomplishments is not the purpose of this chapter. Instead, this chapter is designed to provide the reader some context by which they might judge the credibility of the claims made by others to having been the first to accomplish powered and controlled manned flight. In effect, we need to understand how the Wright brothers were able to succeed when so many others failed so we might better gauge the claims made by and about others to being the first. We will examine those claims in the next few chapters, but for now it is necessary we concentrate on the brothers from Ohio and their role in aviation history.

The Brothers from Dayton

As is often the case with men who have made significant contributions to human advances throughout history, it is easy to forget they often hailed from fairly humble beginnings, and that is especially true for the Wright brothers. Orville (1871-1948) and Wilbur (1867-1912) Wright were two of seven children born to Milton and Susan Wright of Dayton, Ohio. Sons of a traveling minister, it would be fair to maintain they grew up in what might be considered at the time a middle-class environment—a status only a fraction of the population were able to subscribe to at a time when most Americans lived in abject poverty. How this might have affected their outlook on things is uncertain, but it did seem to provide them with opportunities to pursue various interests other young men of their era might not have been able to do. On area of interest that the brothers appeared to have shared from an early age is aviation which came about, at least according to legend, as a result of receiving a toy from their father. It seems that Milton Wright, who occasionally purchased small gifts for his children during his travels, came upon a toy "helicopter" designed by French aeronautical pioneer Alphonse Pénaud (see chapter five) and decided his sons might find the paper, bamboo, and cork device amusing. His instincts proved to be correct, and Wilbur and Orville played with the wind-up toy until it broke, after which they built one of their own (and in doing so, already

The Wright brothers' first glider was a primitive machine when compared to their later efforts but it taught them much about what would be required to build a successful manned glider.

The 1902 Wright Flyer was the most successful of their early gliders in that it enabled them to make turns, an important step in building their first powered machine.

demonstrating that they possessed a high degree of mechanical aptitude).

This seemingly inherent engineering skill was to become the brother's trademark. Even while still in their teens they began putting it to good use by building and operating their own printing press, which they used to produce a local paper as well as print jobs for paying customers. This soon permitted them to earn enough money to move on to their next adventure: Capitalizing on the national bicycle craze, in 1892 the brothers opened a bicycle repair and sales shop, which proved to be so successful that soon they were manufacturing their own brand of bicycle. It was this success that provided them not only the funds, but the mechanical know-how, to begin experimenting with gliders. Even more fortuitous, being successful businessmen in their own right gave them the business acumen they would require to make their study of aeronautics potentially profitable as well—a talent few inventors of the era possessed.

Fortunately, by this time there was considerable information available on aeronautics that they could study, and they spent a number of years developing the technical knowledge that would be required to build a flyable aircraft. They also began closely following the exploits of men like Lilienthal, Chanute, and Langley, and even began corresponding with a number of these men on a regular basis as a means of gaining further knowledge, ideas, and support.

Finally satisfied they had acquired enough information to begin work in earnest, during the summer of 1899 they constructed the first of their gliders—a five foot box kite in the approximate shape of a biplane—to test their theories about wing warping (which they considered to be the key to controlling an aircraft in flight). After working with the small model for a few weeks, they decided to build a larger, manned glider, which they took to Kitty Hawk, North Carolina, to test in the summer of 1900. (Kitty Hawk was chosen largely on the advice of aviator Octave Chanute, who suggested they attempt their experiments on the Atlantic coast because of its regular breezes and soft sandy landing surface. The spot also gave them privacy from reporters.) Unfortunately, the glider proved to be too small to be useful and after attempting a few test flights with it, they soon packed up and headed home to Ohio to rework their basic design.

The next year they returned with a larger glider that proved considerably more successful, though it too exhibited some serious flaws. While it allowed them to make scores of flights over the next two months, some of which reached distances of four hundred feet before gently pancaking into the soft sand, the glider produced only about one-third the lift they calculated it should. Even worse, it sometimes failed to respond to their wing-warping techniques and would turn in the opposite the direction intended. Still, with each test the brothers grew more experienced and incorporated each lesson learned into their next design.

They returned to North Carolina again in 1902, this time with a still larger and much improved glider, with which they would experience their greatest unpowered success. Incorporating a steerable vertical rudder that permitted them to achieve true control in turns for the first time, during September and October of that year they made hundreds of flights (with some biographers claiming they made as many as one thousand glides), the longest of which lasted 26 seconds and covered an impressive 622 feet in distance. What these flights did is not only make them expert pilots, but convinced them they were finally ready to build a powered flying machine, which they set about doing as soon as they returned home to Dayton that fall.

Almost as remarkable an achievement as the flight itself was the ability of the Wright brothers' chief mechanic, Charlie Taylor, to hand-build an aluminum engine in just six weeks. Though badly underpowered, it has to be considered an example of American ingenuity and resourcefulness at its best.

Transitioning from unmanned gliders to a powered aircraft, however, proved to be more difficult than they had imagined. The problem was in finding a power plant both powerful enough to provide the thrust needed yet light enough to lift its own weight (a problem shared by all aviators of the era). At first the Wrights wrote to several engine manufacturers in search of an engine which would meet their requirements, but none were available that could meet their needs. Desperate, they finally turned to their shop mechanic, Charlie Taylor—a man known for his mechanical skills—for help and in just six short weeks he managed to build them an engine cast from a solid block of light-weight aluminum. Though building an engine from scratch would be a remarkable achievement in any era, the engine Taylor built proved to be problematic from the start. First, while it was light, it wasn't very powerful, producing barely 12 horsepower. Second, it was archaic not only by modern standards (the engine had no carburetor or fuel pump; the gasoline was simply gravity-fed into the crankcase through a rubber tube from the fuel tank mounted on a wing strut) but was considered primitive even by the standards of 1903. However, it was all they had and they soon mounted it to their next machine—a slightly modified version of their 1902 glider—and returned to North Carolina in the autumn of 1903 to test it out.

Things did not go well at first as several propeller shafts broke during engine testing, causing weeks of delays, but they were nothing if not persistent. With dogged determination they stuck with it, and by December they finally got the flyer working well enough that they were ready to attempt a manned, powered flight. Pulling their muslin-covered spruce machine out to its launch rail, Orville (who had won a coin toss to be the first to pilot the craft) pushed the throttle forward and launched the plane into a freezing 27 mph headwind. Remarkably, the frail craft left the rail and flew for a few seconds before gently setting back down, thereby making Orville

Probably the most famous aviation photo in history—the first successful flight of the Flyer I at Kitty Hawk, North Carolina, on December 17, 1903. Though the flight lasted only 12 seconds and covered a mere 120 feet, it signalled the start of the era of modern aviation and immortalized the Wright brothers.

Wright the first man officially credited with flying an airplane. Fortunately the brothers had the presence of mind to ensure the achievement was captured on film—the result being perhaps the most famous aviation photo ever taken. (The photo was taken by an employee of the nearby lifesaving station named John Daniels using Orville's pre-positioned Gundlach Korona box camera.) The brothers would make three more flights that day with Orville and Wilbur taking turns at the controls. The most successful flight of the day covered an impressive 852 feet and lasted nearly a full minute before the craft pitched into the sand, breaking a rudder support beam in the process and ending all further flights for the day, but the modern era of aviation had been born.

Of course, while their brief hops through the air were significant and, in a way, even impressive, what would make the aeroplane truly useful was the ability to control it in flight. Just as it was the axle that made the wheel genuinely practical, the Wright brothers had to be able to consistently turn and maneuver the plane in the air for it to be considered a true aeroplane. Fortunately, they managed to accomplish that the following summer when, flying from a cow pasture turned airfield eight miles northwest of Dayton and in a bigger and improved version of the their first plane, the *Flyer II*[39], they made a number of circles and turns, demonstrating that flight was not only possible, but becoming increasingly routine. By 1905 they were consistently making flights of over twenty miles and staying aloft for half an hour at a time, which, while seemingly insignificant by modern standards, was a technological triumph as great for the era as the moon landings would be to future generations.

Their success was not without its risks or sacrifices, however. The Wright brothers may have the distinction of being the first men to have successfully flown a powered, fixed wing aircraft, but they also hold the dubious distinction of being participants in the first heavier-than-air aviation fatality in history as well. The incident occurred on September 17,

39. This is the aircraft currently on display at the Smithsonian Air and Space Museum.

Orville Wright flying the successful Flyer II near Dayton in November, 1904. This flight, in being under complete control, is the one that proved the airplane to be truly practical.

1908 when Orville was demonstrating the flyer for the U.S. Army Signal Corps at Fort Myer, Virginia. Acting as an observer during the flight was Lieutenant Thomas Selfridge, a Signal Corps officer and a man who already possessed considerable experience working with both heavier and lighter-than-air aviation (he had even designed his own aircraft, the *Red Wing*, which had flown earlier that year) making him uniquely qualified to observe the brother's fragile contraption firsthand. Tragically, while the plane was making a series of circles around the field, the right propeller broke, which set up a vibration that caused the split propeller to hit a guy wire bracing the rear vertical rudder. This tore the wire out of its fastening and shattered the propeller, forcing the rudder to swivel to the horizontal and sending the *Flyer* into a nose dive. Orville immediately shut off the engine and put the plane into a glide, but the flyer hit the ground so hard it threw both Selfridge and Orville against the wooden uprights of the framework, fracturing the lieutenant's skull and leaving Orville with a broken left thigh, several broken ribs, and a damaged hip. Selfridge died the next morning without regaining consciousness[40] while Orville was to spend the next seven weeks in a hospital, recovering from his injuries.

Unfortunately, Selfridge was not the last man to die on a Wright-built aircraft. Just a year later, a

Scene of the crash that killed Lieutenant Thomas Selfridge and badly injured Orville Wright near Ft. Myers, Virginia, September 17, 1908. Unfortunately, the Wright flyers were to develop a reputation for being difficult to fly, resulting in the deaths of a number of pilots over the next few years.

[40] It was generally believed at the time that, if Selfridge had been wearing a helmet of some sort, he most likely would have survived the crash. As a result of his death, the U.S. Army's first pilots were required to wear headgear reminiscent of early football helmets.

The Wright Model A was the aircraft that convinced the world that the Wrights had actually flown. It was also the first two-seat aircraft, and the first Wright aircraft in which the occupants sat upright.

French pilot, Eugène Lefebvre, flying in a Wright-built *Model A*, crashed near Reims, France (making him the first person to die while piloting a powered aircraft), and on November 17, 1910, one of the pilots the Wright brothers had hired to demonstrate their aircraft, Ralph Johnstone, died when he was unable to recover from a shallow dive while flying near Denver, Colorado. Their deaths, like that of Selfridge earlier, served to underscore the very real danger inherent not only the early Wright aircraft, but in aviation in general. Unfortunately, death was to be an element of the aviation industry that would not only mar its future development, but remain a part of it to this very day.

Conclusions

While the Wright planes could be dangerous to fly, what the brothers accomplished cannot be underemphasized. Patiently working step-by-step towards their goal, they invested literally thousands of man-hours (and considerable money) into accomplishing what man had been trying to do since the days of George Cayley—fly like a bird. And they did it the old fashioned way: through hard work, ingenuity, and perseverance. In the process they endured brutal weather and bitter disappointment, and exposed themselves to considerable danger, but in the end they made it possible for the airplane to come of age, and for that they deserve our admiration and respect.

But when dealing with the Wright brothers and the history of aviation in general, the burning question that remains to be answered is what would the world look like today had the Wright brothers given up on their experiments? Would we be where we are today, or would the evolution of aviation have proceeded down a different path?

Most likely the answer is a bit of both. Remove the Wright brothers from the equation and you still have others waiting in the wings to take their place—men whose own successes came a few years and, in some cases, just scant months after the Wrights' own. But remove the Ohio brothers from history and perhaps the innovations they introduced are delayed by years, thereby retarding the entire process of aeronautical evolution and leaving us to only speculate what twists and turns history might have made without the key characters being in their rightful places.

Chapter Thirteen: Was Gustave Whitehead the First Man to Fly? 13

While no one doubts that the Wright brothers actually flew at Kitty Hawk that cold December day in 1903, there are some who insist that they were not the first to have taken a powered plane aloft. Of course, most of these claims remain unsubstantiated and so are not accepted by aviation historians as fact, which is as it should be, for establishing who was the first to fly is an important element of the history of early aviation.

As is usually the case with multiple claims to the same title, however, some of these stories have more credibility than others, which should force us to approach the issue with an open mind. After all, history is notoriously unkind to those who work in obscurity, leaving the objective observer to wonder if our knowledge of that history is as complete and accurate as we imagine it to be or whether there may not still be stories out there that remain to be told.

In this chapter we will look at one of the more credible claims made in regards to who was really the first man to fly a powered, heavier-than-air machine. This is the story of a little known German immigrant whom some maintain not only beat the Wright brothers into the air by as much as two years, but may have done so in an aircraft that was, in many ways, superior to the brothers' little fabric and wooden craft. The reason we have chosen to devote an entire chapter to this claim is twofold: First, there is considerable evidence—along with a number of eyewitnesses who were willing to go on public record affirming the prospect—that this man may have truly been the first to fly; and, second, because this inventor's work as a gifted mechanic and a daring aviator deserves to be examined and appreciated for its ingenuity and foresight, especially when weighed in the context of the era in which he lived.

So who was this man and what evidence is there to substantiate the remarkable claims some have made about him?

The Big Man from Leutershausen

By all accounts, even as a boy Gustave Albin Weisskopf (later Americanized to Whitehead, 1874-1927), displayed considerable mechanical aptitude and an even more pronounced interest in aviation, which he demonstrated by spending days studying the way tethered birds flew, experimenting with kites, and even making tiny parachutes out of tissue paper. It is reported that at the age of thirteen he even made himself a pair of cloth wings and jumped from the roof of his home which, not surprisingly, earned him the nickname "der Flieger" ("the flyer"). As it would turn out, the title was to eventually not only follow him the rest of his life, but ultimately prove to be entirely appropriate as well.

At first, however, it did not appear that young Gustav's future was likely to take him in that direction, Losing both his parents a year apart, while still a mere boy of thirteen he found himself parceled off to a relative in Hamburg, where he was find his

Whitehead with one of his engines, probably taken sometime around 1900. Had he concentrated on developing his powerful, lightweight motors instead of trying to build airplanes, he might have become a wealthy man.

options extremely limited. Eventually finding work as a deck hand onboard a sailing ship, he quickly found himself working long days of difficult and often backbreaking work, which was to remain his lot in life for the next several years as he served on a number of vessels in both Germany and Brazil. However, despite the drudgery his life had become, he never lost his interest in aviation: Being at sea for weeks on end gave him time to study the way seabirds flew along with the opportunity to learn as much as he could about wind and weather, all of which would prove useful in later years.

In 1893 Gustave finally managed to return to Germany where, like so many Europeans with limited prospects during that era, he became intent on making his way to America to start afresh. To acquire the funds to book passage on a ship to America, he found work as a mechanic, (which, one might surmise, was where he acquired his mechanical aptitude—a skill that was to serve him well in the future). While he worked towards that goal, however, he kept his fascination with flying alive, and spent much time studying the work of the German glider designer Otto Lilienthal. (There were reports he managed to meet the great aviator and, according to some, that he even helped him with the construction of some of his gliders—though such stories remain largely unsubstantiated. If true, however, it would explain much about how the man acquired his aeronautical acumen at such a young age.) In any case, he finally found the funds to emigrate in 1895, at which time he anglicized his name to Whitehead and began looking for work in Boston, Massachusetts.

Apparently Weisskopf's—now Whitehead's—background in aeronautics paid off and he eventually found work building and flying gliders on behalf of the Aeronautical Club of Boston. It is reported that during that time he built a couple of small gliders of his own design, one of which—a Lilienthal-type glider—he evidently flew with some success. That next year he also found employment building hang gliders, aircraft models, and motors for flying craft in New York City for Horsman Manufacturing, which sold them in their toy store.

By 1898 Whitehead had married and moved to Pittsburgh, Pennsylvania, where he began to work full-time on his two life passions: engines and aeroplanes. As for the former, it turned out that Whitehead had a remarkable aptitude for designing and building his own light-weight engines, which proved to be so good he was able to advertise and sell a number of them. (Many biographers maintain that had he concentrated on simply building and selling his motors, he might easily have made himself a wealthy man.) Unfortunately, his obsession with his second interest—building a flying machine—proved to outweigh his practical senses as he built only enough motors to keep his growing family fed and continue to finance his work on his airplanes, leaving him always just one step ahead of being destitute.

In fact, building an airplane became such a consuming passion of his that over the next few years he spent almost every waking moment and every spare cent he made on the venture, often to the detriment of his family's wishes (and the goodwill of his neighbors, whom he frequently awoke in the middle of the night whenever he tested one of his new motors). It appears, in fact, that he and his noisy motors (this was an era in mechanical history when the muffler was still theoretical) became such a disturbance that he was forced to move his fledgling family to a more remote home near Bridgeport, Connecticut, where he could at last work with a degree of privacy.

The Alleged 1899 Flight

It's uncertain when Whitehead actually built his first airplane. It has been claimed he built a steam-powered monoplane aircraft as early as 1899 and that he actually flew it while he was still living in Pittsburgh, though the story appears doubtful. First, it's unlikely Whitehead would have had the resources to build an airplane at such an early date. While it was known that he had constructed a number of unpowered gliders at that time, a powered machine would simply have been too costly at a time when his engine manufacturing business was just beginning to take off. But the most suspicious element of the story is that it seems to have originated almost entirely from a single source—a friend of Whitehead's named Lewis Darvarich. It was Davarich, apparently, who maintained that in the spring of 1899, Whitehead successfully flew his home-built airplane about half a mile—even reaching the impressive altitude of 25 feet—before crashing into a three-story building. The claim is almost universally rejected even by many Whitehead proponents, however, largely because there is no record of it having occurred (which, considering the magnitude of the achievement and the fact that the machine supposedly crashed into a building, should have brought at least some attention from the local press). Additionally, Mr. Darvarich claimed to have been a passenger in the plane, meaning the plane would have had to have the capacity to carry the weight of two full-grown men at a time when getting even a single man into the air was problematic. Finally, Davarich claims that he was so seriously scalded while stoking the boiler aboard the craft that he required several weeks in a hospital to recover, yet there is no record of a "Mr. Darvarich" to be found in any area hospital records at the time. Taken together then and, considering that Whitehead himself made no claim to having flown in 1899, the entire story appears to be more likely a figment of Darvarich's imagination than a genuine historical event and so is generally discounted by even Whitehead fans.

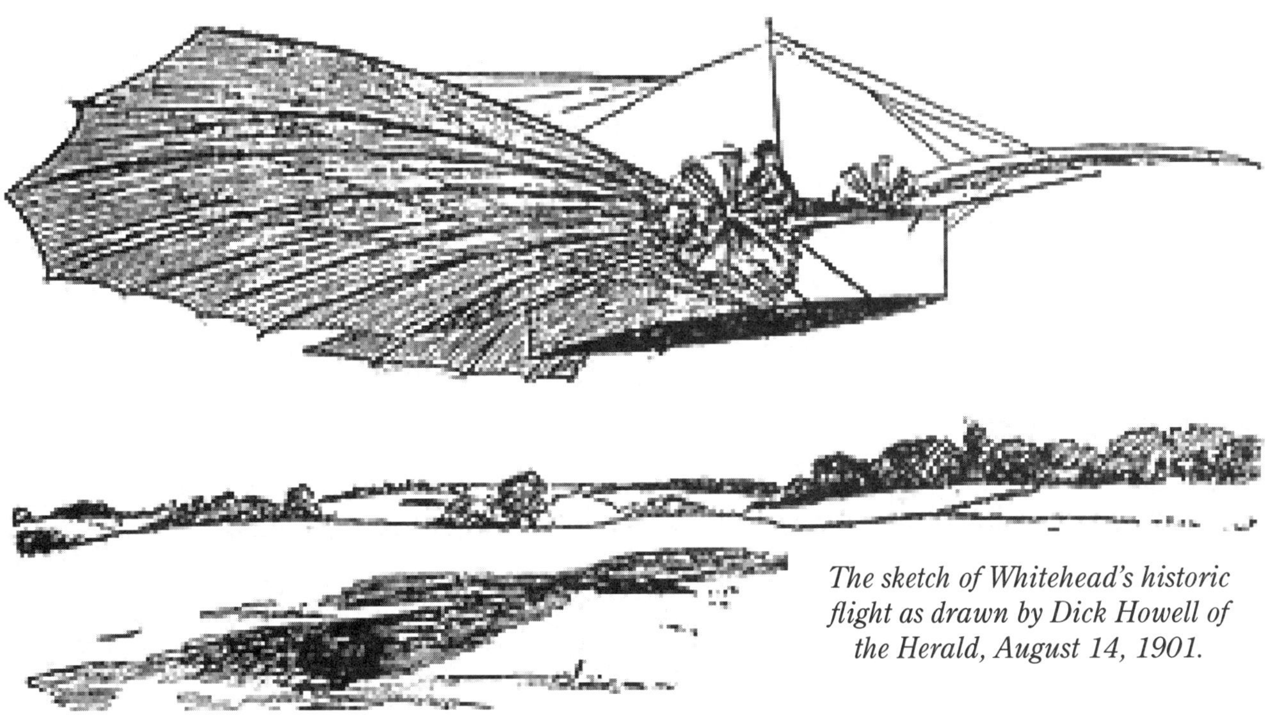

The sketch of Whitehead's historic flight as drawn by Dick Howell of the Herald, August 14, 1901.

"Number 21" and the August, 1901 Flight

However, there is a better substantiated account that suggests Whitehead may have flown a couple of years later, and it is this possible event that continues to generate the most debate to this day.

It apparently occurred—assuming it happened at all—while Whitehead was living in Bridgeport, Connecticut in 1901. It appears that around this time, Whitehead did manage to build his first full-sized, piston-engine powered monoplane, which he simply called *Number 21.* (We will discuss the specifics of this aircraft in more detail later in the chapter.) In any case, according to the *Bridgeport Herald* newspaper and a few witnesses who gave their statements more than thirty years later, it was in this plane that on August 14, 1901, Whitehead supposedly made a powered, controlled flight that covered the remarkable distance of 2,600 feet and reached an altitude of almost fifty feet before making a gentle landing. If true, of course, that would have been an extraordinary achievement, and one that would have not only bettered the Wright brothers first powered flight by 1700 feet, but one that would have preceded their efforts by more than two years!

Unfortunately, as was often the case with Whitehead—and would be for a host of other would-be aviators at the time—no photos of the event were taken, which is astonishing considering the magnitude of the flight and the availability of cameras at the time. The only visual record of the flight that exists is in the form of a rather fanciful sketch made by *Herald* reporter Dick Howell—who also wrote an eyewitness account of the flight.

In the article, Howell reported that the airplane took off from a flat surface without the aid of a catapult, launch ramp, or headwind. The *Herald* also reported that when Whitehead encountered some trees that stood directly in his path: "He simply shifted his weight more to one side than the other. This careened the ship to one side. She turned her nose away from the clump of sprouts when within fifty yards of them and took her course around them as prettily as a yacht on the sea avoids a bar." (If true, this would be the first documented controlled maneuver of a manned, powered aircraft as well.) In the end, the paper reported that the vehicle made a safe landing with the motor turned off.

The 1902 Flights

Had that been the end of it, that would have been remarkable enough, but Whitehead wasn't quite done with his attempts to fly. In 1902 he again claimed to not only have flown again, but to have made two even more spectacular flights in his much improved and more powerful *Number 22,* one of two miles and a second in which he claims to have remained airborne for an unbelievable seven miles!

Unlike the 1901 flight, however, there was no press

Gustave Whitehead with his daughter in front of his "Number 21" airplane—circa 1901. Was this the machine that beat the Wright brothers into the air?

on hand to report on either attempt, leaving it firmly entrenched in the realm of popular legend. In fact, the bulk of the evidence to support the contention that the flights actually occurred came from Whitehead himself in the form of two letters he sent to *American Inventor* magazine, in which he wrote that the flights took place over Long Island Sound near Lordship Manor. Going into some detail about the flights themselves, he wrote that he reached altitudes of up to two hundred feet and that both of them ended with safe landings in the water.[41] Expressing considerable pride in the accomplishment, he wrote: "...as I successfully returned to my starting place with a machine hitherto untried and heavier than air, I consider the trip quite a success. To my knowledge it is the first of its kind. This matter has so far never been published." Whitehead claimed that the flight was witnessed by dozens of people and that a log entry to the effect was supposedly made by the captain of a ship passing through the sound at the time.[42]

Whitehead promised the magazine that he would try to get photos of the plane in flight and send them along, but the promised photos never materialized. This was either because Whitehead was unable to get his aircraft to repeat the feat—if, indeed, he ever accomplished it in the first place—or he simply ran out of the funds to attempt any more powered flights after that—which is not difficult to imagine from the always cash-strapped inventor. In either case, that appears to be the last that anyone heard of Gustave Whitehead's flying machines and the brief excitement he generated quickly faded.

At least, that was case until an article appeared in *Popular Aviation* magazine in 1935 entitled "Did Whitehead Precede the Wrights in the World's First Powered Flight?" Written by Stella Randolph, the article[43] maintained that Whitehead had flown before the Wrights, an assertion which not only rescued Whitehead from oblivion, but inaugurated a running debate that continues to this day.

The Controversy Begins

So did any of these flights actually happen?

It seems destined to be a question that will most likely always remain unanswered. That it might have happened seems both credible and yet entirely impossible at the same time, further compounding the mystery and magnifying the mystique that surrounds this gifted if quirky man from Germany.

Unfortunately, the best case for Whitehead making each of his flights is based almost entirely upon the testimony of witnesses who maintain that they saw him fly his machines some thirty years earlier. Some of these people claimed to have been his helpers or employees, while others were neighbors—including a number who were children at the time—each of whom were willing to go on public record and sign affidavits affirming that they did, indeed, see the man fly when and where he claimed to have flown. These accounts remain the best—and some might argue, the *only*—pieces of evidence, in addition to the *Bridgeport Herald* newspaper report, that substantiate Whitehead's claims.

The problems with the story are obvious. First, the idea that anyone could build an aircraft capable of flying seven miles as early as 1902 is difficult to imagine. Had the claims been that he simply flew for a few hundred feet before gently setting down, it might not be so difficult to accept. Whitehead did, after all, know how to build an airplane and he certainly had the engines capable of powering them. (Consider that Whitehead's 1901 plane boasted a 20 horsepower engine whereas the *Wright Flyer* had only a 12 horsepower engine—and his 1902

41. Whitehead's claim to have landed in the water as well as claiming to having returned to his starting point (on presumably dry ground) is confusing and seems contradictory. However, the man's English was not good, making it possible that he meant he flew out over open water before making a safe landing at his starting point.

42. There is a report that the captain involved was named Brown, and that he not only made a logbook entry about Whitehead flying over Long Island Sound, but that he even photographed the airplane in flight. Unfortunately, later attempts to retrieve the captain's journal to substantiate the claim was unsuccessful when the family who owned it moved out of state and subsequently lost the book. Other photos of this flight—as well as the 1901 flight—were rumored to have existed early in the century, but have yet to turn up.

43. Randolph later expanded the article into a 1937 book entitled *The Lost Flights Of Gustave Whitehead.*

plane had an even more powerful engine.) However, that any of his planes were capable of flying miles at a time strains credulity.

The other problem is that Whitehead had a reputation for exaggeration. For example, he once claimed his aircraft were capable of speeds of 70 mph—a speed no aircraft was to reach until late in the decade. He also claimed the reason he never had photographs taken of any of his machines in flight was because they moved too fast to be photographed. Additionally, failing to deliver on his promise to the *American Inventor* to get photos of his *Number 22* in the air also made people wonder if Whitehead didn't have a tendency to promise more than he could deliver.

Of course, none of these aspects of the man in and of themselves disprove his story. Certainly, he may simply have overestimated his airspeed—easy to do without a reliable means of measuring it—and he may really have believed that his aircraft could not be captured on film. (Photography was still a new and mysterious technology to many people at the time and he may genuinely not have understood what it was capable of doing.) Then there is the abundance of eyewitness testimony, much of it quite compelling, to take into account. Even recorded thirty years after the fact, it seems unlikely people would confuse seeing an airplane in flight as opposed to not seeing one. They may be hazy on the specifics—dates, time of day, location, etc.—but seeing something as unusual as a flying machine in the pre-aviation skies of the early twentieth century is not something one was likely to forget, even after three decades.

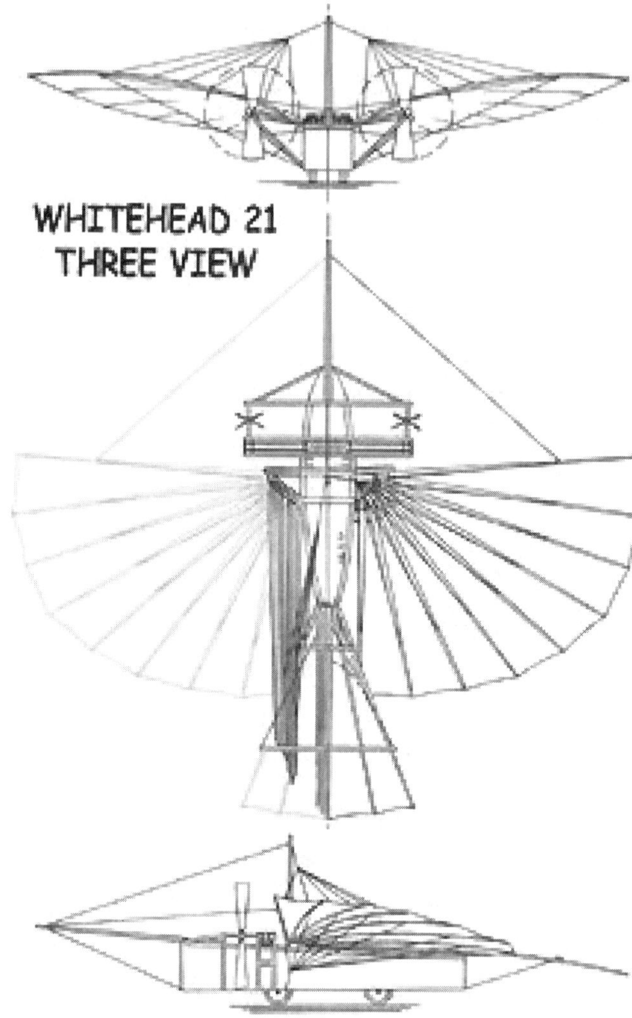

Three-view plan of Whitehead's Number 21 aeroplane reveals a design that possessed all the necessary elements to make it a truly flyable machine, potentially capable of at least a short flight.

Since the evidence both for and against Whitehead's claim to have flown before the Wright brothers appears to cancel each other out, then, we must consider other factors to determine whether there is any validity to his claims, and perhaps the best evidence we might examine is his airplane itself. After all, if it can be determined that the plane he built in 1901 was capable of flight, at least from an engineering standpoint, it stands to reason he may have actually flown it. Fortunately photos of the plane—at least the one he designated as *Number 21*—do exist (though they are all of the plane on the ground), and we have working drawings and other information regarding its construction, so we are not left empty-handed.

So was the Number 21 capable of flying, or are Whitehead's exploits purely a product of his own imagination?

The Aircraft

Regardless of what one may think of Whitehead's story, what cannot be denied is that Whitehead built one racy-looking monoplane. In fact, with its 36-foot long silk and bamboo wings modeled after the shape of a soaring bird's wing, it even appears in many ways light years ahead of the boxy, kite-like configuration of the *Wright Flyer* and other machines of the era. Though the plane may appear fragile from a modern standpoint, it is obvious it also appears to be quite light, which is precisely what would be required were it to have any chance of achieving flight with such a small engine. Its boat-like hull also provides a degree of streamlining quite advanced for its era, and with its great folding wings (similar to those Otto Lilienthal used in 1896), small wheeled landing gear, and broad propeller blades, it was the very image of how a proper flying machine of that time should appear.

In terms of power, it also seemed up to the task. Originally steam-powered, Whitehead later installed a 20 horsepower carbide-acetylene powered engine to drive the plane's two counter-rotating propellers, which should have provided sufficient lift—with a bit of cooperation from a fortuitous headwind—to get the machine off the ground. He even equipped his plane with a second, smaller 10 horsepower engine designed to drive two of the landing gear's four wheels, which was quite a novel and clever approach to getting the craft up to speed quickly. (By using the wheels to propel the craft to takeoff speed, it took much of the burden off the propellers, thereby making for a more efficient use of power to achieve take-off. The concept of using the landing gear as a means of propulsion rather than as just an implement for landing and taking off was actually quite ingenious and one that Whitehead alone among all the aviation pioneers of the era thought of.) Additionally, the later *Number 22* plane was built along similar lines, but in having a 40 horsepower motor instead of a 20 horsepower one—along with aluminum instead of bamboo ribbing in the wings—it was potentially more capable than the earlier craft and, as such, even more likely to have flown.

As far as control is concerned, roll was controlled by the pilot shifting his weight, much as

This rare overhead view of the Number 21 aeroplane does a good job showing off the sleek lines and aerodynamic shape of Whitehead's machine. The fact that this aircraft predates the Wright brother's by two years and yet looks to be at least a decade ahead of their relatively primitive machine demonstrates just how capable an aeronautical engineer Gustave Whitehead really was.

Whitehead in the air in his Glider 1, probably taken sometime around 1900. Whitehead built and flew gliders for a number of years without incident, demonstrating that he was a most proficient pilot.

on a hang-glider, while pitch would be controlled by a tail wing. Yaw—the third force that acts upon an aircraft in flight—and normally controlled by the rudder in modern aircraft—was controlled by differentiating the thrust of the two propellers. As such, the aircraft had all the necessary elements—lightness, power, and control—to make it a true aeroplane in every sense of the word and, as such, entirely capable of controlled flight—at least under ideal conditions and for short amounts of time. In fact, it has been conclusively demonstrated since that a replica of Whitehead's *Number 21*, built in 1986, twice become briefly airborne (when flown into a decent headwind), demonstrating that the aircraft—or at least a modern facsimile of it—was flyable.[44] Of course, this doesn't prove that Whitehead actually flew in either 1901 or 1902, but it does suggest that he well could have.

The only other question is whether it would have been capable making a flight of several miles as Whitehead claimed. While it seems doubtful, if the plane carried enough fuel and had a good tailwind, it is conceivable it might have remained airborne for some time. All things considered, then, if Whitehead wanted to beat the Wright brothers into the air, he certainly had the airplane that could do it.

The Pilot

Assuming that Gustave Whitehead had the airplane capable of pulling off the feat of the century, the next question to be addressed is whether he possessed the expertise to actually fly the thing. In this case, we have good evidence that, indeed, he had.

It is a well-attested fact that Whitehead was an accomplished glider pilot, having built and flown gliders for several years prior to his first alleged powered flights. It was even attested to by his wife and those who knew Whitehead personally that when he lacked sufficient funds to build powered aircraft, he would build gliders, which he apparently did with some frequency. Additionally, photos exist of him flying his homemade gliders (and, by all accounts, for some distance), demonstrating that he was an accomplished pilot, making him easily capable of handling a powered aircraft. As such, we can be reasonably confident that Whitehead had the skills to fly his airplane, if it were capable of being flown at all. The only question that remains, then—and one which will probably never be answered—is did he actually pull it off?

The question, unfortunately, cannot be answered and so must remain a matter of faith. On the one hand, there is no hard evidence of the type the Wright brothers produced to prove that he ever got off the ground. On the other hand, he did build an airplane that appears to have been capable of flight, and we know he knew how to fly it, plus there is at least circumstantial and anecdotal evidence that he did, in fact, beat the Wright brothers into the air by at least a year—and possibly by two. The final verdict, however, pending the introduction of new and compelling evidence, must remain for the reader to determine for themselves.

44. Admittedly this replica was being towed behind a sportscar at the time, but the replica's builder, Andy Kosch, had made a number of unassisted flights in the craft previously, one as far as 330 feet. Additionally, another reproduction flown in Germany in 1998 reportedly flew over 500 meters (nearly 1600 feet) unassisted.

Later Career

Fate appears to have been unkind to Gustave Whitehead after the events of 1902. Without the proper funding to continue his research on aeroplanes (apparently his hobby proved to be quite a financial strain on his family), he went back to doing what he did best: building small, light-weight motors. The curious thing is that he was so gifted in this arena and his engines proved so popular that had he concentrated on marketing them instead of only producing enough to fund his aeronautical research, he might well have become a millionaire (and, as such, would have finally had the resources to pursue his flying experiments full-time). No less a personage than Glenn Curtiss, (see chapter seventeen) wrote about Whitehead (whom he referred to by his German name of Weisskopf), saying:

> *"Weisskopf's ability and mechanical skill could have made him a wealthy man at a time when there was an ever increasing demand for lightweight engines, but he was far more interested in flying. Even so, word of his talent as a machinist spread rapidly. His daughter, Rose, remembers bringing home so many letters with orders and advance payments on engines that she could scarcely carry them. She stated that one day, her father returned 50 orders, for he built only as many engines for sale as he felt would provide him with funds to advance his own work upon airplanes."*

He didn't give up on his aviation experiments entirely, however. In 1911 he designed a small, 60-bladed helicopter, which actually lifted itself off the ground. Probably realizing he needed a much more powerful motor to make the experimental helicopter practical, however, Whitehead apparently gave up on the project and, as far as is known, did not attempt to build any more aircraft the rest of his life. Part of the reason for this may not have been a lack of desire on his part, however: In 1912 Whitehead lost a lawsuit when one of his motors failed to work up to expectations and he saw some of his tools and equipment seized. As a result, Whitehead was finally forced to give up his flying tests entirely and find work in a factory as a laborer while he repaired motors on the side to support his family. Worse, he lost an eye in an industrial accident and suffered a severe blow to the chest when he was hit by a piece of factory equipment, an injury that left him permanently weakened and likely contributed to his increasingly frequent angina attacks.

These setbacks, however, didn't reduce his creativity. Whitehead continued to invent all sorts of useful devices and one time came very close to winning a substantial cash prize offered by a local railroad for a braking safety device. He even constructed an "automatic" concrete-laying machine, which he used to help build a road near Bridgeport. Unfortunately, his marked lack of business acumen ensured that he profited no more from these inventions than he did from his airplanes and engines. Additionally, when he found his discoveries being credited to other people[45] and as others surpassed him in aircraft engine design, Whitehead finally lost all interest in aviation and new inventions and spent the rest of his life working as a mechanic until his death in October of 1927, bringing an end to the life of a most talented man.

Conclusions

So what can we make of Gustave Whitehead and the claims made about him? No doubt about it, the man is an enigma. On the one hand he seemed to have possessed the technical skills to have built a flyable machine and the piloting experience necessary to have actually flown it. On the other hand, there are those who claim he was a man prone to exaggeration and hyperbole, and one whose imagination and enthusiasm may well have outpaced his actual accomplishments. Finally, there are those who are adamant about having seen his aeroplanes flying as early

45. Is it possible that some of this misfortune was a by-product of the prejudice against Germans that was prevalent in the country during World War I? Whitehead never lost his German accent nor did he acquire American citizenship, making him an unsympathetic figure to many people. Could such prejudice also have affected the way history has dealt with the man as well?

as 1901, while there are others who were willing to attest to the fact that nothing he built ever flew.

The truth, as is usually the case with controversial claims of this nature, probably lies somewhere between these two extremes. Clearly, it is undeniable that Whitehead built and successfully flew a number of gliders of his own design, many of which flew quite successfully. It is also a fact that he built a number of small motors that proved to be quite advanced for the time. Bringing these two elements together, then, should not have been beyond his abilities, making the prospect that he successfully built a flyable airplane not an unreasonable one.

The problem seems to be, in large part, Whitehead himself. His detractors suggest that the man was less than honest on many occasions, and it is a fact that he frequently promised "evidences" of his accomplishments that never materialized. Unfortunately, we have only second- and third-hand accounts of what sort of man Gustav Whitehead was in terms of personal honesty. Clearly he was an eccentric personality, as gifted inventors frequently are, and he remained something of an enigma even to those who knew him best (including his own family).

But that doesn't answer the question of whether he fabricated or, at very least, exaggerated his successes. No doubt he probably embellished his accomplishments. He was, after all, only human. But I wonder if those who later came out to discount his stories weren't themselves working from their own agendas—ones perhaps influenced by the rampant anti-German sentiments of the time. No doubt Whitehead also made enemies along the way, as did many of the great inventors of the era. (The feud between the Wright brothers and Samuel Langley being one of the more prominent, though by no means the only, example.) Further, there is the desire by many to ensure that the Wright brothers' legacy remains unblemished. As such, can we be certain that those who challenged the claims made by and about him did so only out of the purest of motives?

Aside from the man and his personal foibles, however, we need to ask a few basic questions. For example, would a man who was so passionate about his experiments that he was willing to sacrifice much in terms of time and money to pursue his projects not have wanted to find out whether his machines would actually fly? After all, he had spent years tinkering with his aircraft in an attempt to perfect them, plus he repeatedly demonstrated that he was not only a capable pilot but was willing to risk his own life to test his creations, so it's difficult to believe he wouldn't have tried to fly at least one of his two airplanes at some point. Additionally, we must ask whether the idea that the *Bridgeport Herald* reporter fabricated the entire August, 1901 event out of whole cloth—and obtained sworn affidavits from a number of alleged witnesses in an effort to maintain the ruse for some unknown reason—truly plausible? That the reporter got all the facts straight or that he quoted Whitehead accurately throughout the article is another question, but that he was reporting on an actual event—as opposed to an entirely fictional one—seems reasonably well established.

At the end of the day we are left with as much a mystery as we started with. All that can be said with accuracy is that Gustave Whitehead, for all his faults, was obviously a gifted mechanic and inventor, and a man with a passion for aviation few could match. Had he the resources—and the business sense the Wright brothers possessed—he might well have taken on the mantle as the "Father of American Aviation", but due to his background and social status—as well as a lack of financial acumen on his part—he was left to toil on his ideas in obscurity. We also suspect that had he been a native-born American who spoke flawless English and came from a family of some standing, history might well have treated him differently. For now, however, we can only speculate about what his rightful place in history should be and debate—ad infinitum—whether he was, indeed, the first man to make an untethered powered flight in a heavier-than-air machine. In the end, we can only let the fickleness of time render the final verdict, as it usually does.

Chapter Fourteen: Contenders and Pretenders 14

The claim to having beat the Wright brothers into the air is an extraordinary one and one that must be approached with a healthy dose of skepticism. Considering that it took the brothers from Dayton almost four years of hard work from the time they first began exploring aeronautics to the moment they finally achieved their first, powered flights makes the idea that some self-taught backyard mechanic with a modicum of aeronautical knowledge and limited funds could accomplish what they did months or, according to some assertions, even years before they did, has to supported by some compelling evidence. (In the previous chapter we made the case that Gustave Whitehead may be one of the few men who might have a legitimate basis for making such a claim, but even the prospect that he beat the Wright brothers is widely dismissed by most aviation experts and remains a source of considerable debate to this day.)

Unfortunately, the fact is that such evidence is badly lacking, even for the best claims. In fact, and to be honest, some of them are difficult to take seriously. However, there are a few of these assertions that are credible enough to force one to wonder if the historical record may not have missed something. As such, and in the interest of being as thorough and open-minded as possible, we would be remiss if we didn't examine a few of these claims.

So who were these contenders and pretenders to the title of being the first to fly? Perhaps the most credible contender we've encountered is a German aviator from Hanover named Karl Jatho.

Karl Jatho

What makes Karl Jatho's (1873-1933) story more credible than most is that unlike so many aviators who worked in secret on their flying machines and supposedly flew them with few if any witnesses being present, Jatho actually tested his machine in front of numerous eyewitnesses, making his one of the better documented attempts at achieving manned, powered flight on record.

Jatho was particularly well suited to be an aviator. As a young man he showed considerable aptitude as an inventor and engineer, both of which would serve him well when he embarked upon a thorough study of aeronautics. Through his exhaustive research, he eventually came to the conclusion that, of all of nature's many wonders, the Zanonia seed—a kidney-shaped seed native to the island of Java known for its abilities to make amazingly long glides—was the most aerodynamically perfect form in nature, making it, he believed, nature's blueprint for a flying machine because of its inherent stability. Deciding to put that assumption to the test, he began constructing a biplane based upon the seed's general shape and within a few months he emerged with what could only be considered to be one of the more unusual machines to grace the pages of aviation history. Christened the *"Motordrachen"* (Motor Dragon), what he built was essentially a large biplane with massive boomerang-shaped wings reminiscent of the Zanonia seed, which was powered by a two-bladed pusher propeller driven by a single-cylinder 10 horsepower Buchet engine. Rolling it out of its

Karl Jatho's unconventional-looking machine being readied for flight. The lack of cambering in the wings gave them almost no lift, practically guaranteed that the craft would never achieve true flight.

specially built hangar in the summer of 1903 before a generally impressed public, Jatho enthusiastically set about beginning a series of test flights with the craft, convinced he was close to achieving manned flight.

Unfortunately, while he patterned the shapes of the wings on the Zanonia seed, he failed to include any significant cambering in them, leaving them essentially flat. Naturally, this gave them almost no lift to speak of, seemingly dooming it from the start. Additionally, his power plant was far too small to turn the craft's large propeller fast enough to produce sufficient thrust, almost guaranteeing failure. As such, the best Jatho could do was taxi the craft along and, on occasion, get it to make a few short hops into the air.

One time, however, either through some fluke of aerodynamics or because of a most fortuitous breeze, one of his "hops" reached an altitude of about ten feet and flew for a respectable two hundred feet—a distance that compares favorably with the Wright brothers first effort a few months later.[46] As a result, many people—especially inside Germany—consider Karl Jatho the first to have achieved manned, powered flight.

But was it a real "flight" in the truest sense of the word? Considering the lack of cambering in his wings, it's likely that his achieving lift-off was more a result of his wings acting like sails rather than providing true lift—a necessary prerequisite for the event to qualify as a true flight. Additionally, and unlike the Wright brothers, he could not replicate the feat, making it more of an anomaly akin to

A front view of Jatho's aircraft. Though some believe it flew in August of 1903, it's more likely its wide wings acted like a sail, briefly lifting it off the ground.

[46] In comparison, the Wright brothers first flight on December 17, 1903 covered a distance of just over 120 feet. A second attempt that same day, however, flew a whopping 852 feet—more than four times farther than Jatho's alleged flight.

Mozhaysky's brief flight in 1882 (see chapter nine). For the time, however, it was an impressive attempt and deserves to be recognized as such.

For some reason, however, Jatho never flew the plane again, which was unusual among aviation enthusiasts, who were known for their doggedness even long after their cause seemed lost. Though he had achieved promising results, he apparently grew frustrated that he could not do better and gave up flying, at least for the time being. Writing in his journal, he stated "in spite of many efforts, (I) cannot make longer or higher flights. Motor weak."

He didn't give up on aeroplanes entirely however, and eventually went on to build a more robust plane six years later, but his unwillingness to follow up on his original design by mounting a more powerful motor or reworking the wing structure ensured that none of his designs would ever fly, thereby leaving him a minor player in world of aviation. Had he pursued his ideas with the same degree of enthusiasm he displayed in building his aircraft, he might well be considered today one of the preeminent aviators of the age rather than merely a footnote in history.

What should Jatho's legacy be then? Like the Wright brother's *Flyer*, his equally underpowered plane did make it into the air in front of witnesses several months before the Wrights flew, giving his proponents fuel to fan the flames of controversy. He also took off from level ground without outside assistance of any kind whereas the *Wright Flyer* required a ramp to launch it and a decent sustained breeze to keep it aloft. As such, if simply leaving the ground to stagger into the air for a few hundred feet in what is really nothing more than a engine-assisted glide for a few seconds constitutes "flight" then Jatho—and, indeed, a half dozen who came before him—could make the claim as being the first. However, if it is defined as powered lift in which the craft responds to the controls of the pilot, then the Wrights (who actually made four successful flights on December 17, 1903) must be considered the first.[47]

That's not to say that Jatho didn't contribute something to the world of aeronautics. He was one of the earliest to argue that lying prone on the lower wing to operate an aircraft was impractical and insisted that the pilot be seated, which was to prove prophetic. (Even the Wrights finally conceded the point and adopted pilot seats in their later flyers). He was also the first to suggest that constructed runways would be necessary for flight to be feasible, which also soon bore itself out. As such, the man, even if he didn't truly fly, must be credited as being a person who clearly understood the difficulties with making flight practical, thereby making him one of the great visionaries of aviation.

Richard William Pearse

Another credible claim to having beat the Wright brothers into the air came from a most unlikely place—the island nation of New Zealand—which, being half a world away from either America or Europe, was about as far away from the center of aviation as it was possible to be in 1903. But that's just where some claim that a simple New Zealand farmer turned aviation enthusiast by the name of Richard William Pearse (1877-1953) made a craft that managed to accomplish what the *Wright Flyer* did, only he did it, according to some accounts, as much as eight months earlier.

What's interesting about this particular claim, however, is that it did not originate by the man himself, but was made by others who testified that they witnessed the event firsthand—which is what makes the claim so problematic.

In any event, there appears to be anecdotal evidence that on March 31, 1903 Pearse achieved a powered, though poorly controlled, flight of 350 yards before crashing into the hedge at the end of the field. (Some maintain he made two such flights that year a few weeks apart, both with similar results.) There are also reports he made two more flights that May, one in which Pearse cleared the thirty foot tall riverbank along the Opihi River and flew nearly 1,000 yards

47. The still unconfirmed prospect that Gustave Whitehead may have achieved controlled flight as early as 1901 notwithstanding.

A replica of Richard Pearse's aeroplane on display at the South Canterbury Museum in Timaru.

along the river before his engine began to overheat and he was forced to land on a dry riverbed which, if true, would have been an astonishing accomplishment at the time. But did it happen?

At least the fact that he had actually built an aircraft is a certainty. Numerous witnesses attested to the fact that Pearse had built a simple monoplane with a 21 foot wing-span powered by a homemade two-cylinder 15 horsepower engine. In fact, many commentators noted that it looked not remarkably different from a modern ultra-light, with the pilot sitting just beneath the engine and wing atop a wheeled tricycle landing gear, where he manipulated a set of small movable inset wing panels, a vertical rudder mounted behind a fixed vertical panel above the flat-section wing, and a movable horizontal surface mounted centrally on the rear wing spar. And if any person captured the "can do" spirit of the age, Pearse was that man. A self-taught inventor who showed a remarkable aptitude for mechanical devices, it was said he built the craft almost from scratch at his family's farm near Waitohi, on the South Island of New Zealand. However, while there is little doubt that Pearse had the ability to build such a craft and that he actually did so, the problem remains in determining just when exactly he built it and, even more importantly, first flew it.

Most aviation historians, while they don't deny that Pearse actually flew, generally agree that the flight took place sometime after 1903—and perhaps several years after at that. The rationale is based upon several statements Pearse himself made as well as other circumstantial evidence. First, it appears that Pearse denied that he ever flew before the Wrights did. (Why he would not admit to it if he had actually done so has never been satisfactorily answered by his proponents.) Second, he claims that his interest in aviation began in 1904—more than a year *after* his alleged first flight. Third, some maintain that his design was reminiscent of Santos-Dumont's *Demoiselle*, (see chapter fifteen), which first appeared in 1909. If Pearse used Santos-Dumont's plane as a template, however, he obviously couldn't have flown before 1909[48]. Additionally, Pearse applied for and received a patent for his airplane design in 1907, four years after his alleged flight. Had he actually flown as early as 1903, why such a lengthy delay?

48. Although it is also claimed that Pearse's design, while similar to Santos-Dumont's machine in some respects, remained uniquely his own.

Finally, there is the question of whether Pearse's machine was capable of flight at all. While his engine may have produced enough power to get the plane airborne, the lack of cambering in the wings (the wings being mostly flat surfaces) would have provided little in terms of lift, which is perhaps why, according to witnesses, the plane seemed incapable of gaining altitude or maintaining level flight. In fact, Pearse went on record as saying that he never flew his first plane[49], making the claims that he did either erroneous or greatly exaggerated. Witness claims to the contrary, then (many made sixty years after the fact), must be considered either highly subjective or simply mistaken as to the actual year—especially in light of the man's own admission that he didn't develop an interest in aviation until 1904.

The other problem remains that no contemporary newspaper record exists of the 1903 flights and what photos were taken of the man flying in one of his machines are undated and of such poor quality as to make them difficult to interpret. Additionally, and apparently unconcerned with posterity, Pearse kept few written records of his work. As a result, he received little formal acknowledgment for his work during his lifetime. (There is some evidence he may have preferred it that way, as the man was known for being something of a loner and for being eccentric.)

While no one doubts that Pearse man was a gifted mechanic and inventor, the claim that he beat the Wrights into the air, then—a claim, by the way, that Pearse never made himself or, as far as can be told, subscribed to personally—must be taken with the proverbial "grain of salt". However, in accomplishing so much with so few resources in a land with little in terms of available technology at the time, he deserves to be applauded for his ingenuity and honored for his daring in being the first New Zealander to build and fly an airplane—regardless of just when that might have been exactly.

The Reverend Burrell Cannon

Another person many believe may have flown before the Wright brothers was a preacher and sawmiller from the small town of Pittsburg, Texas, by the name of Burrell Cannon (1848-1922)—a man who, according to the good reverend himself, believed that God had instructed him to build and fly an airplane—which he called an airship[50]—sometime in the fall of 1902.

Born in Coffeeville, Mississippi, in 1848, Cannon had the good fortune of having a father who was himself known for his creativity and mechanical aptitude. (During his lifetime Cannon's father patented a number of inventions, including a marine propeller, a wind-driven water pump, and even a specialized camera.) Burrell apparently shared his father's penchant for invention and went on to obtain a degree in mechanics at Mississippi College in Clinton, Mississippi.[51] Apparently possessing something of an entrepreneurial spirit, Cannon soon decided to leave Mississippi and head to Texas to make his fortune. Apparently appreciating the nation's need for lumber, there he started a number of successful lumber operations in the state's forested eastern region which made him, if not a wealthy man, at least one capable of providing his family with a comfortable living. He was also, by most accounts, a pretty fair preacher as well, who could often be found preaching at area school-houses and community centers when he wasn't busy running his lumber business (or, eventually, building airships).

It was apparently sometime around the turn of the century when the good reverend first became enthused about the idea of building an airplane. Apparently inspired by a passage of scripture he found in the Old Testament book of Ezekiel that read: *"The appearance of the wheels and their work was like unto the color of beryl and...their appearance and their work was as it were a wheel in the middle of a wheel,"*—which he believed was a de-

49. His claim that he didn't fly could be construed to mean he didn't consider the plane successful, not that he didn't *attempt* to fly it.
50. The term "airship" is usually associated with lighter-than-air craft (dirigibles). Why Cannon consistently referred to his craft as an airship is unclear and, at times, confusing.
51. Apparently he also found the time to become an ordained minister while attending the University.

Replica of the Ezekiel Airship on display at the Northeast Texas Rural Heritage Museum in Pittsburg, Texas. (Used with permission.)

scription of a flying machine[52]—Cannon decided that God was telling him to build the very ship the Old Testament prophet was describing. Obviously not a man to take divine instruction lightly, almost immediately he began working on designs for his craft, which he named the *Ezekiel Airship,* with the same fervor he displayed in his preaching.

Fortunately Cannon was a good businessman and knew that in order to get his "airship" off the ground—so to speak—he needed to raise capital, which he did by selling shares in his "Ezekiel Airship Manufacturing Company" for $25 apiece. Quickly raising the $20,000 (around half a million dollars in modern currency) needed to fund the project, by the summer of 1901 he was ready to begin construction on his remarkable machine. Leasing the upper floor of a nearby foundry and working with three hired assistants, within a year he had completed the machine. Weighing in at just over 850 pounds, the huge, circular-shaped wings were constructed around a rigid metal frame and layered with lightweight sail cloth. (Some have since commented the craft looked to have been a cross between a Conestoga wagon and a massive kite, which is not an inaccurate description.) The power plant, at least according to his notes and eyewitness descriptions, was made up of four interlocking wheels, with two larger wheels working in tandem with two smaller, faster turning wheels. The wheels turned a series of interconnected paddles at a rapid rate, thereby creating vertical and horizontal blasts of air that were supposed to move the machine forward as well as provide upward thrust. Control of the machine was provided by varying the amount of airflow on each set of paddles and by adjusting the angle of the sail. To power the machine, Cannon built a custom-made, four-cylinder, hand-crafted engine at the foundry where the craft was being assembled that he described his being a "gas" engine, though he doesn't specify whether he meant "gasoline" or something more like "natural gas" (which was a possibility as there was no fuel tank aboard. It also appears the "gas" may have been contained within the hollow tubing from which the fuselage was constructed).

Finally ready for its maiden voyage sometime in the fall of 1902—no one seems to recall the pre-

[52] A belief shared by many ancient astronaut theorists today.

cise date for some reason—the machine was towed to a nearby pasture on the edge of town and readied for its one and only flight. According to legend, one of Cannon's employees, a gentlemen named Gus Stamps, apparently climbed onboard the craft, fired up the engine, opened the throttle, and—if the handful of witnesses who were present are to be believed—the craft built up enough speed to actually make it off the ground. Unfortunately, the flight was cut short after about fifty yards when Stamps, who of course had never flown before, grew alarmed by the vibration and noise the engine was making and cut the engines, causing the craft to lurch hard to the left and land hard, badly damaging its undercarriage. (It's also said the plane just missed hitting a group of children who had been sitting on a large wooden fence watching the event, but this may be more anecdotal than factual.)

In any case, Cannon's craft was permanently grounded after that. While on might imagine he would have wanted to make repairs and try again, for reasons known only to himself he decided instead to have his airship loaded onto a flatbed railcar and take it on tour. It is speculated he was going to use it for promotional purposes—perhaps as a way of selling more shares in his airship company—but before he got the chance to do so, a storm blew up and the airship was torn from its railcar and completely destroyed. This unfortunate—and hopefully not diabolically orchestrated event—not only resulted in Cannon losing his machine, but in his backers losing their investment as well (which may be one of the reasons very little was heard of the good reverend for the next ten years).

But Cannon was not done yet. Still convinced that God had ordered him to build a flying machine, by 1910 he was back at it again, selling stock in an attempt to produce a second airship. Remarkably, he managed to not only get investors (he must have been a very persuasive man) but, according to reports, he actually built a second craft. How much it resembled his first craft is uncertain as, in keeping with the good reverends' insistence on privacy, there are no records or photos taken of this second machine. (Additionally, by this time flying was not the novelty it was a decade earlier, reducing the press's interest in any flight he might have made.) In any case, the one time Cannon tried to fly the new machine, (or, more precisely, convinced someone to fly it for he would have been in his sixties at the time),

Close-up of the inner workings of the Ezekiel Airship replica currently on display at the Pittsburgh, Texas Northeast Texas Rural Heritage Museum. The process by which it might have flown remains as mysterious to engineers today as it must have been back in 1902. (Used with permission.)

it crash landed after a few seconds and its bottom was torn out by a utility pole. This, at last, was apparently enough to convince Cannon that God did not want his craft to fly after all, inducing the man to finally give up his "flights of fancy" and retire. He died ten years later at the age of seventy-four, never having stepped foot off planet Earth despite his best efforts at doing so.

So did Burrell Cannon actually beat the Wright brothers into the air? It's hard to say. Unfortunately, there is little of substance to prove that the flight actually happened. Beyond supposed eyewitness accounts and a brief mention of the flight in local papers, the prospect that the *Ezekiel Airship* ever flew must be taken by faith.

That there was an actual aircraft is reasonably certain, however, as one photo taken of it survives and there is a paper trail that exists to prove that Cannon actually built the thing. Whether it was capable of flying or not remains problematic. A static replica of Cannon's plane, which is now on display at the Camp County Museum in downtown Pittsburg, though an intricate and beautiful machine, does not appear to be capable of flight.[53] Of course, since this replica was built without Cannon's original blueprints and based almost entirely upon witness notes and a single photo taken during construction, there's no way to tell how true it is to Cannon's original design, which makes it especially difficult to determine whether his would have flown. While it does appear to possess sufficient wing surface to provide enough lift, the motor seems too small to provide the kind of power required to lift the craft's quarter-ton weight into the air. However, like Mozhaysky's steam-powered monoplane of 1882 (see chapter nine), it may have caught a fortuitous gust of wind that inflated the sail-like wings and sent it skyward for a short distance. Unfortunately, without the original aircraft for engineers to study (it's uncertain what became of the wreckage after it was destroyed or what became of the 1911 prototype), it's difficult to know how it might have flown—much less whether it was capable of flight at all. As such, that Cannon's machine flew must, for the time being anyway, be relegated to the status of local legend, at least until someone comes forward with a photo of it in flight or there is something more substantial to consider. In any case, Cannon's ambitious if unsuccessful attempts at manned flight deserve to be applauded for their ingenuity and bravado if nothing else.

William Frost

There have been many who have claimed to be the first to fly. Some of these claims are plausible and require further investigation, while others are too outrageous to be taken seriously. One name that seems to fit somewhere between these two extremes is that of a Welshman named William Frost (1848-1935), who

For some reason, Cannon didn't take many pictures of the craft, either during construction or after it was finished, forcing us to guess at the particulars of its design. Fortunately, one grainy photo survives, giving us our only visual reference to what must have been a truly unique aircraft.

53. Although there are a handful of experts who have been willing to go on record affirming that it would have been flyable.

some insist was really the first man to fly, though most historians suggest otherwise.

William Frost, the simple Welshman who dreamed of flying and, according to some, may have actually pulled it off.

Few people have ever heard of Frost, which is understandable. As a poor carpenter from the small village of Saundersfoot on the southern Pembrokeshire coast of Wales, he was not a likely candidate for fame. Had things gone just a little differently, however, he might well have achieved just that and more.

According to those who champion Frost's aeronautical claims, sometime in the 1890s the man became smitten with the idea of building a flying machine and, despite his poverty—or, it should be noted, any aeronautical or engineering education—he managed to construct a small and possibly flyable machine. Christened the *Frost Airship Glider* (an exceptionally apt name if there ever was one), what he came up with seems to have been a cross between a vertical takeoff aircraft and a hydrogen balloon. Writer Byron Rogers, apparently relying on the book *"A Pembrokeshire Pioneer"*[54] by Roscoe Howells, gives a description of the flying machine as "part balloon (gas cylinders gave it lift), and part powered hang glider, with foot-pedals operating helicopter-style blades." As for how it was controlled in flight, Frost wrote in his patent application: "The flying machine is propelled into the air by two reversible fans revolving horizontally. When sufficient height is gained, its wings are spread and tilted by means of a lever, causing the machine to float onward and downward. When low enough, the lever is reversed causing it to rise upward & onward. When required to stop, the wings are tilted so as to hold against the wind or air and lowered by the reversible fans. The steering is done by a helm fitted to front of machine." The patent seems to contradict itself slightly in the next paragraph, however, when he writes: "The steering is done by a rudder at both ends."

Such inconsistencies aside, according to some biographers, the man actually managed to get his machine to take flight sometime in September of 1896. Observers claim the curious craft flew about five hundred meters (around 1,640 feet) before crashing into some bushes which, if true, well outdistanced the one hundred twenty feet the Wright brothers would achieve in their first powered flight seven years later. Of course, this is an unfair comparison as Frost's craft—assuming it actually flew—achieved lift through its hydrogen envelope whereas the Wrights' airplane was purely heavier-than-air. Yet considering the lack of resources and engineering knowledge available to the man and the fact that he carried out no preliminary studies before building a full-sized machine, it must be considered an impressive feat and one that rightfully belongs in the public record of aeronautical accomplishments.

Unfortunately, and to Frost's great misfortune,

Somewhat speculative illustration of what William Frost's airship/aeroplane hybrid may have looked like in flight. The craft's vertical lift propeller blades are mounted directly above the head of the pilot, which must have made for a very noisy and uncomfortable ride.

54. Llygad Gwalch Cyf, UK, 27 June 2007.

the event was not recorded except in local memories that, when combined with the lack of photographic or physical evidence, makes aviation historians naturally skeptical that the flight ever took place. Even worse, it seems that during the night following the flight, a violent storm destroyed and scattered the flying machine, thereby eliminating any physical evidence of its existence (not unlike what happened to Burrell Cannon's flying machine a few years later).[55] As such, without any evidence that his machine ever flew, Frost's accomplishment, along with his unusual machine, soon evolved into a sort of local legend, which remains its status today.

As for Frost himself, even though he was awarded a patent for his design in October of 1894, he was unable to pay the renewal fees four years later and so had to allow his patent to lapse. Returning to his trade as a carpenter, he lived out the balance of his life in comparative poverty right up to the time of his death in 1935, his accomplishments as an aviation pioneer going largely unrecognized and unappreciated outside his native Wales to this day.

So was William Frost really the first to fly? While the evidence is purely anecdotal (and, with the death of any potential witnesses since the event, unverifiable), there is no way of knowing. All we know for a fact is that he did patent a design for an airship (Patent No: GB189420431) and that claims were made by multiple witnesses to the effect that he flew a craft based on that design on or around September 24, 1896. Beyond that, it's all a matter of faith.

However, even if Frost did somehow manage to build a small airship, and even if he did manage to get it airborne, it was not a true heavier-than-air machine, thereby seemingly disqualifying it from contention as the first manned powered flight. If it did fly, it was due to the hydrogen gas within the envelope and not as a result of lift achieved by air flowing over the wing surfaces. In fact, the only thing that made Frost's machine stand out from other airships of the time was the preponderance of control surfaces and the prospect that it possessed vertical thrusters that, while a novel concept and even revolutionary at the time, did not make his vehicle an airplane any more than von Zeppelin adding motors and rudders to his dirigibles made them airplanes, thereby rendering the entire question of whether he beat the Wright brothers into the air moot; Frost was flying an early balloon and not an aeroplane, and numerous, manned, controllable balloon flights had already been made by that time.

Lyman Wiswell Gilmore, Jr.

At the time the airplane was being developed, it was a popular perception among the general public that people engaged in such work had to be a little crazy. After all, who else but a mad man would strap himself into what was little more than an oversized kite with a loud and dangerous motor and launch himself skyward in the quixotic attempt to soar with the eagles? While the question of just how sane aviation pioneers of the era were remains a source of debate, there did appear to be one personality who may have given the public ammunition for that contention. That man? A largely self-taught engineer from the state of Washington named Lyman Wiswell Gilmore, Jr., (1874-1951).

Lyman Gilmore was easily one of aviation's more colorful characters.

Easily one of the more eccentric personalities of the era, it's unclear exactly when Gilmore's fascination with aviation began. It was reported, however, that he first attempted to fly in a small homemade glider—towed by a horse—near Red Bluff, California, in July of 1894. Unfortunately, the source of this story, as with many of the assertions surrounding Gilmore's efforts, seems to have been Gilmore himself, who in 1898 described himself as an "aerial Fulton" who intended to open the skies in the same way that Robert Fulton's steam-powered ship had opened the waters.

55. It is unfortunate (and inexplicable) that some pieces of the craft couldn't have been retrieved. This would have done much to prove the machine actually existed since no photographs of it are known to exist.

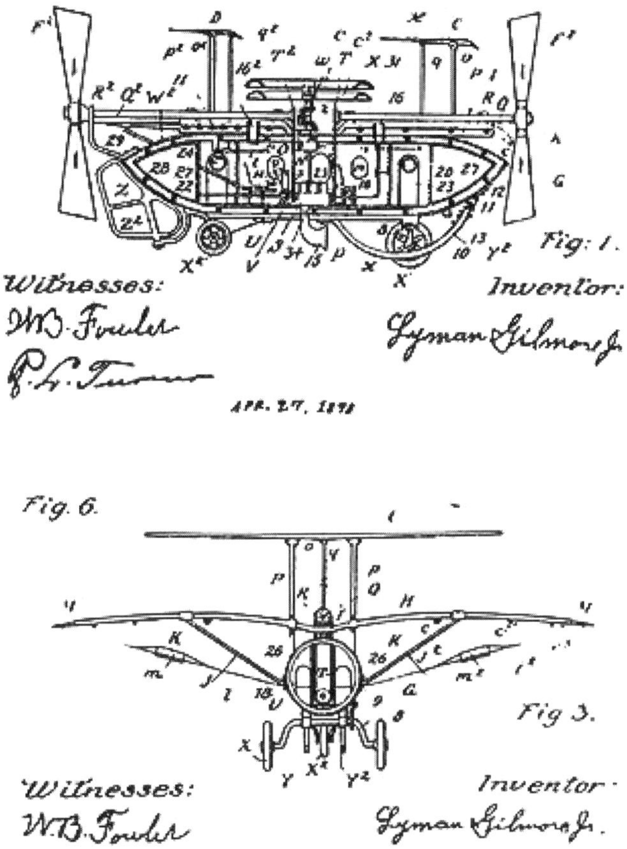

The Gilmore Monoplane design—said to date from 1898 (though the date, in being in another hand, may have been added later—possibly by Gilmore himself). While this design appears more fanciful than practical, at times Gilmore displayed flashes of engineering brilliance throughout his long life.

While he never quite lived up to his own aspirations, Gilmore did apparently apply for and receive a patent for a massive monoplane—which he called his "1898 design"—sometime around that year. (It's not entirely clear whether he actually designed the craft at that early a date as the April 27, 1898 written on it appears to have been done in a different hand.) When he first conceived of his craft seems scarcely relevant, however. What is important is determining if he actually built it and, if he did, whether it was potentially capable of flight.

In examining Gilmore's curious machine, what strikes one straight away is its great size and complexity. Basically a pusher/tractor monoplane with an enclosed boat-like fuselage, the craft looks more like a flying submarine than an aeroplane. It does have a couple of features that were more forward-looking than one might expect, however, such as a tricycle landing gear (which was a rather novel idea for 1898) but for the most part, the design appears more fanciful than practical. For example, it has the rudder attached to the hull precisely as one would see on a boat and its two propellers are ridiculously oversized and lacking any aerodynamic shape.

In any case, many years later Gilmore claimed that in May of 1902 he actually built and flew the plane (or one similar to it), which, powered by one of his patented steam engines, supposedly achieved flight. Unfortunately, no supporting evidence has ever been presented to support this claim (nor is there any to support the contention that the plane even existed outside of Gilmore's fertile imagination).[56] And even if the plane did exist, it is unlikely it could have even lifted its own weight, much less have flown. From a purely engineering standpoint, the plane would have been enormously heavy, and the steam engine Gilmore patented in 1902 was too underpowered to lift such a mass. Further, the lifting surfaces on his design are flat rather than curved, giving the wings almost no lifting capability. As such, it seems likely that the 1902 flight took place only in Gilmore's mind.

Whether Gilmore ever built his "1898 design"

[56] There is a rumor that a photo of the plane sitting on the ground was taken. If true, the authors have been unable to locate it.

airplane is unknown, but that's not to say he never built any aircraft over the course of his life. In fact, he built two planes, both of his own design. The first, built around 1910, was a large monoplane apparently patterned to some degree after Louis Blériot's *Blériot XII* monoplane of 1909. Gilmore even attempted to fly this plane before a large crowd on March 17, 1912, but it proved incapable of lifting its 1,600 pounds off the ground. (This failure appears to have ended the active phase of Gilmore's aviation career and he never built or test flew another aircraft again.)

It was Gilmore's second aircraft, however—built around the same time as his smaller monoplane—that was really interesting. A monster nearly as large as a Second World War era fighter plane (and probably nearly as heavy), the plane was supposedly based on his 1898 design, though with some significant modifications being evident, especially when compared to his original design. For example, Gilmore omitted the two smaller secondary wings from the top of the plane, and the propellers were considerably smaller and more contemporary to those of the time.[57] In any case, like his smaller monoplane, the plane never flew (perhaps Gilmore was enough of a realist to recognize that it was far too heavy to fly and wisely chose to save himself the embarrassment of proving the fact to others) and he kept it, along with his other creation, in a barn for the next twenty-five years, only taking them out on occasion to display to friends. Both planes, unfortunately, were destroyed when the barn caught fire and burned to the ground in 1935, which is a pity as they would have made

A monoplane loosely based on Gilmore's "1898 design" under construction in his garage. Notice the plane's impossibly large size, especially when compared to most aircraft of the era.

Gilmore's completed "1898 design" monoplane (inside the shed) next to his smaller Blériot XII-like monoplane (on right).

[57]. Since there are no available photos of the back side of the aircraft, it's not known if he retained the boat rudder or swapped it for a more contemporary rudder/elevator configuration.

excellent static displays today and likely would have rescued the man from obscurity to some degree.

In any case, the fire apparently destroyed whatever nascent aeronautical ambitions still existed within Mr. Gilmore, for immediately afterwards he left his home in Grass Valley for the gold fields of nearby Nevada City, California, to try his hand at prospecting. Unfortunately, his ability as a miner was no better than his capacity as an aviator and he died penniless in 1951.

So what should Gilmore's legacy be? Clearly, the man was a complex and curious individual who seemed to manifest elements of both genius and madness in his life. It is apparent that Gilmore had the intelligence and engineering acumen to win a number of patents, making him seem to be a man who might conceivably have possessed the skills to build a flyable airplane, possibly even as early as 1902. However, and while a gifted engineer, it also seemed he sometimes had a hard time distinguishing reality from fantasy, making it difficult to take him seriously. Part of the problem appears to have been his unorthodox life style: Gilmore was, by all accounts, a foul-smelling vagabond who never bathed and vowed to never cut his hair or beard. (He believed that if he did so, it would "diminish his strength and vitality.") Gilmore was also known for wearing a trench coat even in the middle of summer in which, local legend has it, he kept all his engineering drawings and technical notes. (It was also said that when he was involuntarily hospitalized for his last illness, the coat containing all his important papers and drawings—which had been sewn into the seams—were burned.)

On the other hand, the man corresponded with other flight pioneers like Samuel Langley and, eventually, the Wright brothers, making it difficult to imagine that such would have been possible had they considered him insane. He was also granted two patents on steam engines, one of which was quite possibly the first radial steam engine in history, which suggests he was a man who possessed a sound and even brilliant engineering mind at times. Gilmore also had a practical side as well, designing one of the first snowplows and even opening the first commercial airfield, Gilmore Airfield, in Grass Valley, California, in 1907. (An elementary school named in his honor sits on the site of the airfield today.) As such, for all his eccentricities, he appears to have been a man capable of innovative thought and practicality as well, which is something the truly insane are generally incapable of doing.

In the end, his long and active interest in aeroplanes should grant him some recognition as a participant in the development of aviation, although (as was also the case with many other early experimenters) his sometimes wild exaggerations and apparent falsehoods have left him without widespread honor. However, Gilmore was certainly one of the more colorful pioneers of aviation who deserves to be acknowledged for his efforts, as misplaced and unappreciated as they may have been.

Conclusions

So what are we to make of these claims that the Wrights were not the first in the air?

It's difficult to say. The proof is always just tantalizingly out of reach. It's almost as though fate enjoys teasing us with just enough evidence to make each claim plausible, only to deny us the proof needed to permit us to render a final verdict. We also suspect there may be something of a nationalist component in many of these claims as well, with each country vying for the honor of "really" being the birthplace of aviation, which makes it even more difficult to gauge the credibility of each claim. As such, it will probably never be possible to determine with any degree of certainty if anyone truly preceded the Wright brothers into the air.

However, the question remains as to whether it really makes a difference if one or more of these claims are true. After all, it was what the Wright brothers did after their first flights that were really important. It was in developing their was, refining them, patenting them, and finally looking for ways to make them commercially and militarily useful that the real value of their efforts lay. That they succeeded in being the first to fly may be debatable, but that they were the first to make flight controllable—thereby rendering it practical—is beyond debate, and that is what they are best remembered for.

Finally, does it really matter who was the first to fly? Flight was something man was destined to accomplish. Whether it was Orville Wright behind the controls, or Gustave Whitehead, or Karl Jatho, or even Lyman Gilmore for that matter, manned flight was inevitable. In that respect, then, perhaps all of these men were the first, if not chronologically, at least in the respect that they were all pursuing the same dream at the same time, and though they all worked independently of each other, each of them was working towards a common goal. This demonstrates like nothing else that the airplane's time had come. Fate was merely content to stand back and watch to see who would be the first to accomplish it.

Chapter Fifteen: The Dawn of a New Era 15

In the aftermath of the Wright brothers' successful flight in December of 1903, it is interesting to speculate as to what might have happened had they failed. What if they had been unable to get their engine to work properly or it proved too underpowered to get them airborne or, worst of all, what if one of their early gliding attempts had ended in tragedy, causing them to give up entirely? Would there have been no age of aviation? Would the airplane have never flown?

Of course not. Without the Wright brothers, the direction aeronautics might have taken may have been significantly altered, but even without their contributions the airplane was inevitable. There were simply too many people working the problem of how to get a man to fly like a bird for the Age of Aviation to not have happened.

But who were these people? While we looked at a number of individuals who challenged the claim that the Wright brothers were the first to fly, there were others who likely would have been the first if the brothers from Dayton had failed. Perhaps the first flight would have taken place in 1905 instead of 1903, or it would have taken place in France instead of the United States, but it would have happened, simply because there were so many determined and gifted inventors, engineers, and enthusiasts out there who were every bit as capable as the Wright brothers proved themselves to be. Therefore, for the sake of thoroughness and because these men deserve to be equally honored for their considerable accomplishments in their own right, it is important we take a look at who these pioneers were and how close they may have been to having been the first men to fly.

Alberto Santos-Dumont

Probably the one man who had the greatest potential to beat the Wright brothers into the air during the first years of the new century was the always interesting Brazilian aviator, Alberto Santos-Dumont (1873-1932).

We looked at some of this man's early work with lighter-than-air airships in a previous chapter and saw how effortlessly he was able to construct and fly small airships around Paris at a time when other lighter-than-air inventors were still struggling to get off the ground. As such, it is not difficult to imagine Santos-Dumont doing the same in regards to heavier-than-air machines, which is, in fact, exactly what he did—though he didn't make this transition until a couple of years after the Wright brothers had already flown.

It's not entirely clear why Santos-Dumont abandoned dirigibles to pursue heavier-than-air craft; it may be he simply grew bored with spiriting about Paris in his tiny airship and was looking for a greater challenge, or perhaps he recognized that fixed-wing, heavier-than-air flight was increasingly grabbing the headlines in the aftermath of the Wright brothers first flight and he wanted to be a part of the excitement. Then again, it may be as simple as the fact that in 1905, French aviation authorities became aware that the rapid development and growing interest in aviation portended an exciting future for mankind and began offering substantial cash prizes for various aeronautical achievements. The first of these were for the first heavier-than-air machines to fly more than 25 meters and 100 meters respectively which, while seemingly rather modest goals from our perspective in the twenty-first century, was considered a tremendous challenged at the time. Undoubtedly, to a man like Santos-Dumont—who had already won the Deutsch de la Meurthe prize in one of his airships (see chapter ten)—such a challenge must have been irresistible, and in 1905 he abruptly turned his full attention to building heavier-than-air machines, apparently abandoning all interest in lighter-than-air vessels in the process.

What the Brazilian engineer eventually came up with was a unique biplane whose overall design continues to fascinate even today. A rather large aircraft for the times, what was most intriguing about it was that it was made up almost entirely of Hargrave cells—a light and tough structure recently developed by the ingenious Australian engineer Lawrence Hargrave (see chapter seven) that allowed for great lift and structural strength with minimal weight—that made it look like nothing so much as a massive box kite. Powered by one of the newest and most powerful airplane engines available at the time—the lightweight but reasonably powerful 50 horsepower Antoinette V-8 engine made by Léon Levavasseur—he named his newest creation the *14-bis*[58] (also known as *Oiseau de proie*—French for "bird of prey"). Beyond its unusual appearance and construction, what made Santos-Dumont's plane even more different from the Wright brothers' aircraft was that it had a wheeled landing gear as opposed to the *Wright Flyer's* skids and could be launched without the aid of a catapult. His plane also had movable surfaces—the precursor to ailerons—that gave him more lateral stability, as opposed to the *Flyer's* dependency on wing-warping. This not only made his craft more practical, but more in line with the direction that airplanes were then evolving towards.

Santos-Dumont's unusual looking 14-bis may have been more box kite than airplane, but it flew.

58. The name came about as a result of weight and balance tests he did on the aircraft when he attached it to the belly of his latest dirigible, the *Number 14.* The test nearly destroyed the airship, but it did provide useful information in regards to what necessary adjustments needed to be made. Because of this, Santos-Dumont named his plane the *14-bis.*

Santos-Dumont behind the controls of the 14-bis, the first airplane to fly in Europe.

Unlike many inventors of the era who were reluctant to test their newest invention in front of large crowds, the flamboyant Brazilian aviator thrived on publicity and so was more than willing to take his creation for its first flight in front of thousands of witnesses—even if it meant that failure would be front page news. And so, on the morning of October 23, 1906, Santos-Dumont climbed into his strange machine and taxied it before a huge Parisian crowd who had come to watch either history being made or a most spectacular crash. Revving up the plane's engine, he pushed the simple throttle forward and quickly built up enough speed to lift the plane from the ground. He was only able to keep it in the air for a few seconds before it came back down hard (damaging the landing gear in the process), but in having flown a distance of distance of some two hundred feet (about sixty meters), the flight was enough to earn him the Deutsch-Archdeacon Prize for the first officially observed flight of more than 25 meters. Even more, it made him the first man to pilot a heavier-than-air machine on the continent of Europe. (It also has the distinction of being the first public demonstration of an aircraft taking off from an ordinary airstrip with a non-detachable landing gear entirely under its own power.)

But the Brazilian aviator was not done yet. Determined to win the 100-meter prize as well, after some quick repairs to the cracked landing gear and after making some minor refinements to the engine and basic structure of the plane, three weeks later he was ready to go for the new record. Once again performing in front of a large crowd of curious onlookers, he struggled to get the *14-bis* to remain in the air long enough to reach the 100 meter (320 feet) mark, but each time he managed to get the machine to lift off, it would quickly lose altitude and land well short of the mark, making it appear as though his machine was simply not up to the task. Santos-Dumont was not a man to accept defeat so easily, however, and he tried repeatedly throughout the day to get the plane to remain airborne long enough to win the prize. Finally, just as the sun was setting, he managed to get his tired plane to stay airborne for an astonishing—at the time—distance of nearly 700 feet (twice the distance needed to win) before he cut power and brought the plane down for a gentle landing. Not only did the flight win him the 1,000 francs prize for distance flying, but he was also recognized by the *Aero Club de France* for setting the first world airspeed record by reaching a speed of nearly 25 mph.

But that wasn't the end of Santos-Dumont's creativity. Moving away from the box-kite configuration of the *14-bis*, in 1908 he designed and built the first of a series of small airplanes which he named the *Demoiselle* ("Damsel" or "Young Lady" in French), which, like the small airships he had built earlier, he designed largely for his own personal use. Considerably different in design to his first plane (and more conventional in appearance), the *Demoiselle* was essentially a lightweight monoplane with a wire-braced wing mounted atop an open-framework fuselage, all of it built around a reinforced bamboo boom for strength. For some reason, though, Santos-Dumont didn't personally fly the first of these planes, but left

Santos-Dumont's most successful aircraft, the Demoiselle, in flight—circa 1909. This was the first plane built for series production, with fifteen being sold. It also had a top speed of over 70 mph, making it the fastest plane in the world at the time.

it to the factory to have it test flown first. This resulted in the plane —designated the *No. 19*—being so badly damaged during takeoff [59] that it was rendered unflyable and was subsequently dismantled. Later variants of the plane—which he flew personally after that—proved progressively more successful, with the model designated *No. 22* even setting a speed record by reaching 75 mph in level flight—a speed at the time as impressive as Mach 1 would be to us today. So successful was his new *Demoiselle* in fact, that the aircraft would go into production to become the world's first series production aircraft. (Supposedly the plane could be constructed in only 15 days at a cost of 7,500 francs though there is no evidence that such was ever attempted.)

Sadly, the *Demoiselle* was to be the last aircraft Santos-Dumont would ever build or fly. Complaining of dizziness and disorientation after a 1910 flight, he was diagnosed with multiple sclerosis and was reluctantly forced to give up flying entirely. Deeply disappointed that his great love—aviation—was no longer open to him, he moved from Paris to the French seaside village of Bénerville (now Benerville-sur-Mer) where he took up astronomy as a hobby. Hounded by local authorities during the First World War, (some of the local citizens who didn't know who he was mistook his German-made telescope and unusual accent as signs that he was a spy who was tracking French naval activity and turned him into authorities), he left France to return to his native Brazil. Becoming increasingly despondent over his worsening illness and depressed by the fact that airplanes were being used in warfare, Santos-Dumont took his own life in 1932 and was buried in the Cemitério São João Batista in Rio de Janeiro—as tragic in death as he was vibrant in life.

Jacob Christian Ellehammer

While the world had long-since conceded the fact that the first heavier-than-air-flight had been accomplished in America in 1903, the title of who was the first man to achieve powered flight in Europe remained more controversial. Although it is generally agreed among historians that Santos-Dumont was the first, there are a few who maintain that the honor really belongs to Danish aviator named Jacob Christian Ellehammer (1871–1946), a man who many believe accomplished the feat some two months before the Brazilian aviator.

Born at Bakkebolle, Denmark, in 1871, Ellehammer began life as an apprentice watchmaker, which is how he developed an understanding of the complex inner workings of miniature devices. Later, he taught himself the principles of the internal combustion engine, knowledge that he later used to design and build his own lightweight and pow-

[59]. This was the aircraft that Belgian aviatrix Hélène Dutrieu, one of the first female pilots in Europe, crashed upon takeoff, largely because it was the first time she had ever attempted to fly. See chapter eighteen for details.

Ellehammer in his unusual flying machine—1906.

erful motors. His motors proved to be so good, in fact, that he began mounting them on motor scooters and selling them, making himself an extremely wealthy young man in the process.

By 1904, Ellehammer, apparently caught up in the aviation enthusiasm of the era and now with enough money from his enterprises to pursue aeronautics full time, he quickly put his considerable engineering skills to use to build a three-cylinder, 9 horsepower, air-cooled radial airplane engine that he attached to a small monoplane of his own design. Unfortunately, the engine proved too underpowered to achieve takeoff, forcing the determined Ellehammer to rework his original engine. It would take him nearly two more years to develop a motor capable of lifting an airplane, but by 1906 he was ready to try again and he mounted his new engine—this one capable of 18 horsepower—to an equally improved version of his earlier, tailless biplane.

What was interesting about Ellehammer's aircraft was not its engine, however, but its unusual design. While the bottom half of the machine was relatively conventional—a diamond-shaped monoplane wing mounted over a reverse tricycle landing gear containing the pilot and engine, it was the top half that was unusual. While the lower wing was fixed, with a movable elevator attached to the trailing edge center section, the upper wing was smaller and nonrigid like a sail. Laying limp upon the top of the lower wing when not in flight, once up to speed it took on the shape of a large kite, giving the machine the appearance of two enormous kites mounted one above the other. While this was a clever way to provide additional lift with minimal additional weight, however, unfortunately it did nothing to help overcome the plane's greatest difficulty: directional control. It seems that while Ellehammer's aircraft had enough wing surface to provide sufficient lift, he had neglected to mount a vertical fin or rudder, giving it no directional sta-

Another view of Ellehammer's unique configuration. The upper wing (or sail) is barely discernable in this picture because it lies flacid until the aircraft picks up speed.

Among the many aircraft Ellehammer built and flew throughout his long career were a triplane (left) and a barely flyable but still impressive—for its time—helicopter (right), making him one of the true pioneers of European aviation.

bility beyond what small degree may have been afforded by its large, steerable tail wheel.

Unlike other aviation pioneers of the era, he also had a unique means of flying his machine. Instead of simply running it up to speed over a straight track, he had it run around a circular runway tethered to a pivoting pole at the center. While this made it easier for him to make modifications and test them without risking the craft getting away from him and being damaged or destroyed, it also made his claim to being the first European to fly more controversial. In any case, after refining and doing a number of unmanned test flights in this way for several weeks, on September 12, 1906, Ellehammer took his place beneath the wings in the craft's pendulum-like seat, gunned the engine to full throttle and, even with his extra weight onboard, managed to get the plane—still tethered to its pivoting pole all—into the air for a few seconds before gently setting it back down.

Though the machine only flew for a few seconds, being that it happened two months before Santos-Dumont's brief flight in the *14-bis*, many people gave Ellehammer credit for the first manned flight made in Europe. However, since the plane was tethered to the ground and did not achieve free flight, it was not officially recognized as such by the *Federation Aeronautique Internationale* and so the accomplishment went largely unheralded outside of his native Denmark. The mere fact that he got off the ground at all, however, was a major achievement at the time and earned him the title of the "first flying Dane."

The flight was far from the end of his Ellehammer's aviation career, however. He continued his aeronautical experiments for a number of years, his most noteworthy accomplishment being the development of an improved biplane in 1908 that was to become the first heavier-than-air craft to fly in Germany. He also experimented with a flying boat, a wheeled monoplane, and, in 1912, he even developed a successful helicopter—the last of his aeronautical achievements. Continuing to aid in the development of Danish aviation until his death in 1946, Ellehammer proved to be so prolific an inventor that by the time of his death, he had more than four hundred patents on a variety of electromechanical devices to his credit, making him in many ways a Danish Thomas Edison.

Traian Vuia

One name it is safe to assume few Westerners have ever heard, especially with when it comes to the early years of flight, is that of Romanian inventor Traian Vuia (1872-1950), a man who, while virtually unknown outside his native Romania, was one of the most celebrated aviators of his time.

Traian Vuia's first flying machine. The sophisticated landing gear and upright seating made this plane more advanced than most aircraft of the era.

One of Vuia's later aircraft. Notice how small his airplanes tended to be, especially when compared to the larger Wright machines.

A mechanical engineer by trade, Vuia began studying aviation early in his career and by 1901 had already designed a flying machine, which he called the "airplane-car". Unfortunately, designing an aeroplane and getting it built are not the same thing, and Vuia found it extremely difficult to find someone willing to bankroll its construction. Rebuffed by numerous aviation enthusiasts throughout France (who perhaps were reluctant to help a Romanian become the first man to fly), and even turned down by the Academy of Science in Paris with the terse comment that building a machine that was heavier than the air was "a chimera," he was forced to rely on his own skills as an engineer to find the funds. Fortunately—and very similar to his American contemporary, Gustave Whitehead—Vuia was a gifted mechanic who was able to design and patent a number of small engines of his own design. While this proved to be a slow method of raising the required funds, it did provide enough to permit him to begin work on his airplane which, in December of 1905—after several years of refining and building his machine—he completed.

Named the *Traian Vuia 1*, Vuia's small, high-wing mounted monoplane was quite sophisticated for the time. The plane was powered by a carbonic-acid gas engine, beneath which Vuia sat centered between four bicycle tires on a light frame. Interestingly, his plane may have been the first to incorporate a car-like steering wheel to affect the plane's various control surfaces, which was a feature that wouldn't be seen in airplanes on a regular basis for many years.

Unfortunately, the aircraft proved to be too heavy to do more than fly a few dozen feet at a time, but the fact that it lifted off the ground at all made Vuia something of a hero in France. Some newspapers even went so far as to claim that he was the first to fly in a heavier-than-air machine (which was a curious claim, coming as it did years after the Wright brothers' first flight). However, and despite the fact that Vuia flew two years after the Wright brothers, some today still consider his to have been the first "true" flight because his machine was able to take off from a flat surface by on-board means without the assistance of an incline, rails, or a catapult, making it many ways superior to the *Wright Flyer.* While that claim may be presumptuous, the fact that he made his attempt in March of 1906—months before either Santos-Dumont or Ellehammer—may have given him some right to the title of being the first European to fly—though the accomplishment remained unacknowledged by aviation authorities at the time (just as it does to this day).

In any case, in tribute to the man and his accomplishments—he later went on to test a number of aircraft over the decades, including a pair of experimental helicopters that contributed much to the development of vertical take-off aircraft in the early twentieth century—Romania's second largest airport is named after him.

Aurel Vlaicu at the controls of his advanced but difficult-to-fly Vlaicu II in 1912.

Aurel Vlaicu

Another Romanian inventor who was a contemporary of Vuia but was, in many ways, even more successful, was Aurel Vlaicu (1882 –1913). He was probably less well known, however, as he came along later than Vuia and had an extremely short career. However, in many ways, his aircraft designs were far superior to his colleague's at the time and may have even proven revolutionary had he lived long enough to develop his ideas to their fullest extent.

Like Vuia, Vlaicu too was a trained engineer and inventor, having earned his degree from the Technical University of Budapest and the Technische Hochschule München in Germany in 1907. Unlike his Romanian countryman, however, his fame came not in being the first to fly—he didn't construct his first manned aircraft, the *Vlaicu I,* until 1910—but in the precision and control his designs exhibited. So advanced were his airplanes, in fact, that his second aircraft, the *Vlaicu II,* won several cash prizes while competing against other well-known aviators of the day, besting the competition in the areas of precision landing, projectile throwing, and tight flying around a pole at the Aspern Air Show near Vienna in 1912.

Sadly, the world never had a chance to see what the gifted aviator might have accomplished as he was killed while flying his beloved *Vlaicu II* over the Carpathian Mountains in 1913. Interestingly, he is buried at the Bellu cemetery in Bucharest—the same cemetery where his colleague, Traian Vuia, would one day join him.

Louis Blériot

While few people had heard of Aurel Vlaicu or Traian Vuia, almost the whole world knew who Louis Blériot (1872 –1936) was. In fact, at one time he was probably the most famous Frenchman in the world and was considered by many then, just as he is today, to be France's greatest aviator, inventor and engineer.

Born in the village of Dehéries near Cambrai in 1872, Blériot studied engineering at the École Centrale Paris. Like most engineers of the era, however, he didn't study aeronautical engineering—which was not a recognized engineering field at the time—but mechanical engineering, which turned out to be more immediately practical and, in Blériot's case, financially lucrative as well. Using his considerable engineering acumen to design and build things like automobile headlights and acetylene head lamps, by the time he was in his

A grim-faced Louis Blériot at the controls of his Blériot XI on the morning of his historic crossing of the English Channel, July 25, 1909.

Blériot standing with his wife next to his battered Blériot XI after it crash landed in the English countryside.

mid-twenties he was already a wealthy man, which is what made it possible for him to begin pursuing his other interest—aviation—full time.

He started in 1900 by towing gliders along the Seine River in an effort to learn as much as he could about flight dynamics before he began tinkering with different designs. Though his earliest attempts at aeroplane design proved to be impractical and, in some cases, even dangerous, by 1907 he had learned enough that he was able to build the world's first true monoplane, the *Blériot V*. Unfortunately, the plane proved dangerously unstable and required a skilled pilot to control it, but Blériot eventually fixed this stability problem in his later, successful *Blériot XI* model.

It would be in the *Blériot XI* , in fact, that he would achieve his greatest fame. In 1909 the London *Daily Mail* offered a thousand-pound prize to the first man to make a successful crossing of the English Channel in an airplane. Of course, the channel had already been crossed by air over a century earlier by Jean-Pierre Blanchard in a hydrogen balloon and repeatedly since by other balloonists (including at least one woman balloonist), but the feat had never been attempted in a heavier-than-air machine. Additionally, Blanchard had made the crossing from the English side of the channel so, perhaps in a belated desire to reclaim French pride, Blériot decided he would make the crossing from the French side in his little monoplane. As such, on the morning of July 25, 1909, he strapped himself into the cockpit of his fragile little plane and took off from Les Baraques (near Calais), intent on reaching the English coastline at Dover, twenty-two miles away.[60]

At first the flight appeared headed for tragedy when the plane, powered by a 25 horsepower, 3-cylinder Anzani radial engine, barely had enough power to clear telegraph wires at the edge of the cliff near the runway. Blériot was a good enough pilot, however, that he deftly managed to avoid the wires and soon he was winging its way out over the chan-

60. While 22 miles (total flying distance was probably closer to 30 miles) may not seem that far by modern standards, it was a considerable distance to fly for an airplane in 1909. Additionally, though flights of a several dozen miles had been made by this date, none of them had been attempted over open water, making crossing the channel an especially risky proposition at the time.

nel. After his close-call on takeoff, things appeared to be going better for the aviator as he flew along at a speed of approximately 40 miles per hour some 250 feet over the frigid waters of the Atlantic, but soon things began going wrong. Encountering stormy weather over the always unpredictable channel, Blériot soon lost sight of any landmarks and for the next ten minutes he flew practically blind, largley lost over the open sea and without a ship in sight to rescue him should his engine fail and he plunge into the chilly waters below. Fortunately, however, his luck held and he eventually found a break in the weather through which he spotted the English coastline and soon he was over dry land.

Completing the treacherous crossing wasn't the end of his difficulties, however, for no sooner had he reached the coast when he was hit with powerful gusts of wind that caused him great difficulty in keeping on course. Even worse, he encountered a series of rain squalls that threatened to stall his engine. Desperate to set down while he still had some control, he finally located a suitable field, cut his engine, and glided in for a crash landing. Though he seriously damaged his landing gear and shattered the propeller upon hitting the ground, the plane remained largely intact and Blériot walked away from the crash uninjured.

Naturally, the flight made Blériot an immediate celebrity and the premiere pilot in all of Europe, as well as something of a *cause célèbre* in Europe for the rest of his life, which he immediately used to his advantage to secure the loans required to expand his aircraft factory. Blériot's aircraft firm soon became the largest in France and he went on to design and build hundreds of aircraft throughout his long career. (Probably the best known aircraft he produced was the famous *SPAD S.XIII* fighter, which became the premier fighter in the French air force during World War I.) Perhaps he is best remembered, however, for being one of the first men on hand to welcome Charles Lindbergh moments after he landed at Le Bourget field in Paris in May of 1927, after having completed his gruelling thirty-six-hour transatlantic flight. When the two men shook hands, they could claim to be the only two men in history who, though separated in age by thirty years, had each become famous by crossing large bodies of water.

Conclusions

The point to be appreciated here is that the development of heavier-than-air flight was not attributable to any single person, but was the product of a conglomerate of inventors, thinkers, engineers, and aviation enthusiasts, all working around the same time. Each built on the successes of others or refined what had previously been attempted, sometimes even rediscovering the same ideas others had already uncovered years before. Of course, this often resulted in redundancy and even, on occasion, futility, but all of it was necessary to the process of learning the hard lessons the world of aviation needed to learn if it was ever to achieve the maturity required to make the world of easy and safe air travel possible for us today.

Yet each of them shared the dream of flight and pursued that dream with the same passion and determination that is the hallmark of all great inventors throughout history. To these men, then—both those who succeeded brilliantly and those who failed miserably—the world owes much.

CHAPTER SIXTEEN: AVIATION ODDITIES 16

The history of early aviation was not only about early triumphs and dismal failures, but innovative ideas and unconventional thinking as well. It was the different and, often, bizarre ways aviators of the era came up with to solve the many issues inherent in keeping a linen and wooden aircraft in the air that made it such an interesting time in history. While some may consider exploring these unusual ideas a departure from the theme of this book, we believe an examination of some of these always interesting and often remarkable ideas is necessary, if only to appreciate the rich history that aviation brings us as well as underscore the fact that there is always more than one way to solve an engineering problem.

Of course, none of the problems early aviators faced in the first decade of the twentieth century was greater than the question of how to make an airplane stable in flight. Flying was an extremely dangerous pursuit at the time, not because of the speeds—which were mild even by the standards of the times—but because of the tendency for aircraft to want to nose over or roll onto their side with a minimum of provocation. This chapter, then, is a look at some of the more novel approaches to addressing this problem undertaken by a small cadre of unconventional designers, many of whom looked to nature for a solution, and in so doing created a line of aircraft the world had never seen before or is likely to be seen again. Though they are considered oddities of their era and were largely unappreciated at the time, they are deserving of examination—as are the men who designed and flew them—if only to appreciate the extraordinarily diverse road aviation took during its long and torturous evolution.

John William Dunne

John Dunne, an aviation pioneer with a most unusual resumé.

Forgoing the common notion that aircraft required a fixed, vertical rudder surface to maintain stability, there were those inventors who dared to push the envelope of what was considered practical by designing—often independently of each other—a type of plane that flew—and, in many cases, flew quite well—without a rudder. One of the earliest and perhaps most successful of these innovators was a man who would later be better remembered not for his work as an aviation pioneer, but for his work in the field of parapsychology!

At first glance, British Army officer John William Dunne (1875-1949) did not seem to be the type of person destined to become one of the early pioneers of aviation, especially as the man had no formal technical or engineering training. In fact, he was to become more famous for his later writings in parapsychology—a subject on which he wrote several books concerning his theories on precognitive dreams and the nature of time—than he was for his airplanes. Yet somehow this author became known for designing a number of inherently stable, tailless aircraft that ultimately proved to be remarkably successful.

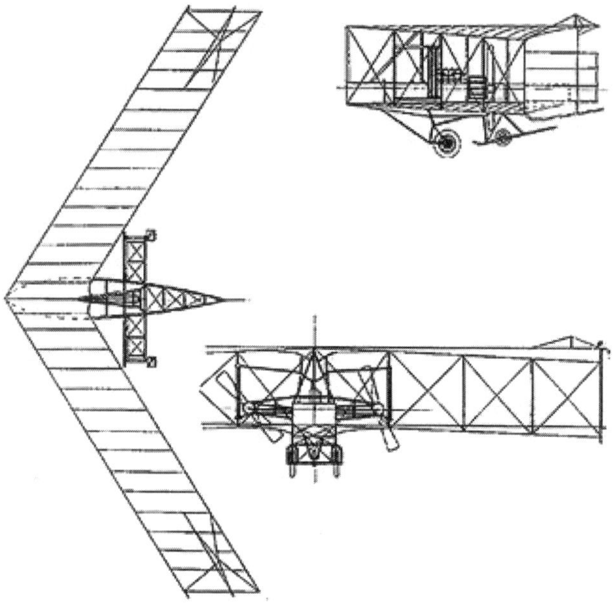

Dunne's D.5 biplane, his first successful tailless design. The unique V-shape design made his aircraft among the most aerodynamically stable in the world.

An RAF version of Dunne's tailless biplane proved to be so easy to fly, instructors felt it lulled student pilots into the mistaken belief that flying was safe. It also lacked the speed and manueverability neeeded to make it a good aerial fighter.

So how does a man known for theories regarding parapsychology become a successful airplane designer? Apparently, it was all the result of reading a book.

From an early age, Dunne was fascinated by science, especially after reading Jules Verne novels as a boy. Once, at the age of thirteen—and perhaps overly influenced by Verne's writings—he dreamed of building a flying machine that needed no steering and would right itself regardless of wind or weather. Why he had that particular dream—being so specific and precise as it was—he had no idea, but as he grew older, he began to study the idea more seriously. Eventually, through his readings, he learned about the Zanonia seed—a large "V"-shaped seed native to Indonesia. What was unusual about the seed was that it was capable of being carried great distances by the wind due to its amazing flying qualities, leading Dunne to conclude that an aircraft built around its basic shape would be extremely stable. Unfortunately, he lacked the resources to actually build such an aircraft and probably would have given up on the idea entirely had not fate intervened in a most un-

usual way. Unbeknownst to Dunne, a curious chain of events were about to unfold that would one day make his tailless airplane a reality.

It all started in 1901 when Dunne was serving with the British Army. While stationed in South Africa during the Second Boer War, he came down with typhoid and was sent back to England to recuperate. While contracting a serious illness is usually not considered fortuitous, in Dunne's case it turned out to be exactly that, for it was while he spent the next few months slowly regaining his health that he found the time to undertake a serious study of the science of aeronautics. Revisiting his earlier idea of using the Zanonia seed as a type of template, he eventually came up with the design for a tailless, V-shaped winged airplane he believed would be inherently stable. Further, encouraged by a family friend by the name of H.G. Wells—another noted science fiction author of the period—Dunne set about enthusiastically testing a number of small models with his unique configuration.

Eventually Dunne was well enough to resume his duties in the British Army and was prepared to return to South Africa when fate again intervened. He was diagnosed with a weak heart, thereby making him unfit for combat duty and effectively ending his career as a field officer. Dunne remained in the army, however, though confined to less physically demanding roles. Aware of his interest in aeronautics and in a rare stroke of good sense, the British army assigned him to its army balloon factory in South Farnborough, which proved to be an ideal billet for a man with Dunne's intelligence and ingenuity. Even more fortuitous, the base was under the command of Colonel John Capper, an experienced engineer and a man who saw in Dunne's rudderless aircraft designs a great deal of promise. Even better, as it was evident that the airplane showed considerable military value, the British army became interested in acquiring one for study and turned to Dunne who, with Capper's guidance and support, began to work on designing Britain's first military airplane.

Before building a full-sized aircraft, however, Dunne decided to build a series of small gliders to refine his V-shaped wing design—which he frequently referred to as "an arrowhead minus a shaft"—first. Working on a number of progressively larger models, by the spring of 1907 he had built a machine large enough to carry passengers, which he began flight testing under the utmost secrecy near the village of Blair Atholl in the Scottish Highlands. While the unpowered craft at first seemed reluctant to take to the air, after a series of modifications it finally became airborne, making a flight of around eight seconds before slamming into the ground. (Unfortunately, it landed hard enough that it slightly injured Colonel Capper, who had been riding along with Dunne on the flight. He wisely chose to observe the flights from the ground after that.) While the craft didn't quite live up to expectations, it did—at least briefly—demonstrate the stability Dunne considered so essential, making the flight a success and inspiring him to continue to work on a number of both gliders and powered aircraft over the next two years, several of which proved marginally successful. None of them proved to be able to fly for more than a few thousand feet or stay aloft for more than a few minutes, however, and eventually the British army council decided to discontinue funding Dunne's experiments, thereby ending his work for the military.

While that should have been the end of the story, Dunne was not so easily dissuaded. Determined to

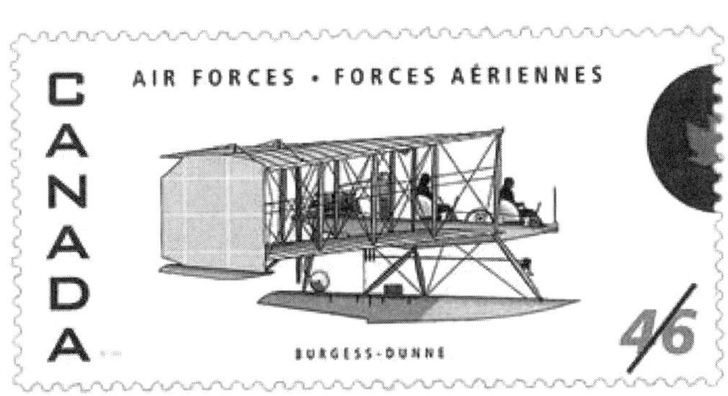

While the Burgess-Dunne seaplane that was sold to the Canadian Aviation Corps in 1914 didn't quite make it into the skies over France, it did make it onto a stamp issued by Canada in 1999.

John William Dunne at the controls of his D.5 biplane at Eastchurch, England — 14 June 1910.

see his tailless airplane concept through, he left the army and formed a small company of his own, the *Blair Atholl Aeroplane Syndicate*, to continue his experiments free of government oversight and interference. The result was that he now had the opportunity to rework his original designs, which soon paid off when he built the *Dunne D.5* in 1910—the airplane that was to validate his tailless airplane theories.

A vast improvement over his earlier designs, like his previous models, the *D.5* was a tailless biplane, with sharply swept back wings. It also had a boat-like nacelle that housed the pilot—with additional room for a passenger—while an engine located at the rear of the nacelle drove two pusher propellers. The key to the success of the *D.5,* as it would eventually prove to be for all of Dunne's designs, were in the twist and the camber designed into its swept wings, which gave the plane a remarkable degree of inherent stability.

Intent on proving just how stable his design was, on December 20, 1910, Dunne took off from Eastchurch, the site of many previous successful flights, in front of crowd of onlookers that included none other than Orville Wright, who had been visiting England at the time. Making a number of slow circles around the base, Dunne quickly demonstrated that the plane was so inherently stable that it could be flown for an extended period of time without handling the controls. To emphasize the point, Dunne was seen scribbling furiously on a piece of paper while the plane flew along under perfect control. It was an impressive display of both the plane's stability as well as Dunne's confidence in his design, though it very nearly ended in tragedy when due to a momentary lack of attention, the plane just missed hitting a windmill.

Although Dunne had proven the stability of his tailless airplane, the *D.5*, his tailless design turned out to be relatively inefficient compared to more conventional aircraft of equal horsepower. It also turned out that the inherent stability so coveted by Dunne was incompatible with the handling characteristics desired by military pilots, who valued speed, maneuverability, ease of handling, and superior performance to basic stability. In fact, it turned out that the ease with which the plane could be flown was even considered dangerous for student pilots as it lulled them into assuming that flying was safer than it really was. As a result, while the Canadian Royal Flying Corps did eventually purchase one of Dunne's planes for evaluation, it was becoming increasingly obvious that the military's interests lay in other, more conventional designs.

Dunne continued to work on his designs for another three years, however, until ill health finally forced his retirement from aviation, at which point he sold the rights to his designs to an American aircraft designer and went on to pursue his parallel interest in parapsychology (an interest which may have been induced by the aviation dream he had as a boy). However, Dunne's ideas would not fade away any time soon, for the man he sold the rights to was W. Starling Burgess, a highly successful aviator and designer in his own right, who later went on to produce a number of successful land and seaplane variations based off, in part, Dunne design.[61]

[61] It was one of these planes that was sold to the Canadian Aviation Corps in 1914 to become the first aircraft in that organization's history.

After two years of trial-and error testing, the Boland brothers produced this tailless biplane, which first flew successfully in 1910. The jibs used for turning the airplane can be seen on the wing tips.

Assessing Dunne's impact on aeronautical history is difficult. While his tailless design never caught on, some of his theories regarding inherent stability proved to be valid and many designers have since benefited from his far-sighted experiments. England's G.T.R. Hill and America's John K. Northrop, two of the more renowned designers of tailless aircraft, often referred to Dunne when discussing the basic problems of maintaining stability and control in a tailless configuration. In fact, the Northrop YB-49 flying wing bomber of the late 1940s and today's modern B-2 stealth bomber are two examples of tailless aircraft that may have been a by-product—at least to some extent—of Dunne's previous experiments.

In the end, Dunne's aircraft were successful because of his brilliant use of aerodynamic innovation in wing design, while his determination to prove the stability of his tailless design and the devotion he showed in building his aircraft made him a true pioneer of aviation. Had he attempted to apply his ideas to the civilian aviation market in the years following World War I—where inherent stability and ease of handling would have doubtlessly found a more favorable reception—he might well have become an extremely successful aircraft pioneer.

The Boland Brothers

While the Wright brothers are undoubtedly the most famous set of siblings in aviation history, they weren't the only brothers who were working out the problems inherent to flight during the first decade of the twentieth century. Three New Jersey brothers, James, Frank, and Joseph Boland—who incidentally also ran a service garage for bicycles and motorcycles just as the Wright brothers had—began investigating the possibility of building an airplane as early as 1907.

What is unique—and some might say rare in history—is that the three brothers turned out to posses the perfect combination of ability, organizational skill, and business sense to make it all work. Oldest brother Joseph (1879-1964) had the mechanical skills required to transform ideas into tangible, workable form; brother Frank (1880-1913) supplied the enthusiasm, ingenuity, and self-taught flying ability; while youngest brother James (1882-1970) brought the business acumen so often lacking in ventures of this sort, giving them the perfect combination of creativity and practicality to lay the foundation for a successful company. The problem for the Bolands, however, was that they didn't seem to be particularly interested in learning what other aviation pioneers had already learned and so the brothers didn't do much by way of research, instead preferring to build their tailless airplanes without drawings, reference to what aeronautical studies existed at the time, or advice. Naturally, this resulted in a number of early set-backs, but by not simply following what other inventors had done before, it also allowed them to come up with unique solutions in regards to controlling their aircraft. Even more remarkable is that it appears the brothers stumbled upon the idea of the tailless airplane concept at almost the same time that John Dunne in England was working on his tailless aircraft designs, even though there is no evidence the men ever met or even knew each other. Their approaches weren't identical, however; the Bolands' planes maintained a straight wing form whereas Dunne's had a v-shape, but in the respect that both Dunne and the Bolands placed their lateral control surfaces on the tips of the wings rather than at the back of the fuselage like most aircraft, they were very similar in both function and form. Also like Dunne's designs, the Bolands' plane lacked anything even remotely like a rudder or elevators—which the brothers considered to be inferior to wing warping and ailerons to achieve stability. They preferred instead to leave the handling and maneuverability issues to a patented system of lateral controls called a jib. In essence, lateral control was provided by elliptically shaped surfaces mounted between the outer ends of the top and bottom wings. Each jib was pivoted on an axis from the lower front strut to the upper rear strut and was movable inward in one direction only. The operation was similar to steering an automobile: a control wheel was turned in the desired direction, the jib on that side was pulled in, and the aircraft banked and turned. In essence, what the Bolands did was simply move the vertical control surface of a centralized rudder to the outer edges of each wing. Control in elevation was provided by a curved control surface located in front of the wings that responded to the movement of the control column; pushing forward caused the craft to dive; pulling it back induced it to climb. The real beauty of the system was that in providing more than a single jib, it gave an aircraft greater lateral control and introduced a degree of redundancy in its control surfaces that could keep the plane airborne and stable even if it lost one of its jibs, as opposed to a traditional design in which a plane would quickly lose control were its single rudder damaged.

As the Bolands continued to test and improve their designs, word of their success spread. Orders for aircraft and engines began to arrive, as did both the curiosity seekers and the more serious investors. In 1911, their unique jib control design had come to the attention of the Wright brothers who, always on the lookout for patent infringement on their aileron design, sent Wilbur to investigate. Paying an impromptu visit to the brothers' workshop to determine if the jib control infringed on their patents, he was quickly satisfied that they did not and even went on to praise the Bolands for their original work which, coming from one of the most renown aviators of the era, was quite a compliment.

The Bolands continued to work on their unusual planes for the next few years, constantly making minor improvements to their designs. The Bolands' dreams came to a tragic end, however, when on January 3, 1913, Frank Boland was killed during an exhibition flight in Trinidad. As sole pilot and chief enthusiast of the group, his death was a crushing blow to the company, and though Joseph quickly took over control of the venture, the brothers had lost much of their direction and enthusiasm for the venture. Though they did manage to come up with a tailless flying boat in 1913, by the next year they were ready to turn control of the company over to others, and in 1914 the newly formed *Aeromarine Plane and Motor Company* of Keyport, New Jersey, took over the exclusive manufacturing rights of Boland airplanes and engines. Though Joseph and James remained with the new organization for a time, it seemed their unique jib control system was unable to compete with the simplicity of the traditional rudder and elevator tail assembly and little more was heard of the Boland tailless airplane again.

Preston Watson

Another man who was to make a small but important contribution to the evolution of flight was a Scotsman named Preston Watson (1880-1915), who would be known for two things: possibly being the first man to have flown an airplane (beating the Wright brothers in the process) and for building a most unusual type of airplane. The first feat is merely a legend but the second is a fact and one of the more interesting engineering configurations ever seen. First, let's look at the legend.

Some claim that Watson built and flew a small monoplane sometime in the summer of 1903 near his home of Dundee and, later, on the banks of the Tay River near Errol, Scotland which, if true, would have made him the first man to fly in controlled flight. Unfortunately, there seems to be little evidence to suggest the man did any such thing. The only accounts of the feat were made by his brother, James, in 1953—a full half century after the fact—a claim that was later discredited by renowned aviation historian Charles Gibbs-Smith in his book *The Invention of the Aeroplane*[62]. Further, in an article published in the December 1955 issue of the magazine *Aeronautics*, James himself maintained that he had been referring to a glider his brother flew in 1903, not to a powered aircraft, which would have been far more likely, especially as Preston would have only been twenty-two years old at the time. As such, and despite the insistence of Watson's fans, there is no solid evidence to support the claim that he flew before the Wrights—a claim which, it should be noted, Watson never made himself.

While the claim that Watson was the first to fly remains unsubstantiated—and is likely to remain that way—what cannot be disputed is that he did later build a number of unusual aircraft—three in fact—one in 1909, one in 1910, and a third in 1913, (of which only the last two got airborne under their own power). Of course, by 1910 manned flight had lost much of its novelty, making his successful flights of only passing interest. What made Preston Watson noteworthy, however, was not his ability to fly, but the unique configuration of his airplanes: Watson built all his aircraft with a rigid lower wing and a second, smaller upper plane—sometimes called a "parasol plane"—which could be tilted or rocked independently to either side and so cause the machine to bank to the right or left, permitting him to dispense with the need for a movable rudder to correct side-slip. Additionally, the tail of the plane was fashioned like a box kite, which also helped to support the machine in the air. While odd looking, this

One of Preston Watson's unique "parasol" planes with its movable upper wing. The ability of the upper wing to swing freely made it one of the most unusual aircraft of its era, but also remarkably stable as well.

62. Taplinger Pub Co, UK, June 1966.

design actually turned out to be a more structurally sound method of control than the twisting and wing warping methods employed by the Wright brothers. (In fact, it proved to be such a superior solution to the problem of lateral control and stability that it won Watson a French award for improved stability in an aircraft.)

While it appears that the belief that Watson flew prior to the Wright brothers is more a by-product of Scottish nationalism than aviation fact, no one can doubt that the man was an innovative thinker—as evidenced by his unique parasol wing concept—and a man of apparently considerable personal bravery as well. This was especially well demonstrated at the start of the First World War, when Watson, although thirty-four years of age, volunteered for service with the newly formed Royal Naval Air Service. Tragically, just two months after obtaining his commission, he lost his life when the service plane he was piloting exploded in mid-air over Eastbourne, England, ending the life of one of aviation's most gifted minds and courageous souls. That Watson—who will always be remembered affectionately by his countrymen as the first "Flying Scot"—deserves to be recognized as a pioneer of aviation, then, is not in doubt—regardless of what date he first flew an airplane might have been.

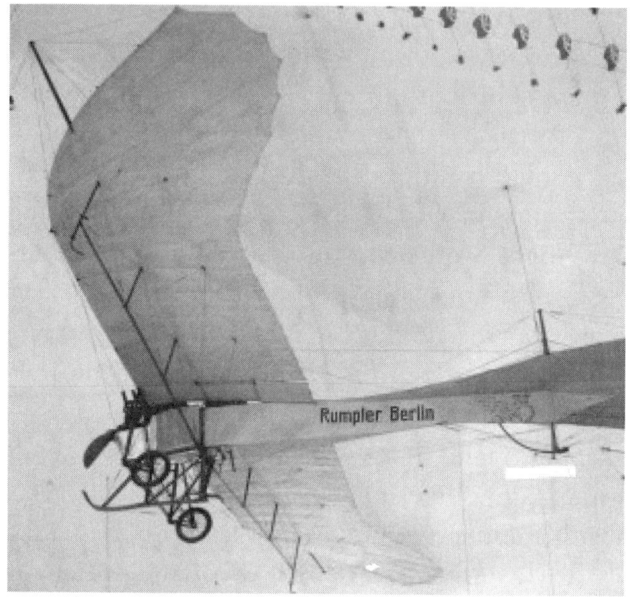

Igo Etrich's mass-produced Taube monoplane. Few aircraft looked more like a bird than Etrich's masterpiece. However, they proved to be too fragile to handle the rigors of aerial combat and were soon replaced by sturdier and more powerful aircraft. This one hangs in the Old Aeronautical Hall of the Deutches Museum in Munich, Germany.

Ignaz "Igo" Etrich

It's curious how the same idea seems to occur to different designers at the same time, often with little or no collusion between them. Such seems to be the case with the tailless airplane concept and the Zanonia flying seed, one or both of which were driving ideas behind the designs of Karl Jatho, John Dunne, the Boland brothers, and, finally an Austrian flight pioneer, pilot, and fixed-wing aircraft developer by the name of Ignaz "Igo" Etrich (1879-1967).

Born on Christmas Day, 1879 in Trutnov, Bohemia, Ignaz—called "Igo" for short—Etrich showed an early interest in aviation, especially after he came in contact with the works of Otto Lilienthal. Fortunately, young Igo was blessed with a father who shared his interest in aviation and together they built a laboratory for developing aeroplanes. At first they simply experimented with some gliders his father had acquired, but after Etrich read a book about flying seeds, he became convinced—just as Jatho, Dunne, and the Bolands had been—that the Zanonia seed was nature's most inherently airworthy design. Working from that premise, in 1903 Etrich developed his first gliders based loosely on the seed's basic shape. Working with noted Austrian aviation inventors Franz Xaver Wels and Karl Illner, two men who themselves would eventually become very important for the future development of aircraft in Europe (in 1906 Illner would be the first Austrian to fly an Austrian-built glider), together they built a tiny pilotless plane called the *Praterspatz* (German for "Sparrow of the Prater")—also known as the *Etrich 1*—in an effort to try out his design. Due to insufficient power provided by the 24 horsepower motor, however, the little plane was unable to achieve flight, even after the men tried different configurations and power plants. Deciding that simply adding a more powerful motor to the Zanonia wing design was not the solution, Etrich did what most successful inventors did and changed course. Again turning to nature for the answer, he simply added a bird-shaped tail to his design, the result being the creation of the passenger-carrying *Taube* (Dove) in 1910. This new design, this time aided by a more powerful and reliable 65 horsepower motor, proved to be the solution and Etrich flew the plane successfully a number of times. As if to remind the aviation community that flying was still far from routine, however, during one of his flights, Etrich nearly lost his life when his *Taube* crashed tail-first to earth, nearly breaking his spine. It was a lesson the man fortunately took to heart and he never flew again, turning all flying duties over to his skilled partner, Karl Illner.

Despite the aircraft's occasional fickleness, however, the *Taube* became the prototype for a class of aircraft Etrich and his partners would build in a bewildering array of versions for both civil and military use over the next few years. In fact, between 1910 and 1914, six hundred of these aircraft would be produced by no fewer than 54 manufacturers in 137 different configurations, making it nearly impossible to find any two that are identical. All were easily recognized, however, by their distinctive Zanonia-shaped wings and dove-like tail, and all possessed the inherent stability that had originally attracted Etrich to the Zanonia design.

Shortly after the start of World War I, however, it quickly became apparent that the *Taube* was too slow to be an effective fighter plane and the aircraft were withdrawn from front-line combat units by the German air force. However, it proved to be an effective training plane, and was used throughout the war in that capacity, often becoming the very first aircraft most German fighter pilots flew.

The *Taube* wasn't the end of Igo's contribution to aviation either. In 1912 he founded his Etrich-Fliegerwerke (Etrich flying works) company in Liebau, Germany (Lubawka, Poland today) where he designed the first passenger plane to have a totally enclosed cabin, the *Luft-Limousine*. After the First World War, Etrich went to the newly-founded nation of Czechoslovakia and built another plane: his *Sport-Taube*[63] which, according to legend flew faster—despite being equipped with only a 40 horsepower engine—than the Czech military planes of the time. While this may be but a bit of hyperbole, if true, it certainly does not speak well of the combat capabilities of the fledgling Czech air force at the time.

Igo remained active all his life, living to the ripe old age of eighty-seven before finally passing away at his home in Salzburg, Austria in 1967. In his death, however, he left a lasting legacy through his contribution to the evolution of aeronautics in Germany in the years leading up to the First World War and beyond.

Conclusions

While to many these unusual aircraft may be considered mere oddities or, at best, but footnotes to history, they demonstrated the spirit of innovation that was such a big part of the early years of aviation. It was that spirit that many maintain has been lost today in a world of computer-aided drafting and 3-D modeling that takes all the guesswork and, perhaps, the human element out of aircraft design. It's difficult to imagine today—with every design being carefully tested in virtual space before it ever reaches an airport tarmac—that men a century ago trusted their very lives to unproven and sometimes even fanciful ideas. Theirs was a world that refused to recognize convention or listen to the "experts" who knew best how to build a "real" airplane; instead, they trusted their instincts and, in doing so, succeeded while their detractors merely watched from the sidelines and debated among themselves why their ideas were preposterous. It also underscores the degree to which aviation was a product of human ingenuity and personal courage, two elements that often seem to be missing from modern endeavors today.

63. This aircraft can still be seen today on display at the Technischen Museum in Prague.

CHAPTER SEVENTEEN: THE REMARKABLE WORLD OF GLENN CURTISS 17

While the contributions made by the Wright brothers to the world of aviation is considerable and well documented, there was another man whose contributions, though later than those of the Wrights, were in many ways as important. In fact, it could be maintained that without his later work both on aircraft and their power plants, the advances in aviation during the first decades of the twentieth century would have been slowed considerably, making him, in many ways, as important a figure to the world of heavier-than-air flight as any pioneer of the twentieth century. The man's name was Glenn Hammond Curtiss (1878-1930), and since no history of the early years of aircraft development would be complete without examining his amazing career, it will be necessary for the sake of thoroughness that we do so.

Born in Hammondsport, New York, Curtiss didn't start out developing aircraft, but more-or-less fell into it through a series of associations and coincidences. As a young man, Curtiss worked as a Western Union bicycle messenger, which was when he developed the strength and skills to become a powerful bicyclist. He soon put those abilities to the test by entering a number of bicycle races, where he quickly developed a reputation as a fearless racer. His success on the track proved financially lucrative as well, and he soon acquired enough money in winnings to open his own bicycle repair and retail shop. (It's curious how many American aviators started out working around bicycles before trying their hand at aviation. It almost appears that many of the technical aspects of bicycle design had inherent aeronautical equivalencies, making the transition from bikes to planes a seemingly natural progression.)

But Curtiss' interests were not limited purely to powerless bicycles. By the turn of the century, internal combustion engines were becoming smaller and more available to the general public, and Curtiss soon began working on the idea of motorizing his bicycles. Purchasing a number of engines, he mounted them to bicycle frames and in doing so created some of the earliest motorized bikes in America, which proved to be especially popular with the buying public. But more than the financial success his new venture brought him was the fact that, in being naturally mechanically inclined, he soon mastered the intricacies of the internal combustion engine and began designing and building his own lightweight and powerful motors. This combination of power and mechanical aptitude soon led Curtiss to begin building some of the first motorcycles in America, which quickly developed a reputation for being some of the finest and fastest motorcycles in the world. As such, in a matter of just a few short years, Curtiss went from being a simple bicycle shop owner to a very wealthy man, all because of his gift for machinery.

More than simply a designer and builder, however, Curtiss was also a daredevil who enjoyed pushing the limits. To that end, in 1903 he set a motorcycle

Heavily retouched composite illustration of Glenn Curtiss on a custom-built, eight-cylinder motorcycle of his own design. The bike earned him the title of the "Fastest Man in the World," when it reached the astonishing speed of 136 mph at Ormand Beach, Florida in 1907.

land speed record of 64 miles per hour—a remarkable feat at a time when the fastest trains could barely manage 60 mph. Four years later he would break his own record and earn the title of the "Fastest Man in the World" when he reached the astonishing speed of 136 mph at Ormand Beach, Florida on a custom-built, eight-cylinder motorcycle of his own design. While he would hold the title only a short time (it would be broken in 1911 by an automobile), it was one he would carry the rest of his life.

During this period, Curtiss' reputation as an engine designer brought him to the attention of the aviation community, which was always on the lookout for reliable, lightweight, and powerful engines. To that end, by 1904 Curtiss had become the chief supplier of engines for a Californian aeronaut named Tom Baldwin, who was working hard on building the first successful dirigible in America.[64] Later, Curtiss was approached by the famous inventor Alexander Graham Bell—an enthusiastic proponent of aeronautics in his later years—and asked to provide a suitable engine for his heavier-than-air flight experiments. Bell soon came to regard Curtiss as "the greatest motor expert in the country" and invited him to join his Aerial Experiment Association (AEA), an offer which Curtiss gratefully accepted.

Through his association with Bell and the AEA (he had even been appointed "director of experiments" by Bell for the group), Curtiss soon learned he had an aptitude for designing aircraft as well, and over the next two years, he had a hand in designing and building several prototypes for the group. Eventually Curtiss became a primary designer, his first aircraft being a biplane named the *June Bug*. In keeping with his daredevil persona, Curtiss insisted on undertaking most of the plane's proving flights himself, and on July 4, 1908, he even flew the little plane a distance of one kilometer to win the *Scientific American* Trophy and its $2,500 purse.[65]

But the ambitious Curtiss was not done yet. Intent on setting or breaking every aviation record possible, he purchased the *June Bug* from the AEA and used it as the basis for his *Curtiss No.1*, which was to be the first of his production series of pusher aircraft. With it, he entered the world's first air meet,

The Curtiss-piloted June Bug winning the Scientific American Trophy, July 4, 1908.

[64] Baldwin was an interesting fellow in his own right. Known as the "Father of the Modern Parachute" for the many leaps he made from balloons, he not only built and successfully flew a small airship, but built the U.S. Navy's first dirigible, the *DN-1*. He also trained a large number of pilots prior to World War I, many of whom became America's first combat pilots. Among his more renown students was the famous aviator Billy Mitchell.

[65] For this flight and for other achievements, Curtiss received U.S. Pilot's License #1 from the Aero Club of America, officially making him the first certified pilot in America. It's a mystery to this day how he somehow beat out both Wilbur and Orville Wright for the honor, both of whom had been flying for years before Curtiss ever touched an airplane.

A serious-looking Glenn Curtiss seated in one of his early pusher-type planes, looking every inch the way one would imagine an early twentieth century aviator should look.

the Grande Semaine d'Aviation flying contest at Rheims, France where, in August of 1909, he went on to win the overall speed event and, with it, the Gordon Bennett Cup. For this he was honored by being awarded French pilot's license no. 2 (French aviator Louis Blériot, who flew the English Channel in 1909 but finished behind Curtiss, had been awarded license no. 1), which was quite an achievement for an American at the time.

Curtiss, along with several demonstration pilots under his employment, spent the next few years crisscrossing North America in an effort to introduce aviation to a curious public and promote his products. (During this period, he even found time to train the first woman pilot, Blanche Stuart Scott—see chapter eighteen.) He also continued to win more competitions and prizes; in May 1910, he flew the 137 miles from Albany to New York City to make the first long-distance flight between two major cities in the U.S., for which he won a $10,000 prize and was awarded permanent possession of the *Scientific America* trophy.

All this attention soon got Curtiss noticed by the American military, which was curious as to the military applications of the airplane. As such, by 1910 Curtiss and his team began putting on a series of demonstration flights designed to illustrate his aircraft's versatility as a weapons platform. These included performing simulated bombing demon-

The Curtiss A-1 Triad—the U.S. Navy's first seaplane and one of Curtiss' most successful designs.

strations and the feasibility of shooting at targets on the ground from an aircraft. One of Curtiss' pilots, Eugene Ely, even demonstrated the potential mobility of the airplane by taking off from a platform mounted on the deck of a U.S. Navy warship, thereby initiating the age of naval aviation and marking the beginning of a relationship between himself and the Navy that would last for decades.[66]

Curtiss eventually went on to specialize in waterborne aircraft (seaplanes) for both the Navy and the civilian market, many of which proved to be spectacularly successful. One of his best known designs, the *A-1 Triad*, proved to be so popular that not only did the U.S. Navy buy them, but so too did the navies of Russia, Japan, Germany, and Britain. It was during World War I, however, that Curtiss built the aircraft for which he would become most famous—a rugged two-seat reconnaissance biplane and light bomber designated the *JN-4*, but known better as the *"Jenny."* While thousands of the hearty little planes would be sold to the armed forces of the United States, Canada, and Britain, they would become better known for their post-war exploits as the plane of choice for most "barn stormers"—men (and few women) who used the airplane to entertain and awe appreciative crowds across the country. In fact, for many Americans during the 1920s and 30s, the *"Jenny"* was the first airplane many had ever seen or ridden in, making the plane in many ways responsible for introducing aviation to the general public.

Curtiss went on to oversee the design and development of many aircraft, particularly his line of sea planes and a number of extremely fast racing planes. In fact, Curtiss' racing planes, by then being flown by some of the finest pilots in the world, set a number of air-speed records during the 1920s, even winning the much coveted Schneider Cup in 1923 and 1925. With his untimely death in 1930 at the age of just fifty-two, the world lost one of its most successful aviators along with one of its most outgoing and gifted engineers.

Conclusions

It's difficult to imagine what the world of aviation would have looked like during the first decades of the twentieth century without Glenn Curtiss' contributions, but what's especially interesting to consider is what role he might have played, had he lived, in aircraft development during the Second World War. In an era when new innovations were being introduced literally every week, Curtiss' contributions would have undoubtedly been significant and, potentially, even game changing. Considering his love for speed, it's even conceivable he might have taken the lead in developing the jet engine years before the Germans began experimenting with them on their next-generation jet fighters. It's not hard to imagine how that alone might have radically changed the nature of the air war over Europe in World War II.

But perhaps even more important than his physical contributions to aeronautics was the mystique he brought to aviation. While it is difficult to gauge just how extensive was the impact he had on the general public, to get some idea, it is believed by many that Glenn Curtiss was the inspiration behind the fictional character of Tom Swift—the central character in a series of juvenile science fiction and adventure novels that first appeared in 1910. Most of the series, written by various authors but published under the pseudonym Victor Appleton, focused on Tom's inventions, a number of which have anticipated actual inventions. In general, the books portrayed science and technology as wholly beneficial in their effects, and the role of the inventor in society was treated as admirable and heroic, which was a generally accurate assessment many people of that era held towards science in general and aviation in particular. It's not a certainty, but it's likely such would not have been possible without men like Glenn Hammond Curtiss to inspire such confidence in the future.

Quite a tribute for the "Fastest Man in the World".

[66] In being the first to illustrate the many waterborne uses of the airplane and for training the Navy's first aviators, he was further bestowed with the title "Father of Naval Aviation."

Chapter Seventeen: The First Ladies of Flight

18

Before ending this look into the early history of aviation, we would be remiss if we didn't take a moment to recognize the contributions of women aviators as well, many of whom demonstrated the same degree of resourcefulness, courage, and determination as their male counterparts. Unfortunately, when people think about famous women pilots in history, the discussion usually begins and ends with Amelia Earhart and her aviation exploits during the 30s, most specifically her crossing of the Atlantic and—almost—circumnavigating the globe. What few people realize, however, is how the daring aviatrix was the beneficiary of a number of woman who had gone before her and were instrumental in laying the foundation that made her amazing success even possible. While we have already mentioned a few of these daring women aviators—Sophie Blanchard probably being the most famous—there are a number of others whose bravery or, some might say, foolhardiness, matched that of their male colleagues.

So who were these early matriarchs of the skies? While it would be impossible to list them all or go into great detail about any one of them, there are a few names that rise to the surface that are especially worthy of note, not just for their willingness to risk their lives for the advancement of aviation, but for their courage in going against the sensibilities of the era and challenging the patriarchal world they found themselves born into. In doing so, they succeeded in changing the way society was to view the ever evolving role of women, which was accomplished, at least in part, thanks to that infernal machine, the aeroplane.

A Few Firsts

While the widow of ballooning pioneer Pierre Blanchard—Madame Sophie Blanchard—was well known for her exploits during the early years of the nineteenth century, as well as for being the first woman to die in a balloon accident (see chapter three for details), she was not the first woman to ascend in a hot-air balloon. That honor goes to Elizabeth Thible, who made the ascent in a Montgolfier-built balloon along with the pilot, a Monsieur Fleurant, from Lyons, France, in June of 1784 (just a mere eight months after de Rozier had made the first manned ascent). Perhaps the most notable aspect of the flight was that Madame Thible, dressed as the Roman goddess Minerva, made the ascent while singing two duets from Monsigny's *La Belle Arsène*, a celebrated opera of the time.

It wouldn't be until 1798, however, that a woman would fly a balloon solo. While one might assume that honor would go to the aforementioned Madame Blanchard, in fact it went to Jeanne-Genevieve Labrosse (1775-1847), the wife of the well known French inventor Andre-Jacques Garnerin (1769-1823). (Garnerin himself was quite a noted aviator of the day, as well as the inventor of the frameless parachute. He was especially known for his tendency to test his devices by affixing an unopened parachute to the top of a basket slung beneath a hot-air balloon, ascending to several thousand feet, and cutting the lines and letting the basket free fall in the hopes—fortunately always realized—that his revolutionary creation would open. Apparently, showmanship was as popular in the eighteenth century as it is today.)

In any case, Madame Labrosse, apparently demonstrating the old adage that somewehere there is a perfect match for everyone, shared her husband's enthusiasm for such daring feats—as well as for aeronautics in general—and convinced him to let her become the first woman to solo in a balloon. Obviously a man well ahead of his time, he permitted her to take to the air solo, which she did on November 10, 1798[67]. After that, the duo became something of a sensation, with both of them making parachute jumps across Europe. Interestingly, however, even though Madame Labrosse was the first woman to both fly a balloon solo and use a parachute, for some reason she never achieved the fame of her contemporary, Sophie Blanchard. (Why this was is uncertain, though it may have something to do with the sheer number of solo ascents Blanchard made—67 in all—and the fact that she lost her life in a balloon whereas Madame Labrosse was part of a husband and wife team that retired before their luck ran out.)

After Madame Labrosse's retirement, however, it would be nearly sixty years before a woman would make it into the record books again. This time she would not be French, however, but an American by the name of Mary Myers—who would eventually come to be known as "Carlotta, the Lady Aeronaut." Her claim to fame, however, was not that she was the first female balloonist in America to make a solo flight, but that she set a world altitude record by soaring to height of 21,000 feet over Franklin, Pennsylvania. What made the feat made even more remarkable was the fact that she performed it without using oxygen equipment.

"Carlotta" and her husband, Carl, continued to perform well into the 1890s, making literally hundreds of ascents without a serious mishap, but perhaps even more noteworthy than their flights was the fact that throughout the 1890s the couple ran what was called the "balloon farm" near Frankfort, New York. It was there, where, during the summer months, they kept as many as a dozen half-inflated balloons tethered to the grounds, each carrying a tag that listed the balloon's ascents, good performances, and bad habits. Unfortunately, with their deaths, "Carlotta" and her and her husbands' balloon farm were soon forgotten and remain a little known footnote to this day. Nonetheless, theirs remains a story of courage, skill, and romance that deserves to survive the eclipse of ballooning and needs to be remembered for the remarkable accomplishments this unique couple gave to the world of aeronautics.

Aida de Acosta

In the modern era of aviation, one of the first women known for her skills as a self-taught aviatrix was an American socialite named Aida de Acosta (1884-1962), who was, as far as can be determined, the first female to not only fly in a dirigible, but to pilot one solo. But what made it even more remarkable was that she didn't fly just any old dirigible, but had the rare privilege of flying one built and operated by the famous Brazilian aviator Alberto Santos-Dumont.

It all happened in the summer of 1903. Acosta, traveling in Paris with her mother, caught her first glimpse of a dirigible and became determined to meet the craft's maker—the by then legendary Santos-Dumont. Somehow managing to charm the man into showing her how to operate his personal dirigible, the *No. 9*, after taking just three lessons, he let her fly solo while he rode below on his bicycle, waving his arms and shouting advice. Apparently mastering the intricacies of flying a dirigible in record

[67] There is some evidence she may have been accompanied by a British woman named Miss Henry on this flight, which, if true, would make it not a solo flight. However, it would still constitute the first balloon flight without a man present.

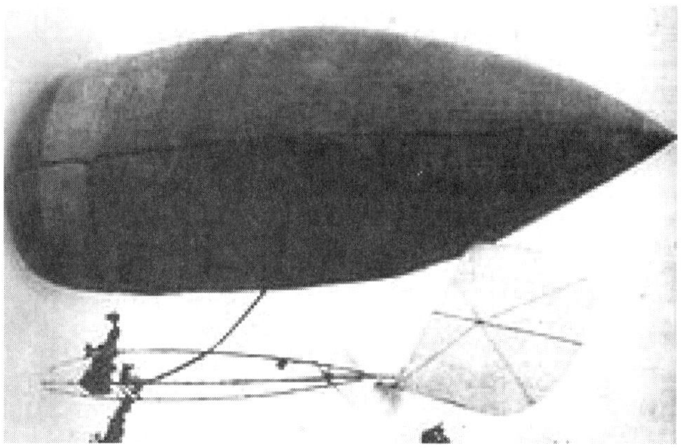

The charming Miss de Acosta managed to be the only person—male or female—to convince Brazilian aviator Santos-Dumont to let them operate his personal dirigible. She is shown here making her way to a local polo match in the man's tiny airship in the summer of 1903.

time, she flew the *No. 9* to the polo field at Bagatelle at the northern end of the Bois de Boulogne where she expertly landed the little ship, disembarked and watched some polo with Santos-Dumont. After a few minutes, she climbed back into the gondola and flew the machine back to Neuilly St. James, the entire trip lasting just over an hour and a half.

Of course, she did this without the knowledge or approval of her family, who were appalled when they learned about the flight. Certain that no man would marry a woman who had done such a thing, they managed to hush it up until she was safely married, and it remained a closely guarded secret for years afterwards.

So why did the Brazilian let her fly his dirigible when he wouldn't let anyone else touch it? Biographers of Santos-Dumont have speculated that there may have been a romantic relationship between them. However, while he may have been smitten with the woman—which may be the reason he let her fly his airship in the first place—there is no evidence she returned his affections. Nonetheless, it was reported that Santos-Dumont, a lifelong bachelor with no known romantic ties, kept a photograph of Acosta on his desk next to a vase of fresh flowers for rest of his life, suggesting that even in the world of aeronautics, love reigns supreme.

E. (Emma) Lillian Todd

While there were several women who have been recognized as being the first female pilots, none were considered inventors or builders of aircraft—a discipline that seemed uniquely suited to men. There was one exception, however, and that was the talented E. Lillian Todd (1865-1937), the first woman in the world to design and build an airplane.

Left: Emma Todd's biplane prior to its first and only flight in 1910. At right is Miss Todd at the controls, though she would never be afforded the opportunity to fly it herself.

Todd started life in a more conventional manner, of course. Working as a patent clerk and typist before eventually becoming a member of the first Woman's Law Class of New York University in 1890, she probably would have gone on to be one of the first female lawyers had she not possessed a most unladylike aptitude for building things. A self-taught inventor who grew up with a love for mechanical devices, she started by designing and building small mechanical and aeronautic toys, largely for her own amusement. By 1906, however, she had become proficient enough that she decided to build a large model of an aeroplane of her own design for an aero show at Madison Square Garden. Her model not only attracted national attention (probably because of the novelty of it being designed by a woman) but it even caught the notice of philanthropist Olivia Sage, the widow of financier and politician Russell Sage.[68] Mrs. Sage, perhaps intrigued by the idea of a female aircraft designer, became Todd's patron and even gave her $7,000 to build a full-size aircraft, which she began working on in the fall of 1908.

Working with the Wittemann brothers of Staten Island, Todd's small, two-seat monoplane was reminiscent of a *Wright Flyer* though with a few design features uniquely her own. Unfortunately, she never got the opportunity to fly her own airplane, for when she applied for a permit to do so, it was denied (most likely for sexist reasons). Her plane still flew, however, in November of 1910, though with a man—Didier Masson (an employee of Glenn Curtiss at the time)—at the controls. Deciding that she didn't want to take on the conventions of the day (even though there were several female pilots by this time), she gave up aeronautics completely and took a position with Mrs. Sage, which allowed her to pursue various philanthropic pursuits she decided to be more fulfilling than building airplanes.

Elise Raymonde de Laroche

While the question of who was the first female to actually pilot an aircraft (as opposed to a balloon or dirigible) is a source of some debate, the general consensus appears to be that the honor belongs to a French woman named Elise Raymonde de Laroche (1882-1919).

Born the daughter of a Parisian plumber in 1882, as a young woman de Laroche worked as an actress. Eventually, however, she became interested in ballooning and aeronautics in general and, to that end, she soon acquired her balloonist license. Already an accomplished balloonist, in 1909 she was approached by noted French aviator Charles Voison[66] who urged her to pursue fixed-wing aircraft, even offering to teach her himself. Not surprisingly, the intrepid Madame de Laroche took him up on his of-

French aviatrix Raymonde de Laroche was the first licensed female pilot in the world and the first woman to fly a fixed-wing aircraft solo. She also went on to achieve several women's altitude and endurance records over the course of her short career.

68. Russell Sage (1816-1906) was a financier, railroad executive and Whig politician from New York who had amassed a fortune, which he passed to his second wife, Margaret Olivia Slocum Sage, upon his death. She used the money for philanthropic purposes, endowing a number of buildings and institutions with his name, including Russell Sage College.

69. Charles Voisin (1882-1912) was an early aviation pioneer who, along with his more famous brother Gabriel, became among the first Frenchmen to fly.

Scene of the July, 1910 crash that nearly cost the "baroness" de Larouche her life. She would survive the crash and, though badly injured, would be flying again just two years later.

fer it soon became evident that she had a natural aptitude for flying. After riding along on a few flights and learning to taxi the plane herself, on October 22, 1909, she decided she was ready to solo. Turning the plane into the wind, she revved up the engine, quickly picked up speed, and was soon airborne, flying a distance of some 300 yards before gently and expertly setting the plane back down. Even more impressively, the next day she repeated the feat, but this time she circled the field twice, keeping the plane under perfect control the entire time, and landed with a skill and precision her instructor—engineer M. Chateau, along with the smitten Charles Voison—could only marvel at. Not surprisingly, a few months later she was awarded a pilot's license by the Aero Club of France, making her not only the first woman to fly solo, but the first licensed aviatrix in the world.

De Laroche—erroneously labeled a "baroness" by the press at the time (a title that stayed with her the rest of her life)—became quite a popular attraction after that, going on to fly in exhibitions at the Heliopolis in Egypt as well as at Saint Petersburg, Russia; Budapest, Hungary; and Rouen, France. She soon learned just how dangerous her new passion was, however, when, in July of 1910, her aircraft lost power while participating in an air show at Reims and she crashed hard into the ground. She was so badly injured in the accident that there was a question of whether she'd ever be able to walk again—much less fly—but just two years later, the "baroness," demonstrating the indomitable spirit that was to become her trademark, was once again flying, much to the delight of the French people.

However, de Larouche appeared to be endowed with an inordinate share of bad luck. No sooner had she recovered to the point where she could start flying again when in September of 1912 she was badly injured in an automobile accident (an accident which, ironically, took the life of her mentor, Charles Voison). Remarkably, even that wasn't enough to keep the strong-willed aviatrix grounded; a year later she was not only flying again, but had even won the Aero Club of France's Femina Cup for making a nonstop flight of over four hours—a feat that would have been impressive for any pilot of the era, either male or female.

Unable to fly during the First World War by law, the guns had barely fallen silent when she was already back in the air. This time, however, she was intent on becoming a test pilot and, to that end, she set two women's altitude records—one for an astonishing 15,700 feet—and the women's distance record of 201 miles. Unfortunately, her penchant for misfortune struck for the last time the following month when, while she was co-piloting an experimental aircraft, the craft went into a steep dive during its landing approach and crashed, killing both her and the pilot instantly. As a tribute to her courage, the people of France erected a statue of her that stands today at Le Bourget Airport in France—a fitting tribute to one of aviation's most indomitable spirits.

An imposing-looking Miss Scott behind the controls of one of Glenn Curtiss' early planes.

Blanche Stuart Scott

While it is generally conceded that the first woman to fly a fixed-wing aircraft was the French aviatrix Raymonde de Laroche in October of 1909, the question of who the first American female to do so is less clear. Probably the best candidate for the nomination belongs to Blanche "Betty" Stuart Scott (1885-1970), who may have accomplished the feat in the fall of 1910.

Something of a tomboy from Rochester, New York, Scott first achieved national attention when she became only the second woman to drive an automobile across the United States. The publicity from the stunt brought her to the attention of aviation pioneer Glenn Curtiss, who agreed to provide her with flying lessons (making her the only woman ever to receive flying instruction directly from Curtiss himself). Perhaps unprepared for her to fly solo, however, Curtiss had attached a limiter on the throttle of her airplane to prevent it from gaining enough speed to become airborne while she practiced taxiing on her own. On September 6, 1910, however, either the limiter broke or a gust of wind lifted the biplane and the plane became airborne and flew a short distance before she gently set it down. (Some sources have suggested that Miss Scott, apparently frustrated at the slow pace of her lessons, may have secretly removed the limiter herself, though this has never been substantiated.) Though her flight was short and possibly unintentional, nonetheless Scott is credited by the Early Birds of Aviation Association[70] as the first woman to solo in an airplane in the United States.

Scott subsequently went on to become a professional pilot, making her debut as a member of the Curtiss exhibition team at an air meet in Fort Wayne, Indiana, in October of 1910. Her exhibition flying earned her the nickname "Tomboy of the Air," and she became an accomplished stunt pilot known for flying upside down and performing "death dives" from 4000 feet. In 1911 she also became the first

[70] The Early Birds of Aviation Association is an organization devoted to the history of early pilots. The organization was started in 1928, with membership being limited to those who piloted a glider, gas balloon, or airplane prior to December 17, 1916. The original organization dissolved once the last living member died in 1998. The current organization is devoted to collecting and publishing biographies on those who met the 1916 deadline.

woman in America to fly long distance when she flew sixty miles around Mineola, on Long Island in New York. In 1912 she accomplished another first by becoming the first female test pilot, flying prototypes for another early airplane designer, Glenn Martin. She retired from flying in 1916, however, because she was bothered by the public's interest in air crashes and was frustrated with an aviation industry that refused to permit women to become mechanics or engineers.

Scott never lost her love for flying, however, and on September 6, 1948, she became the first American woman to fly in a jet when she rode with aviation legend Chuck Yeager in a T-33 Shooting Star. In a tribute to her history as a stunt pilot, it is said that Major Yeager treated her to some snap rolls and a 14,000 foot dive.

The diminutive Belgian aviatrix Helene Dutrieu may have looked frail, but she was a daredevil of the first magnitude who set numerous records during her short career.

Helene Dutrieu

Another woman who is frequently credited with being the first to fly was the Belgian aviatrix Helene Dutrieu (1877-1961), though it's more likely she was the second to do so—the first being Madame de Larouche in October of 1909.

Helene Dutrieu did not start out life in a manner that suggested that she was destined to be anything but a traditional young woman of her era. Born in Tournai, Belgium to a Belgian army officer, she had to drop out of school at the age of fourteen to earn a living, thereby suggesting that her future held little promise of adventure. However, Dutrieu didn't accept that and took a very unconventional path—especially for a woman—by becoming a professional cyclist. Racing for the Simpson Lever Chain team in 1895, within a few months she had won the women's speed track cycling world championship in Ostend, Belgium, (earning the nickname "La Flèche Humaine"—"The Human Arrow" in the process) and making quite a name for herself. She quickly followed that up by winning the Grand Prix of Europe in August 1898 and in November of that year she won the Course de 12 Jours in London, England, demonstrating that she was a consummate athlete despite her small size. On top of being an extremely successful stunt cyclist, she soon went on to become the first professional female motorcyclist, an automobile racer and even a stunt driver. As such, it is not at all remarkable to imagine her trying her hand at aviation, which she did when she was asked by the Clément Bayard factory in France to be the first to pilot its new ultralight aeroplane, the Santos-Dumont-designed *No.19 Demoiselle* monoplane. Unfortunately, the opportunity almost cost her her life when, never having flown an aeroplane before, she lost control of the machine during take-off and crashed, abruptly putting an end to her immediate flying ambitions.[71] (She also just missed out being the indisputable first woman to fly, for had she managed to get the plane in the air, she would have flown

[71] That she crashed was not all that surprising; not only had she never flown before, but the *Demoiselle* aircraft—with its temperamental Antoinette engine—was considered an especially difficult plane to fly, making such an outcome practically assured. What is difficult to understand is why the factory or Santos-Dumont would entrust the plane's first test flight to anyone—man or woman—who had never flown before. Clearly such an oversight was done for publicity purposes, making it a miracle it didn't end in tragedy as well.

a full year earlier than any other woman aviator.)

Dutrieu, however, wasn't one to give up that easily. Determined to be a pilot, she took flying lessons until she mastered the airplane. Soloing in April of 1910, she thus became only the second women to do so and the fourth female to earn her pilot's license.

Afterwards, and like many of her contemporaries, she went on to compete for numerous aerial records as well as accomplish a number of other firsts. She was best known, however, for her competitive spirit; battling fellow female pilots for the coveted Coupe Femina (Femina Cup) trophy for distance and endurance flying, she won it several times over the next two years before losing it for good to "Baroness" Raymonde de Laroche in 1912. But perhaps her greatest achievement was in winning the Coppa del Re (King's Cup) in Florence, Italy in 1911. In that event, she not only won the coveted trophy, but beat out fourteen male pilots to do so, thereby earning her the nickname "Girl Hawk." It was a title she proudly carried with her for the rest of her life.

Once World War I started, however, and since women were not allowed to fly during the war, Dutrieu's flying days came to an abrupt end. She did, however, offer her services as an ambulance driver in France and even became the director of the Campagne à Val-de Grâce military hospital, where she oversaw the recovery of thousands of allied servicemen injured during the war.

After the war she decided not to return to aviation but instead became a journalist and married, effectively ending any further flying adventures. However, in recognition of her accomplishments, in 1953 she was awarded the Médaille de l'Aéronautique (French Medal for Aeronautics).

Though she died in 1961, her legacy lives on to this day through the Coupe Hélène Dutrieu-Mortier (Hélène Dutrieu-Mortier Cup), an award that gives a cash prize of 200,000 francs to the French or Belgian woman pilot who makes the longest nonstop flight each year.

Bessica Medlar Raiche

Another woman frequently credited with being the first to fly—at least in America—is Bessica Medlar Raiche (1875-1932) of Mineola, New York. What's interesting about her accomplishment as compared to the others, however, is that she and her husband actually built the aircraft she soloed in, and she flew it without any previous flying instruction or aviation experience of any kind.

They built the plane—a Wright-type flyer that used bamboo and silk instead of a heavier canvas covering to save weight—in their living room over the course of several years, with the final pieces being assembled in the couple's back yard in the summer of 1910. Obviously unconcerned with the fact that she had no training to fly such a craft, on September 16, 1910, the couple dragged the craft to Hempstead Plains on Long Island, where—at least according to the Aeronautical Society of America—she supposedly made the first solo airplane flight by a woman in the United States. (There's no word on why her husband wasn't similarly tempted to try his hand at flying.) Of course, there is some debate about this as Blanche Stuart Scott had flown solo earlier that same month, but her flight was less well documented and arguably unintentional. When later asked about Scott's claim to being the first, Raiche said: "Blanche deserved the recognition, but I got more attention because of my lifestyle. I drove an automobile, was active in sports like shooting and swimming, and I even wore riding pants and knickers. People who did not know me or understand me looked down on this behavior."

It's not clear if Mrs. Raiche ever flew again, but it is well known that, now convinced their product could fly, her and her husband went on to build two more airplanes as part of a successful cottage-industry aviation engineering business. Perhaps the couple's greatest contribution to aviation was that they were innovators in the use of lighter weight materials in aircraft construction, and introduced the idea of using piano wire to replace heavier iron wire in aircraft controls and struts, thereby reducing the overall weight of the aircraft—an important consideration when it comes to maneuverability.

The Raiches' didn't stay in the aircraft business for long, however. In 1915 the couple had a daughter and moved to Newport Beach, California, where Mrs. Raiche went on to pursue careers as a dentist and a physician before dying of a heart attack in 1932.

Marie Marvingt

While a number of women were known for their courage and sheer audacity during the early years of the century, few could complete with the French athlete, mountaineer, balloonist, and aviatrix, Marie Marvingt (1875-1963).

Born in Aurillas, France, young Marie had a very untypical childhood—especially for a girl. Encouraged to participate in sports by her father, by the time she was an adult, she had excelled at water polo, horseback riding, boxing, martial arts, fencing, shooting, tennis, golf, hockey, football, a variety of winter sports, and mountaineering. In a rare display for the era of independence and self-reliance, in 1890, at the age of fifteen, she canoed over 400 kilometers from Nancy, France, to Koblenz, Germany. To top that off, in 1899 she also earned her driving license, becoming one of the first women in the world to do so.

Obviously, for such a woman, aeronautics seemed the next logical step, and she began to pursue it shortly after making her first ascent in a balloon. By 1909 she had not only soloed in a balloon, but was to become the first woman to pilot a balloon across the North Sea and English Channel from France to England.

But it was in fixed-wing aircraft that Marvingt was best known. On November 8, 1910 she became only the third woman in the world licensed as an aeroplane pilot and the only woman ever licensed to fly the hard-to-handle *Antoinette*[72] monoplane. Despite the plane's bad reputation, however, she made over nine hundred flights in the craft without once "breaking wood"—a popular euphemism at the time for a crash or hard landing resulting in damage—a record unequalled at the time. She also participated in many air shows and in December of 1910, while competing for the Coupe Femina (Femina Cup), she set the first official women's flight records for duration and distance by staying aloft for 53 minutes and flying a total of 42 kilometers.

It is what she managed to accomplish during the First World War that was most remarkable, however. As was the case for all female aviators at the time, she was not allowed to fly during the war. Marvingt wasn't one to allow her gender to keep her from doing her patriotic duty, however, and so she disguised herself as a man and made her way to the front where, with the help of a French infantry lieutenant, she briefly served as a common infantry soldier. Soon discovered and sent home, the determined Marvingt somehow managed to find her way back to the action, but this time with the Italian army in the Dolomite Mountains along the Italian-Austrian border. Though initially serving as a Red Cross nurse, she eventually tired of that and found her way back into the air, managing to even fly actual combat missions. For one of these mis-

The remarkable Marie Marvingt, one of the truly astonishing women of aviation.

[72] The Antoinette monoplane was a French-built, high-speed monoplane known for its difficult handling characteristics and unforgiving nature. Only the most capable pilots were permitted to fly it.

Marie Marvingt at the controls of the Antoinette monoplane—1911. Notice the plane's fairly substantial size.

sions, she even received the Croix de Guerre for bombing a German military base in Metz, making her, as far as can be determined, the first woman to participate in a military combat mission by air.

After the war she worked as a journalist, war correspondent, and medical officer with French Forces in North Africa. Her greatest passion, however, and one she pursued until her death in 1963, was her design for an air ambulance—a dream that was finally realized when a fully-organized air ambulance service was established during the African and Middle Eastern Colonial Wars of the 1920s.

Marvingt never gave up her love for flying, however, and in 1955, on her eightieth birthday, she had the honor of becoming the first woman to break the sound barrier as a passenger in an American fighter jet over France. Obviously not accepting the common wisdom that one needs to slow down later in life, that same year she pursued her helicopter pilot's license and in 1961, at the age of eighty-six, even cycled an astonishing 176 miles from Nancy to Paris.

Marie Marvingt's greatest legacy, however, comes from the fact that she is officially recognized as the most decorated woman in French history, having won numerous prizes for her sporting achievements and aerial accomplishments, as well as for her work in aviation medicine. In doing so, it could be easily argued that she packed more into a single lifetime than most people could in three.

Harriet Quimby

While other women flew before she did, probably no aviatrix was as famous in her day or had as much influence on the role of woman in aviation as did Michigan native Harriet Quimby (1875-1912).

Educated as a journalist, it was while she was working as a theater critic for *Leslie's Illustrated Weekly*[73] that she first became interested in aviation. While attending the Belmont Park International Aviation Tournament on Long Island, New York, in 1910,

[73] A popular American literary and news magazine published between 1852 and 1922.

she met Matilde Moisant and her brother John, a well-known American aviator who, along with his brother, Alfred, ran an aviation school. Intrigued at the prospect of learning to fly herself, she asked Matilde's brother to give her lessons but he was killed in a plane crash a few months later before he could teach her to fly. Fortunately, Matilde's other brother, George—who was also a pilot—agreed to teach both Harriet and his sister to fly and on August 1, 1911, Quimby become the first American woman to receive a pilot's license.[74]

After receiving her license, both her and Matilde began touring with the Moisant International Aviators, performing at numerous flying exhibitions. Understanding the power of good publicity, Quimby quickly created a look for herself by donning an elegant purple satin flying suit. With her elegant looks, her and her trademark flying suit quickly captured the public's attention while she used her considerable abilities as a writer to chronicle her adventures in articles for *Leslie's Weekly*, sharing with the public her love of flying.

Eventually determined to be the first woman to cross the English Channel in a heavier-than-air machine, in March of 1912 she left for England with a letter of introduction to Louis Blériot, the famous French aviator who had made the first crossing of the channel himself only four years earlier. Somehow managing to convince the dubious Blériot to lend her one of his planes, on the morning of April 16, 1912, she took off from Dover on a one hour flight to Calais, where, despite the fact she'd never flown the plane before and the weather was less than ideal, she landed safely on a beach near Hardelot-Plage. In an incredible example of poor timing, however, her accomplishment did not receive the attention it deserved as the papers were filled for the next few weeks with news of the sinking of the *RMS Titanic*, pushing her exploits to the back page. In any case, the feat, combined with her good looks and the novelty of being the first li-

Harriet Quimby proved to be one of the most popular aviatrixes of her day whose life had a profound impact on the role of women in aviation.

74. George's sister Matilde got her license a few days later to become the second licensed female pilot.

censed American aviatrix (it's not clear when Blanche Scott—who had been flying a year earlier—received her license) made her an even bigger celebrity, and overnight her image began appearing in magazines and on posters around the world, selling everything from chewing gum to soda pop.

Her fame and career as an aviatrix was short-lived, however. Just a few months after her daring channel crossing, tragedy struck. While flying her two-seat *Blériot XI* at an air show near Squantum, Massachusetts, for reasons still unknown, the aircraft unexpectedly pitched violently forward and Quimby, along with her passenger, show organizer William Willard—neither of whom happened to be wearing seat belts at the time—were ejected. Falling from a height of some 1,500 feet, both died instantly upon hitting the water, thereby bringing to a tragic end one of the most promising careers in aviation. (Ironically, after having ejected its passengers, the lightened plane righted itself and made a fairly intact unassisted landing in the mud, where it was subsequently recovered and restored to flying condition.[75])

While her career as an aviatrix lasted just eleven short months, Quimby left behind a legacy as a woman ahead of her time. Even though she was not a self-proclaimed suffragette, her independence and sense of adventure inspired many women and helped to pave the way for other female pilots in the future. Her death also left a great hole in the history of what might have been but for the sometimes cruel whims of fate.

Ruth Bancroft Law

Though not the first to fly, few women pilots were as well known in the 1910s as was Ruth Bancroft Law (1887-1970), one of the pioneers of endurance flying and a vocal activist for women being allowed to fly military aircraft in wartime.

Sister of the famous pioneer movie stunt man and parachutist Rodman Law (1885-1919), Massachusetts native Ruth Law decided to pursue aviation as a career shortly after her marriage. (Fortunately, her husband, Charles Oliver, was apparently a man ahead of the times as well, for he enthusiastically supported his wife's unconventional profession.)

It wasn't easy at first, however. When she first attempted to enroll in the flying school operated by the Wright brothers, she was turned down by Orville Wright, who believed women were unfit to fly. He did agree to sell her a plane, however—a *Wright Model B* (for a whopping $7,500—a fortune at the time) and, after finding a man willing to overlook her gender and teach her to fly (in this case, the very fine pilot Phil Page), on August 1, 1912, Ruth soloed and acquired her pilot's license, becoming only the sixth woman to do so.

Once she had obtained her license, she very soon afterwards began appearing at air shows all over the country, where she enjoyed thrilling the crowds with her death-defying maneuvers. In fact, it was during one of these demonstration flights—this one before a large crowd at Daytona Beach in 1915—that she announced she was going to do a "loop the loop"—a dangerous maneuver she had never attempted before. Much to the consternation and relief of her husband—who also happened to be her manager—she was successful and landed to tumultuous applause. Another feat, and one that made her even more famous, occurred in November of 1916, when she smashed the then existing woman's cross-country distance flying record of 452 miles by flying from Chicago to New York state, a distance of 590 miles, nonstop.[76] The accomplishment made her a national

[75] This aircraft may possibly be the one seen on display today at the Old Rhinebeck Aerodrome in Red Hook, New York. A curious coincidence about the tragedy is that Blanche Scott—the first American aviatrix—happened to be present at the event and watched Quimby and Willard fall to their deaths from her own aircraft circling nearby.

[76] While to modern sensibilities a flight of 600 miles might not seem that impressive, it must be remembered she performed this feat in an open aircraft that left her completely exposed to the wind and cold. Considering that aircraft at the time could barely manage 40 mph, this made her flight an endurance test some twelve hours in length which, under the circumstances, would be a considerable accomplishment even today.

The indomitable barn stormer and endurance flyer Ruth Law set a number of records for long-distance flight--some as far as 600 miles–all in an open airplane capable of barely 40 miles per hour.

figure and got her an invitation to the White House to dine with President Woodrow Wilson, making her, for a time, one of the most famous women in the world.

When the United States entered World War I in 1917, Ruth used her popularity to campaign for allowing women to fly military aircraft in combat, but her efforts proved unsuccessful. Upset that women were blocked from flying based purely on their gender, she wrote an article entitled "Let Women Fly!" for the magazine *Air Travel,* in which she argued that success in aviation should prove a woman's fitness for work in that field. Not surprisingly, her pleas fell on deaf ears and women would not be allowed into aerial combat roles for the balance of the century. (However, Law and other female American pilots were allowed to fly in bond drives and recruiting tours, and she was even allowed to wear the uniform of a noncommissioned Army officer, achieving at least a symbolic victory.)

After the war, Law returned to flying full time and even established a flying circus to travel around the country, entertaining crowds wherever she went. In 1919 she set a new altitude record of 14,000 feet and even tried her hand at wing walking. It all came to an abrupt end in 1922, however, when, at the urging of her husband Charles, who had apparently grown increasingly uncomfortable with the danger her flying exploits entailed, she announced her retirement and never flew again, perhaps demonstrating that, once again, love conquers all.

Conclusion

There are other women who contributed to the early years of aviation as well—women like Hilde Hewlett, the first Englishwoman to earn her flying license and become a highly successful aviation entrepreneur; Marjorie and Katherine Stinson, the first flying sisters; Alys McKey Bryant, the first Canadian aviatrix; Adrienne Bolland, the first woman to fly over the Andes mountains, and, of course, Bessie Coleman, the first African-America to earn a pilot's license—but brevity does not permit us to list them all. We apologize for such an oversight, but at some point every author is forced to draw the line on whose contributions were significant and whose were merely redundant and let the chips fall where they may.

The point of this chapter is not to simply point out the largely unknown history of female aviators prior to Amelia Earhart, but to demonstrate that taming the skies had an important feminine element to it as well. Additionally, it's important to recognize how much aviation may have done to change perceptions of the role of women in regards to pursuing traditional male roles. It's no coincidence that just a few short years after these women made their extraordinary flights, challenging the status quo in the process, that women in much of Europe and North America acquired the right to vote. It could easily be argued that were it not for these determined aviatrixes demonstrating that they could compete with men on an equal basis, endure the rigors and physical hardship of flight, and even best them at their own records, it may have been decades before the majority of the world's nations would be convinced that women should also be afforded the same right as men to elect their leaders. Further, without the aeroplane, it would have made it increasingly difficult for women to push their way into other, less traditional roles and find equal footing with their male counterparts. As a result of their efforts, today few people consider women CEOs and police officers or even commercial pilots a novelty, while the idea of woman fighter pilots and space shuttle commanders is becoming not only accepted but increasingly commonplace—all due, in large part, to the first aviatrixes of the twentieth century.

CONCLUSIONS

Undoubtedly there are many names we have left out of this work, the omission of whom some will find inexcusable, for which we apologize most profusely. It was difficult to not include everyone we came across during our research, but at some point one must decide whose contributions were important to the process and whose were merely interesting. The point of this work remains to recognize that even in omission, the daring and brilliance of all the men who first made human flight possible cannot be over-appreciated. Whether from a fragile basket suspended beneath a balloon filled with volatile hydrogen or from the confines of a tiny seat slung below a pair of bamboo and silk covered-wing being pushed through the air by engines little more powerful than those seen on a modern outboard motor, they are all to be honored and appreciated for the sacrifices they made and the daring they exhibited, which we hope we've accomplished in this work.

The second purpose of this book is to force us to consider the question of who, exactly, deserves to be credited with being the first to fly a heavier-than-air craft. Though we made the statement earlier that in the end, it really doesn't matter, we suspect most readers will still want to know the position of the authors, which we are only too happy to oblige.

The weight of evidence suggests that several men may have managed at least a few seconds of untethered, manned and powered flight prior to December 17, 1903. Gustave Whitehead, Burrell Cannon, Karl Jatho, Richard Pearse, Felix du Temple and even Aleksander Mozhaysky all have excellent claims to being the first. It should be noted, however, that success is not just about accomplishing some great feat as much as it is about being able to prove it. The Wright brothers kept careful records documenting each of their flights, maintained working blueprints on their craft, and, most important of all, had both eyewitnesses and photographers on hand to attest to and document each of their accomplishments. In other words, they recognized that their word was not enough, nor were the often uncertain memories of eyewitnesses recalling an event years or even decades later sufficient. They understood that the world would demand proof of their accomplishments and they dutifully provided it. Had the others named above taken the same precautions, they might well have achieved the recognition their proponents are so keen to bestow upon them now.

We can't very well blame the Wright brothers for the failure of those who may have preceded them from documenting their own accomplishments. Whitehead had only to ensure someone with a camera was present at his alleged flight in 1901 to have made his way from obscurity into the history books, but he didn't do it. Jatho, Pearse, Cannon each had that same opportunity but failed to take advantage of it and so their obscurity is, in some ways, a consequence of their own desire for secrecy or, more likely, simply a failure to appreciate the importance of what they were doing. In effect then, their lack of fame was their own fault; it strikes us as being unfair to hold the Wrights responsible for it.

But then this work was never primarily about determining who was the first to fly as much as it is about the process humanity took in its progression towards breaking the bonds of gravity. As such, it's not so much about individual accomplishments as it is about the process itself. Additionally, it's less about the ingenuity these men exhibited than it is about their bravery and determination. The men and women mentioned in this book risked much to realize their dreams—sometimes even their own lives—and that's what this work was meant to appreciate.

In a way, however, it's also a tribute to the machines they flew as well—those remarkable, fragile, frail, barely airworthy craft that were able to overcome every obstacle thrown at them and somehow manage to claw their way skyward, if only for a few brief seconds. Most of these plucky little craft have long since been destroyed, dismantled, or otherwise lost, but the age of discovery and inventiveness they represent will never die. As such, both the machines and the men that flew them remain phantoms of the skies, for while the material substance of their accomplishments have long ago perished, the spirit they left behind remains forever, which is the greatest tribute any of them can be afforded.

AVIATION TIMELINE

MAJOR MILESTONES in AVIATION HISTORY thru 1912*

450 BCE	Chinese engineer, philosopher, and inventor Lu Ban invented a wooden bird, which may have been a large kite or an early glider.
200 BCE	Earliest known use of kites in China, though they probably have been around much longer.
400 BCE	The Greek philosopher, Archytas, supposedly designed and built a bird-shaped, apparently steam-powered model named "The Pigeon" which is said to have flown some 200 meters while suspended from a string.
180-234	The Kongming lantern, a prototype hot-air balloon, was known in China from ancient times. Its invention is usually attributed to the general Zhuge Liang, who is said to have used them to frighten enemy troops.
875	Muslim astronomer, musician and engineer Abbas Ibn Firnas is said to have built and flown a rudimentary glider near Cordoba, Spain.
Circa 1005	English Benedictine monk Eilmer of Malmesbury allegedly flew a glider from the tower at Malmesbury Abbey.
16th Century	Leonardo da Vinci designed a number of flying machines, some of which were more practical than others. He never built or flew any of his designs, however.
1670	Italian Jesuit priest Francesco Lana de Terzi published a work that suggested lighter-than-air flight would be possible using copper foil spheres that contained a vacuum that would be lighter than the displaced air.
1709	Portuguese priest Bartolomeu de Gusmão built and tested a number of small, unmanned hot-air balloons in front of the king of Portugal, with one of them rising to the ceiling of the hall of the Casa da Índia in Lisbon.
1714	Christian mystic and philosopher Emanuel Swedenborg sketched out what is widely recognized today as the first published drawing of a practical flying machine, the *Daedalion*.

*Early dates are approximations.

1783 4 June: The first of a series of hot-air balloons designed and built by the Montgolfier brothers successfully ascended. It was unmanned but it demonstrated that lighter-than-air flight was possible.

19 September: A Montgolfier-built balloon successfully ascended with a sheep, a duck, and a rooster onboard, making them the first official aviators.

21 November: Frenchmen Pilatre de Rozier and the Marquis d'Arlandes became the first humans to fly when they ascended in a Montgolfier-built balloon in Paris.

31 December: French inventor Jacques Charles and his co-pilot, Nicolas-Louis Robert, became the first men to ascend in a hydrogen-filled balloon.

1784 15 July: The Roberts brothers built and flew a proto-dirigible designed by Frenchman Jean Baptiste Meusnier, though it proved to be uncontrollable in flight.

1785 7 January: Frenchman Jean-Pierre Blanchard—along with American Dr. John Jeffries of Boston—became the first balloonists to successfully cross the English Channel in a hydrogen-filled balloon.

15 June: Pilatre de Rozier and a companion were killed while attempting to cross the English Channel in a hybrid hot-air/hydrogen balloon, making them the first men to die in a balloon crash.

1793 15 May: Spanish inventor Diego Marín Aguilera possibly flew a glider from the castle at Coruña del Conde in northern Spain, gliding nearly 1000 feet before crashing.

1799 English scientist Sir George Cayley etched the design for the first airplane—complete with a fuselage and wings—onto a small silver disk designed to illustrate how the forces that govern flight operate.

1809 George Cayley made several successful flights with a large, unmanned glider of his own design.

1819 6 July: Jean-Pierre Blanchard's widow and female balloonist, Madame Sophie Blanchard, was killed when her balloon caught fire and crashed during an exhibition at the Tivoli Gardens in Paris, making her the first woman to die in an aeronautical accident.

1825 Norwegian inventor Hans Navrestad may have built and successfully flown a manned glider over Lake Navrestad.

1848 English inventor John Stringfellow successfully flew a small steam-powered glider 30 feet.

1849 George Cayley successfully test flew a triplane glider with a ten-year-old boy at the controls.

1852 24 September: French engineer Henri Giffard built and successfully flew a small, steam-driven dirigible, becoming the first to demonstrate that a powered airship could be steered.

1853 George Cayley's largest glider, piloted by his own coachman, allegedly flew some 900 feet at Brompton Dale, England.

1856 December: French sea captain Jean-Marie le Bris built a machine that may have briefly flown as it was being towed behind a horse.

1863	1 June: American inventor Solomon Andrew glided an airship over Perth Amboy, New York by making use of a system of inclined planes to steer the craft, all without a motor. He carried out several more flights later that year, all of them apparently successful.
1866-1869	Polish inventor Jan Wnek made a number of successful manned glides over a three-year period, most of them over Austro-Hungary.
1868	John Stringfellow's most successful steam-powered glider, a triplane design, successfully flew at London's Crystal Palace.
1872	German engineer Paul Haenlein built and flew a small semi-rigid airship that ran using the combustible hydrogen gas contained inside its own envelope.
1874	French Naval Officer Felix du Temple briefly achieved manned, powered flight in his aluminum monoplane, making him possibly the first man to fly.
1882	July?: Russian naval officer Aleksanedr Mozhayski built and briefly flew a massive steam-powered aircraft a few hundred feet.
1884-1886	American aviation enthusiast John Montgomery successfully flew a series of manned gliders that incorporated such modern features as ailerons.
1884	9 August: Two French army officers, Charles Renard and Arthur Krebs, made the first of several successful flights in their electric-powered airship, the *La France*.
1890	9 October: French designer Clement Ader may have briefly flown his powered airplane, the *Eole*, near Paris, France.
1891-1895	German aviation pioneer Otto Lilienthal made over 2,000 successful flights in a series of gliders he designed and built himself.
1893	Australian inventor Lawrence Hargrave designed the box kite, which was to serve as the basis for many later successful airplane designs.
1895	British inventor Percy Pilcher successfully flew a number of gliders in England.
1896	May through November: American inventor Samuel Langley successfully flew a pair of powered, unmanned aircraft distances of up to 4,000 feet.
	10 August: Otto Lilienthal dies while demonstrating one of his gliders.
1897	3 November: An aluminum-skinned airship designed and built by Croatian inventor David Schwarz briefly flies over Tempelhof airfield in Berlin before crashing.
1899	30 September: Percy Pilcher is killed while doing a demonstration flight in one of his unpowered gliders.
1900	2 July: German inventor Count Ferdinand von Zeppelin flew his first airship, the *LZ-1*, briefly over Lake Constance near Friedrichshafen, Germany.
	October: Brazilian aviator Alberto Santos-Dumont won the prestigious Deutsch de la Meurthe prize by flying around the Eiffel tower in a small dirigible.
1901	14 August: Inventor Gustave Whitehead allegedly flew his monoplane over Bridgeport, Connecticut, which, if true, made him the first to achieve controlled, manned flight.

1902	American inventor Burrell Cannon allegedly flew an airplane near Pittsburg, Texas,
1903	31 March: New Zealand farmer Richard Pearse is thought by many to have briefly flown a small airplane near Waitohi, New Zealand.
	August-November: German inventor Karl Jatho may have achieved manned flight in his biplane aircraft, including a brief hop of 200 feet.
	17 December: The Wright brothers achieved the first historically accepted, powered, manned flight at Kitty Hawk, North Carolina.
1906	12 September: Danish pilot Jacob Ellehammer made a short flight in a tethered aircraft, which some people consider to be the first manned flight in Europe.
	23 October: Brazilian aviator Alberto Santos-Dumont is recognized as the first man to fly in Europe when he lifts his biplane, the *14-bis*, from a runway in Paris.
1909	25 July: French aviator Louis Blériot flew a monoplane across the English Channel, flying from Calais to Dover in just 37 minutes.
	22 October: Raymonde de Laroche became the first woman to fly solo in a powered, heavier-than-air craft. She was also the first woman in the world to receive a pilot's license.
1910	14 November: Eugene Ely, flying in a Glenn Curtiss aircraft, becomes the first pilot to fly from the deck of a ship.
1912	16 April: Harriet Quimby is the first woman to fly solo across the English Channel.

IN MEMORIUM

While many men—and a few woman—were killed while pursuing aeronautics, there are a few who stand out, either for their bravery or for being a significant pioneer. While this list is far from complete (more than a score of aviators died between 1910 and 1912 alone, most while learning to fly), these are the individuals named in this work who went on to make the ultimate sacrifice in pursuit of their dreams. They may be gone but their sacrifices will never be forgotten.

Pilâtre de Rozier, 1785

Pierre Romain, 1785

Jean-Pierre Blanchard, 1809

Marie Madeleine-Sophie Armant (aka Sophie Blanchard), 1819

Jan Wnek, 1869

Joseph Croce-Spinelli and Theodore Sivel, 1870

Otto Lilienthal, 1896

Friedrich Hermann Wölfert, 1897

Percy Pilcher, 1899

Daniel Maloney, 1905

Lt. Thomas E. Selfridge, 1908

Eugène Lefebvre, 1909

Ralph Johnstone, 1910

John Moisant, 1910

John J. Montgomery, 1911

Eugene Ely, 1911

Harriet Quimby, 1912

Aurel Vlaicu, 1913

Frank Bolland, 1913

Elise Raymonde de Laroche, 1919

ADDENDUM
THE MYSTERIOUS SACRAMENTO LIGHTS

While this book has dealt largely with those inventors and inventions known to history, there is one event from the early days of aviation that needs to be recounted here, not because it had a significant effect on the history of aviation, but because it is such a curious story that we thought the reader might enjoy it. The event in question is known by several names, though it is most commonly referred to simply as the "Sacramento lights" event of 1896-97.

For those unfamiliar with this incident, it all began in November of 1896 when a mysterious light was seen flying low in the sky over the capital of Sacramento, apparently under manned control which, of course, would have been an unusual sight in the pre-aviation skies of 1896. Whatever was creating the strange light apparently made a slow pass over the city and headed out of sight, leaving the citizens somewhat perplexed as to what it could have been.

Normally such an incident would be dismissed as some kind of astronomical or meteorological anomaly and promptly forgotten, but the light reappeared a few days later, again flying slowly over the city. This time, however, it was even seen to be moving against the prevailing wind—suggesting it was not a free flying balloon—while some witnesses claimed that the light was seen to be emanating from beneath a large, cigar-shaped body, suggesting it was some kind of an airship.

Of course, airships were not unknown in 1896, but the speed at which the craft moved, especially *against* the wind, made it doubtful it could have been a dirigible of the era (which tended to be fragile, slow-moving affairs at the time). Additionally, there were no reports of anyone at work building such a craft, which seems inexplicable, especially considering how newsworthy such a project would have been. But what really made this second visitation remarkable was that it wasn't limited to the Sacramento area. In fact, if accounts at the time are to be taken seriously, the same light—and accompanying craft of some kind—was seen later that same evening over a hundred miles away in San Francisco!

In a remarkable chain of events worthy of a Jules Verne novel, the craft was not only supposedly spotted over the bay city just a few hours after it was seen over Sacramento (suggesting it had the capacity to fly considerable distances at a time), but it was even seen to shine what appeared to be a searchlight on a herd of seals lounging on the appropriately named Seal Rocks—sending them scattering into the sea—before moving off to the northeast and being lost from sight. It was even said to have been seen by no less than the secretary of state of California and the Mayor of San Francisco, making it one of the best observed aerial phenomenon in modern history.

Quickly dubbed an "airship" by the press—probably due to reports of its great cigar-shaped body as well as for lack of a better term—airship fever soon took hold of the state, with newspapers all over California publishing accounts of the mysterious craft appearing over cities all over the West Coast, capturing and holding the public's fascination for weeks. Then, almost as abruptly as the sightings had

This somewhat fanciful drawing of the Sacramento airship first appeared in the November 22, 1896 edition of the San Francisco Call newspaper, igniting a fire storm of controversy over the ensuing weeks and prompting reports of mysterious lights from all over the West Coast.

begun, they came to an end just three weeks after the first sighting. It was almost as if whomever—or whatever—had been responsible for the mysterious lights had tired of their game and left as suddenly as they had appeared, leaving millions of Californians to ponder what the thing was and who might have built it.

But that was not the end of the mysterious visitor. After a two month hiatus, reports of the same craft—or one very similar to the one spotted over California—began coming in from the Midwestern states—most noticeably Nebraska and Kansas. It was almost as if the craft was leisurely making its way across the nation's remote farming belt, eliciting numerous reports from all over the country and being written up in many of the largest and most prestigious newspapers in the region. It even supposedly appeared in the early dawn skies over Chicago in mid-April of 1897, where it allegedly had its picture taken by a man with a small box camera.

Then, just as anticipation as to who might be behind the mysterious craft was growing, once more the reports abruptly ended, again leaving nothing in its wake but a mountain of newspaper clippings and rampant speculation. Though reports of the mysterious vessel would continue to trickle in for a few more weeks, by late spring the great "airship flap" of 1896-97 was over and the world was left with one more mystery to ponder.

So what was it that so many thousands of witnesses saw? Clearly something had been spotted over central California that late autumn of 1896, and later that next spring in the skies over the Midwest, but what could it have been? Researchers have uncovered as many as 1500 individual newspaper accounts of the ship between November, 1896 and May of 1897, but to this day no one is certain what the bright object (or, perhaps, objects, for there may have been more than one) might have been.

Most skeptics simply assume it was all the result of media-inspired mass hysteria fueled by a public so enamored with the idea that there was some mysterious "airship" flying about that every stray light or unusually bright appearance of the planet Venus was reported to be the craft. (Some debunkers even go so far as to maintain it was a hoax devised by the newspapers of the day to boost sales or simply to bring a bit of "color" to the ordinariness of the news of the time.) In more modern times, some in the UFO community—noting that modern UFOs are sometimes described as being "cigar shaped"—have suggested the "airships" were early appearances by extraterrestrials.

Most witnesses at the time, however, thought the vehicle neither imaginary nor extraterrestrial, but believed it to be a very terrestrial machine—a powered balloon of some kind or, even more precisely, a dirigible of considerable sophistication built being flown by parties unknown. Most dismiss this explanation today, of course, as being inconsistent with the technological capability of the time, assuming that a world still in its industrial infancy incapable of such an accomplishment. This book has catalogued an impressive array of men and machines built during that time, however, making the prospect that the "Sacramento lights" could have been an example of a nascent technology not at all outlandish.

Of course, as is usually the case with phenomena of the kind, descriptions of the craft varied widely and seemed to run the gamut from fairly conventional airship designs (which one farmer described as being "...cigar shaped, about 200 feet long and 50 feet across at the widest point, gradually narrowing to a point at both ends") to fantastic aircraft-like configurations complete with flapping wings and bird-like rudders. Some even claimed to have encountered the airship's crew—who, by most accounts, appeared to be quite human—with a few lucky witnesses saying they were even given tours of the fantastic craft.

But despite the variety of descriptions provided, there was also a remarkable degree of unanimity among many of the witnesses as to the vessel's general size and capabilities. For example, most reported the craft was cigar-shaped and roughly 100 to 200 feet in length, and described it as having a single, bright white light slung beneath it (though a few accounts claimed dimmer red lights were seen as well). What is even more interesting and, in some ways, more telling about the many reports, however, were their geographic consistency. While airship sightings were reported all over the country (and sometimes over different states on the same day), the earliest and most reliable reports had the craft moving generally in an easterly direction, cutting a fairly well-defined swath through the skies between Nebraska and Illinois as if making its way slowly towards the eastern seaboard. In effect, it seems that, while much of what was reported about the mysterious ship was fanciful, erroneous, or just plain nonsense, there was a core of reliable reports that possessed a degree of internal consistency that implied that something real was being spotted by many of the citizens of the Midwest.

As this book aptly demonstrates, since airships had been under development in Europe and America prior to and even contemporary to the airship flap of 1896-97, it is likely the general public of the era would not only be very favorably inclined towards accepting such a vehicle as being possible but, considering the pace of scientific and technological progress of the time, inevitable. In fact, there was little debate at the time whether the airship was a man-made vessel of some kind, making it difficult to understand why modern skeptics so casually dismiss the idea. Clearly, the people of the era accepted them as evidence of a nascent technology and, while intrigued as to whom might have been behind their construction, most did not find the possible existence of such a vessel particularly extraordinary. In fact, the main question on most people's mind at the time was not *what* it was, but *who* was behind the fantastic machine?

While it must appear unlikely to modern sensibilities that someone actually had the wherewithal in 1896 to build and fly such a seemingly advanced airship, what if it were true? In effect, what if history has simply missed something extraordinary and the citizens of the late nineteenth century were witnesses to one of the most important engineering and aeronautical feats of the age—an event that for some reason has been lost to history?

Let's play a little game of what-if and consider the possibility—if only as an exercise in logic. Let's pretend, for the sake of argument, that the remarkable airship was real, and that it had been built in secret by parties unknown, and that it proved to be capable of flying extraordinary distances—at least for the time. Who could have pulled off such a feat of engineering? Further, considering the immense investment such an endeavor would require, how could he have done it in secret—and, even more curiously, why would its builder not want the world to know about his magnificent machine?

Unfortunately, we have very few clues to work

from, but we are not completely flying blind here. If we consider the basic elements and setting of the story and take at least some of the earliest accounts of the airship seriously, we can begin to deduce a number of things from them and, perhaps, begin to construct a profile and motivation for our airship builder. We may not be able to produce a name, perhaps, but we may be able to at least point to some intriguing possibilities.

Of course, it's difficult to imagine that any inventor—even one working in great secrecy—could have pulled off such a feat of engineering without leaving a tangible clue of having done so. Such seems too much to ask, especially when one considers the intense scrutiny other aviation pioneers of the time endured in their efforts at conquering flight. Additionally, such a person would have had to overcome a number of technological hurdles as well—some of which would have been formidable even to many of the finest inventors of the era—making the prospect even more difficult to accept.

If the story is true, however, that means we would be dealing with an inventor who might have been years ahead of his time, both in terms of airship design and the ability to construct the power plant needed to propel it. Fortunately, building an airship at that time would not have required any new technologies be created in order to make it work; it would only have been necessary to take the various technologies available at the time, perhaps refine or modify them, and finally bring them all together in an efficient manner. As such, an "inventor" would not be required so much as the services of a gifted engineer—a commodity that would have been in far greater abundance in 1890s America than were inventors. (Further, an engineer would have been more likely to have received formal training in various disciplines as opposed to most inventors—who were often self-taught "tinkerers"—and so could be counted on to be a bit more grounded in his ideas.)

Obviously there's no way of knowing who this person might have been, but then it's not really necessary that we do. What's really at issue here is if such a man did exist, how could he accomplish so much in secret? In effect, how could he have kept the construction of something as substantial as an airship so well hidden from the press and the public for so long?

Or did he? Contained within the earliest reports of the airship is an interview with a San Francisco attorney named Collins, who claimed that he not only knew who the airship inventor was, but even where his "secret" base of operations was. In a particularly detailed and matter-of-fact interview, Collins claimed that he was approached by a man in November of 1896 who asked him to represent his interests in patenting his airship. Collins maintains that the man had come from Maine seven years earlier in order to conduct his experiments without danger of interruption near the little town of Oroville, some fifty miles north of Sacramento, where he was working on something that would "revolutionize the world."

Unfortunately, in the ensuing weeks the mysterious inventor failed to make an appearance and all references to him soon faded from the public discourse (along with, it stands to reason, much of Collins' credibility), leaving skeptics safe in their pre-

The "airship" sightings weren't confined to California. In April of 1897 the craft was supposedly seen flying over Chicago, Illinois, as this engraving from the era illustrates.

sumption that Mr. Collins had an extraordinarily gifted imagination. But is it really as simple as that? Obviously, while there's an excellent chance Collins was making the whole thing up, there is also some reason to wonder why he would do so. Certainly he was risking his reputation and legal credentials in doing so, potentially damaging his law practice if it were demonstrated he was lying, forcing us to wonder what would have induced Mr. Collins to fabricate the story. Certainly, for a pure fabrication it appears to be unusually detailed. (Why Maine, for instance, or Oroville? Why so many details at all, if it was merely a joke?) As such, we would be foolish to not at least entertain the notion that there may well have been a kernel of truth to Collins' story.

Since we are playing devil's advocate, let's presume that Collins was telling the truth (exaggerated and sensationalized as it may have been). The first question we might ask is why he would have picked Oroville as the scene of his alleged inventors activities. What is it about that place in particular that would entice him to choose it?

In retrospect, Oroville would have been a choice locale. For one, it was located off one of the main rail lines running through the state, giving a potential inventor easy access to materials and supplies from the East Coast. However, it was also remote enough that any "airship" construction could be easily hidden from public view. With San Francisco close by, it also gave him access to the largest sea port on the West Coast at the time as well as access to something even more important: someone to finance his project.

Consider that in 1896 San Francisco was one of the wealthiest cities in America, with the largest number of millionaires per capita of any city west of the Mississippi. As such, there would have been any number of wealthy industrialists, mine and business owners, or speculators with the resources to finance something on the scale of our airship—an important consideration when imagining the cost such a project must have incurred. As such, with this ideal combination of investment capital, accessibility to goods by both rail and sea, generally mild weather, and a still largely remote location, central California in 1896 would have been one of the most perfect places in the nation from which to build an advanced airship in relative secrecy.

But why would either the inventor or the investors want it to be built in secrecy? Wouldn't they want publicity in order to attract additional investors?

Not necessarily. There were two very good reasons for wanting to keep his airship under wraps. First, if word got out about what he was doing, that would put the entire operation under the microscope of the press, who would have essentially camped out around the airship—possibly hindering construction. Of even greater concern would have been their presence during testing of the ship, when even the smallest failure or most trivial setback could easily scare away potential investors. Secondly, however, and perhaps even more importantly, would be that constructing the vessel under the watchful eye of the general public would have opened the door to patent theft as competitors and opportunists looked for ways to take advantage of the situation. Such would not have been a petty concern, either, especially when one considers that the last half of the nineteenth century was a time of tremendous competition, especially among inventors. In fact, the drive to be the first to the patent office was almost cutthroat in nature, with stolen ideas and even sabotage not unheard of. Were unscrupulous competitors or another inventor working on a similar design (remember that David Schwarz's aluminum airship was being built around the same time) legal wrangling—perhaps encouraged by jealousy on the part of competitors—could have grounded the airship more securely than the strongest gale force winds. One only need consider the fierce legal battles that had been fought between George Westinghouse and Thomas Edison over their alternating current versus direct current designs to appreciate the spirit of unhealthy competition and, on occasion, outright thievery that marked the early years of the Industrial

Revolution. In such an atmosphere, it's not difficult to appreciate why someone would want to keep their ideas closely guarded and work quietly to secure the necessary patents.

While this scenario may lay out a reasonable premise from which to explain the initial sightings in California, how would they explain the later sightings in the American Midwest that next spring? Would it make sense that our airship inventor would have moved his whole operation east of the Rockies after just a few test flights over California and, if so, why?

This is the most mysterious element of the story and, in many ways, the most difficult to explain. Clearly it wouldn't make sense to move the operation out of California without some greater agenda being in play, but what could it be? There is one possibility to consider: If we accept that the 1896 sightings were, in fact, a man-made dirigible being tested, we must ask ourselves just why it had been built in the first place. After all, building an airship would have required a huge financial investment, which would suggest that there had to be an equally large payoff in the offing to make it feasible, but what could that have been?

It had to have been more than to simply determine if lighter-than-air flight was possible. LTA technology had already been demonstrated decades earlier, so just building a larger version of an airship for the sake of progress would not have been worth the great expense. There had to be some larger, ulterior motive, to make it worth the effort. The only answer that seems to make sense is that this individual—or possibly a consortium of investors—was planning to kick off what could only be described as revolution in travel. Like Solomon Andrews and his airship company idea in 1866 (referred to in chapter six), I suggest this man had hit upon exactly the same idea. In effect, could the airship have been a prototype for what would one day be a fleet of airships that together would form the nucleus of the world's first transcontinental air service?

It seems fantastic, but consider the prospect from the perspective of the late nineteenth century. At that time travel across the country was done either by rail or by steamship, both of which were expensive, time-consuming and often uncomfortable modes of traversing the vast distances between cities. A train from New York to San Francisco, for example, was an arduous journey under even the best conditions, with frequent stops, cramped quarters, and often noisy conditions being the norm, with the prospect of derailment or even catastrophic disaster always a possibility. A steamship offered more comfort, of course, but was hugely expensive and, in the decades before the Panama Canal was built, required a journey of several weeks duration (and one that involved rounding the treacherous and usually storm-tossed Cape Horn). An airship service, on the other hand—assuming the technology could be perfected to make it feasible—would have made traversing the vast distances across America much easier. Not only would such a service make transcontinental travel a journey that could be measured in days (and ultimately hours) rather than in weeks, but it would have made its investors wealthy beyond their wildest dreams. Of course, it would have been financially risky, but when one considers that Germany's Ferdinand von Zeppelin accomplished that very thing with his massive airships a little over a decade later, the prospect doesn't seem as far-fetched as first presumed.

Obviously, this fleet of airships would not have been able to cross the entire continent in a single hop (at least not at first) but airships capable of flying two or three hundred miles at a time would not have been out of the question. With further refinements, larger versions might well have been able to make transcontinental flights, possibly within as little as ten years, thereby making crossing the entire country a relatively quick and even pleasant experience and ushering in the age of aviation a full decade earlier than it would otherwise have emerged.

Naturally, to accomplish this grandiose plan would require considerable investment capital, especially over the long term. Our engineer/inventor and his consortium of financial backers may have possessed sufficient resources to finance the prototype airship, but ultimately they would need more investors if they were to eventually oversee a fleet of passenger-carrying airships. To get them, however, would not be easy. Investors are not by nature anxious to put their money into risky schemes (of which an airship service would certainly qualify), so to get investors of the era to be more forthcoming with their dollars would require that they demonstrate airship travel to be practical, safe, and reliable. And what better way to do that than by flying an airship from one coast to the other—a feat that at the time would have seemed more like magic than simple engineering?

As such, with this scenario in place, it's possible that the flights over California in November and December of 1896 were test flights of a prototype airship that would eventually serve as a template for a fleet of much larger passenger-carrying vessels in the future. The later flights over the plains, then, may have been designed to manifest growing anticipation as the vessel progressively moved eastward via a series of short flights across the nation's midsection, effectively telegraphing its next appearance. Flying over Chicago in early April, then, ensured that "airship fever" would be at its peak and set the stage for its eventual arrival on the East Coast sometime later that summer. If all went according to plan, by the time it sat down near the financial and cultural center of America to throngs of enthusiastic onlookers, it would be a media sensation and an attractive investment for every man with some investment capital on the planet. In fact, it's difficult to see any other way an airship could have been practically built and flown in 1896, considering all the obstacles such a venture would have presented.

This may have been another reason behind the secrecy then. Until the prototype proved itself, anonymity would be necessary. That way, if the venture proved a failure, the entire affair could be quietly disbanded and financial losses kept to a minimum. So the only question that remains then, is what happened? Assuming the "Sacramento lights" were sightings of a genuine airship, and the craft was being tested in secret in anticipation of being eventually unveiled to a increasingly excited public, why didn't it succeed?

The question of why reports of the mysterious airship ended so abruptly in late April of 1897 remains among the greatest mystery of the entire affair. Did something happen that ended the craft's journey, and if so, what was it? Did the inventor determine, even after several apparently successful flights, that the ship was simply too fragile to make

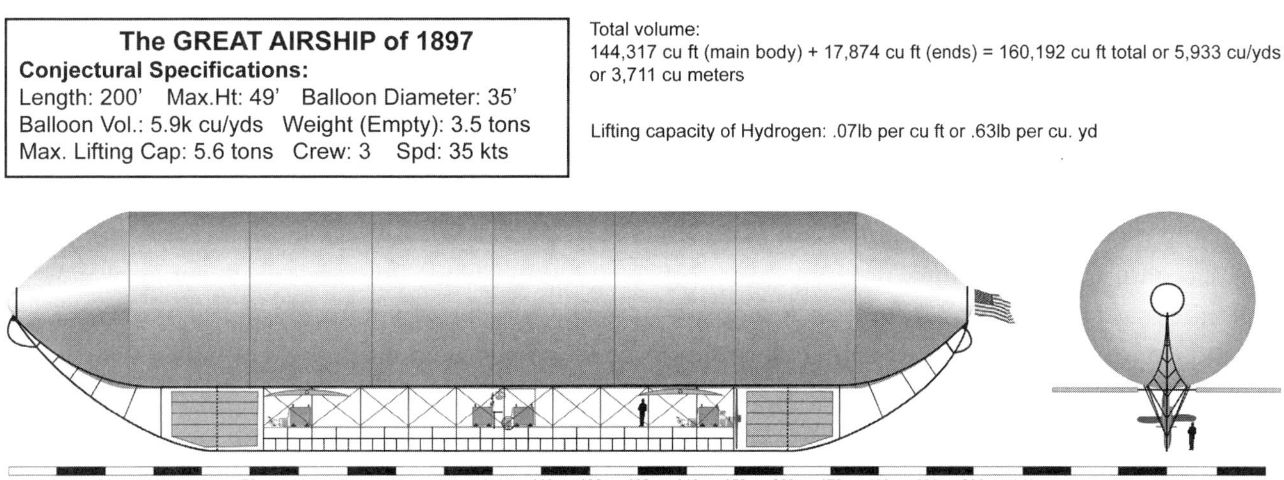

A highly speculative concept of what the great airship of 1897 may have looked like, based in part on eyewitness accounts and on the technologies available at the time. Could something like this have been responsible for the wave of "airship" sightings across California and the Midwest in 1896 and 1897?

the journey and called the whole thing off? Or did our financiers suffer some economic setbacks that forced them to pull the plug on the entire operation before it left them completely destitute?

In the context of what had transpired over the preceding months, neither option makes sense. The ship had been repeatedly tested over California and had already proven itself to be a robust vessel capable of flying under all sorts of conditions, making the chances of its designer suddenly deciding it was not up to the task inexplicable. Had the vessel proven dangerously unstable or irrevocably underpowered, such would have been discovered long before it reached the Midwest, not months later when it was on the eve of making its grand appearance. Additionally, even if the ship did have problems, it's hard to imagine how anyone with the determination and genius to construct such a ship in the first place wouldn't have wanted to press on and see the adventure through to the end. To walk away from the entire project just as it was on the verge of success simply doesn't make sense from either an engineering or human perspective.

And as for the ship's backers suddenly running out of money, this seems even more unlikely. Even if they had incurred considerable debt as a result of the venture, pulling the plug on it would have been the quickest way to ensure financial ruin. Were dwindling finances truly a growing and increasingly insurmountable problem, the only way they could have solved the problem was to complete the flight in the hopes of attracting new investors, thereby bringing a new infusion of capital into the venture.

As such, there is only one option left for us to consider, and that is the unpleasant prospect that something happened to our airship that was so sudden and catastrophic that it ended its cross country adventure in mid flight and brought the entire affair to an untimely end. But what could it have been?

Of course, we don't know, but could there be a clue contained within the many thousands of stories that were in circulation at the time? If so, there is one we need to pay special attention to. It came from a small Kalamazoo, Michigan paper dated April 15th, 1897, that reported that the vessel had exploded near the city with "...a report like that of heavy ordnance," and that the sound of an explosion was "immediately succeeded by a distant sound of projectiles flying through the air". This, of course, makes it sound as though the mysterious craft had come to a fiery end. It's interesting that the story broke around the same time sightings of the airship began to diminish dramatically. Could this be evidence that the Kalamazoo paper's story may have been more than just another tall tale but may have, in fact, contained some basis in fact? In essence, could the still largely experimental airship have exploded over Michigan, killing its designer and tragically ending its maiden flight as the story implies?

Is it possible that something could have touched off the hydrogen contained within the ship's envelope? While it largely depends upon what sort of power plant the craft used, there's any number of things that could have brought the vessel down. Heat from the motor, a stray spark, static electricity, a bolt of lightning—even a crewman smoking onboard, as inconceivable at that might seem to our modern safety-conscious sensibilities—must all be considered.

However, there is a problem with such a scenario: If the gas was somehow ignited and the ship crashed in flame, one would imagine that eventually some of the larger pieces of the ship—its charred motor for instance—would have been identified, yet to date not as much as a rivet or bent girder has been found. Of course, it could have gone down in dense forest or in a lake (Michigan being home to a number of fresh water lakes large enough to easily swallow an airship whole), leaving nothing to be found. Eventually, time and elements would have done the rest, erasing all evidence of the ship having ever existed (a fate which has overtaken many a missing aircraft and ship over the centuries).

While such a scenario is unpleasant to consider, it would account for the abrupt and dramatic end of the airship flap of 1897 and explain why nothing was heard of the mysterious vessel again. Obviously, the loss of the only prototype and its brilliant designer would have been irreplaceable, forcing the vessel's backers, perhaps already tapped out financially and too disheartened to start again from scratch, to cancel the whole venture, their dream of engineering an airship fleet dashed. Eventually the whole tragic affair would be forgotten and the mysterious airship and its crew relegated to the world of legend. It would have been a terrible tragedy, of course, and an incalculable loss to aviation, but such would have been the inevitable end to a noble but ill-fated experiment.

Unfortunately, there's no way of knowing whether the mysterious airship of 1897 really suffered such a tragic end or if the press and public—despite the many months of excitement the story had elicited—simply lost interest in the story and went on to other things (as unlikely as this may seem). The only fact we can take away from the whole affair is that for a few brief months during the final years of the nineteenth century, something unusual happened in the night skies over a large portion of America that captured the public imagination like few things had before or since. That it ended as suddenly as it began seems as unlikely as the idea that an airship could over-fly the prairie states of America in the nineteenth century, but ended it did—not with a bang (unless we accept the Kalamazoo account of a fiery explosion as accurate) but with a whimper, leaving us with one of the great aviation mysteries of our time.

Like Amelia Earhart's Lockheed Electra lost beneath the vast expanses of the Pacific Ocean, I suspect the mysterious airship of 1897 will remain a timeless puzzle that will never be solved, and perhaps that's the way it should be. After all, sometimes the greatest pleasures in life come not in solving our mysteries but in permitting them to continue to intrigue us, and therein may lie the real attractiveness of the "Sacramento lights."

U.S. PATENT OFFICE BLUEPRINTS -- 1844-1904

It is customary in a work of this nature to maintain a certain serious tone, as though history is something of a somber, no-nonsense affair to be read only through the prism of humorless objectivity. However, in doing so, it is our contention that some of what makes history enjoyable is lost, reducing its attractiveness to future generations. We, on the other hand, have chosen to throw caution to the wind and present these patent plates in a more lighthearted manner than purists may be comfortable with, not in an effort to humiliate or otherwise denigrate their originators, but in an effort to make what could only be considered dry engineering blueprints more interesting and, if possible, even fun.

Of course, many of these drawings set themselves up for some good-natured lampooning. The reader will find no lack of presumptuous would-be inventors and eccentric personalities, many of whom wear their lack of engineering acumen, aeronautical knowledge, or common sense like badges of honor, reflected in their drawings. (This tendency is more easily perceived in the descriptions that accompanied each illustration—many of which are often undecipherable and, on occasion, even inadvertently hilarious—but space would not permit us to include these with the drawings.) It's our contention that men whose hubris is such that they refuse to take the science behind aeronautics seriously and blindly foist their ridiculous concepts upon an unsuspecting public make themselves prime candidates for satire, and we are only too happy to oblige.

Not all of these ideas are nonsensical, however. There are a few that show early insights into practical applications that may not have been recognized for their value at the time but can be appreciated in hindsight. There are even occasional glimpses of genius in some of the renderings, and when we come upon such an example, we point that out as well.

All of the drawings in this collection, however, are imbued with the sort of imagination that exemplified an age when most believed technology was the savior of mankind and that anything was possible with a bit of determination and hard work. To many of the originators of these patents, flight had an almost magical quality to it which gave them permission to let their imaginations flow freely with the most bizarre notions. While most of what they came up with was woefully impractical and some of it even downright dangerous, it's still both interesting and, at times, even amusing, to look at their ideas and shake our heads in wonderment. Hopefully the reader will take this tongue-in-cheek examination in the spirit in which it is intended and will find in it the ability to laugh at the follies of those who came before us and, in doing so, perhaps find reason to laugh at ourselves in the process.

These plates are presented in generally chronological order so the reader may get a sense for how aeronautical ideas have evolved over the last half of the nineteenth century.

We apologize up front for the poor quality of some of the drawings. Many have been repeatedly copied before being committed to digital format, resulting in a considerable loss of clarity and the acquisition of background clutter. Every attempt to clean the scans up has been made but even Photoshop can do only so much.

SIDE VIEW

Fig. 2.

Fig. 1.

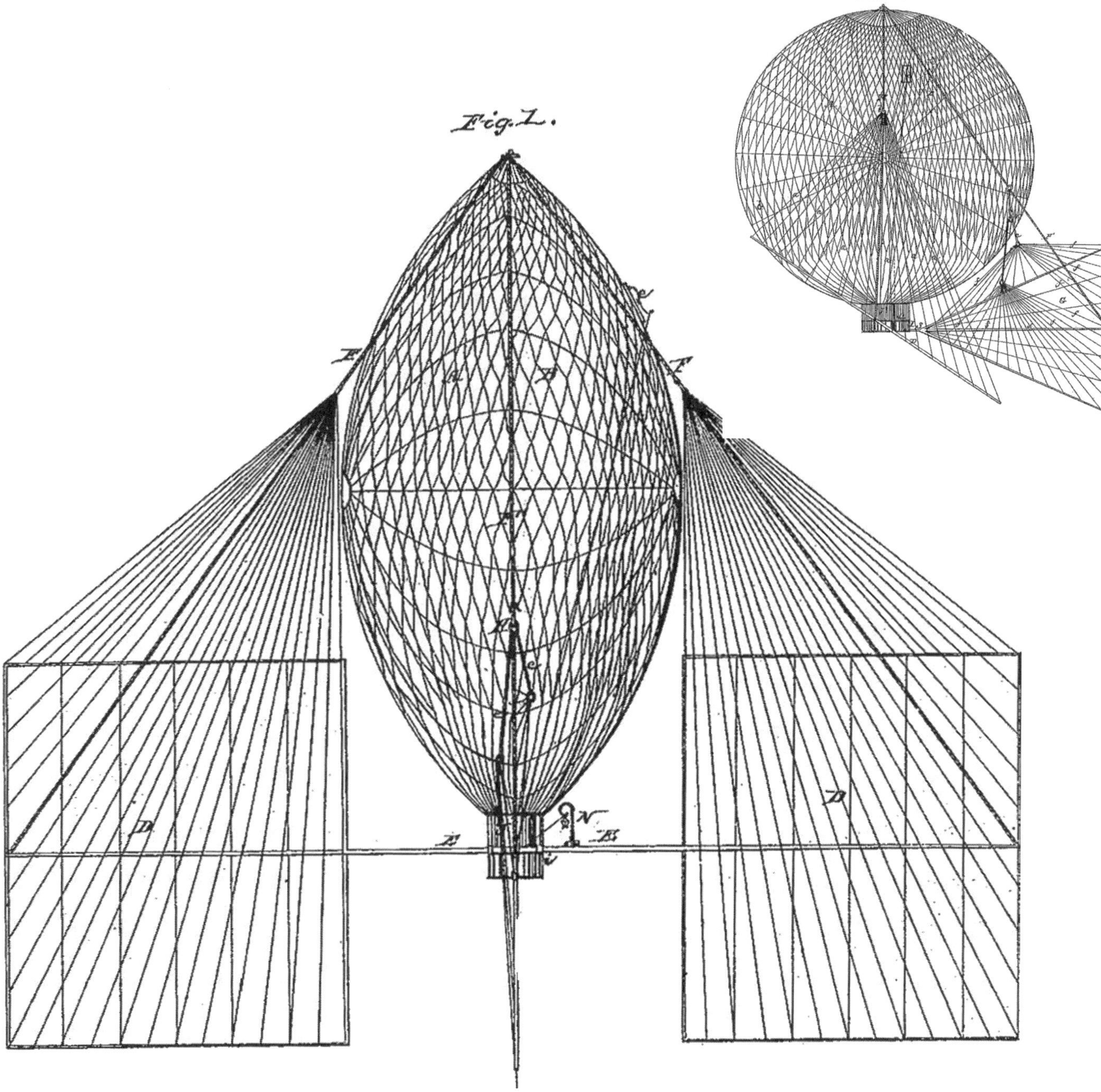

Plate 1: Muzio Muzzi's Flying Machine—October 16, 1844: *This is a balloon—and a rather forward-looking one at that considering the design dates from 1844—that is to be steered by means of "inclined planes which turn upon shafts or axles." Apparently the resistance of the atmosphere against the planes is supposed to turn an ascent or descent into horizontal flight, though it's not quite clear how, exactly, this was to be done. It's possible he's talking about simply using fixed sails—which he refers to as "planes" to guide the ship along, but this, too, is not entirely clear. Of course, none of these features would help the balloon maintain a direction; it would simply be blown along by the wind as all balloons are, though it certainly would make good time if it were assisted by sails.*

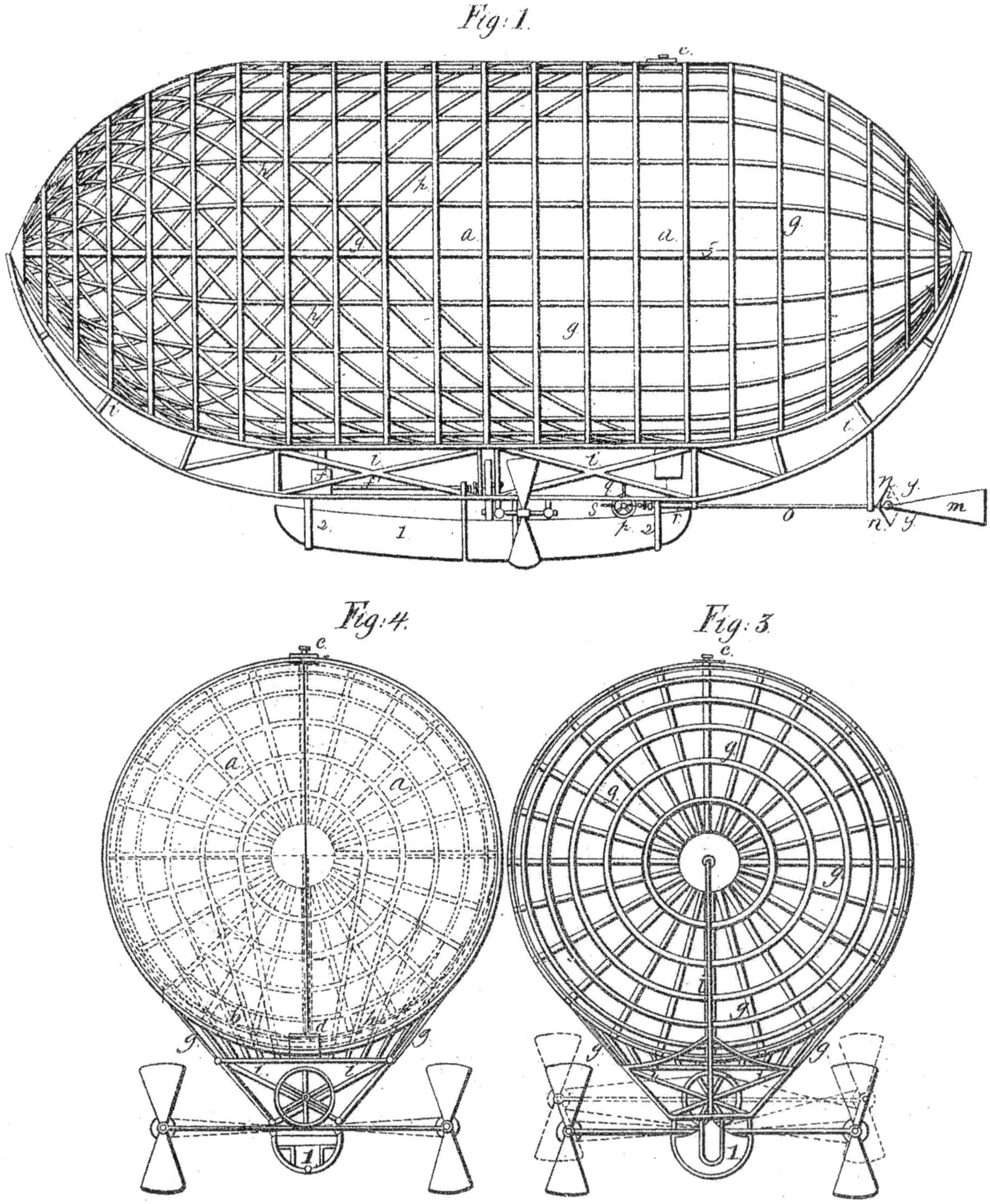

Plate 2: Hugh Bell's Balloon "with appendages"—March 26, 1850: *This fully rigid airship design with its cigar-shaped envelope, bow-to-stern keel structure, and a pair of propellers on the sides was actually quite advanced for the time. He also spends a great deal of time discussing an internal membrane that traverses the length of the ship, dividing the envelope roughly in half—another very forward-looking concept that predates von Zeppelin's compartmentalized airship by half a century. Unfortunately, Mr. Bell doesn't explain how the craft is propelled, though it seems there is a handle that one can crank by hand is so inclined, or he affords one the option of using steam power. In either case, with a powerful enough engine this would have proven to be an extremely practical design.*

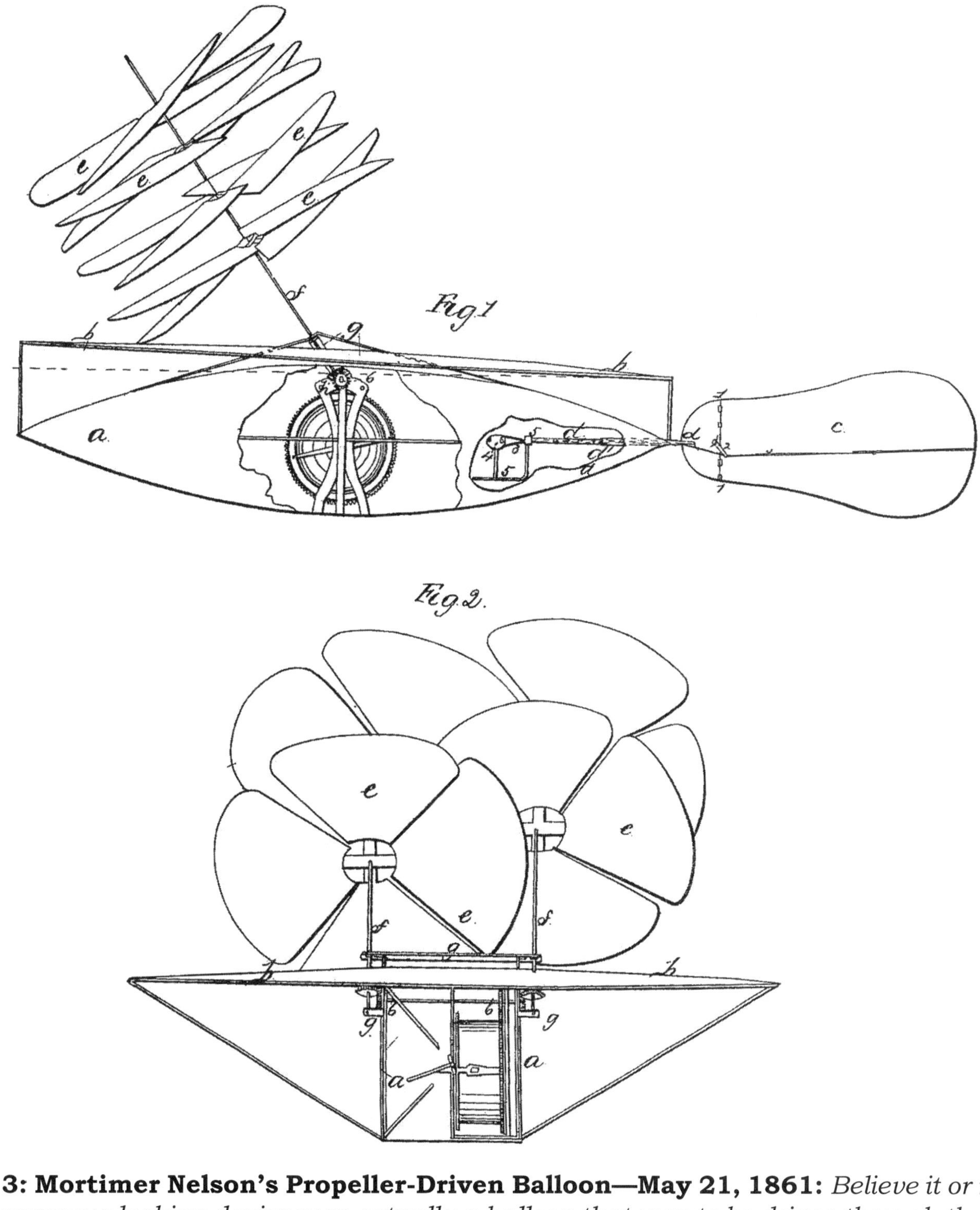

Plate 3: Mortimer Nelson's Propeller-Driven Balloon—May 21, 1861: *Believe it or not, this dangerous looking device was actually a balloon that was to be driven through the air with the assistance of four massive rotating fans hooked to a crankshaft that was to be turned by an unnamed power source. This was to provide enough thrust to push the craft in a horizontal direction, aided by a barely attached but massive ship-like rudder at the stern designed to steer the vessel in whatever direction its frightened pilot desired. Clearly the vessel lacked enough balloon—which appears to have been more of an afterthought in any case—to lift its own weight. The massive propellers would certainly have made Mr. Nelson's craft an intimidating sight had he been able to make it workable, however.*

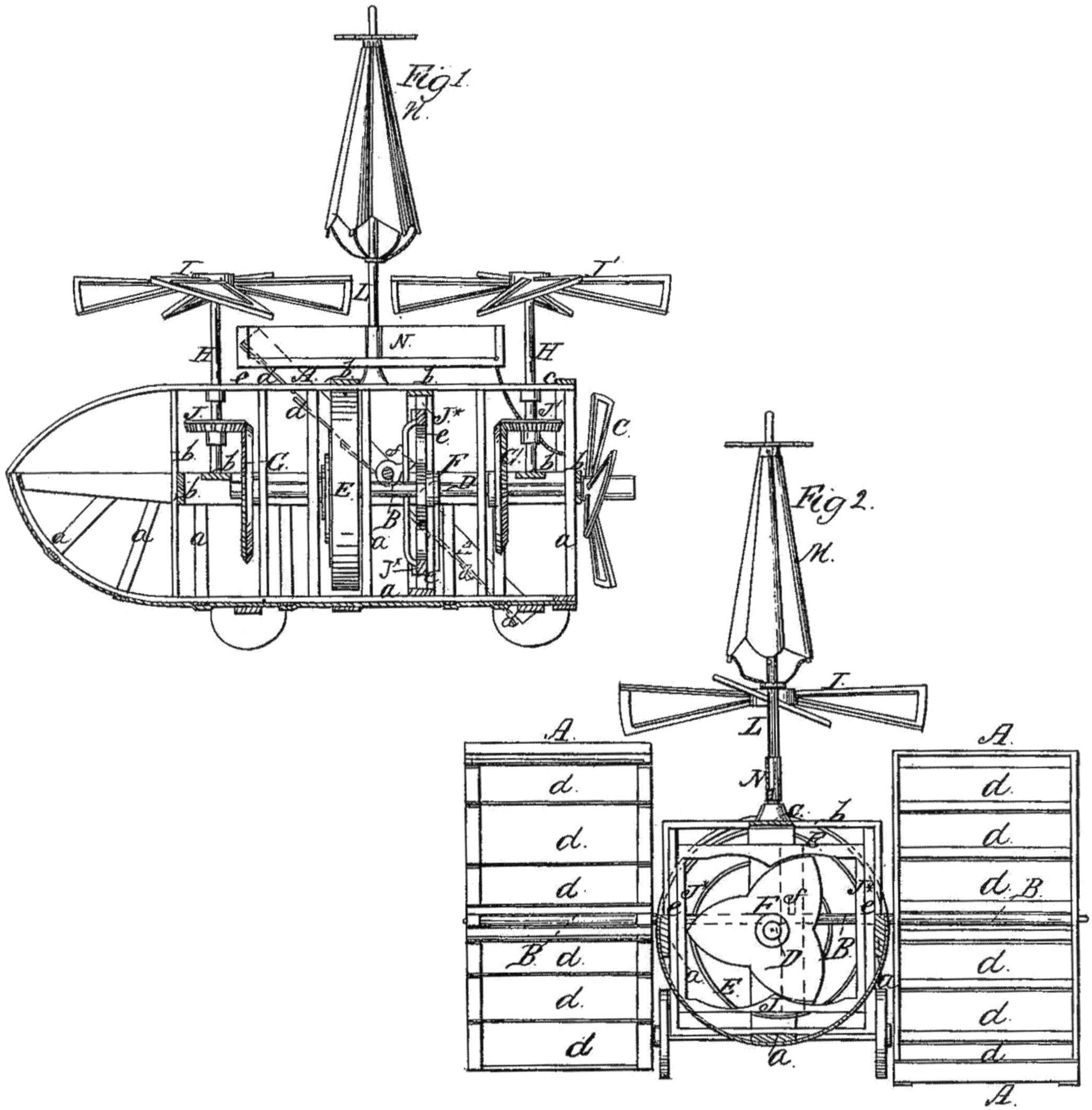

Plate 4: W.F. Quinby's Improved Apparatus for Navigating the Air—November 26, 1861: *Looking like a cross between a submarine and an army tank, this was actually a boat-shaped hull attached to a series of propellers and what appears to be a pair of massive, side-mounted paddle wheels. In fact, the craft is really nothing more than one big set of spinning props and oscillating wheels (which Quinby is certain would make the craft ascend.) It could either be driven by "a small steam engine, by manual labor, or by any other suitable means"—although he doesn't seem to leave any space inside the craft for an engine or crewmen. It's not clear whether the pointed shape at the top of the monstrosity is a closed umbrella or another type of propeller, either of which would only serve to underscore the futility of the entire design.*

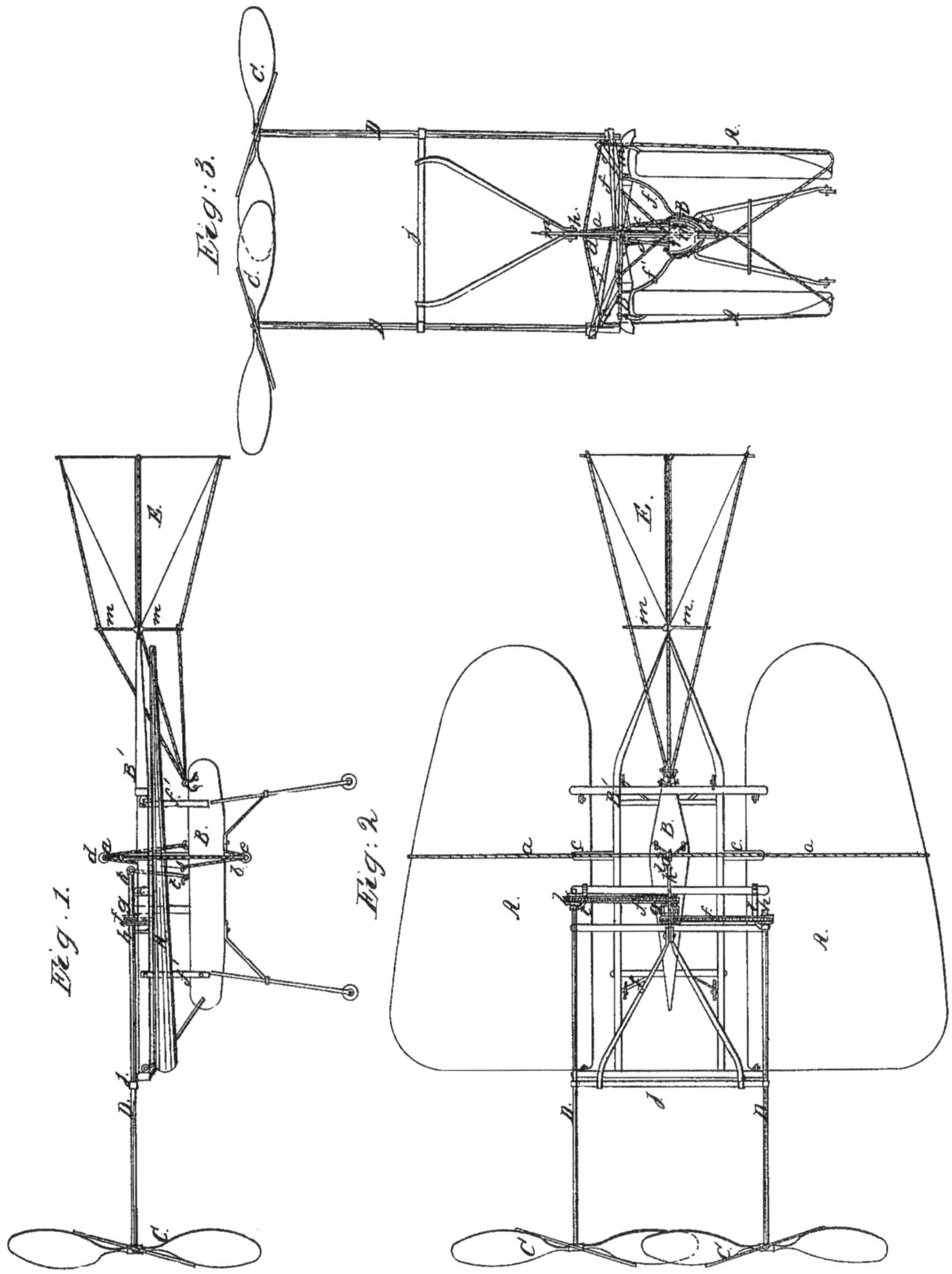

Plate 5: Luther Crowell's Aerial Machine—June 3, 1862: *This delicate-looking machine, looking a bit like a cross between a kite, a beanie, and one of Robert Goddard's rockets, is actually much larger than it appears. Lift is provided by filling the wings with hydrogen "or some other suitable gas," making this the first design in which the lifting agent resides within the confines of the control surfaces (the wings), which would be the equivalent of making the envelope of the Goodyear blimp into one giant aileron! At some point, the wings were to fold back into themselves to make descent easier (which it certainly would do). And don't worry about the fact that the huge propellers seem to overlap: they are counter-rotating. One interesting feature is that in figure 3, it appears that the propellers are rotated vertically, thereby making this design the first tilt-rotor aircraft in history!*

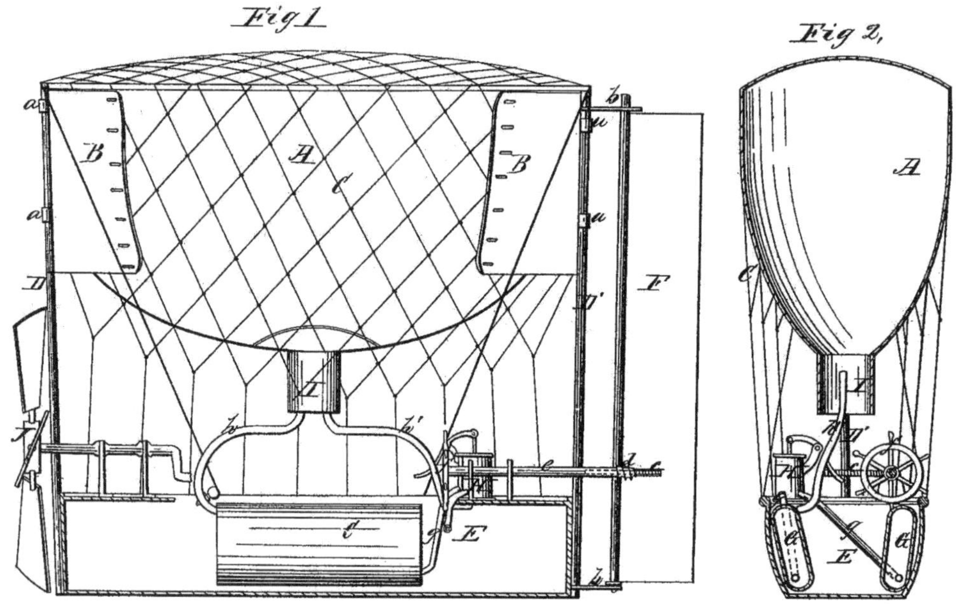

Plate 6: James Connell's Air Balloon—November 17, 1863: *Imagine creating a balloon in the shape of a huge ship, with a tractor propeller on the front and a huge rudder (letter F) on the back, and you have James Connell's brainchild. The most mysterious part of this contraption was figuring out how it flew and what powered its oversized propeller; apparently, it uses compressed "gas" of some kind (possibly magic gas), contained in chambers in the gondola/hull and connected by pipes to the pump at the base of the envelope, to both provide lift and, presumably, to turn the propeller (though this is not specifically articulated). Without even the hint of a motor of some kind (or even the mention of one), it's hard to see how the vessel was going to go anywhere.*

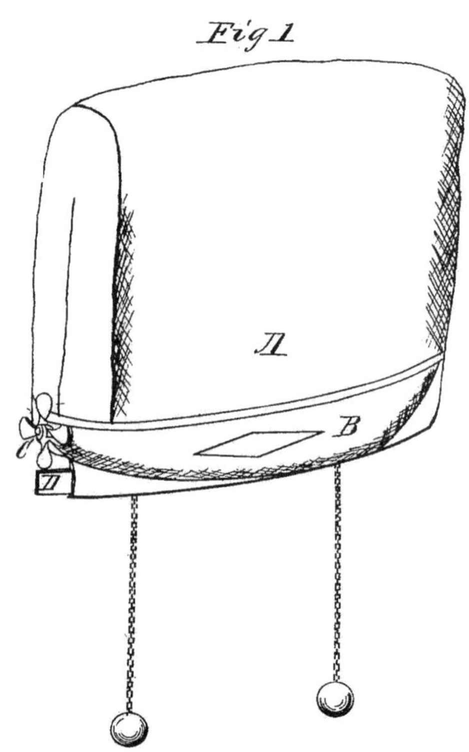

Plate 7: A.G. Wright's Improved Balloon—December 27, 1864: *Who says engineering drawings have to be complex? This knapsack-shaped device is about as simple as they come. (Actually, the envelope is supposed to be boat-shaped but Mr. Wright's—no relationship to the brothers from Dayton, Ohio—drawing skills leave much to be desired.) A.G.'s big contribution to the world of aeronautics was having the hull attached directly to the envelope instead of suspended underneath by cables, making it, in his words, "more compact." Wright is also very sparse when it comes to details about the interior layout of the hull or the means of propulsion, telling us only that it "contains the requisite machinery for the purpose of driving a propeller." It also has a hydrogen-making machine inside that apparently creates the stuff out of thin air. (Actually, that would have been the device he should have patented, as it would have made him a wealthy man!) I'm not sure what the dangling chains out of the bottom are for; perhaps to turn on lamps.*

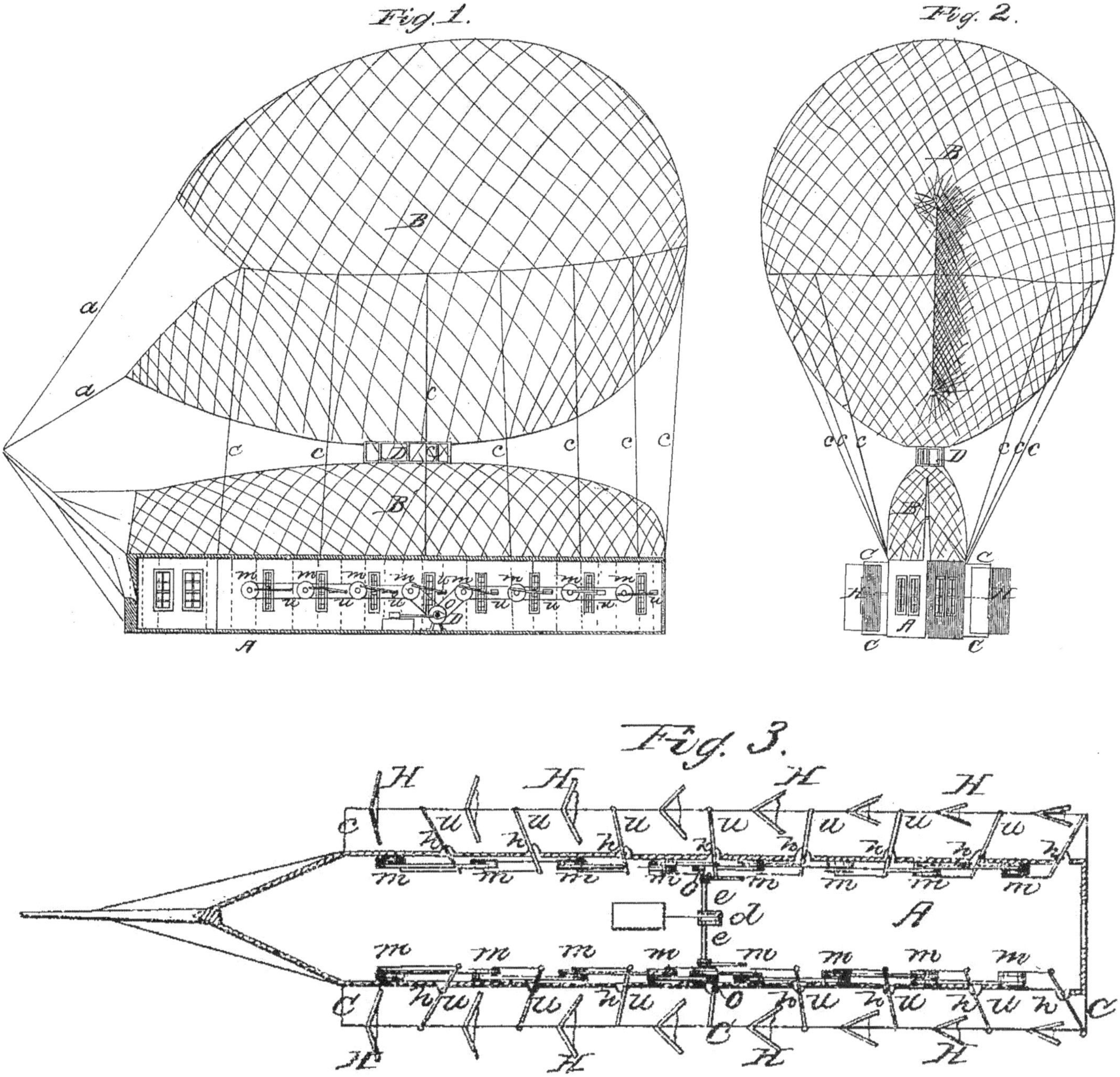

Plate 8: Francis Just and August Koellner's Aerial Car—October 10, 1865: *Reminiscent of a Roman galley suspended beneath a balloon, Just and Koellner's "aerial car" was designed to move through the air by means of a series of tiny wings, each flapping in sequential order so as to reduce the surging motion that would inevitably result were all the wings made to stroke at the same time. To power this behemoth (the hull being fifty feet long by nine feet wide and high), they propose the use of a "small caloric engine of from two to three horsepower." There's also a strange mini-balloon attached to the top of the hull, designed, we are told, to "prevent the balloon from affecting the position of the car in case of violent wind or squalls" (though it's not entirely clear how it might do that. We're told it's the most important feature of the design, however.)*

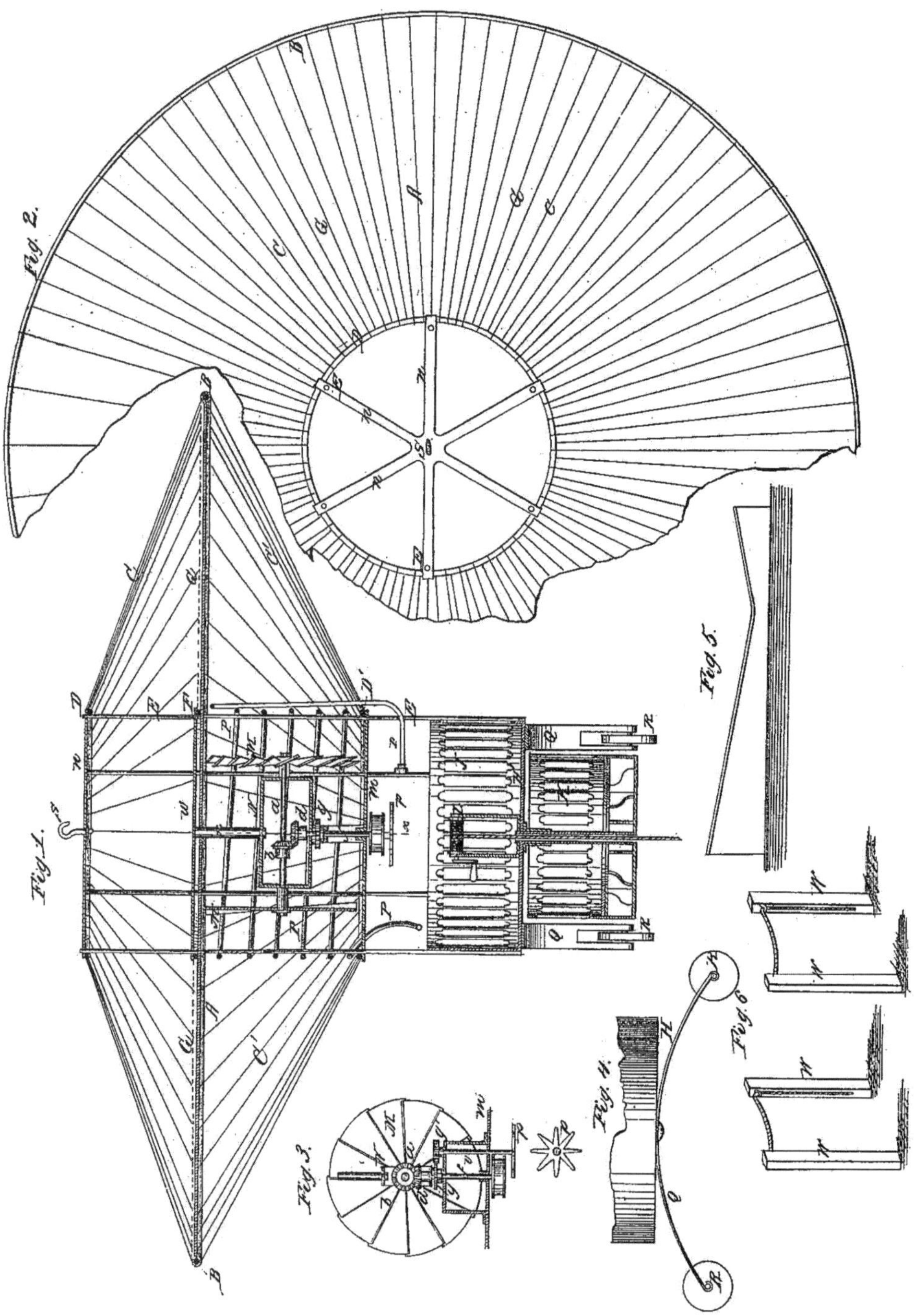

Plate 9: John Wootton's Flying Machine—May 22, 1866: *Looking a little like one of Hitler's unrealized secret weapons from the end of World War II, Wootton's "flying machine" was basically a parachute pulled tautly over a wooden frame, which is then pushed into the air by a massive, steam-powered unidirectional propeller. At first I thought this might be an early vertical take-off and landing (VTOL) craft, but the patent informs us that the entire monstrosity is designed to roll down an inclined track to a ski-jump at one end and rise—with the assistance of the violently twirling propeller—majestically in the air. Stopping the thing was even more novel: it has a hook attached to the top of the chute that is designed to catch on cables strung between poles, much like a fighter jet's arresting gear does during a carrier landing. I doubt if the results would be as hoped, however.*

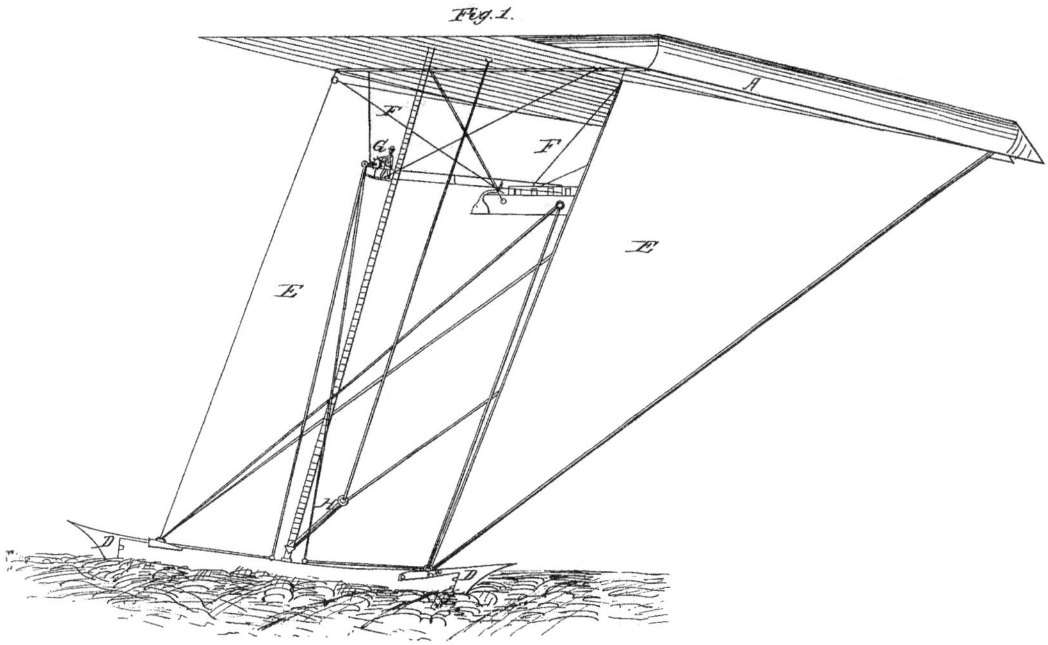

Plate 10: Chester Sykes Marine Balloon—September 11, 1866: *This is actually an interesting idea: tether a boat to a series of pencil-shaped balloons and fix massive sails between the boat and the balloons for directional control. Brilliant! Actually, it's really just a glorified sailboat with a hydrogen-filled sail, but it has the advantage of not needing a power plant (it's wind-driven) and of being somewhat steerable. The only problem is getting the wind to cooperate; in calm conditions it would be dead-in-the-water. In a gale it would be uncontrollable and may even pull the hull into the air before it has an opportunity to be dashed to pieces on a nearby rock outcropping. A good idea with danger written all over it.*

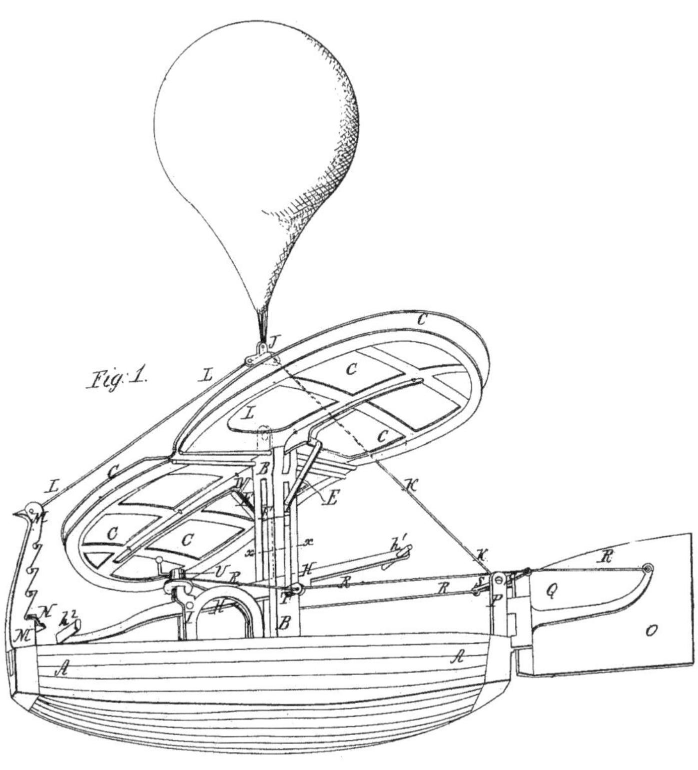

Plate 11: J.A. Elston's Aerial Machine—August 13, 1867: *In yet another attempt to make a boat fly, J.A. Elston of Missouri came up with the novel approach of creating a pair of massive, flapping bird wings that, along with the tiny balloon, should be enough to counter the forces of gravity. As far as can be determined, the wings flap through the use of a seesaw-like hand-pump, which sounds a bit fatiguing to say the least. It is clear from the plethora of boat-shaped designs that inventors of the era failed to appreciate the negligible effect even a massive rudder would have in determining direction in the air, especially when there is no force directed against it. Yet they keep trying...*

Fig:1

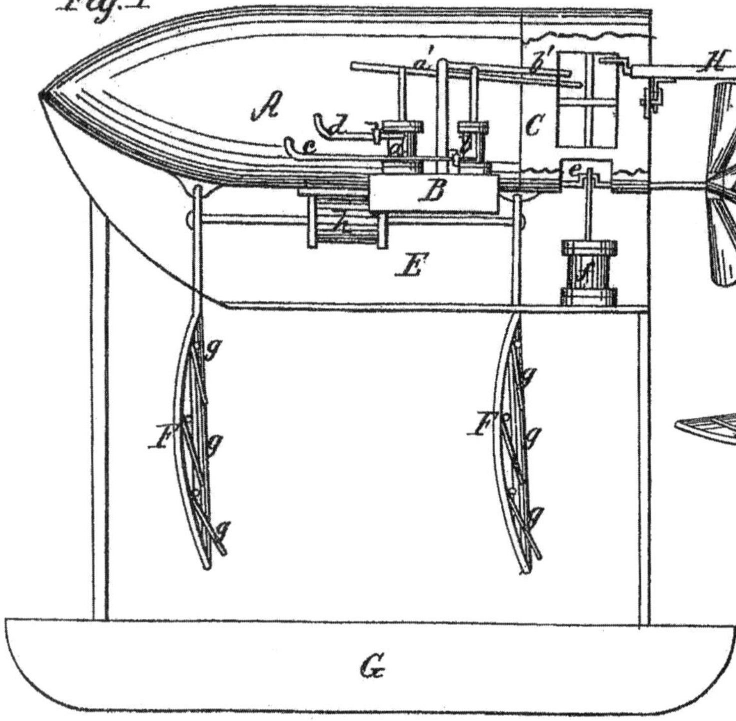

Fig: 2.

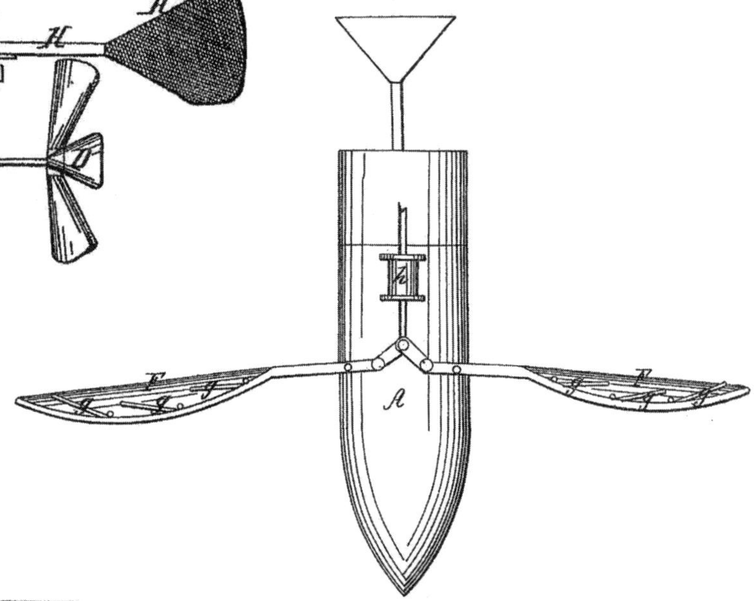

Plate 12: Onofrio Abbruzzo's Aerial Car—July 21, 1868: *As far as I can determine, this is essentially a flying bullet that can be propelled by means of a turning screw, scoop-shaped wings (attached either below or to the sides), or, when all else fails, "war rockets" fired from a canister slung below the envelope. What keeps the machine in the air is some sort of mysterious "fluid" that is pumped into and out of the "sheet iron" clad bullet-shaped hull (which also, by the way, houses the engine and fearless pilot). Steering is afforded through the always ubiquitous rudder. So confident was Mr. Abbruzzo in the lifting power of his vehicle that one could even attach a second, massive passenger carrying gondola (letter G) beneath the main one (letter E) if desired. The strange device shown in figure 3 is some sort of controllable parachute, suggesting that the ship had a much appreciated escape device installed in the off chance that something might go wrong with Mr. Abbruzzo's baby.*

Fig:3

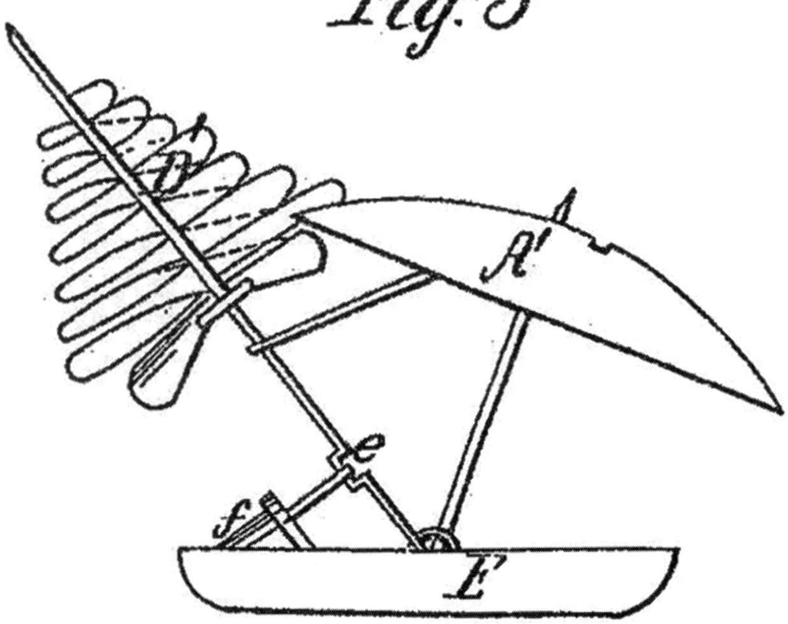

Fig. 1.

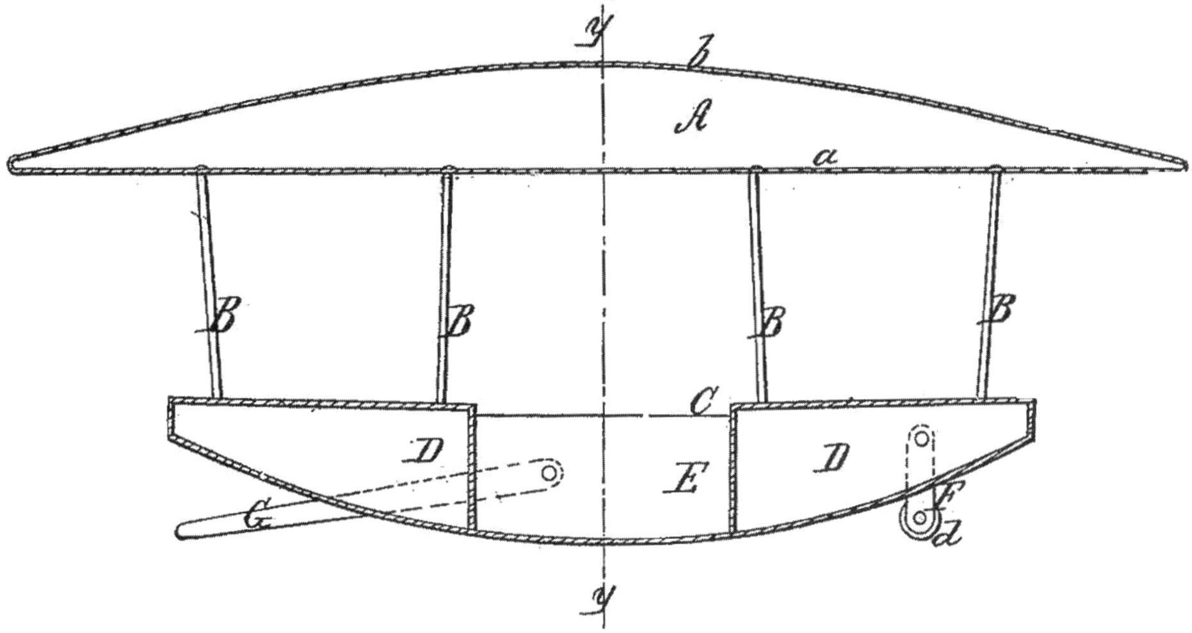

Fig. 2.

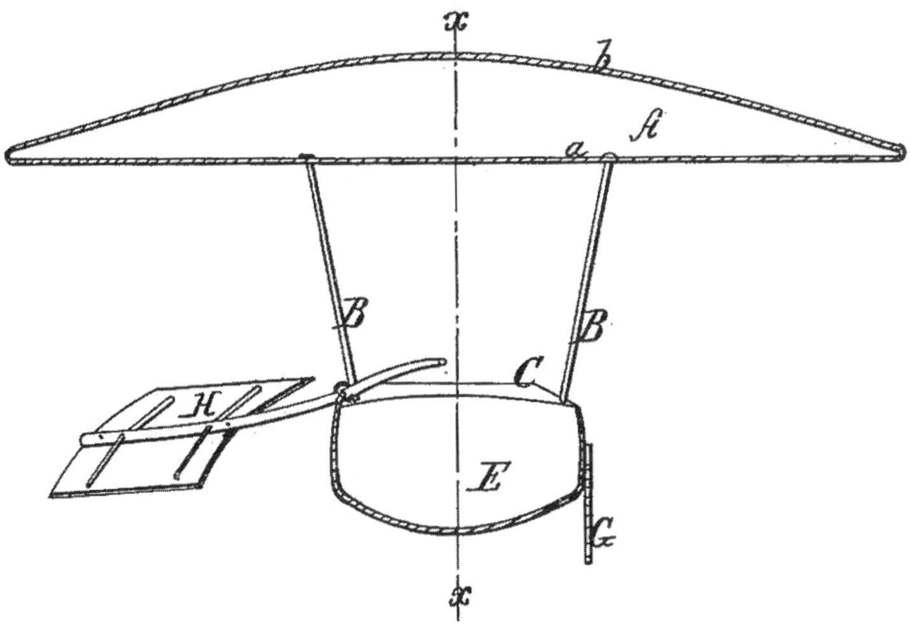

Plate 13: Zaphna Stone's Improved Aerial Navigator—May 12, 1868: *This dainty little gazebo-looking device is really nothing more than a small balloon designed to be paddled around in the sky. Perhaps recognizing that the tiny hydrogen-filled concave envelope was a tad too small to lift the craft, Stone had the gondola cordoned off into thirds and filled the ends with more hydrogen, making it the first design in which the lifting agent was contained in both the envelope and inside the gondola at the same time. Propulsion was provided by tilting the craft in the desired direction of travel and furiously pumping the paddle (letter H) until it jetted you to your destination. Primitive even by nineteenth-century standards.*

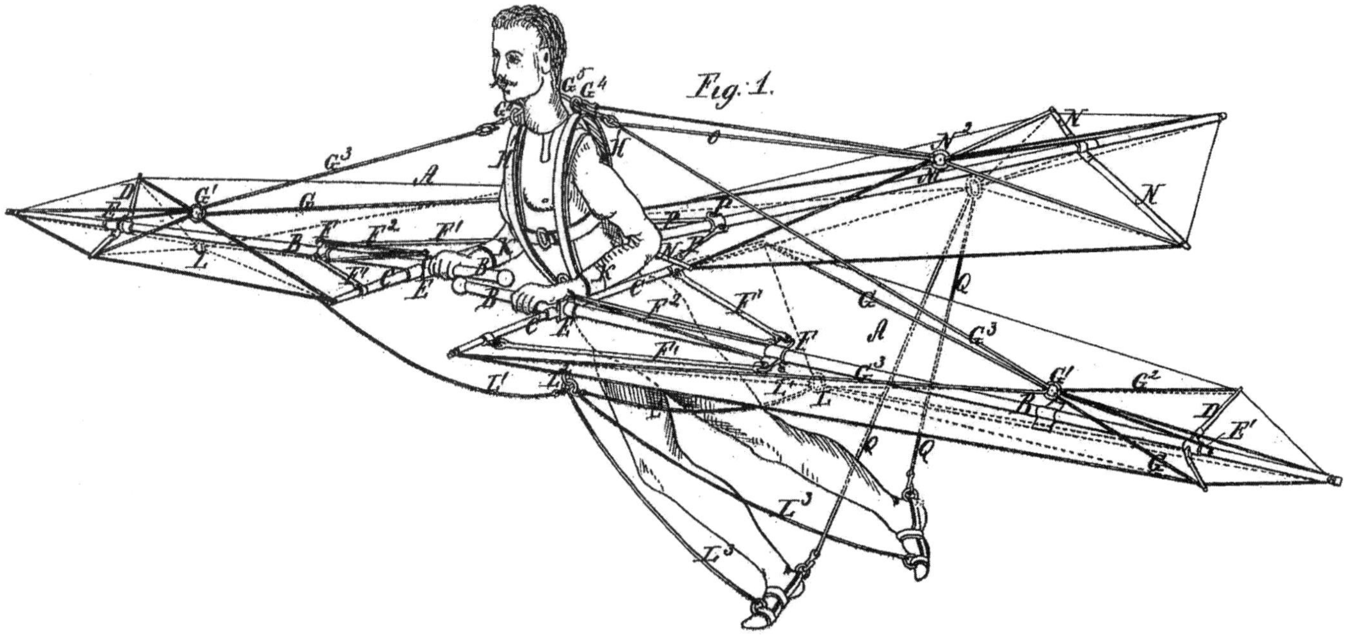

Plate 14: W.F. Quinby's Flying Machine—October 5, 1869: *Apparently seeing little success with his earlier device (see plate 4), the always creative W.F. Quinby is back for a second try. This time, however, he's returned to the basics, creating a lightweight set of wings that anyone with a death wish could easily operate. It appears that no part of the anatomy is spared the rigors of flight with this device, with the neck—usually considered a sensitive part of the body prone to decapitation when affixed to multiple cables—doing double duty. (We can see this better illustrated in figure 2 lest there be any uncertainty as to what Mr. Quinby had in mind.) Another unnecessary and frightening solution from the drawing board of one of the finest engineering minds of Wilmington, Delaware.*

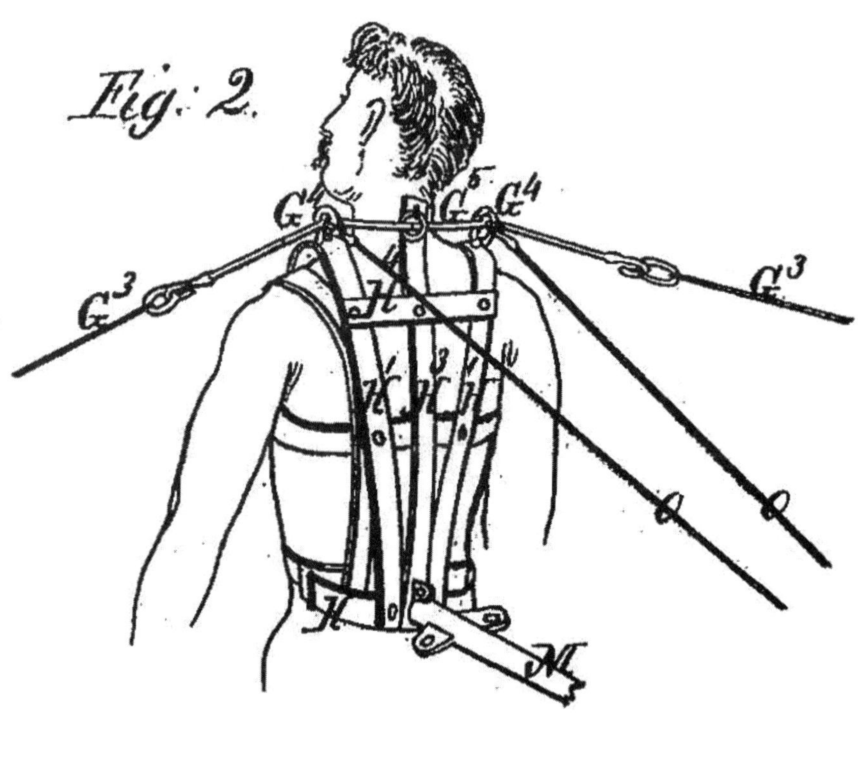

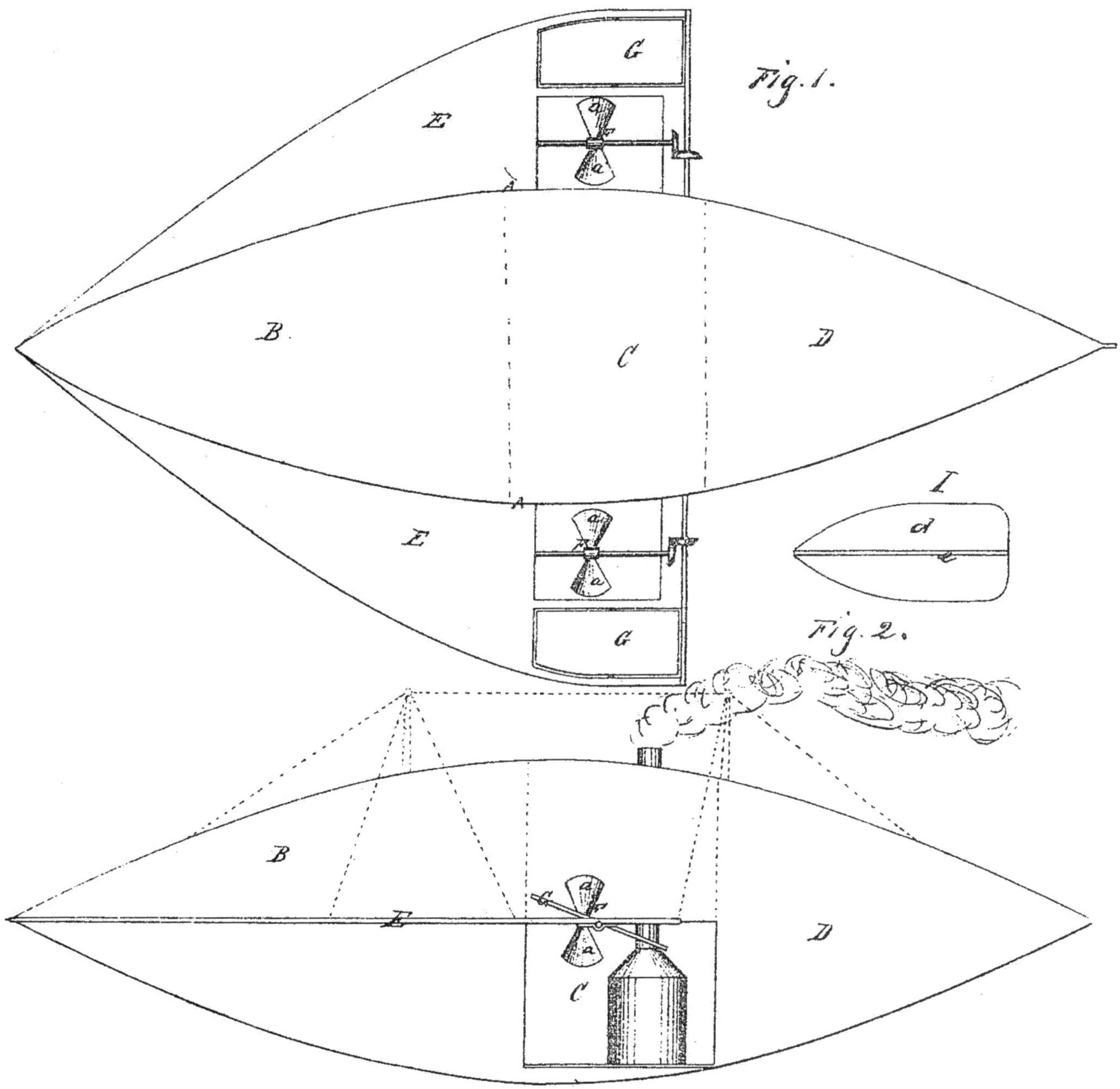

Plate 15: Frederick Marriott's Aerial Steam Car – November 23, 1869: *Straight from the Buck Rogers School of Dirigible Design comes this masterpiece of engineering. Basically a pointy-ended balloon with screws attached to the sides, notice that the elevators are mounted outside the airflow of the screws, thereby ensuring minimal control, while the four-vained rudder or tail, described in some detail in Marriott's description, is nowhere to be seen. (Apparently after figuring out the complexities of the side-mounted screws and elevators, he was too exhausted to figure out exactly where or how to place the rudder, being content merely to describe it.) Notice the steam boiler is contained inside the envelope sandwiched between two hydrogen-filled compartments. Presumably the pilot is also sealed into this death chamber as well, where he will have the best view of the spectacular explosion that will inevitably occur once the steam engine reaches the prescribed temperature.*

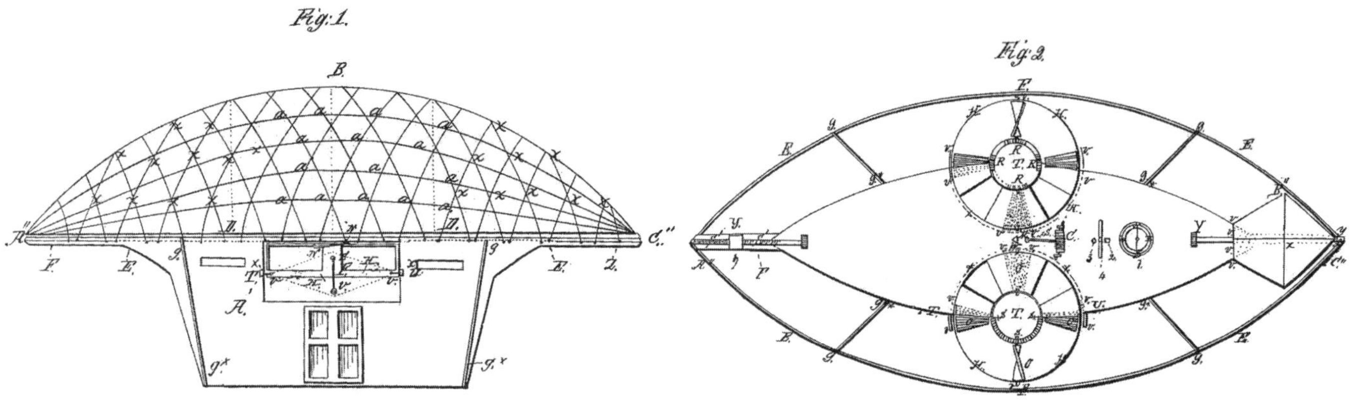

Plate 16: Martin Braun's Aerial Ship—February 8, 1870: *A design that would warm the heart of the late Buckminster Fuller, this half-balloon/ half VTOL craft actually incorporates a few fairly futuristic features such as a compartmentalized envelope—it's divided into thirds—and multi-directional propellers, both ideas well ahead of their time. It also has a number of instruments inside its quaint cottage-like boat hull such as a barometer, a thermometer, and something called a "spirit-level"—which I assume is required to ensure that one's spirit remains level. Not an unrealistic design but one that lacked the technological know-how to make it workable.*

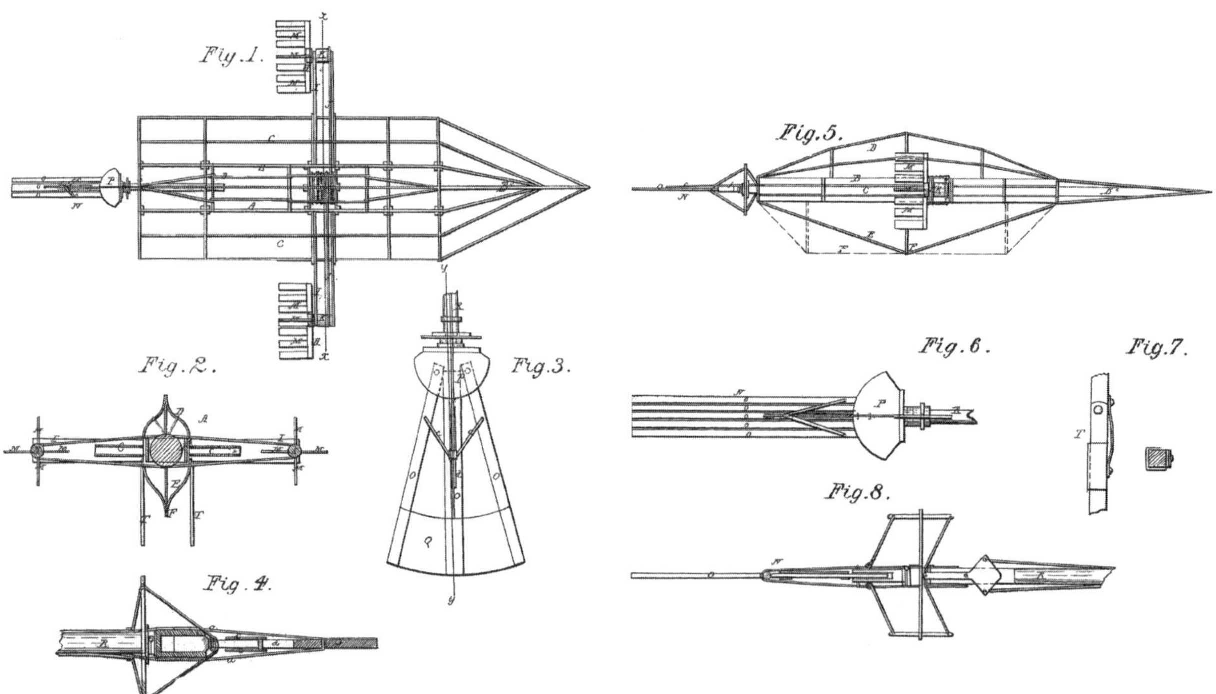

Plate 17: A.P. Keith's Improved Aerial Car—March 1, 1870: *Looking more like a collection of lawn darts than an "aerial car", the idea here was to take streamlining to never-before-attempted extremes, which the designer has clearly accomplished. Though it's difficult to see where engines, crew, and passengers might be located, we are assured by Mr. Keith that there is, "sufficient space for the reception and arrangement of devices generating the motive or actuating-power for paddle or other driving-wheels to act upon the air, and the accommodation of operators and passengers, together with storage for provisions, baggage, etc, according as may be desired." Obviously, he couldn't say that if it weren't true.*

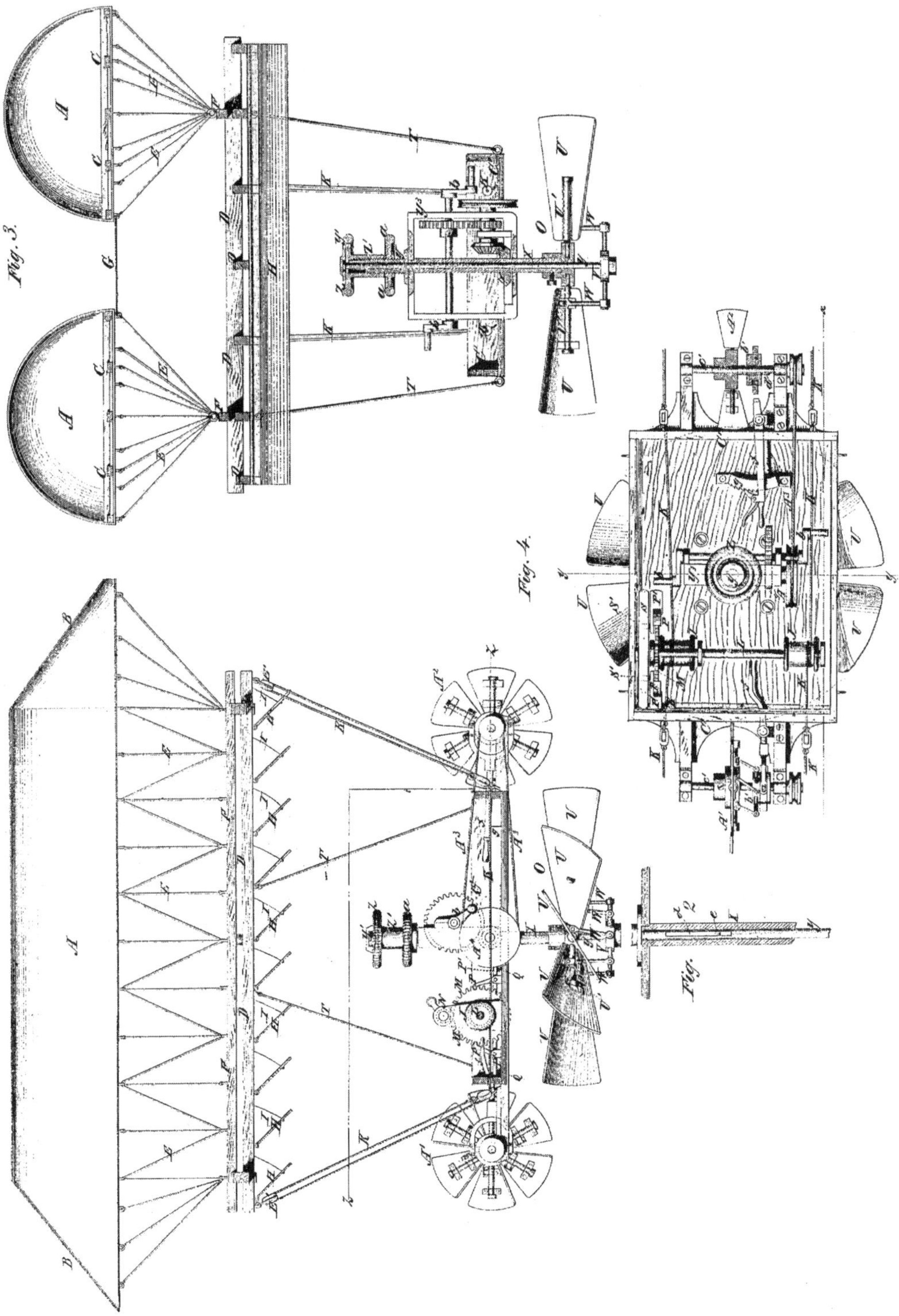

Plate 18: Edward Oakes' Aerial Car—August 25, 1870: *There can be no doubt that whatever it is Mr. Oakes has designed here, he is a draftsman of the first order. In fact, this fairly pedestrian design for a ceiling fan suspended by balloons looks so technically sophisticated, I was tempted to invest in it myself. Unfortunately, the tandem balloons would be far too small to lift the monstrosity off the ground, and even if you could, I'm not sure who would want to be near those massive spinning blades. Still, a fine example of draftsmanship when such precision was often sadly lacking in patent drawings of the era.*

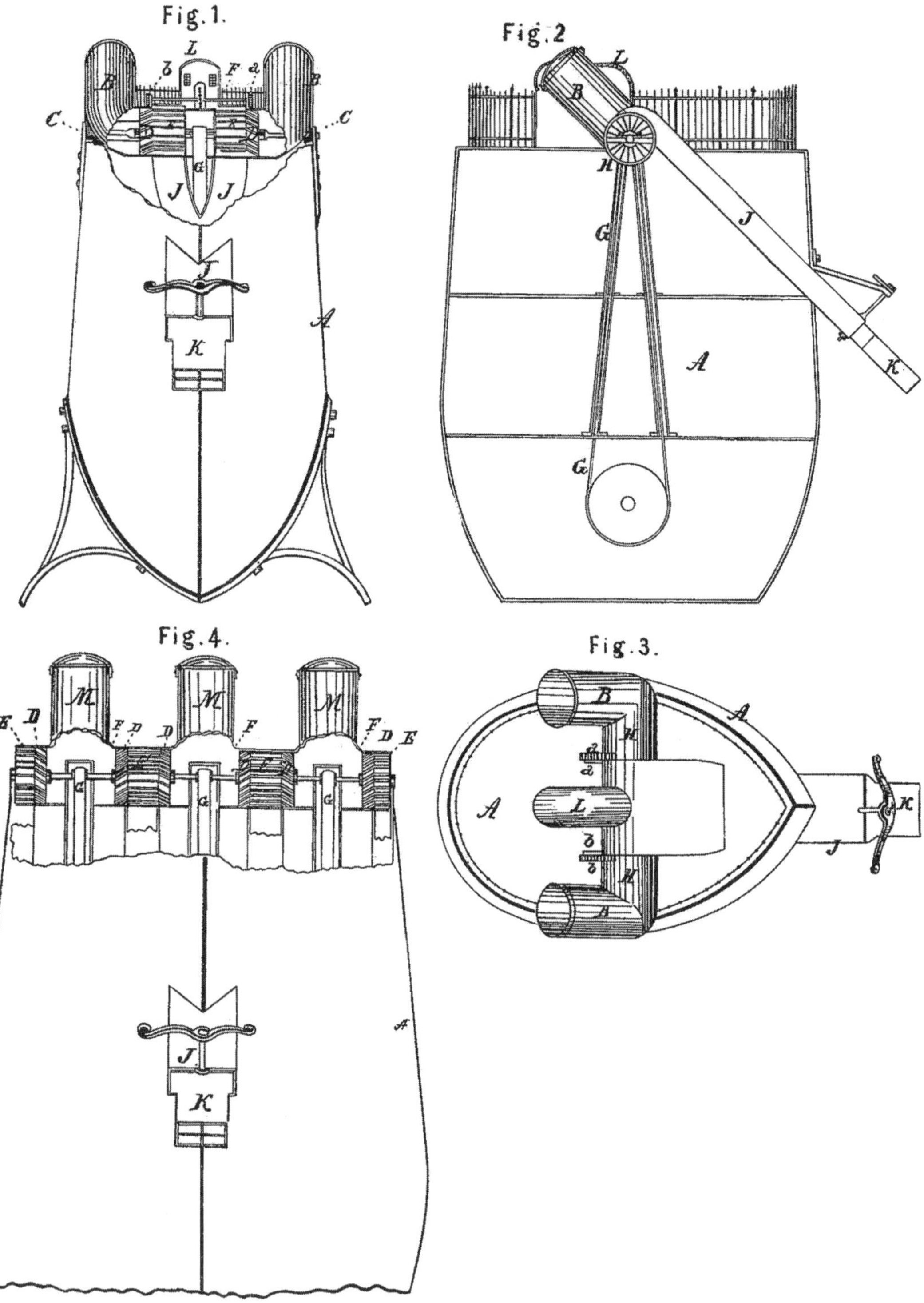

Plate 19: Isaac Forbes' Aerial Car—July 16, 1872: *Looking a little like a 1930's-era Maytag washing machine with pistons, we are assured that it is merely a device that maintains lift through the use of "helical fans, adjustable induction-tubes, a discharge-tube, and rudder." Basically, it's an early helicopter that is lifted through the use of massive fans and propelled through the air by tilting some elbow joints in the opposite direction of intended travel (sort of like a modern Harrier jet), aided by the always useless rudder. Apparently steam-powered, the amount of energy required to produce such an effect was still over a century away, though you can't blame Mr. Forbes for thinking big.*

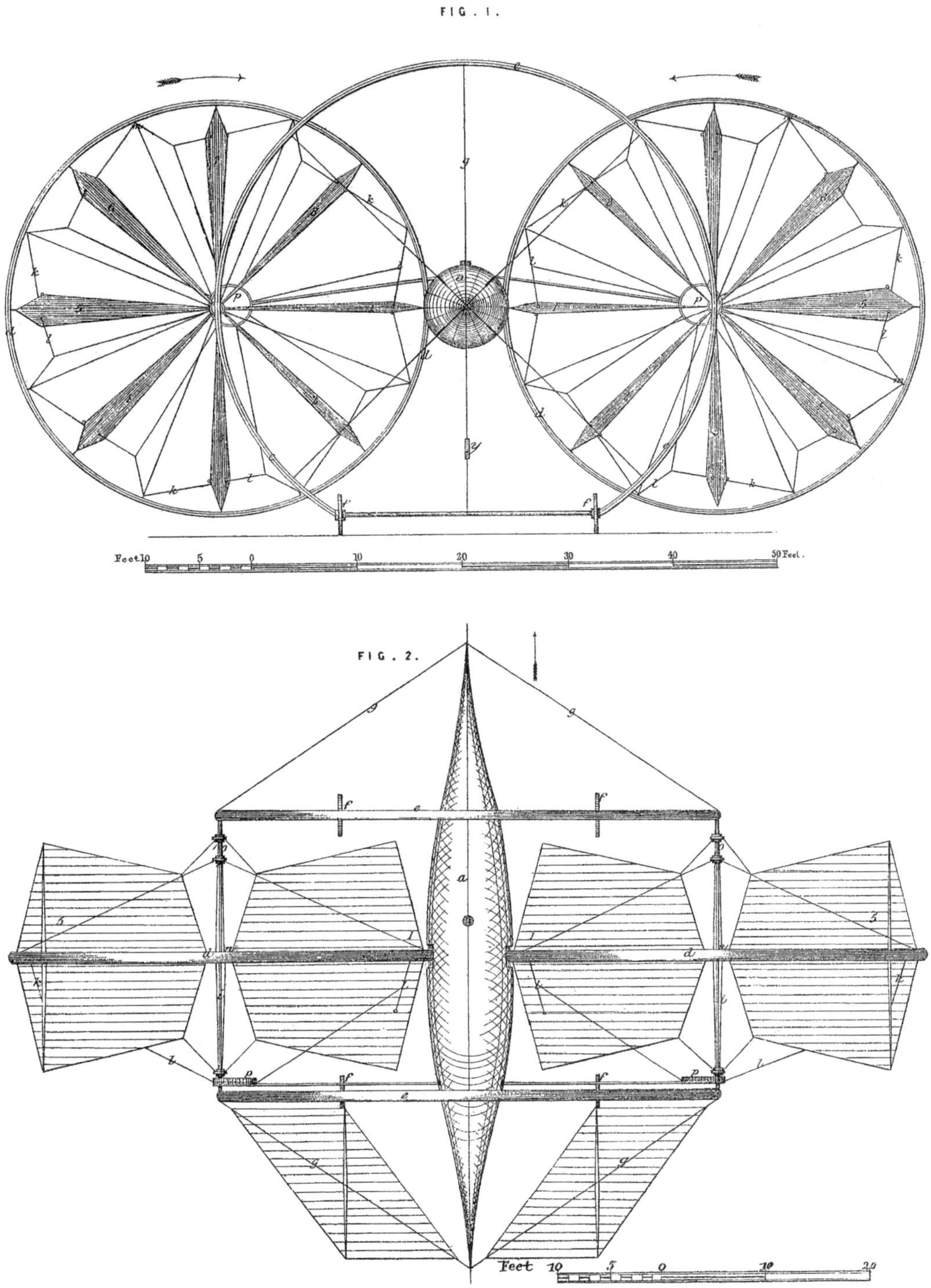

Plate 20: Moy and Shill's Apparatus for Aerial Navigation—November 26, 1872: *This elegant craft is really nothing more than a small, cigar-shaped hull surrounded by massive iron rings, two of which house a series of spinning canvas blades designed to change angle so as to either provide lift or propulsion. Think of it as two giant fans attached to a cucumber and you get the idea. This is one of the few drawings to include a scale bar, which tells us that, while the propellers are nearly forty feet in diameter, the central control car—which also contains the vessel's steam engine—is barely eight feet in diameter, making it a tight and decidedly uncomfortable squeeze for however many passengers and crewmen were unfortunate enough to book passage.*

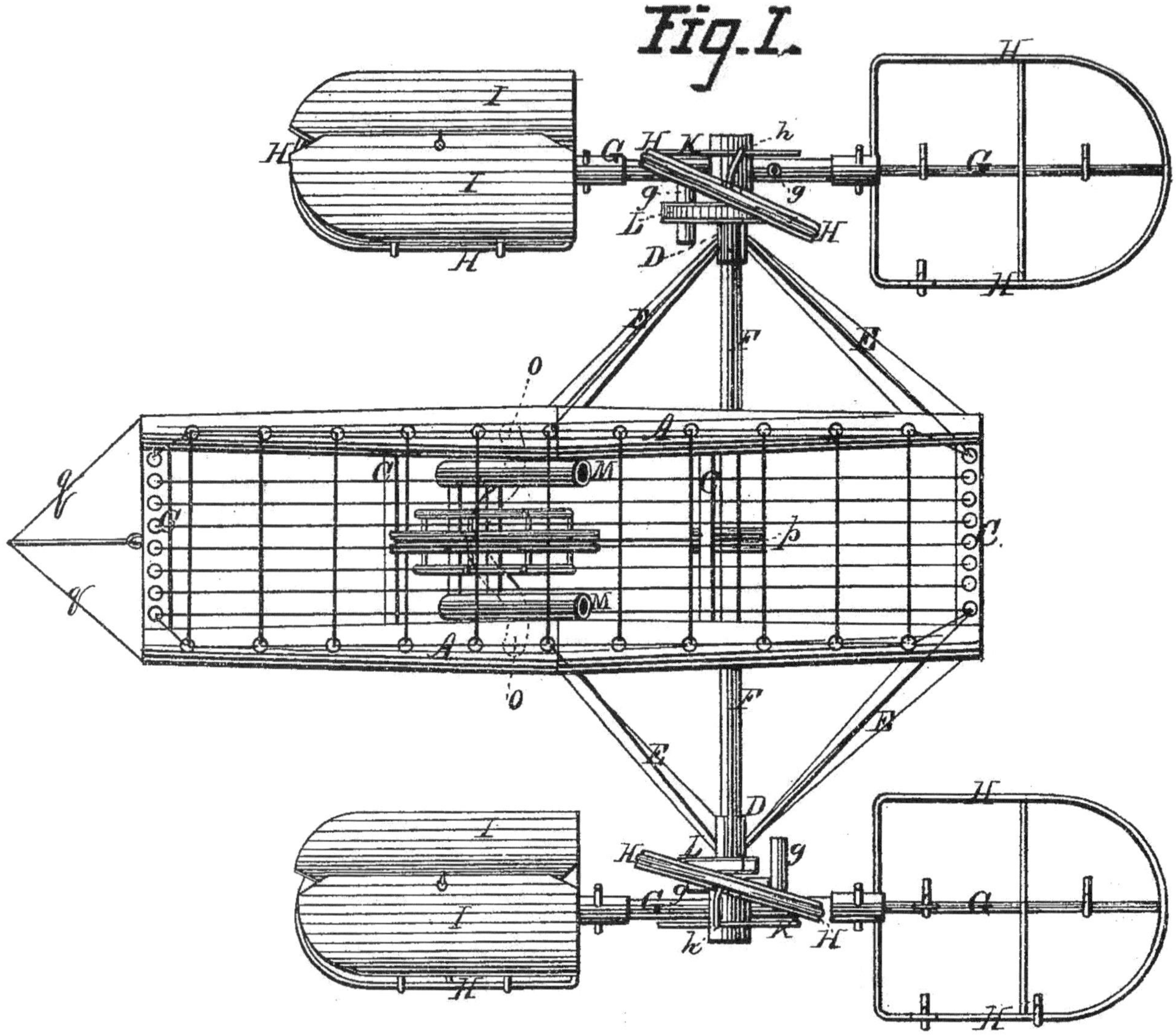

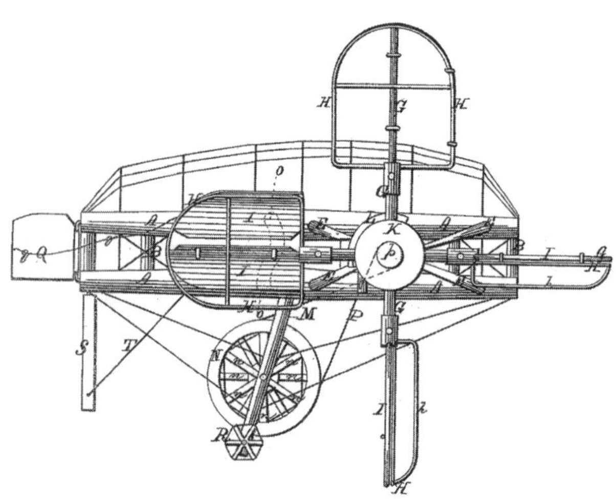

Plate 21: Thomas Green's Apparatus for Navigating the Air—August 12, 1873: *This machine, at least according to its inventor, "is simple, of light weight, strong, and enables the entire strength of a man to be exerted in such manner as to elevate and propel him through the air with great ease and rapidity." Apparently all a person had to do was furiously pump a pair of pedals to make the four-bladed side props spin "with great rapidity" thus easing its transition through the air. More likely it would result in a heart attack after about four minutes of flightless activity. One tiny design flaw is that the blades of the propeller extend below the tiny ground wheels at the bottom (designated R in figure 2), making it difficult to ascertain how the blades wouldn't be dashed to pieces when the machine is started.*

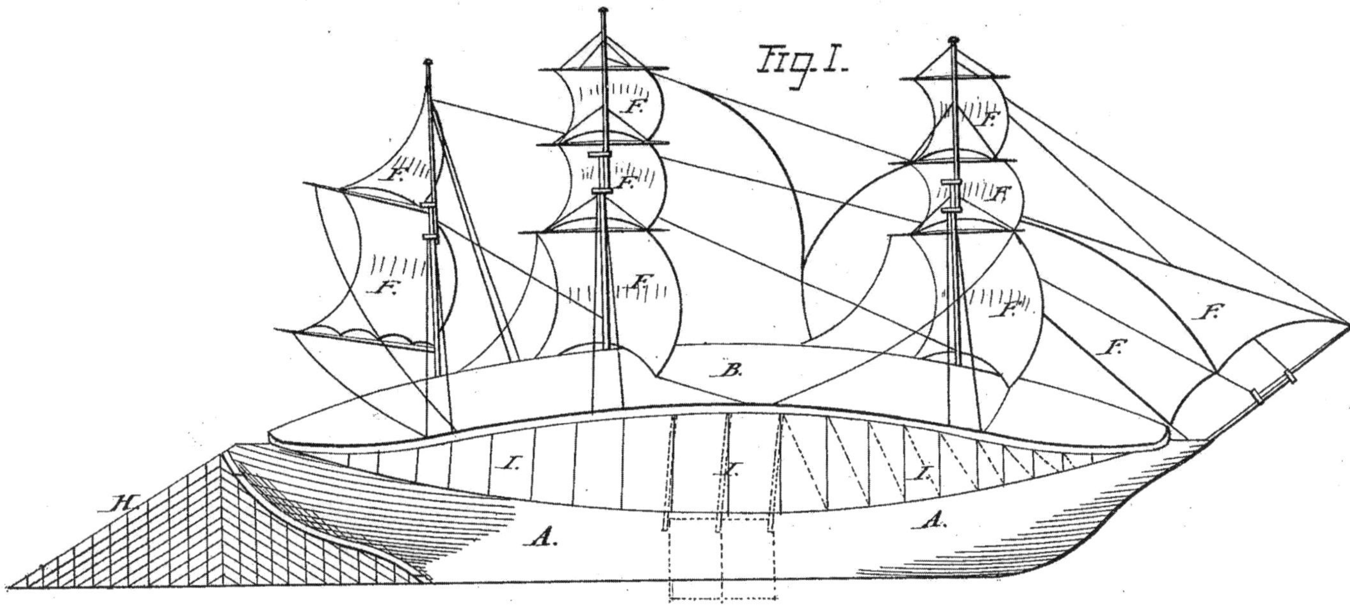

Plate 22: Otto Francis' Apparatus for Navigating the Air—February 10, 1874: *Surely the most elegant airship of all time, this so-simple-why-didn't-I-think-of-that approach to the problem of proper aerial navigation had been sitting under our very noses all along: why not simply take a three masted schooner, fill the hull with bags of hydrogen, outfit it with a fine set of canvas sails and attach the world's largest rudder to it, and simply set sail? Sometimes the best solution is so obvious we miss it—although it certainly didn't escape the notice of Mr. Otto Francis of Grand Rapids, Michigan.*

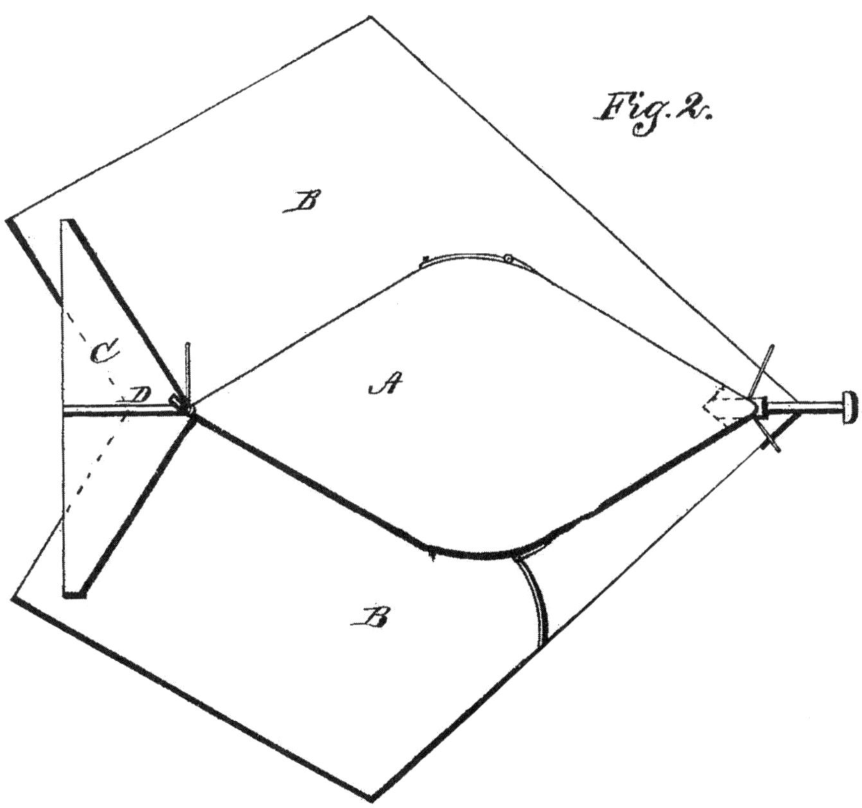

Plate 23: Daniel Rhone's Aerial Boat—June 1, 1874: *Basically a rigid-hulled, hydrogen-filled boat with rudders and a tail, Mr. Rhone's theory is that the "wings are to create resistance to the air, so that the boat in rising and falling will be compelled to take a forward course, on the principle that force always expends itself in the direction of least resistance." Actually, the direction of least resistance would be straight down, which would most likely be the direction this simple machine would take were anyone foolhardy enough to try to fly it.*

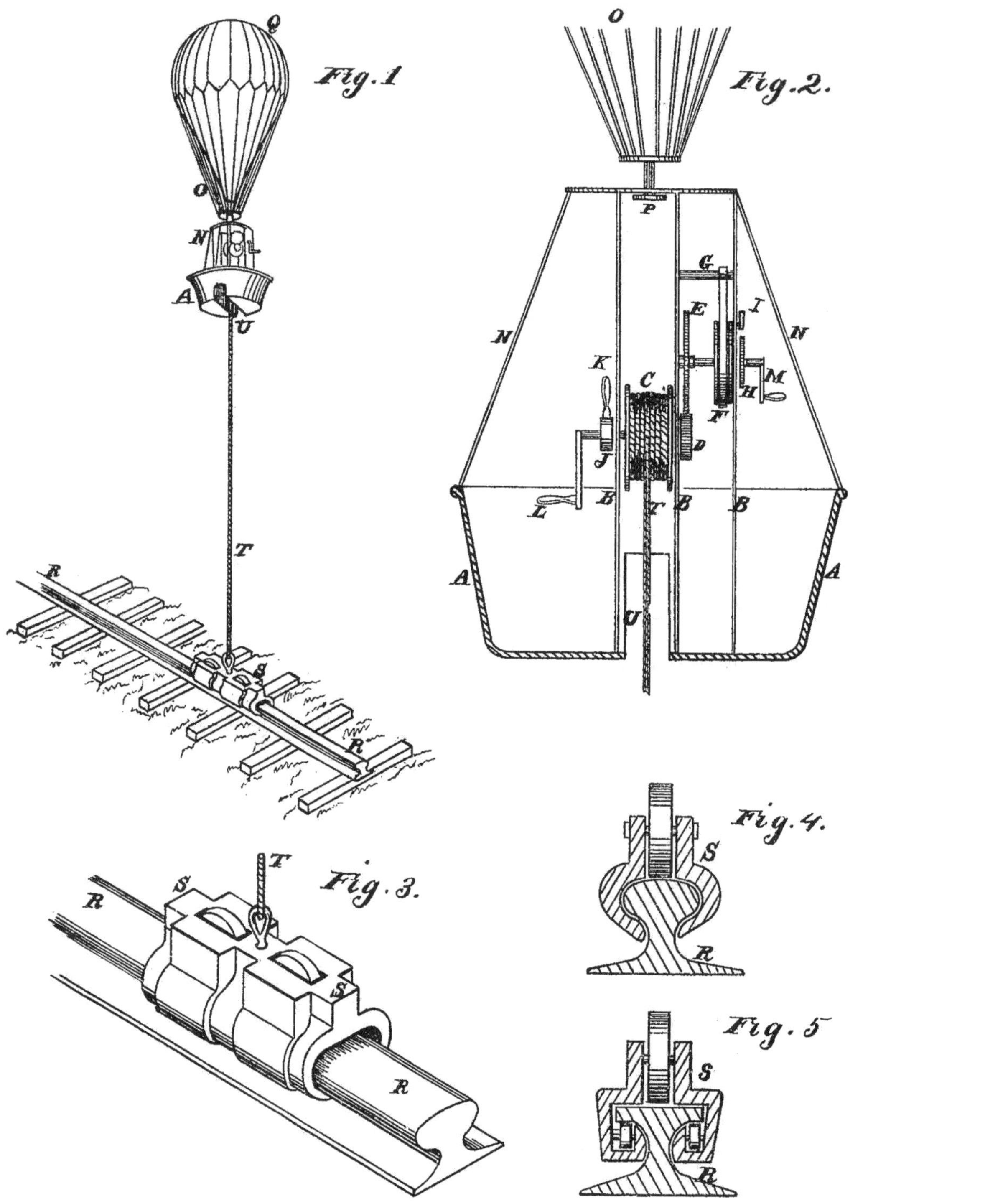

Plate 24: Charles McGlashan's "Guiding" Balloon—April 15, 1874: *For the intrepid balloonist who doesn't care to be pushed about according to the vagaries of the wind currents, this tethered balloon can't be beat. Firmly attached to a wheeled rail attachment, all the balloonist has to do is just follow the rail lines anywhere his heart desires, raising and lowering himself to skirt telegraph wires, trees, and the occasional church steeple as necessary to avoid unfortunate accidents. Okay, so not the most daring way to accomplish flight (and practically useless over water, lakes, or anywhere without a railroad line) but then not everyone is a daredevil. The question of what to do should a train happen along and demand use of the track is not addressed in Mr. McGlashan's description.*

Fig.1.

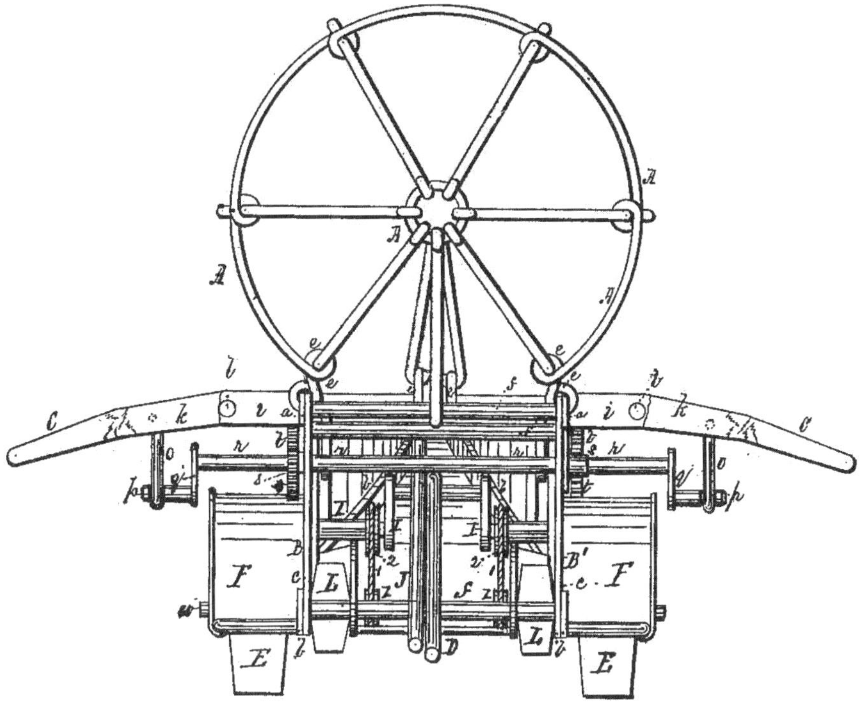

Fig.2.

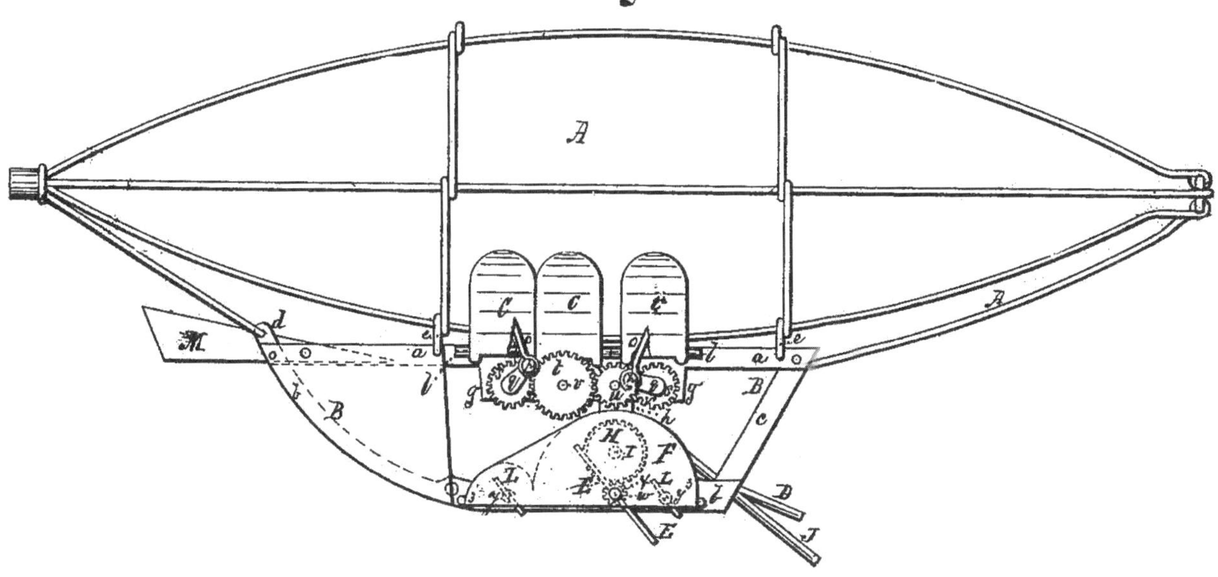

Plate 25: Micajah Dyer's Apparatus for Navigating the Air—September 1, 1874: *This robust example of aeronautical engineering is basically a simple hydrogen-filled cigar-shaped balloon with steam-powered flapping wings or paddles designed to provide forward propulsion as well as assist with lifting itself off the ground. It also has a set of small paddle wheels (letter E) at the bottom to do all of the real forward propulsion work. Unfortunately, the cam and rod system would be so heavy—even without an attached steam engine (curiously not shown)—and unwieldy that it would take a balloon ten times the size of the one shown just to lift it off the ground, with or without flapping wings.*

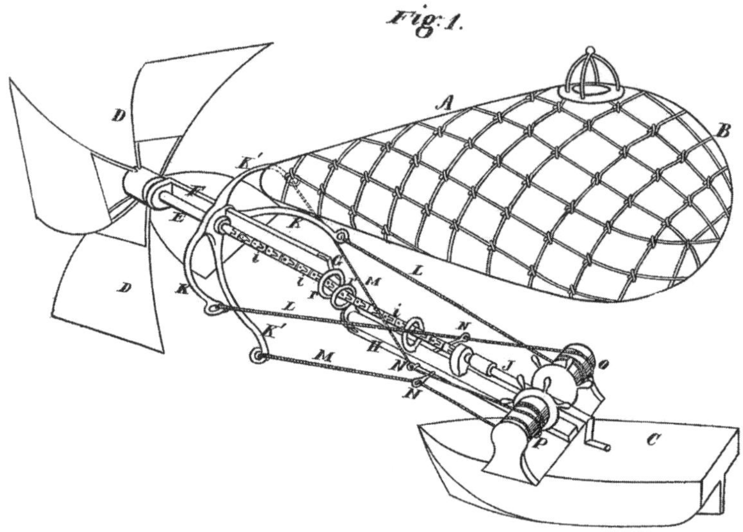

Plate 26: Peter Fernandez's Aerial and Marine Gyrator—April 6, 1875: *At last, someone has found a way of doing away with that pesky rudder by replacing it with a massive gas bag attached to an oversized propeller turned by a hand crank. Not only that, but this will work equally well for a boat in the water as for one in the air, making it a truly revolutionary design. And even if it doesn't improve steering or get you airborne, it is a sure bet that other boaters will give you a wide berth if they see you coming at them with this contraption, thereby clearing the sea lanes of unwanted nautical competition.*

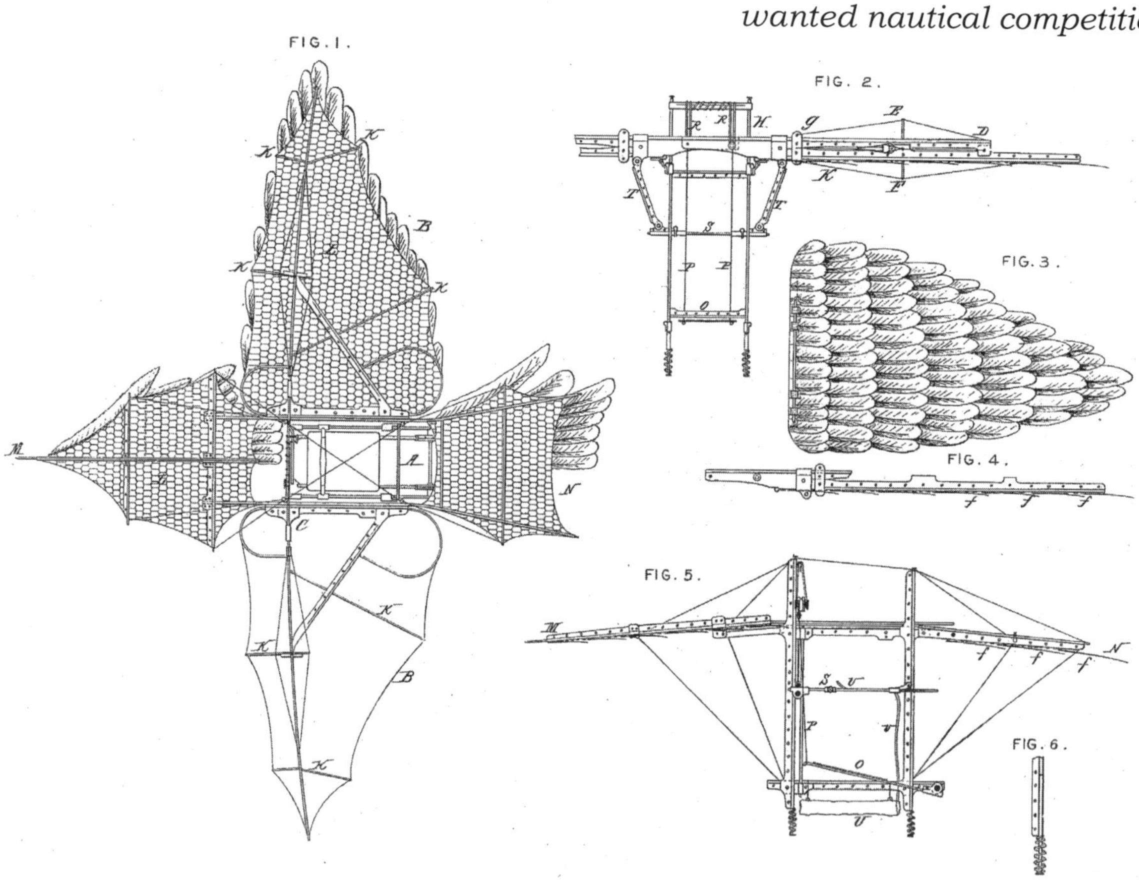

Plate 27: Francois Lamboley's Flying Machine—July 19, 1876: *Apparently Mr. Lamboley took the avian world seriously by creating a man-powered flying machine that really looks like a bird—complete with real feathers (from which species is not stated.) All one needs to do is flap and stomp and jump furiously on a bewildering array of springs and levers and they too may slip the surly bonds of gravity and soar among the eagles. Not recommended for those who are allergic to feathers or who are unusually susceptible to gravity.*

Fig. 1.

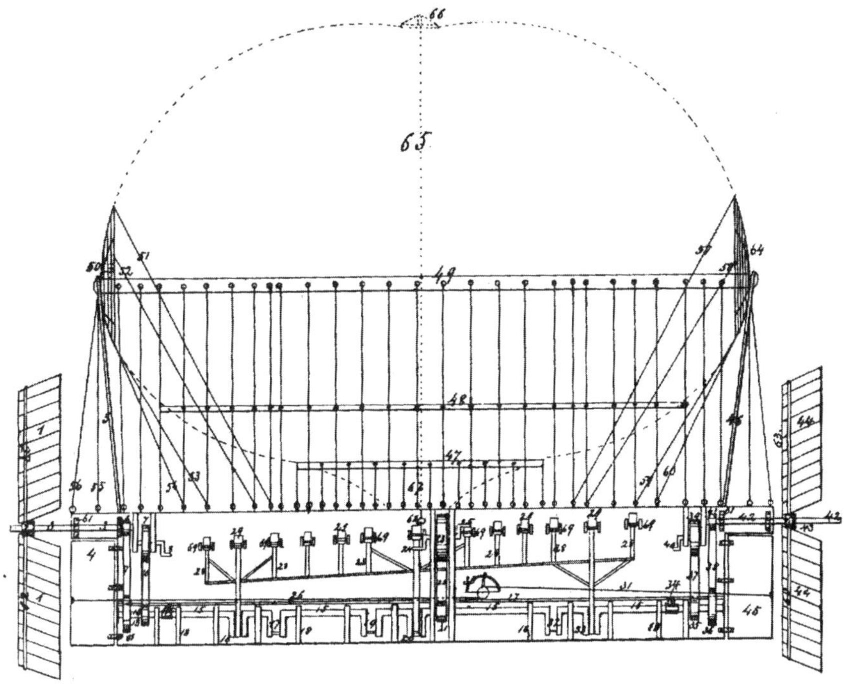

Fig. 2.

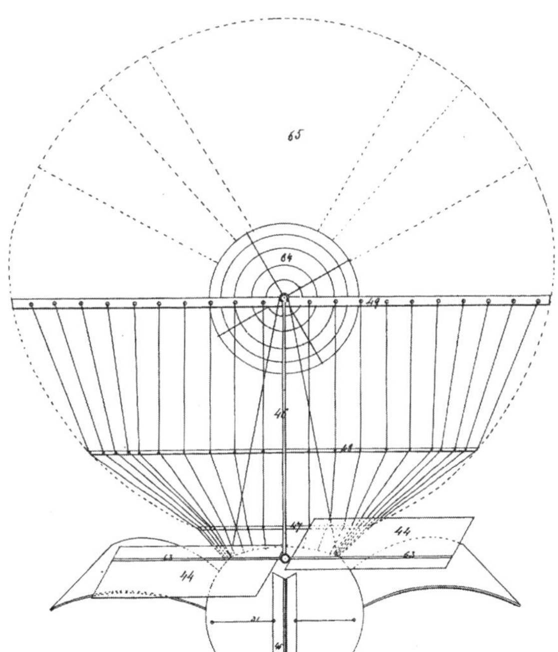

Fig. 3.

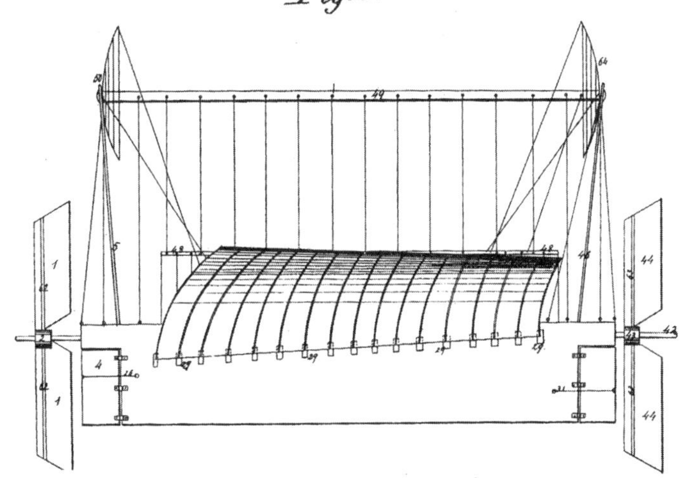

Plate 28: Frederick Schroeder's Steerable Airship—July 1, 1875: *Another misguided but well-meaning effort to marry massive propellers, tiny misplaced rudders (mounted inside the propellers where they are certain to do the least good), and flapping wings to a hydrogen-filled balloon with again no appreciation of the weight-to-power-to-lift ratio. It would take decades for inventors to learn that they couldn't turn boats into airships simply by adding balloons and flapping wings.*

Plate 29: Melville Murrell's Aerial Navigator—April 16, 1877: *In yet another attempt at achieving flight through human locomotion, Tennessean Melville Murrell introduces us to his "aerial navigator", a machine that truly is as dangerous as it looks. One interesting design feature is that the wings are made up of a series of pivoting slats that automatically open when being pushed upward and so meet little air resistance and close when pulled down, providing maximum resistance (and, it is presumed, lift). Merely launching the craft would probably have proven an even greater challenge than flying it, one might imagine.*

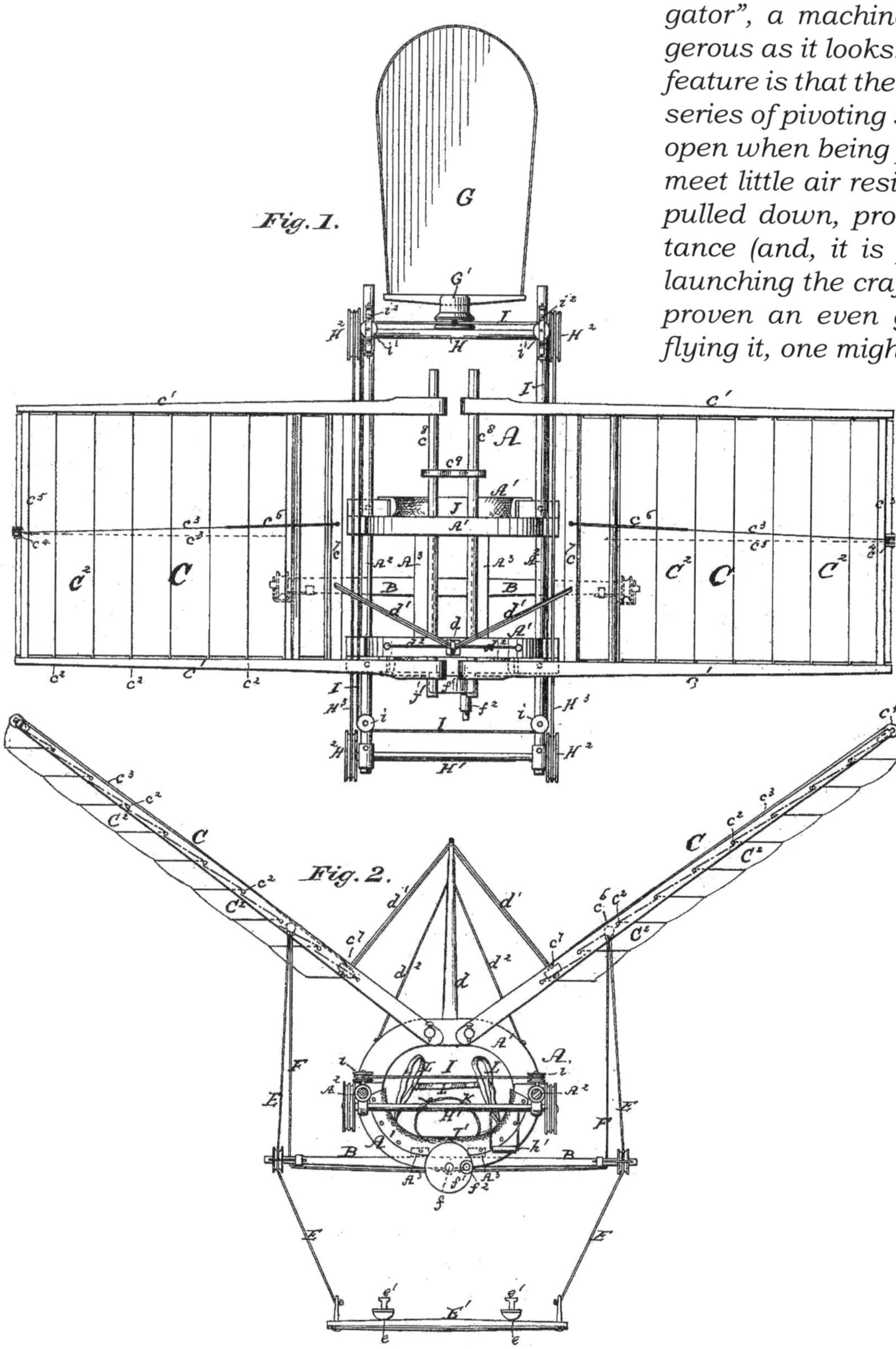

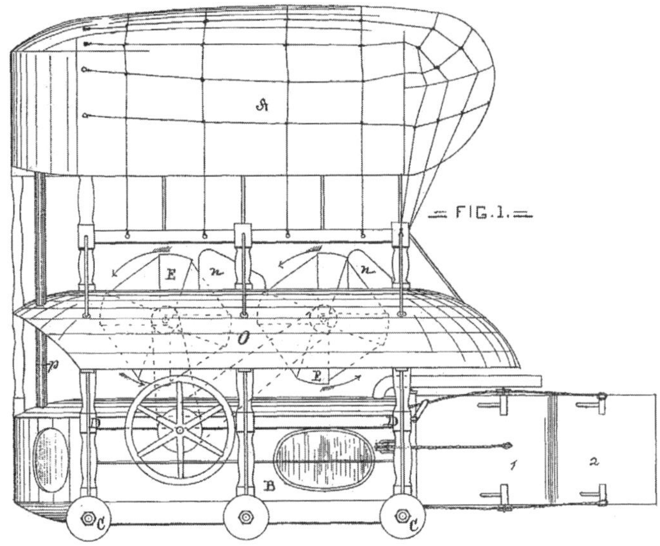

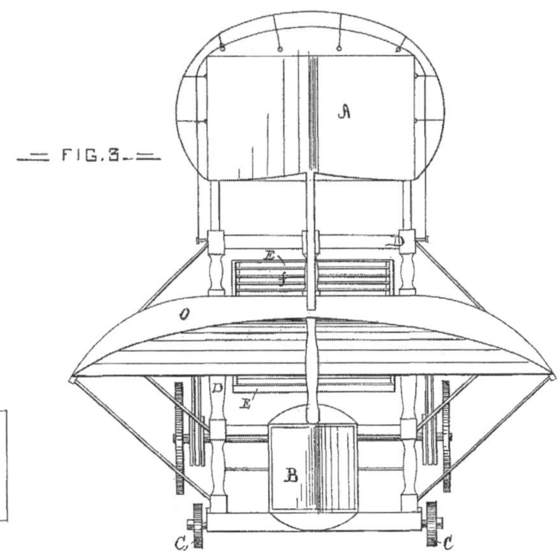

Plate 30: Fritz von Ehren's Aerial Vessel—January 15, 1878: *This strange-looking craft incorporates the novel idea that the steam driven fans between the balloon (which can be either hydrogen-filled or heated with hot air from the boiler) and the enclosed gondola will compress the air to the point that the craft will rise majestically into the air. In fact, von Ehren tells us the balloon isn't really even necessary and he added it only "...for the relief of the machinery, as well as to guard against the latter becoming accidently out of order, and to overcome public prejudice." (Apparently, in 1878 the public was prejudiced against the idea of flying machines not being firmly affixed to even a token balloon, a point which von Ehren thoughtfully took note of.)*

Plate 31: Charles Ritchel's Airship—March 12, 1878: *In this example of kinetic flight, the rotor (letter "W") at the end of the helicopter-looking gondola is capable of being pointed either straight back for forward propulsion or to either starboard or port, thereby providing lateral motion, while a second propeller (letter R) is pointed straight down, providing redundant vertical lifting power in case the canister-shaped balloon didn't have quite enough lift of its own (or to cushion the inevitable out-of-control descent). The idea that all these devices could be powered through sheer muscle power alone demonstrated that few people at the time had any idea what it took to even turn the blades, much less produce enough thrust to make them useful. Of course, this was before most people had ever ridden a bicycle, after which such foolish ideas quickly lost their appeal.*

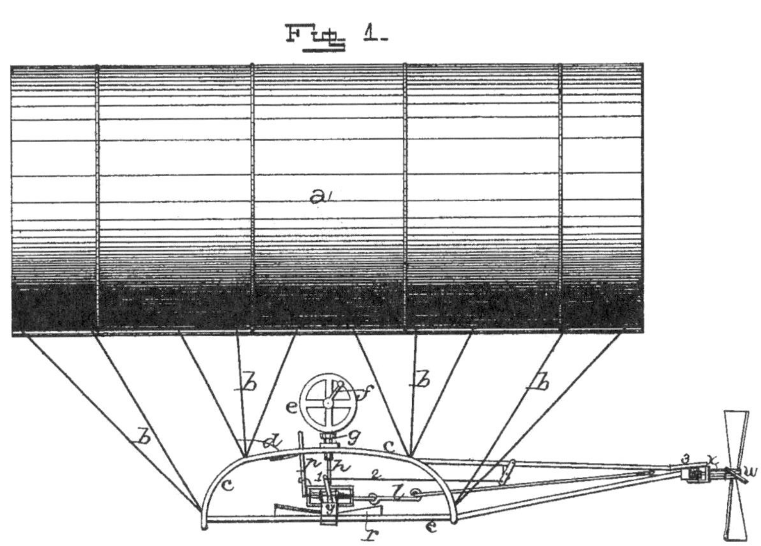

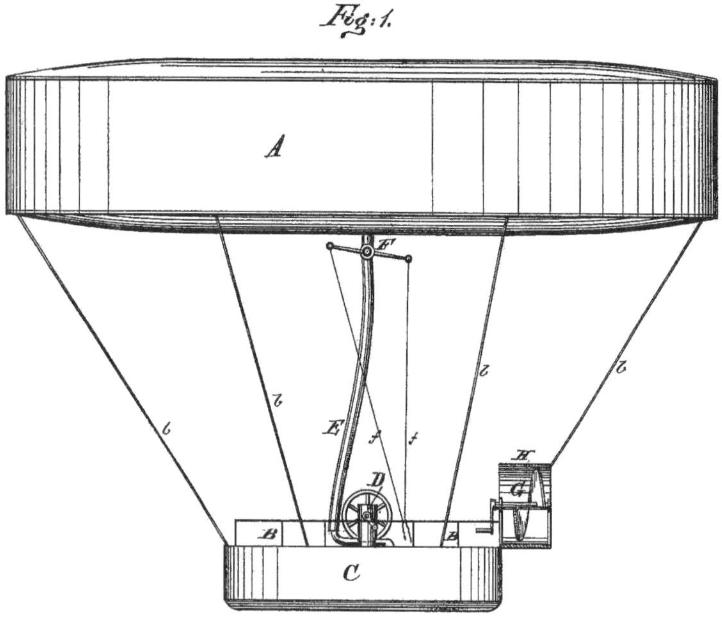

Plate 32: William Pearson's Improvement in Balloons—April 23, 1878: *This is ballooning at its simplest: a lozenge-shaped envelope with a simple pipe to allow for gas to be added or extracted according to whether one wanted to ascend or descend, and a hand-cranked propeller (which looks to be more useful as a meat grinder than a propeller) for forward thrust. Pearson believed that one could get by with a much smaller balloon if they kept the gas under pressure, and that this gas could be easily pulled out of the pressurized balloon and kept in a tank on the gondola floor to be reused later. A simple, elegant, and utterly unworkable concept.*

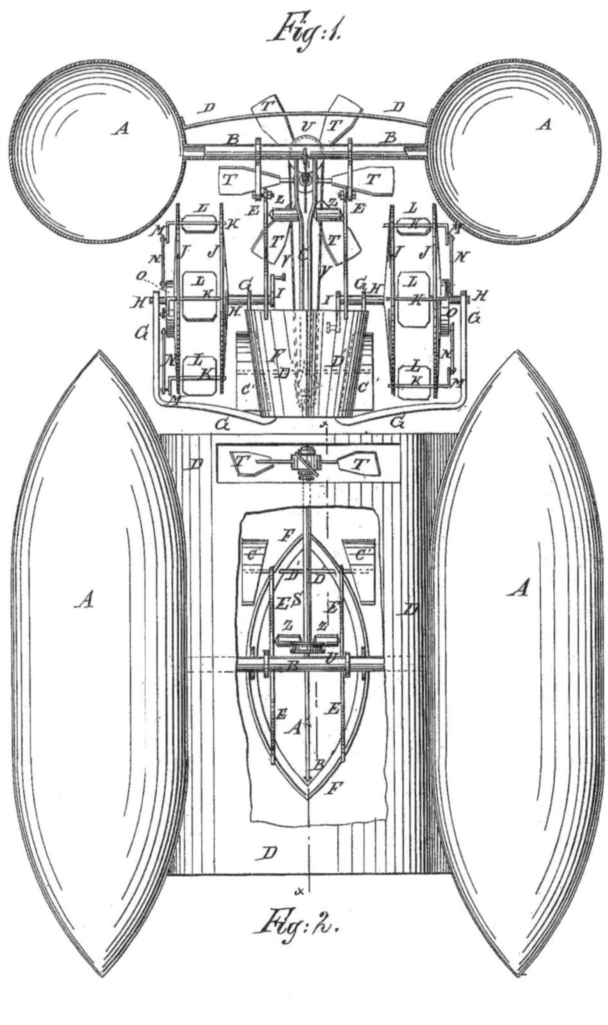

Plate 33: John Cameron's Airship—August 28, 1878: *Another typical example of a hazard to navigation so popular during this era. The only real surprise about this tandem balloon airship is Cameron's note that the hydrogen gas in the envelope is to be sent down to the engine so that "the temperature of the gas may be raised or lowered as may be required to correspond to the temperature of the surrounding air." Unfortunately, heating hydrogen in an "engine-furnace" sounds like a surefire recipe for creating a bomb, making it fortunate Mr. Cameron lacked the resources to actually build such a craft. It's curious that even at this late date, some treated hydrogen as though its lifting power was a by product of heat rather than a natural characteristic of it being a lighter-than-air gas.*

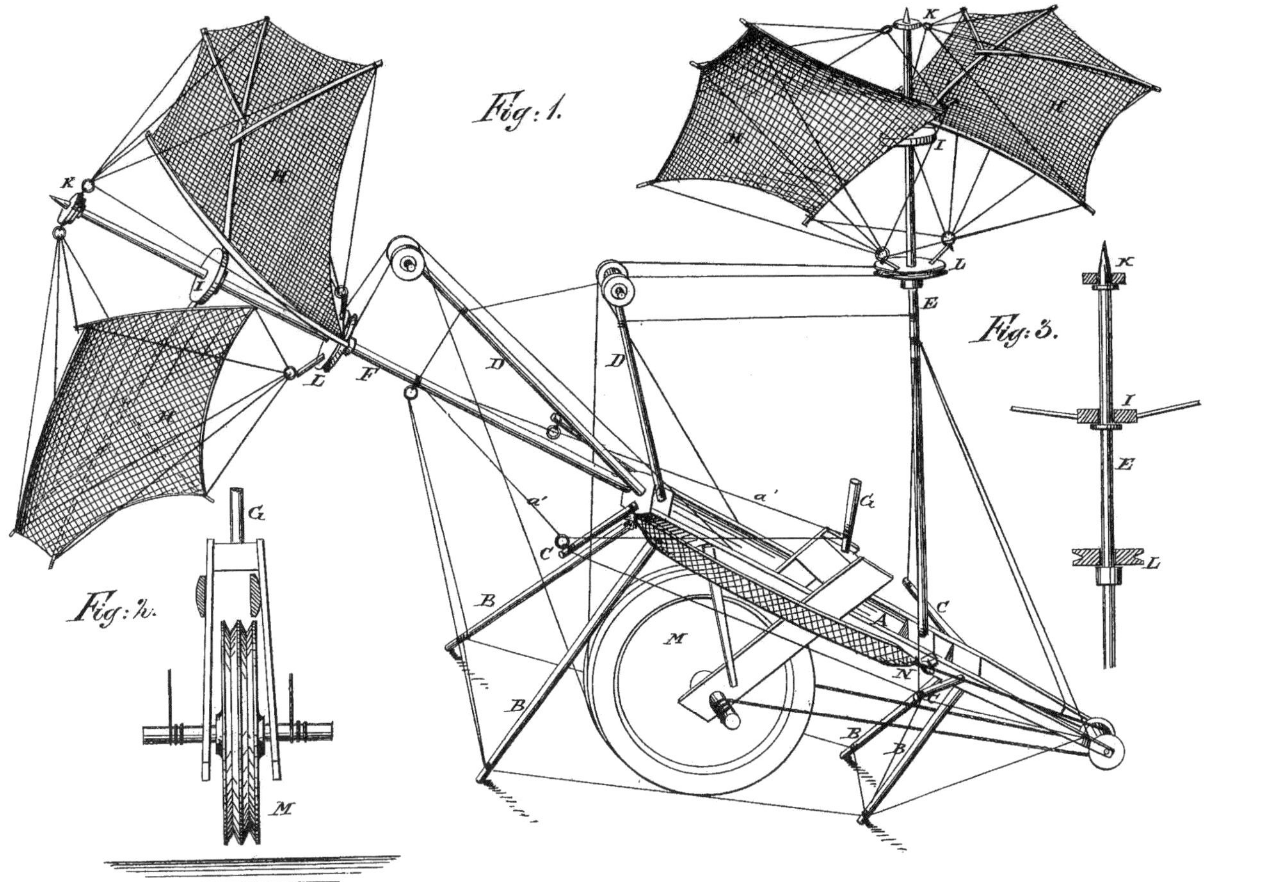

Plate 34: W.F. Quinby's Improvement in Aerial Ships—January 16, 1879: *After a decade of much appreciated silence, the prolific Mr. Quinby of Wilmington, Delaware, is back with yet another remarkable and long-anticipated addition to aviation. Looking a bit like a weaving loom attached to a pair of spinning parasols, the principles of flight behind this creation are not entirely clear—as Mr. Quinby makes clear in his description—nor is it apparent where the pilot sits/stands/lays within its maze of cables, pulleys, and wheels. If nothing else, we can applaud the man's determination to conquer gravity in most unconventional ways.*

FIG.1

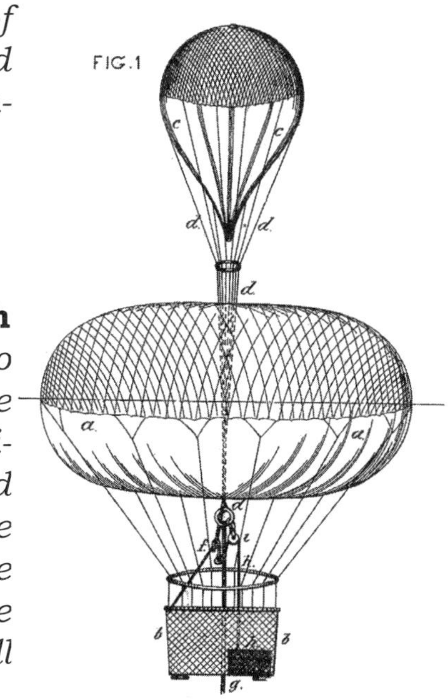

Plate 35: Count Antoine Apraxine's Aerial Balloon—March 25, 1879: *In what must be considered a truly novel way to get a balloon to ascend and descend without having to valve off precious hydrogen. His solution is to fill the upper, spherical balloon with hydrogen while filling the lower, donut-shaped balloon with regular air. The idea is that the balloons can be either tethered together or apart; when tethered together, the lifting balloon will make the whole affair rise; when apart, the air-filled, non-lifting balloon, now free of the lifting balloon, will permit the whole thing to gently descend. Brilliant, no? No.*

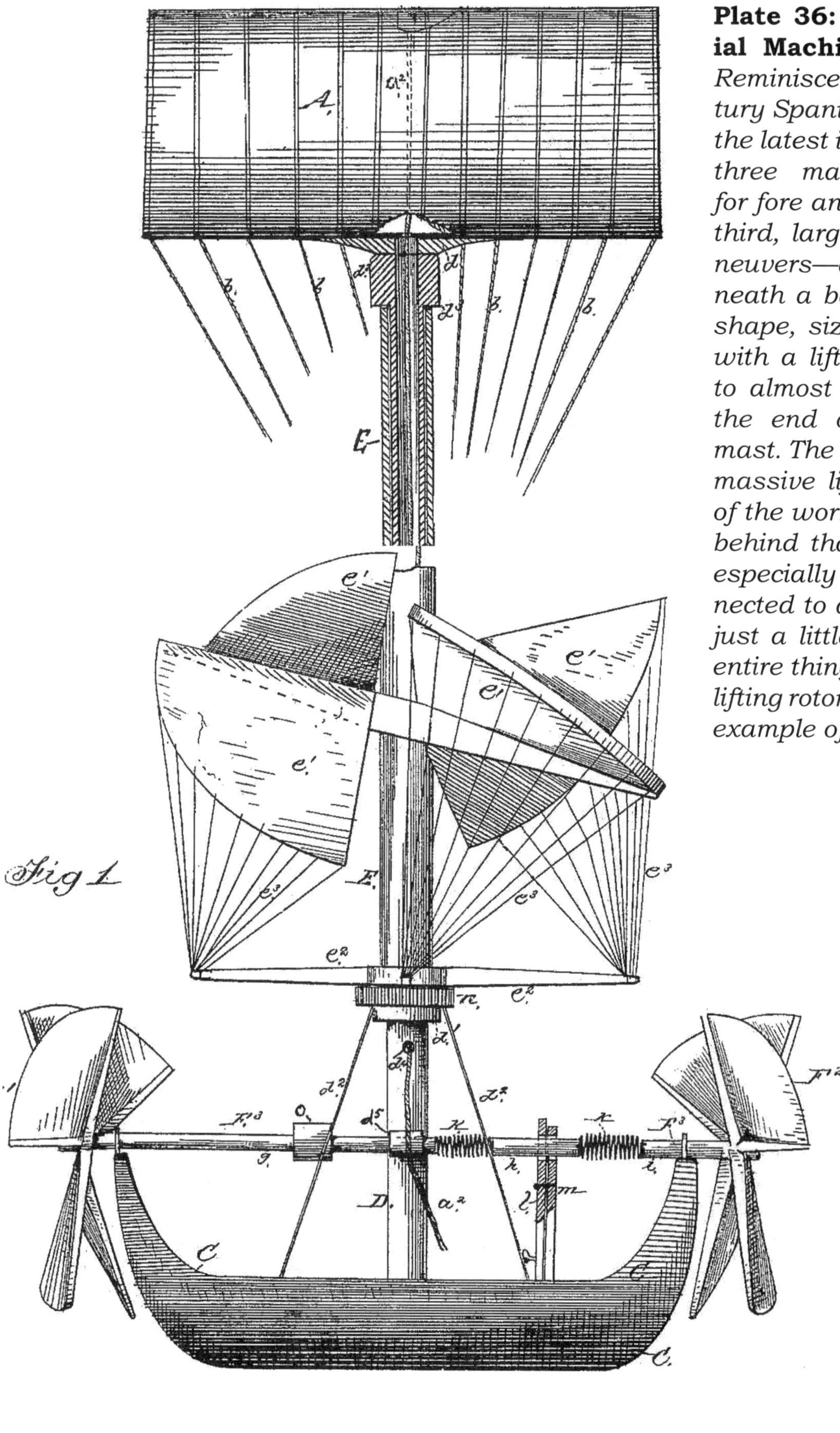

Plate 36: Henry Bagley's Aerial Machine—April 22, 1879: *Reminiscent of a sixteenth century Spanish galleon on steroids, the latest in flying boats featured three massive propellers—two for fore and aft propulsion and a third, larger one for vertical maneuvers—all neatly housed beneath a balloon of "any desired shape, size, or construction, but with a lifting capacity sufficient to almost lift the machine..." at the end of an especially long mast. The idea here was to let the massive lifting rotor do the rest of the work, though the rationale behind that is somewhat fuzzy, especially as you're already connected to a balloon that, if made just a little bigger, could lift the entire thing into the air without a lifting rotor. Clearly, another good example of over-engineering.*

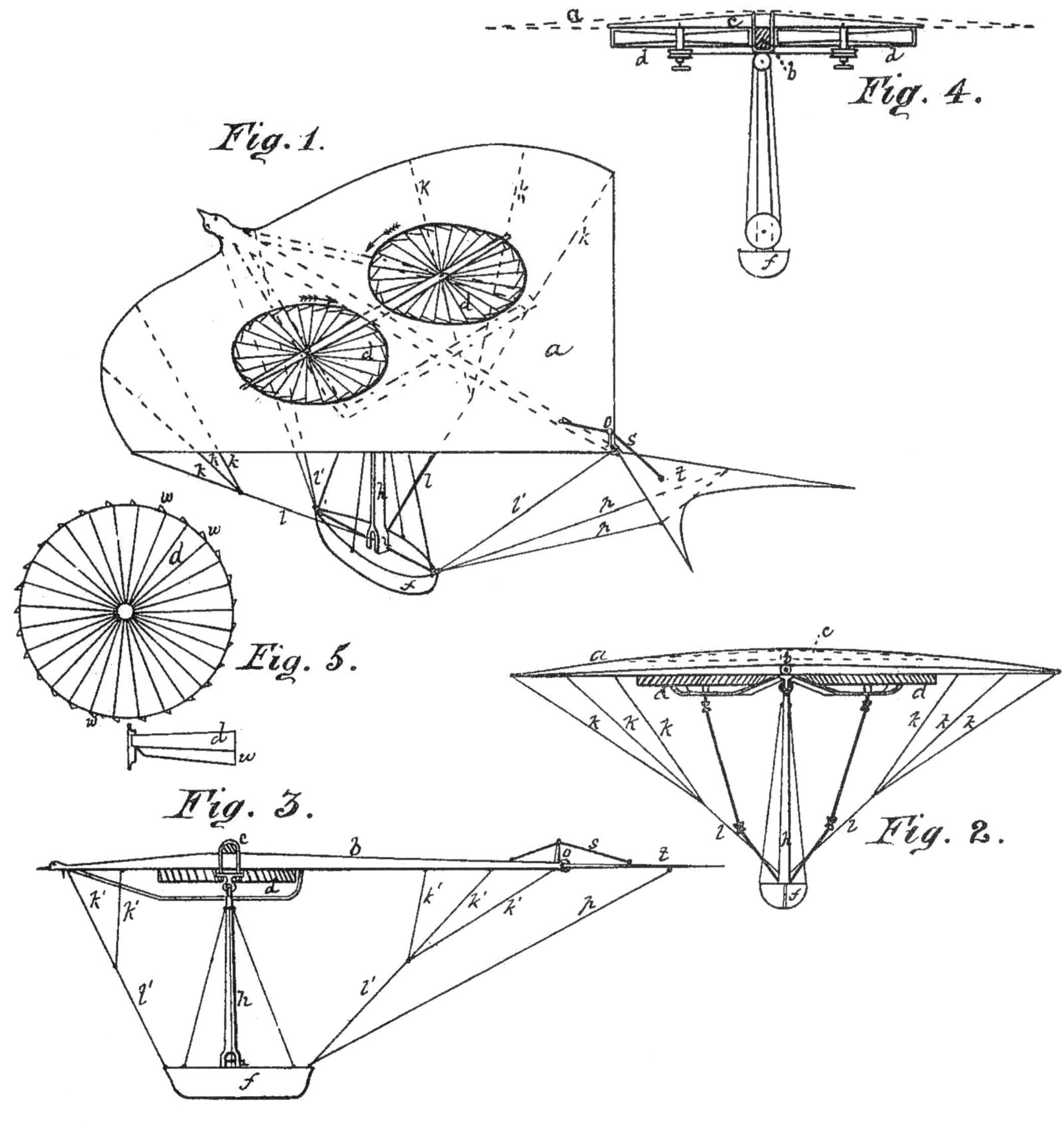

Plate 37: John Greenough's "Aerobat"—October 14, 1879: *This somewhat futuristic-looking device is actually a massive kite—which Greenough calls a "poiser"—held aloft by two massively overworked counter-rotating turbofans—which he calls "wind wheels". Curiously, though the craft has a tail, there is no indication of how forward propulsion is induced as there is no horizontal propellers apparent. The best part of the design is the whimsical bird head at the front of the "poiser", complete with eyes. Mr. Greenough was nothing if not artistic.*

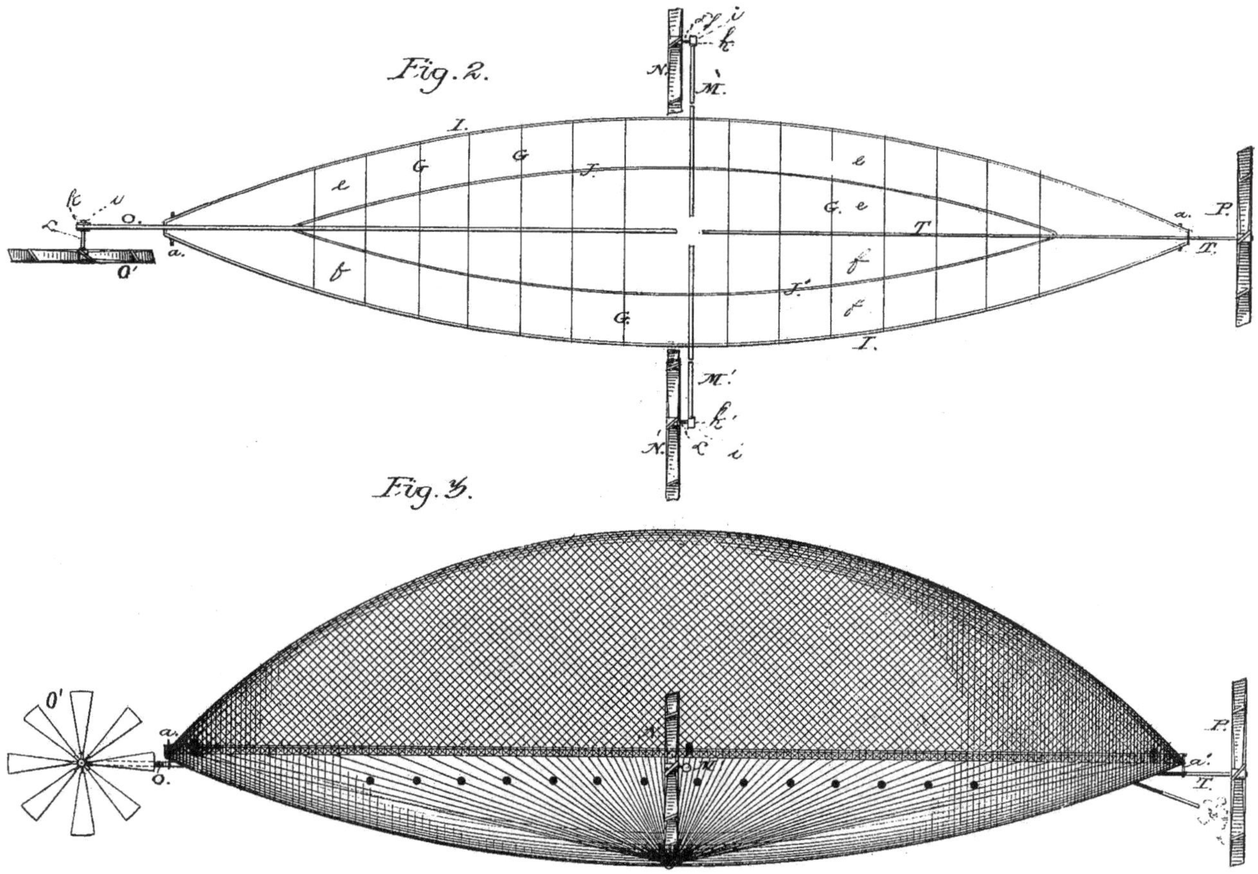

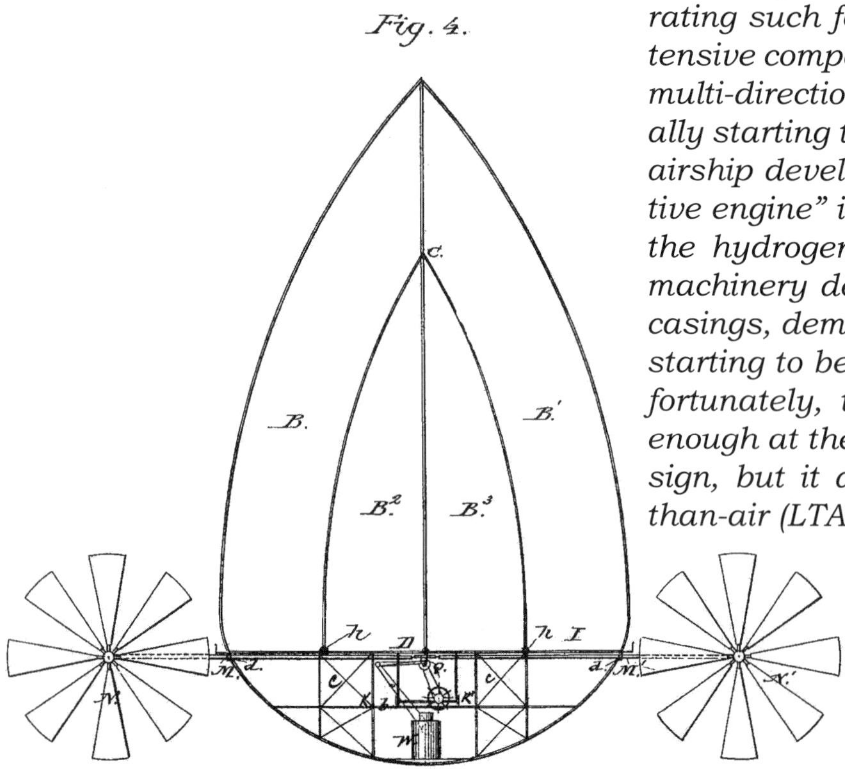

Plate 38: Albert Blackman's Vessel for Aerial Navigation—January 6, 1880: *In incorporating such forward-looking innovations as extensive compartmentalizing of the envelope and multi-directional propellers, this design is actually starting to reflect a better understanding of airship development and practicality. The "motive engine" is even placed some distance from the hydrogen-filled envelope, separated by a machinery deck, and surrounded by fire-proof casings, demonstrating that safety issues were starting to be taken increasingly seriously. Unfortunately, there wasn't an engine powerful enough at the time to make this a workable design, but it does foresee the future of lighter-than-air (LTA) development.*

Fig. 1

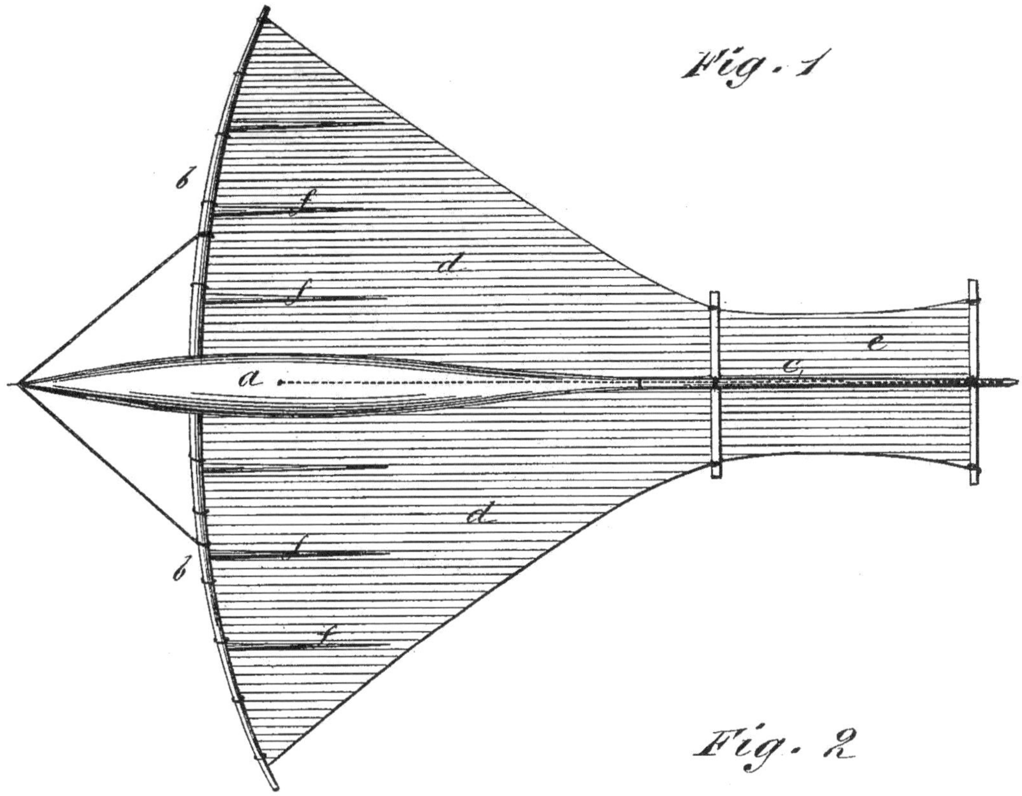

Fig. 2

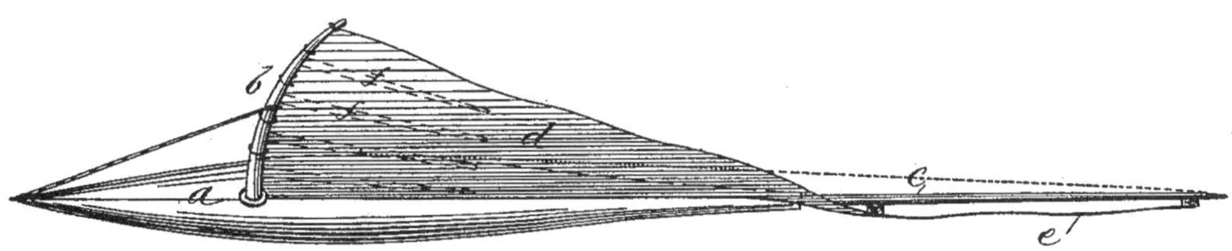

Fig: 3

Plate 39: Frederick Brearey's Aerial Apparatus—January 23, 1880: *In yet another attempt to fly by flapping one's wings, Englishman Brearey has taken it up a notch by making the machine large enough to carry multiple passengers and some kind of engine (which he fails to describe) to flap the wings. Lacking a rudder, the craft, if built, would be extremely unstable while the flapping of the wing surfaces would produce little or no appreciable thrust. There is also no camber to the wings forward to provide lift, making it a near certainty that, even on a good day, this machine would remain firmly earthbound, its flailing appendages making it look much like a beached stingray gasping out its final moments of life.*

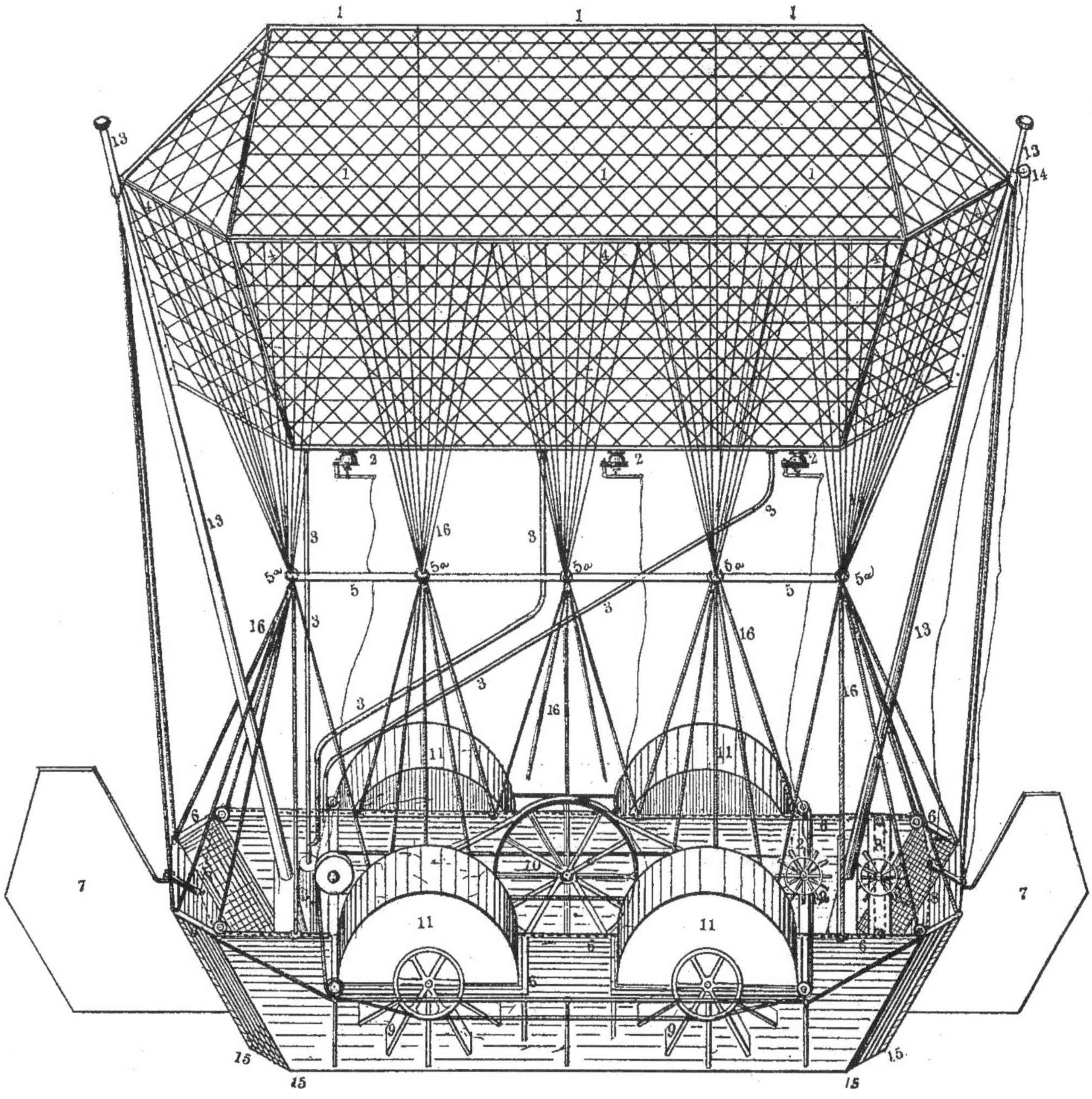

Plate 40: Charles Sullivan's Flying Machine—November 30, 1880: *Can you guess that Mr. Sullivan spent an inordinate amount of time around riverboats? Obviously convinced that a riverboat can be easily adapted to flight and that paddle wheels would do as good a job in the air as they do in the water, Mr. Sullivan's creation, while quaint and very much in keeping with the era, would have been, of course, utterly unflyable. What I find interesting about the patent description is his claim that he is the first to propose compartmentalizing the envelope (an idea that had been suggested a number of times previously) and his reference to a hydrogen gas producing machine that he suggests had already been invented, which, if true, would be a genuine breakthrough in aeronautics. (Apparently the knowledge of how difficult it was to produce hydrogen gas was not well understood at the time, with a number of inventors imaging that it could be easily done in transit.)*

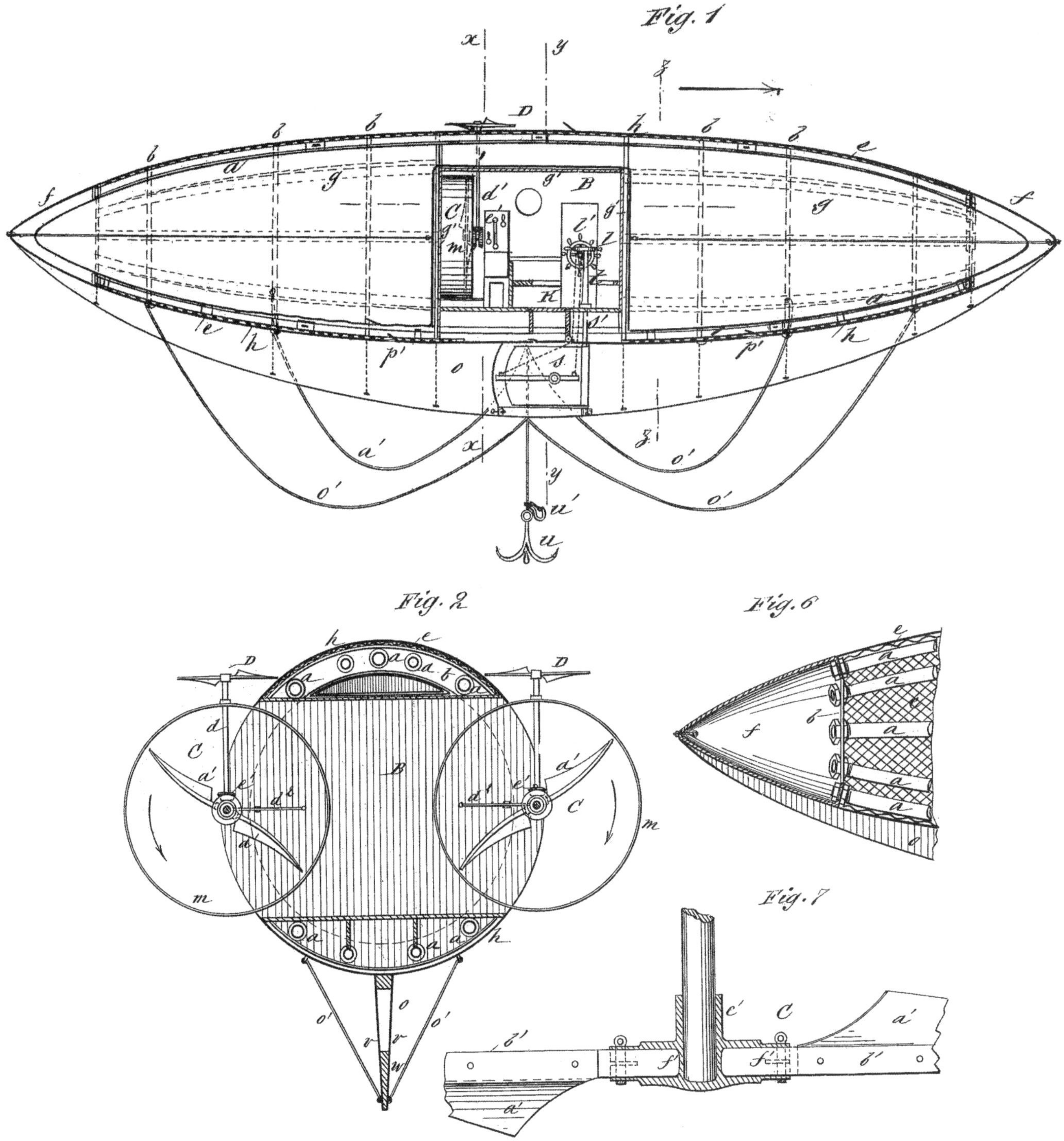

Plate 41: John Mackenzie's Aerial Machine—December 21, 1880: *Obviously heavily influenced by Jules Verne's novels, this device by Scotsman John Mackenzie aptly demonstrates that with enough determination and enthusiasm, even a submarine can be made to fly. This metal-skinned hydrogen balloon is obviously far too small to lift its own weight, but the overall sleekness and the scimitar-shaped recessed props make it a very cool looking vessel indeed. Note the dangling anchor—evidently necessary to winch the machine down to the ground when landing—and how well it lends itself to the overall nautical theme Mr. Mackenzie was evidently striving to achieve.*

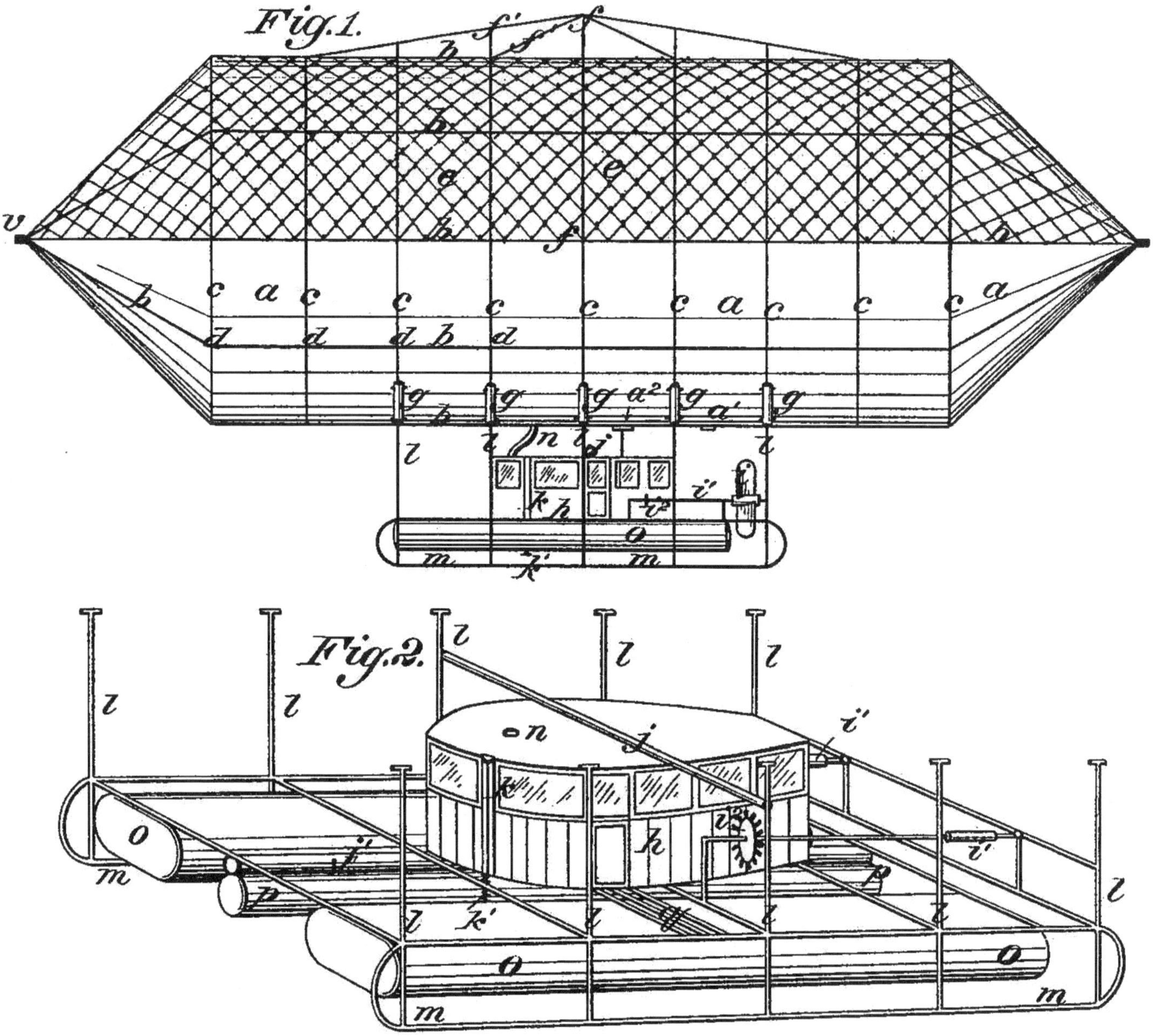

Plate 42: Edward Pearse's Machine for Aerial Navigation—January 11, 1881: *Somewhat reminiscent of a houseboat tethered to a balloon, Mr. Pearse has devised an ingenious means of propelling his craft through the air: the gondola is surrounded by a series of shock absorbers that can be raised or lowered in such a way that it will induce the propeller to point in the exact opposite direction that one wishes to go, making the craft easy to launch in the desired direction. Apparently, the undescribed power plant housed in the glass-enclosed "wheel house" is so powerful, the slightest change in orientation will send the craft hurtling through the air, mandating that it have a pilot of great deftness and hair-trigger reflexes at the helm.*

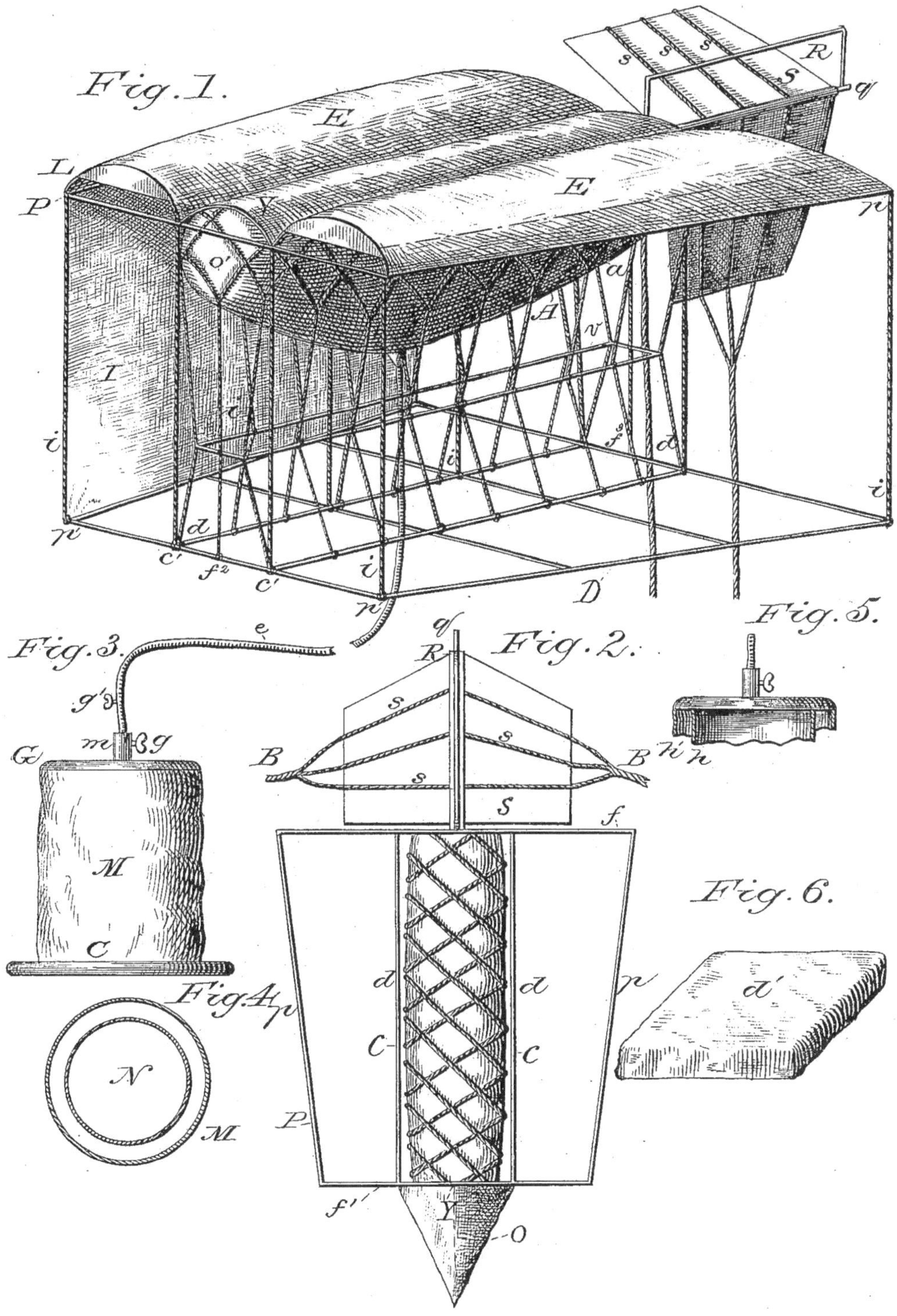

Plate 43: Edward de Jongh's Aeronautical Apparatus—September 13, 1881: *Supposedly this bewildering array of apparently unrelated objects are all part of some sort of tethered balloon that one could easily attach to their sailboat or wagon to make it airworthy. Mr. de Jongh's description of how all of this works is rather fuzzy, but since it's nearly impossible to make sense of anything, I'm prepared to suspend disbelief. I will say that it does appear that in figure 6 (above) Mr. de Jongh may have inadvertently patented the first canvas baseball plate.*

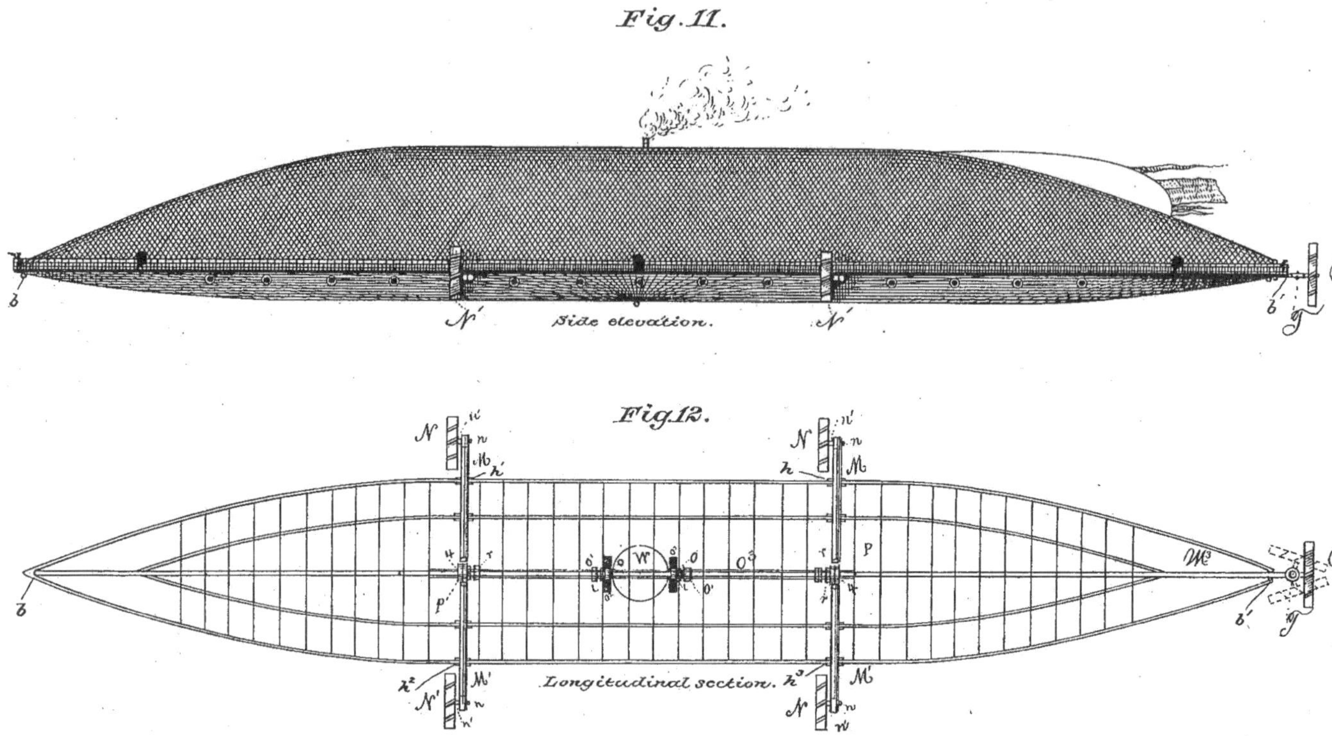

Plate 44: Albert Blackman's Improved Vessel for Aerial Navigation—December 6, 1881: *Compare to plate 38. Same idea, only the balloon has grown in size and there are four side propellers instead of two. Also, the nose-mounted propeller has gone away.*

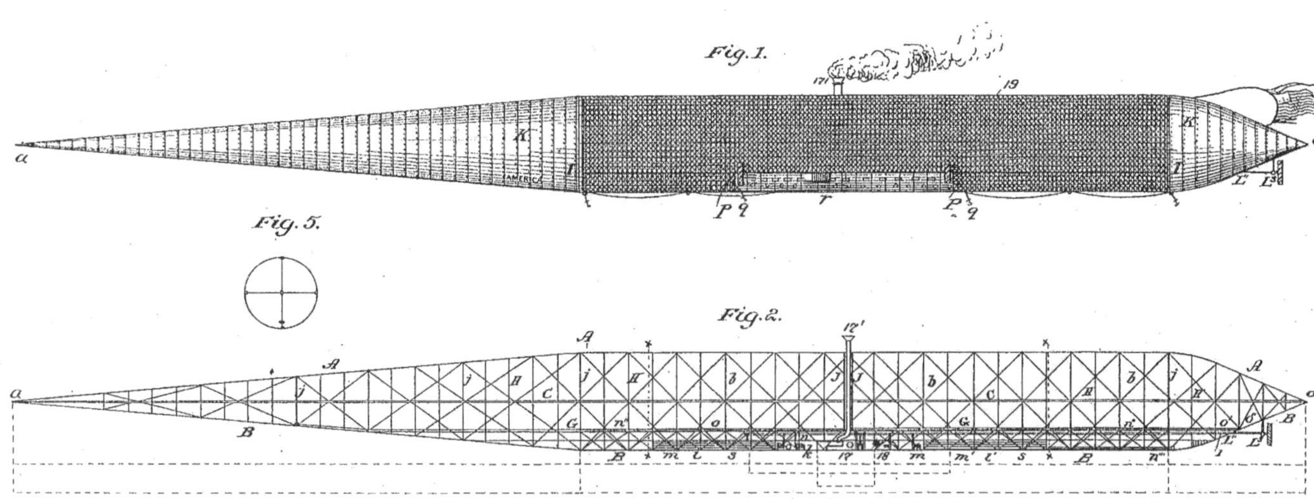

Plate 45: Albert Blackman's Even More Improved Vessel for Aerial Navigation—June 13, 1882: *The prolific Mr. Blackman's third revision to his original airship design (see plates 38 and 44) continues to grow, this time to truly gargantuan proportions that would rival the great airships of the 1930s in size. Clearly finding his inspiration in all things pointy, by this time his machine has gotten so big it is too large to build—much less power or maneuver, but at least it's beginning to look like a real dirigible—or at least a very sharp stick. Unfortunately, Mr. Blackman still hasn't solved his venting problem and continues on insisting on sending the smoke from the boilers through an especially long chimney out the top of the ship. It can only be assumed he understands the potentially difficult relationship that burning embers and hydrogen have with each other.*

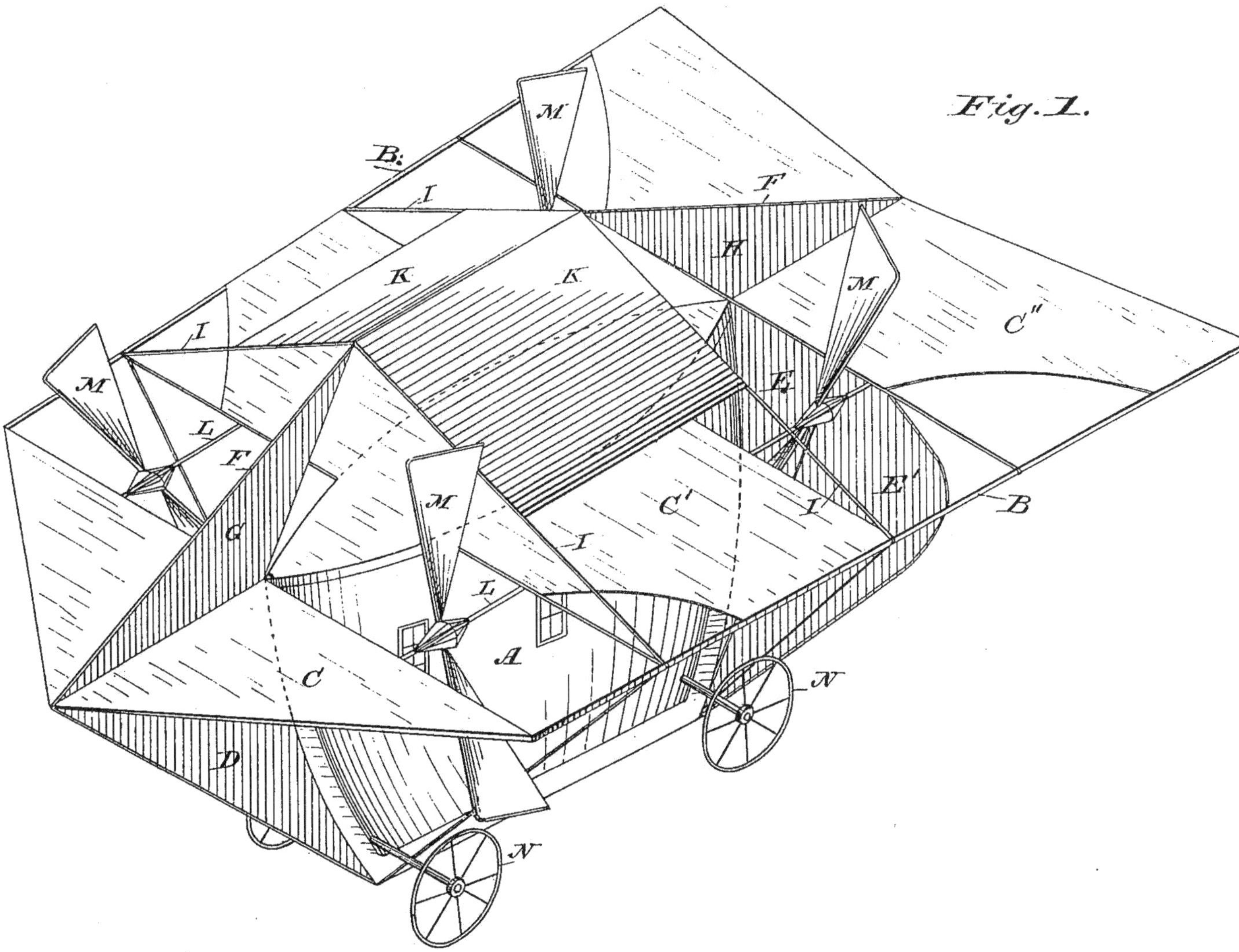

Plate 46: William Krueger's Airship—January 31, 1882: *Despite all visual evidence to the contrary, this device, which Mr. Krueger calls an "airship", is actually a heavier-than-air craft designed to take off and fly much like a regular airplane would—at least, if an airplane could fly without any lifting surfaces of any kind. Mr. Krueger gave his craft four fine propellers (which could be turned either by hand or by any other "source of power according to the size of the machine and the power required to drive it") along with plenty of control surfaces, but in omitting wings, he gave it no capacity whatsoever for flight. Curiously, Mr. Krueger seems lackadaisical in regards to the craft's final shape or design (or even such small details as to whether the landing gear has four wheels or three), preferring to leave that up to the craft's intrepid builder to decide for himself.*

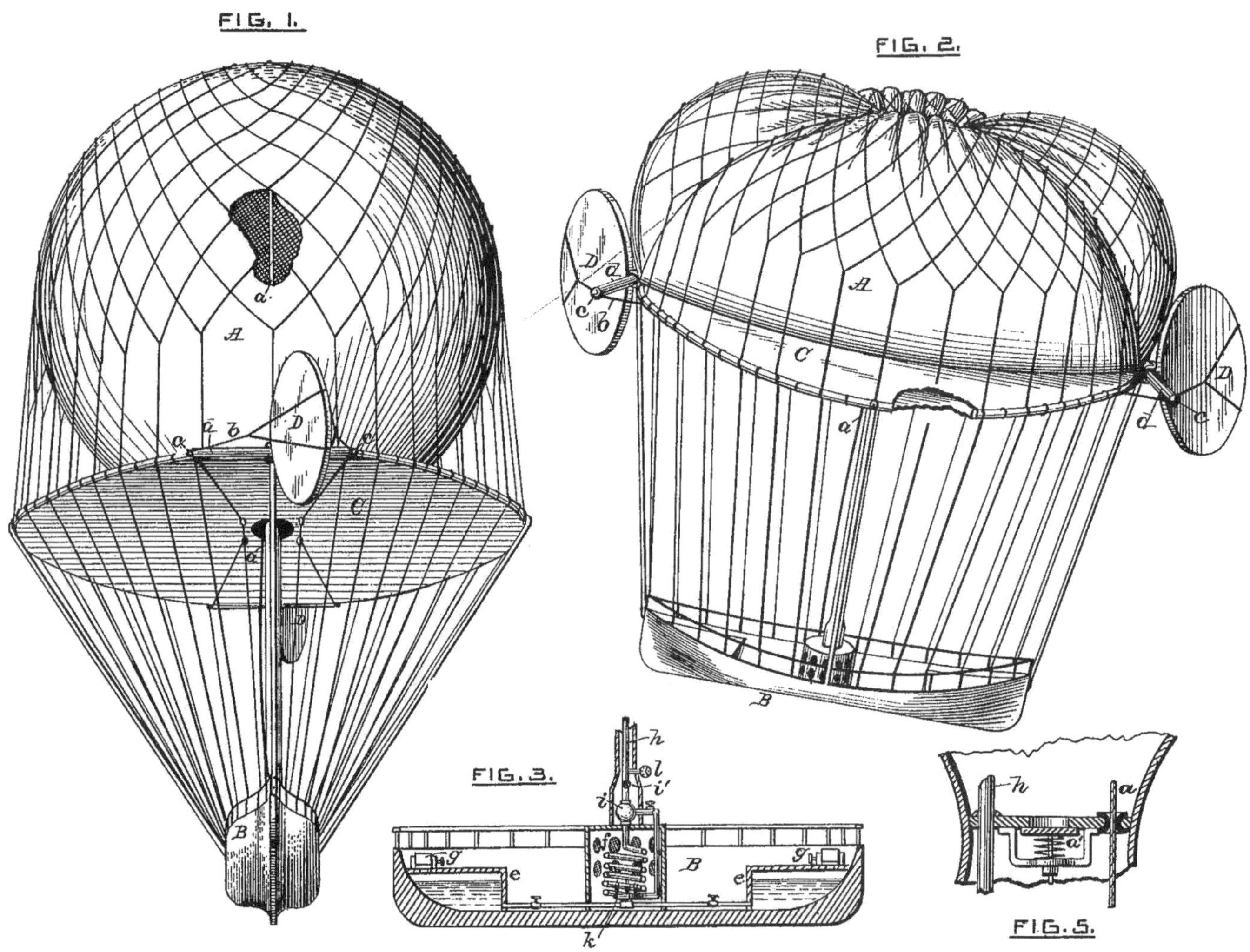

Plate 47: Carl Petersen's Airship—April 11, 1882: *This curious-looking and yet strangely cheery device, complete with what look like massive mouse ears (rudders), is the brainchild of San Francisco inventor Carl Petersen, who is the first man to fully describe how hydrogen gas is produced by burning gasoline (which, I assume, was news to science community at the time). He even tells us that when heated, the hydrogen by product of the burning gasoline becomes even more buoyant than before, making me wonder if Mr. Petersen ever attempted to test that hypothesis. In any case, he also appears to be enamored with the idea of "reefing" the envelope, by which he means compressing the top of the balloon when valving off gas in anticipation of a descent, thereby forcing the unnecessary surplus out the bottom of the envelope rather than out the top. This is to prevent the interior of the balloon from becoming "contaminated" with regular air and make the craft more economical. I suspect, however, it would do neither.*

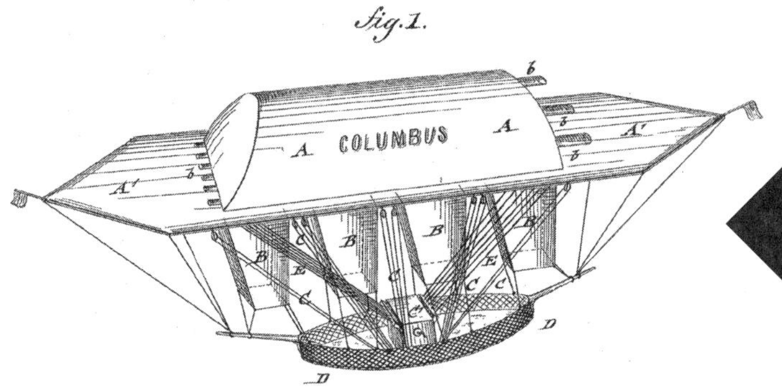

Plate 49: Robert Downton's Aerial Ship—September 12, 1882: *This truly frightening machine made maximum use of the craft's lifting agent by filling the spinning propeller blades with hydrogen, forcing them to do double-duty as both lifting agents and propelling agents. It's not clear just how fast these blades are churning when at full throttle, but it's a safe bet the unfortunate operator of the machine would have been wise to remain extremely still in the midst of this whirling, slicing, and dicing menagerie of death. Also note Mr. Downton's fondness for cams, gears, and flywheels of all shapes and sizes. Obviously a man ahead of his time.*

Plate 48: Charles Fest's Airship—August 29, 1882: *Looking a little like an airworthy version of a Civil War-era ironclad, this is actually a well disguised hot-air balloon, the big difference between it and regular hot-air balloons—besides its unusual shape—being the means by which hot air is introduced into the envelope. Instead of simply shooting hot air upwards directly into the envelope, fluids are heated in the gondola and the heat transferred through the body of the ship via wicks. Thus by turning certain wicks on or off, one could control the rate of ascent or descent of the balloon. Excessive heat was to be vented out lateral pipes (letter B) to not only prevent overheating and fire, but to also serve as a means of propulsion to assist the attached paddle wheels (not shown). Would it have worked? Of course not, but it's a fun idea.*

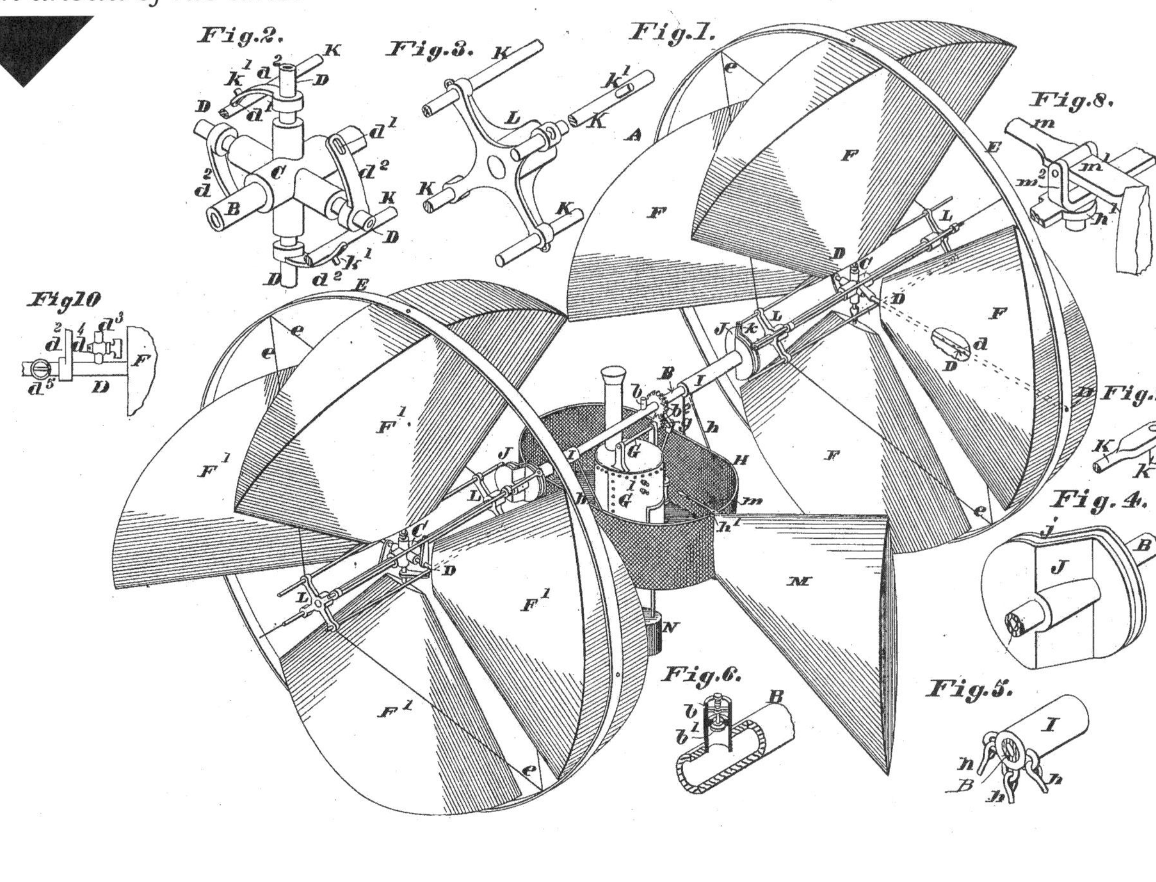

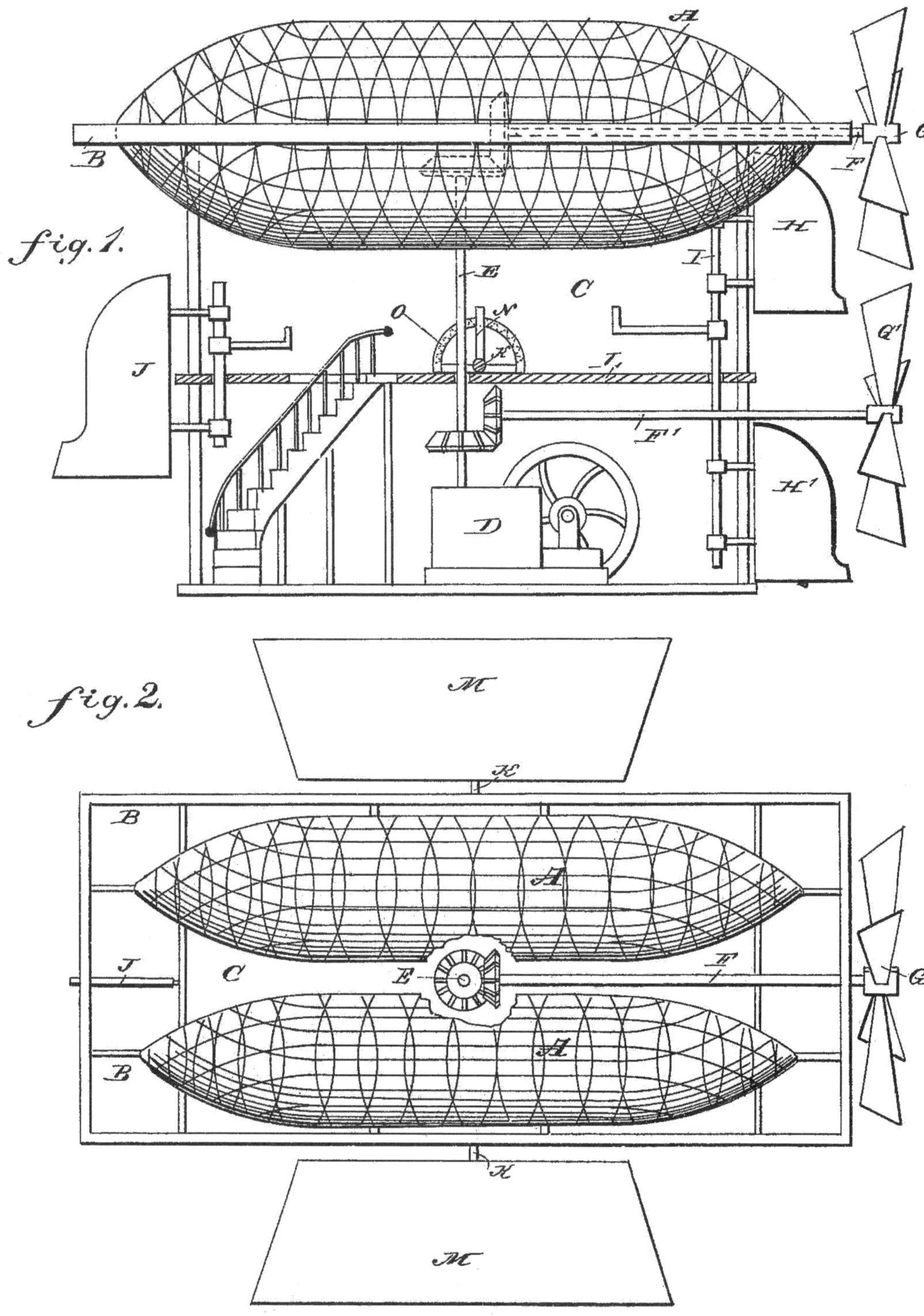

Plate 50: Joel Ray's Aerial Vessel—January 2, 1883: *One can't help but admire Mr. Ray's no-nonsense approach to dirigible design: two balloons encased in a metal frame with two pusher propellers, three rudders, and two massive wings, all being spirited through the air by some unspecified and doubtlessly badly overworked "motor of suitable construction" (mysteriously hidden from view within the box marked D in the drawing). The quaint spiral staircase at the front of the machine, however, does a fine job of adding a much-appreciated touch of elegance consistent to the fashion sensibilities of the era.*

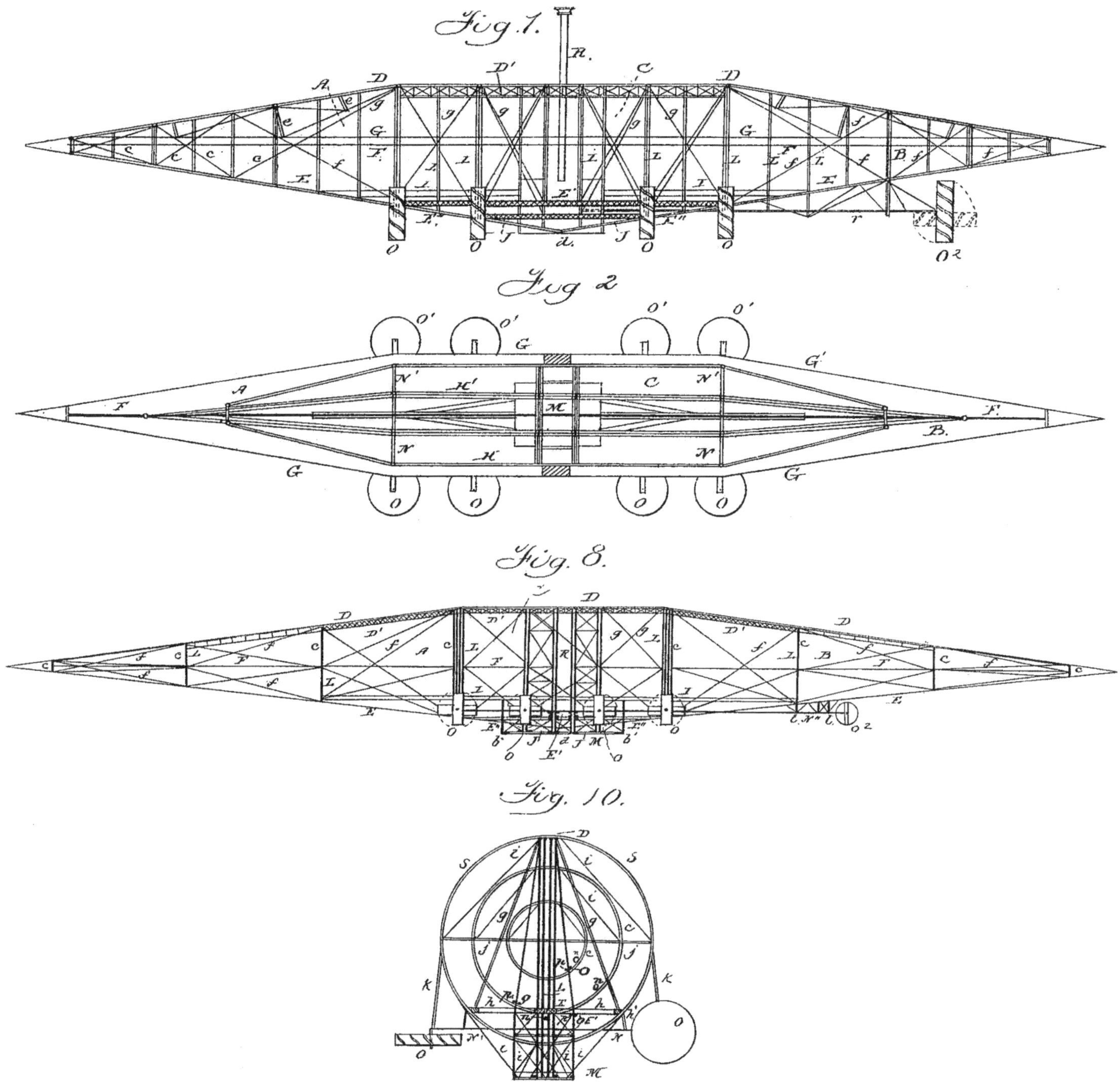

Plate 51: Eugene Falconnet's Aerial Ship—January 23, 1883: *In a truly massive ship Count von Zeppelin would be proud of, Nashville inventor Eugene Falconnet's great ship includes many of the innovations the German count would later incorporate into his own zeppelins twenty years later, most notably among them an interior, metal skeleton. Of course, such a vessel was beyond the technology of the era to build, and it would have been ungainly to fly without control surfaces to steer it, but it demonstrated that even at this early date, men were not afraid to think big...and pointy.*

Fig. 1.

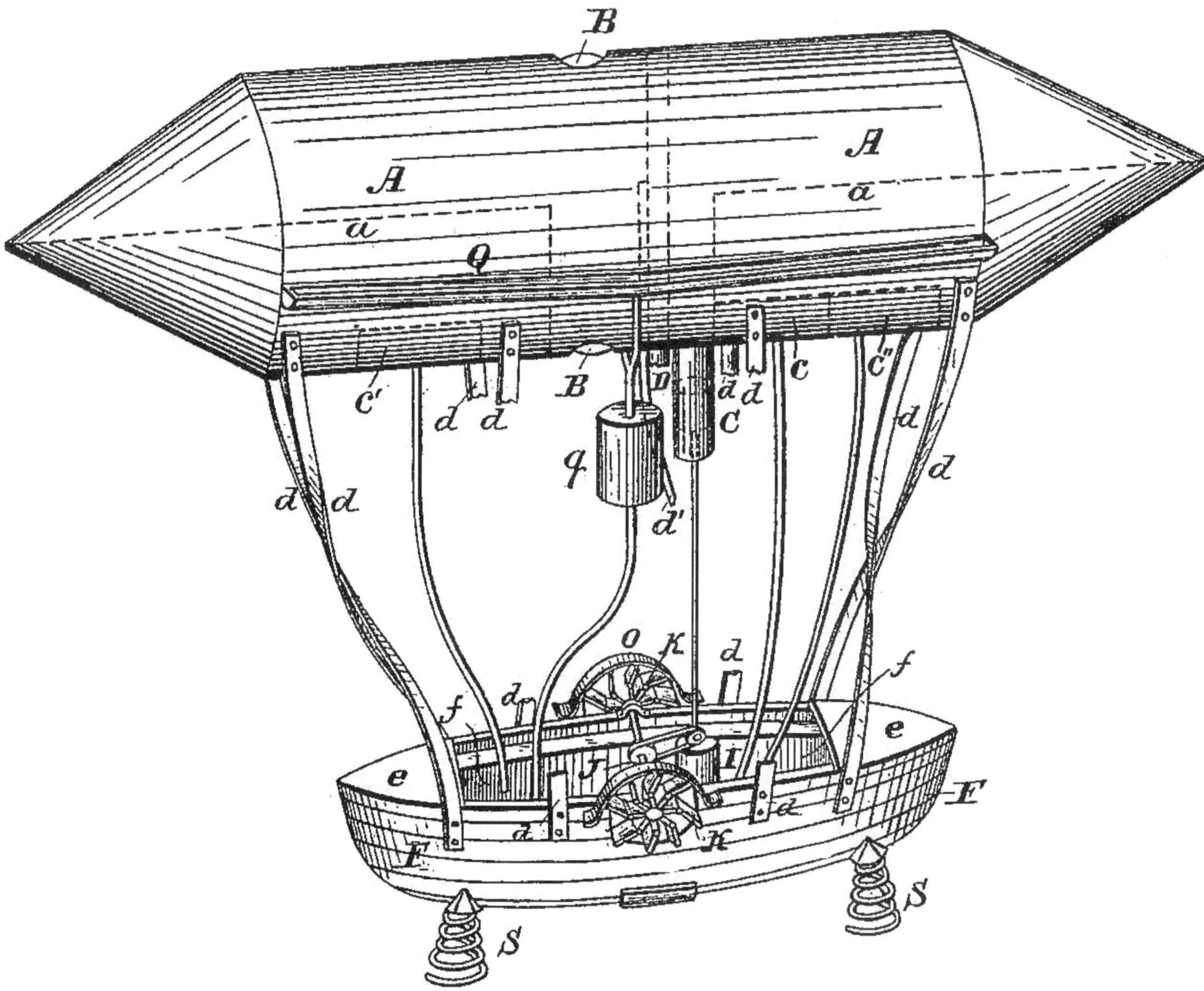

Plate 52: John Emsley's Balloon—April 17, 1883: *In yet another variation on the flying dinghy theme so prevalent in the nineteenth century, inventor John Emsley is intent on figuring out a way to eliminate the need for valving off gas or dropping ballast as a means of controlling vertical flight. I'm not certain how all of his innovations work on his all-metal balloon, but they look impressive—especially the oversized springs at the bottom of the gondola in case these innovations fail. One curious aspect of the craft is his inclusion of two glass portholes (figure B) that permit the pilot to look straight up through the middle of the balloon. While he tells us this is to allow for the inspection of the envelope's interior, he doesn't say what we are inspecting for, as the gas is invisible.*

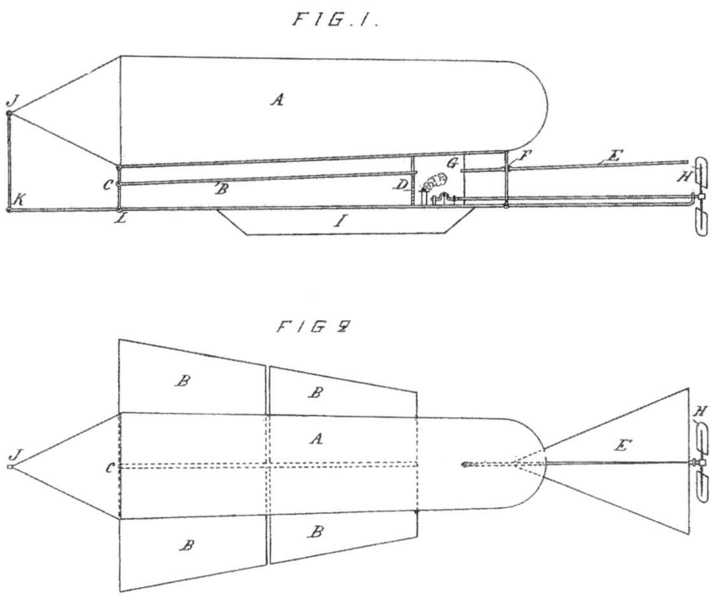

Plate 53: Julio De Souza's Apparatus for Aerial Navigation—July 10, 1883: *Patent drawings don't get much simpler than this. Believe it or not, Mr. De Souza, one of Brazil's unheralded aviation pioneers, designed this around the idea of a bird, although it's possible some of the craft's more bird-like capabilities may have fallen along the wayside. Basically this is simply a hydrogen-filled balloon with wings and a massive tail (which he calls a rudder but is, in fact, actually an elevator). The tiny propeller at the end is more of an afterthought, as most of the craft's forward momentum is going to be achieved by essentially gliding through the air on "bird-like" wings.*

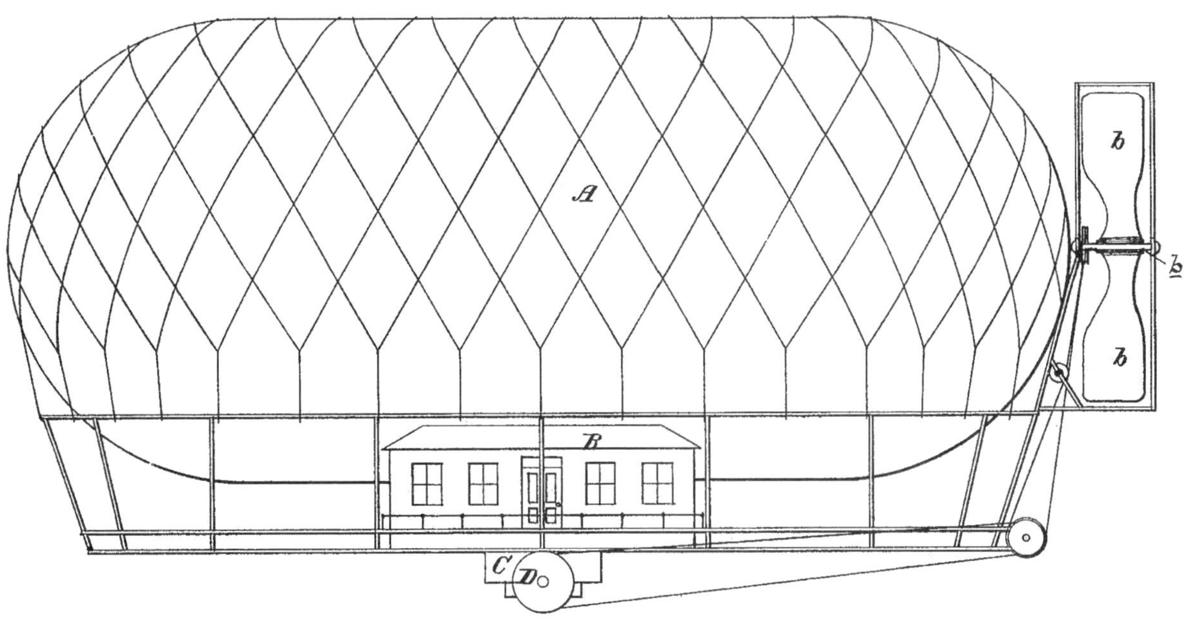

Plate 54: Auguste Debayeux's Aerial Machine—July 31, 1883: *Frenchman Auguste Debayeux has decided that the best way to propel a balloon through the skies is not to have a pusher fan bully it against the wind, but to use a fan to create a vacuum—what he calls "rarefaction of the air"—at the intended direction of travel that will compel the machine to move towards the less denser air. In effect, he proposes essentially sucking the balloon forward by creating a type of vortex in front of it. Having done all the difficult work in figuring out how to rarefy the air, he generously leaves such details as how to steer or power the craft (he recommends steam unless it's small enough that it can be hand cranked) to later designers to wrestle with. In fairness, however, this design does include a very nice cottage, complete with a picket fence and darling curtains.*

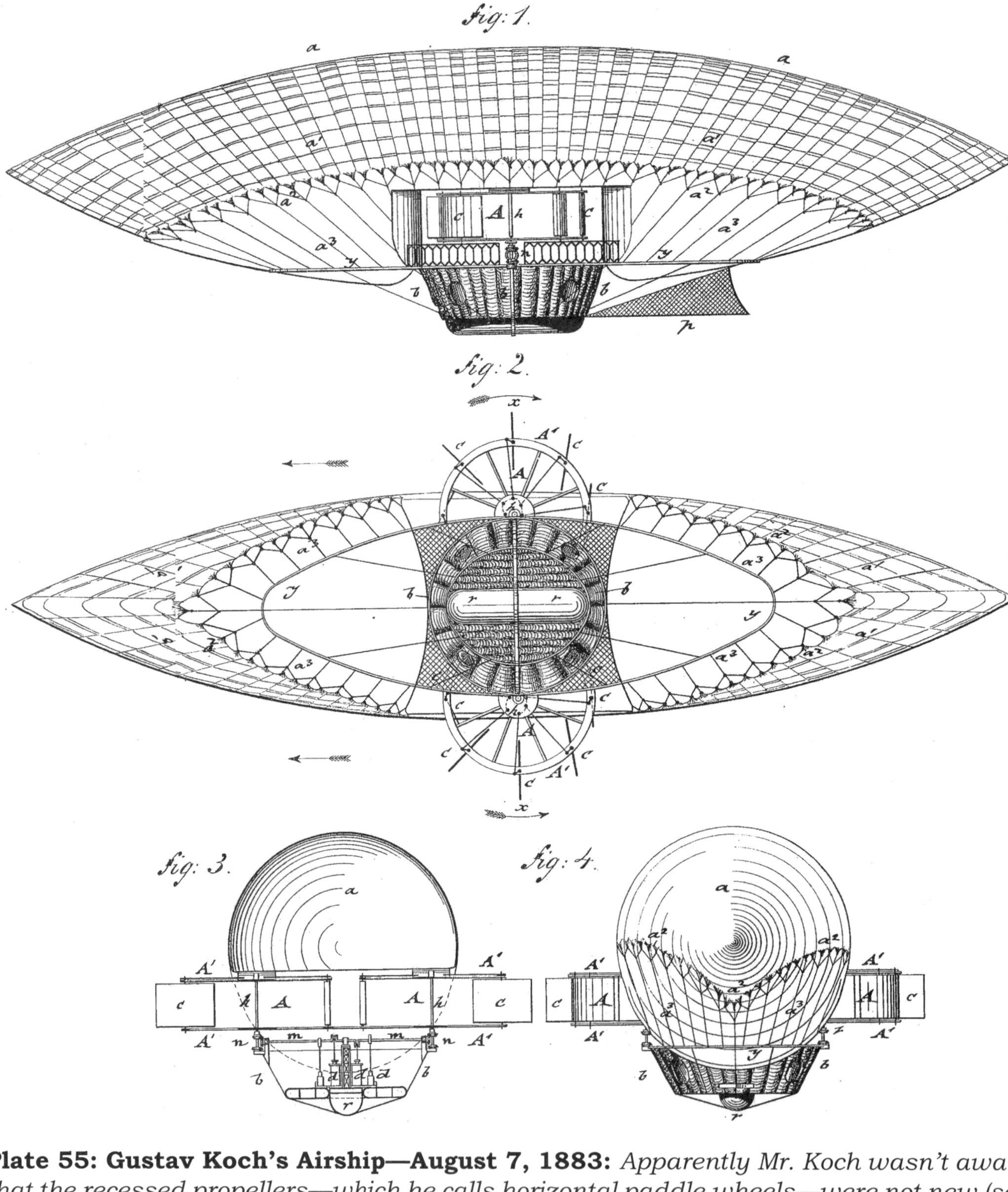

Plate 55: Gustav Koch's Airship—August 7, 1883: *Apparently Mr. Koch wasn't aware that the recessed propellers—which he calls horizontal paddle wheels—were not new (see plate 39), his certainly are beefier. While they may have actually been useful in vertical flight, how they might assist in steering the ship horizontally remain unclear, however. One interesting innovation is the silk ribbons used in lieu of the traditional netting to affix the gondola to the balloon that extend across the bottom of the envelope so as to provide a smooth, aerodynamic surface. They also make an effective parachute in case the balloon should burst, though he doesn't say whether he has tested that hypothesis personally.*

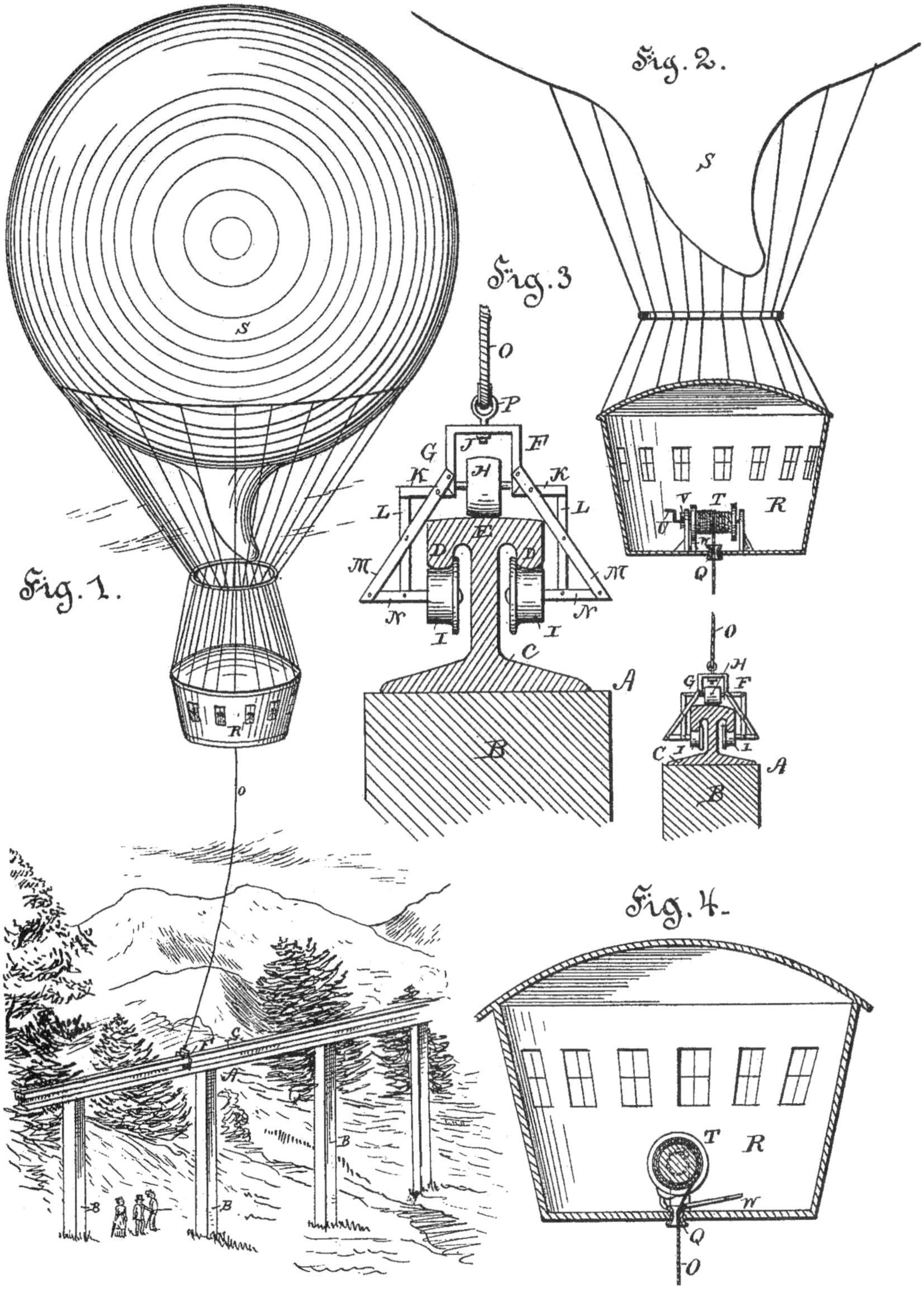

Plate 56: David Olmstead's Aerial Railway—October 30, 1883: *In an obvious rip-off of Charles McGlashan's "Guiding" Balloon" design of 1874 (see plate 24), David Olmstead has simply taken it to the next level by creating a network of monorails as opposed to attaching his balloon to existing rail lines, as Mr. McGlashan recommended a decade earlier. While this eliminates the problem of what happens should a train come along, Mr. Olmstead does not tell us how he proposes to get the aerial tram to move down the rail against the wind or on a calm day, apparently content to leave such details to others to ponder.*

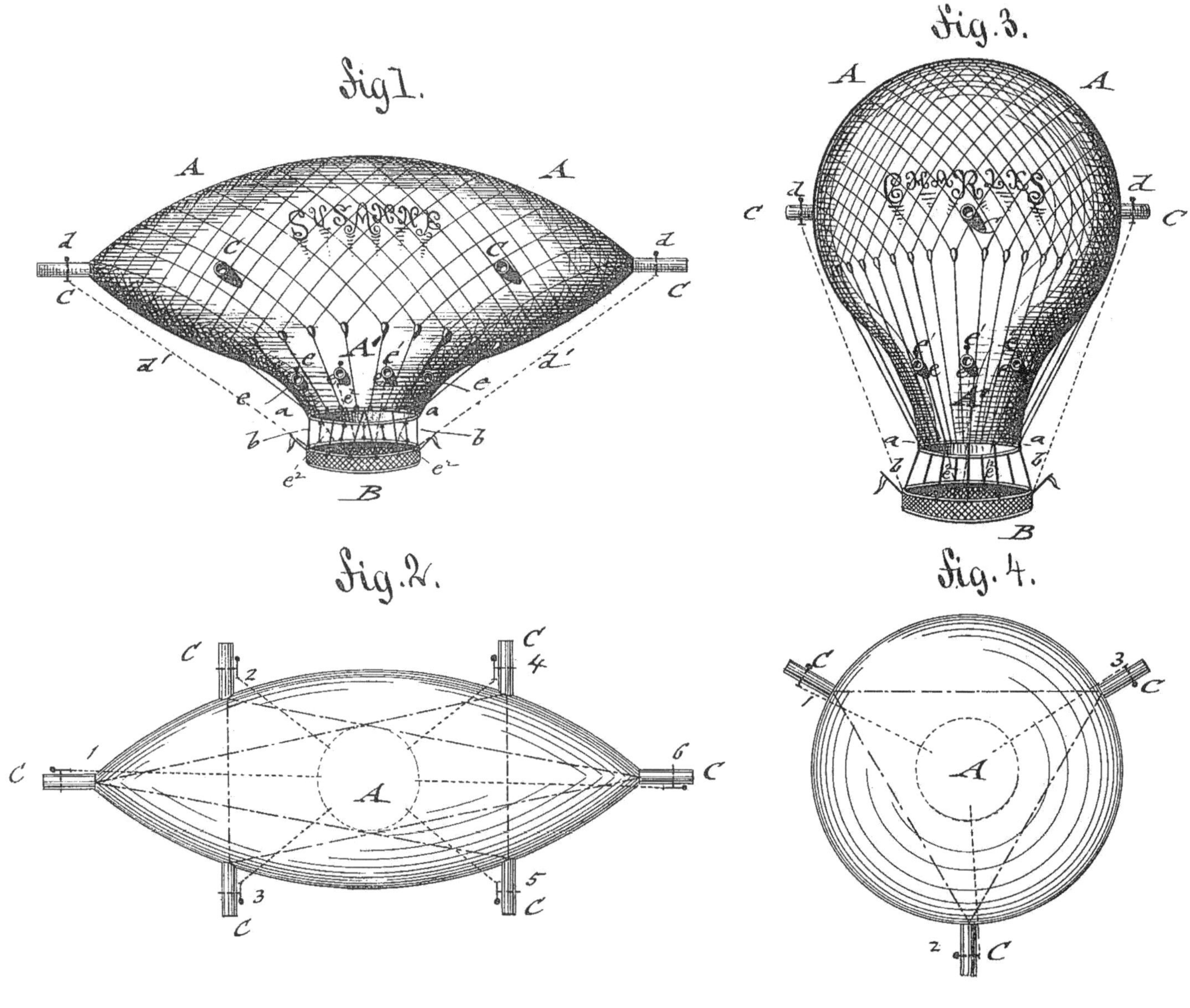

Plate 57: Charles Fest's Airship—March 18, 1884: *Philadelphia's own Charles Fest has thoughtfully provided two versions of his revolutionary airship design in which the balloons are simply maneuvered about by releasing super-heated air through a series of pipes that transverse the center of the envelope. This is made possible by the "great pressure" hot air creates inside the skin of the balloon, which must be valved off immediately lest the balloon simply explode from the great pressure. Note that these are his and her balloons, the spherical one with the name Charles inscribed upon it and the oblong one named Susanne (whom we presume to be Charles Fest's beloved wife)—a quaint reminder that romance still had its place in aeronautics.*

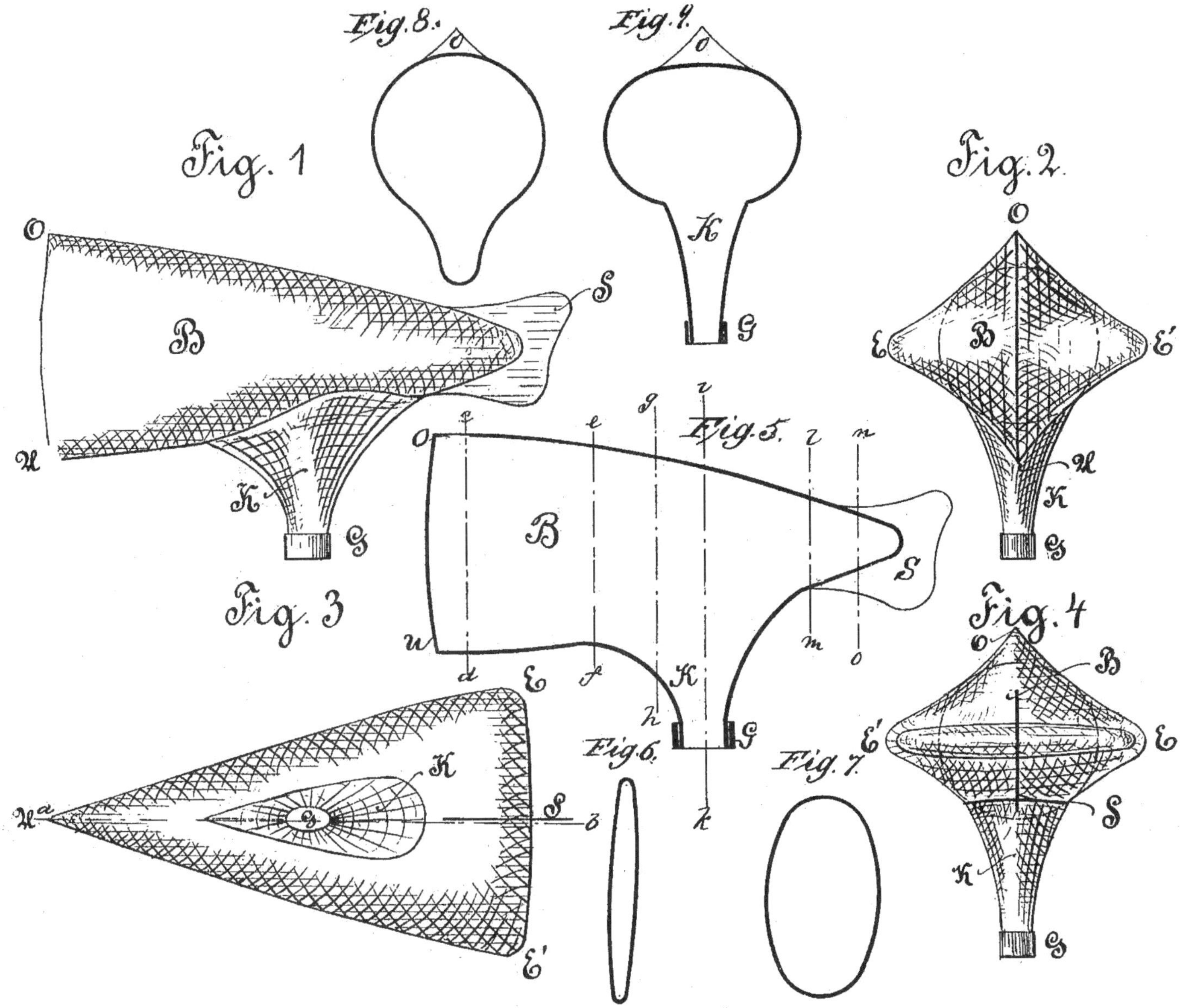

Plate 58: George Wellner's Wedge-Shaped Air Balloon—December 2, 1884: *Looking a little like a Macy's parade balloon with a medical condition, this interesting design is simply a wedge-shaped hot-air balloon, which Mr. Wellner imagines will be pushed through the air due to the wind acting upon the balloon's flat back end while its sharp prow cuts through the air like a knife.The problem, of course, is that once the wind got hold of the balloon, it would act upon the largest surface—the sides—and blow it sideways, completely impervious to the efforts of its doomed pilot to steer it or otherwise have any ability to control its flight.*

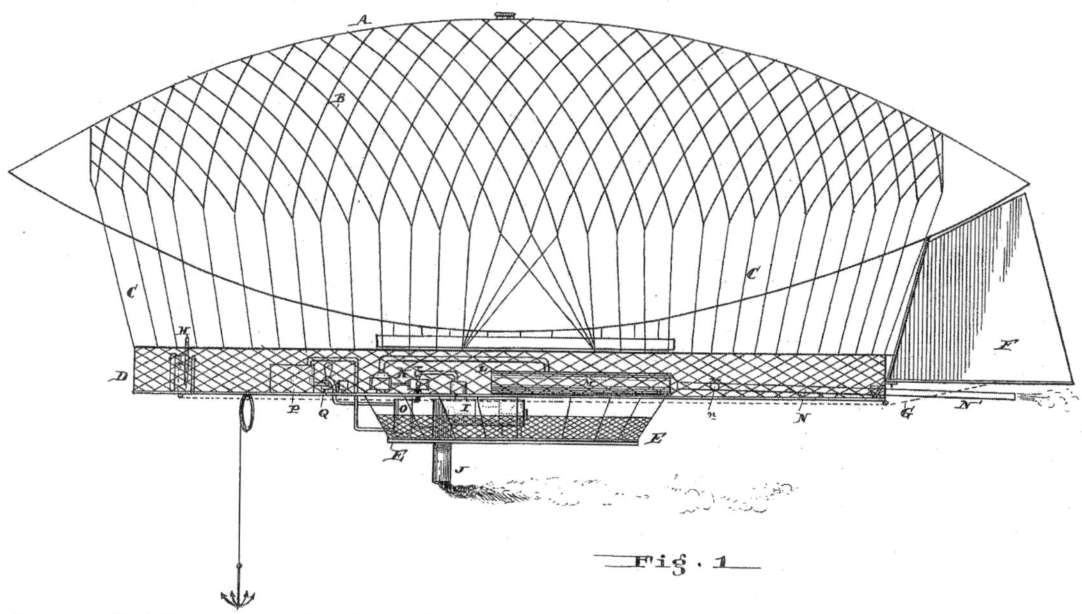

Plate 59: Russell Thayer's Aerial Ship—December 9, 1884: *In what may be the first attempt at creating a dirigible powered by jet propulsion, Russell Thayer of Philadelphia suggests that steam kept under very high pressure could be jetted out the back of the ship, creating the necessary thrust. A peculiar innovation is the downward pointing smokestack coming off the boiler; apparently unconcerned that smoke tends to rise, it is hoped that Mr. Thayer has devised some sort of blower to direct the smoke out the bottom or his gondola will quickly become a smoking car.*

Plate 60: Frederick Brearey's Aerial Apparatus—June 16, 1885: *Mr. Breary, apparently finding his earlier flapping-wing creation impractical (see plate 39), is back with a new and improved version; the main difference—or the only difference for that matter—being that this one has four wings rather than just the two in his earlier design. They still flap up and down like a bird, however, much to the distress of the pilot, who must possess the strength of Hercules to make the thing go more than a few feet. This is a good example of why the successful pioneers of aviation always started with fixed-wing unpowered gliders before moving on the powered flight; it was the only way they could learn what sort of airfoil provided the proper lift—a small detail Mr. Breary apparently was loath to address.*

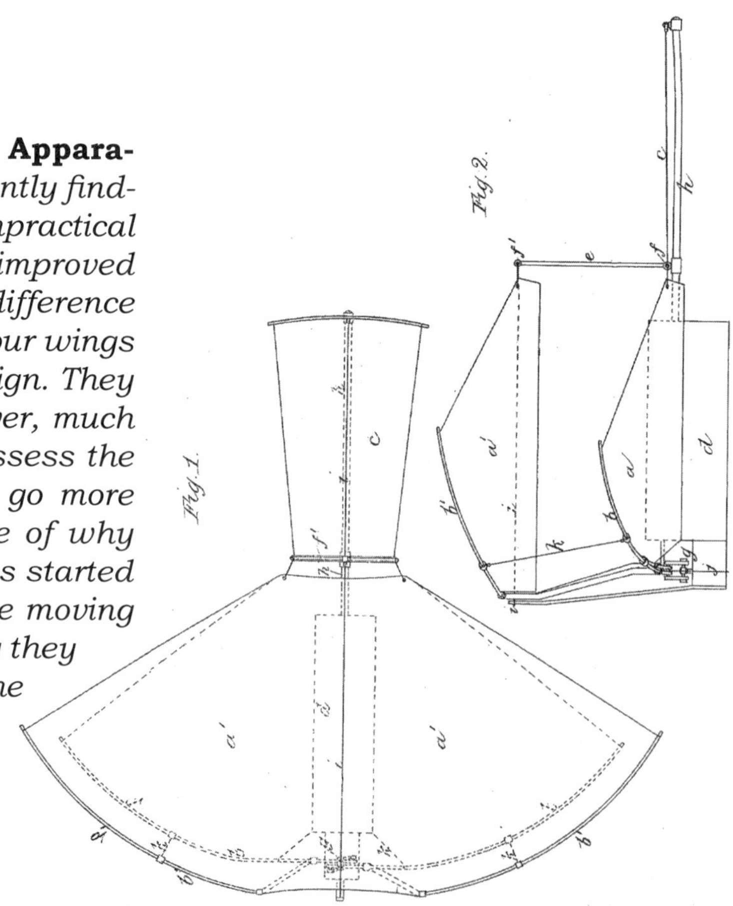

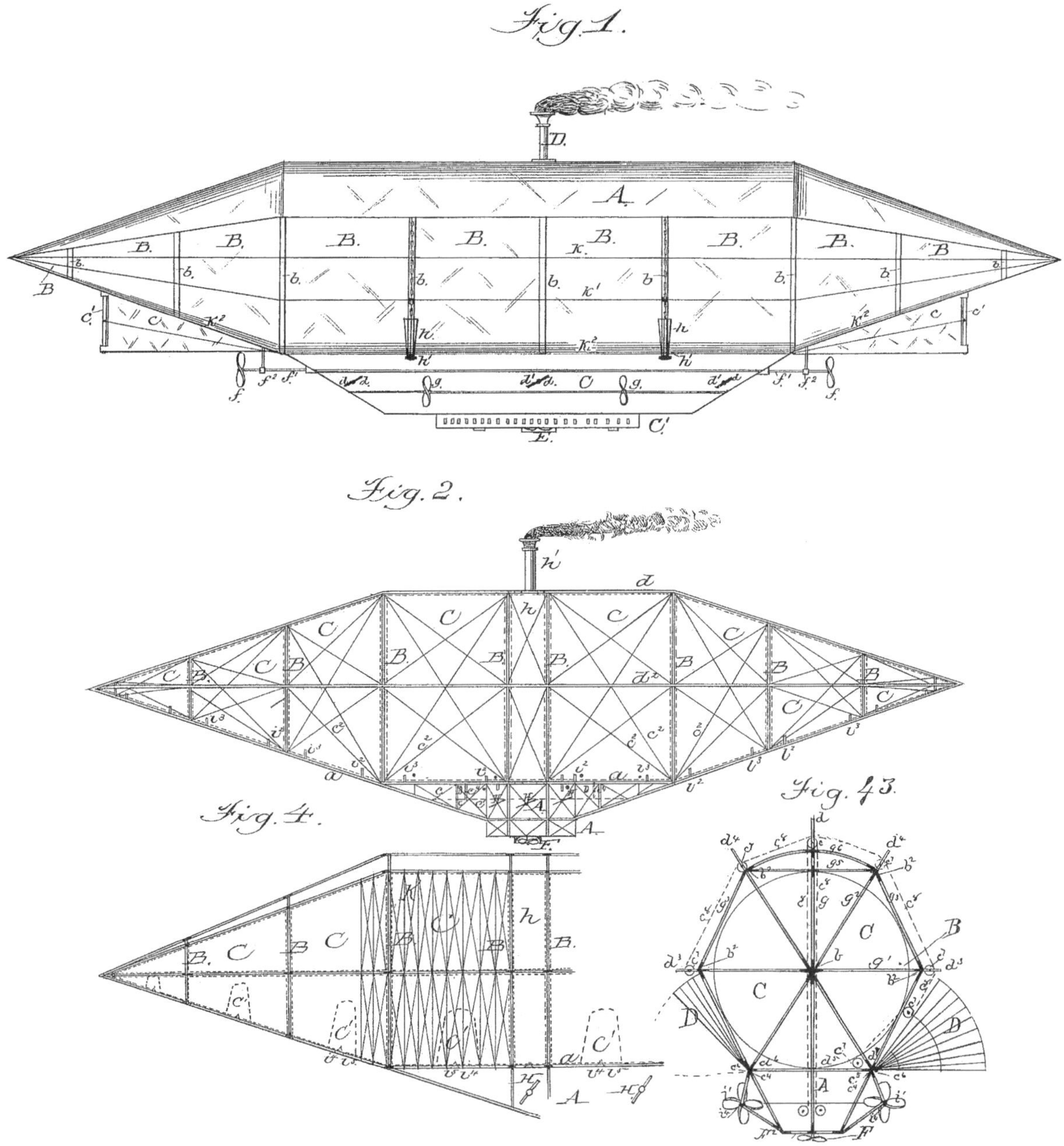

Plate 61: Eugene Falconnet's Aerial Ship—February 10, 1885: *Having designed one of the largest airships ever (see plate 51), Mr. Falconnet is back again with a new and improved, but no smaller, version. (One gets a sense of scale of the ship by comparing the size of the envelope to the gondola at the bottom.) The main difference between this design and his earlier airship is that this one has a greater diameter, and he has installed rudders (a feature his earlier design lacked). Additionally, the propellers are considerably smaller and look more like ship's screws than propellers, but otherwise his "improved" version isn't far removed from his earlier efforts. (Actually, the man submitted five new patent drawings on this date, each of them essentially variations on the same theme, making him easily the most prolific inventor of the decade.) An interesting—and one would think problematic—addition is also the centrally-mounted smoke stack that belches smoke from the top of the hydrogen-engorged envelope, which some—though apparently not Mr. Falconnet—might consider it to be a fire hazard.*

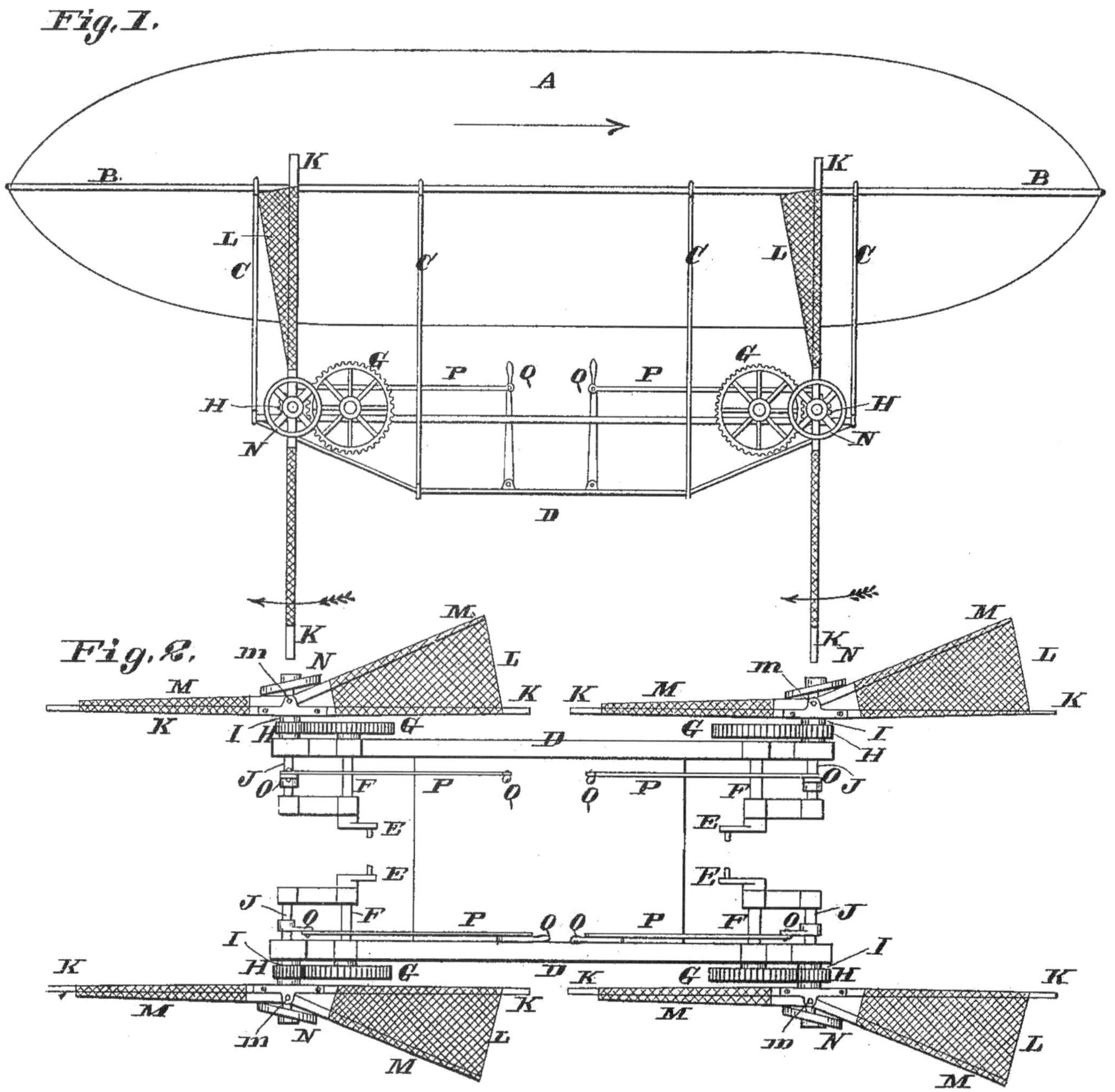

Plate 62: Anton Spier's Airship—June 9, 1885: *Another typical concept of the era, this airship is distinguished by its four large, collapsible fabric wings, each of which can be made to turn independently of the other for maximum maneuverability. It's not entirely clear why they are collapsible other than it has something to do with exerting lesser or greater wind resistance to make the wings more efficient for propulsion, although without an adequate power plant to turn the cams—a problem rarely addressed by most designers of the day—it would all be a moot point.*

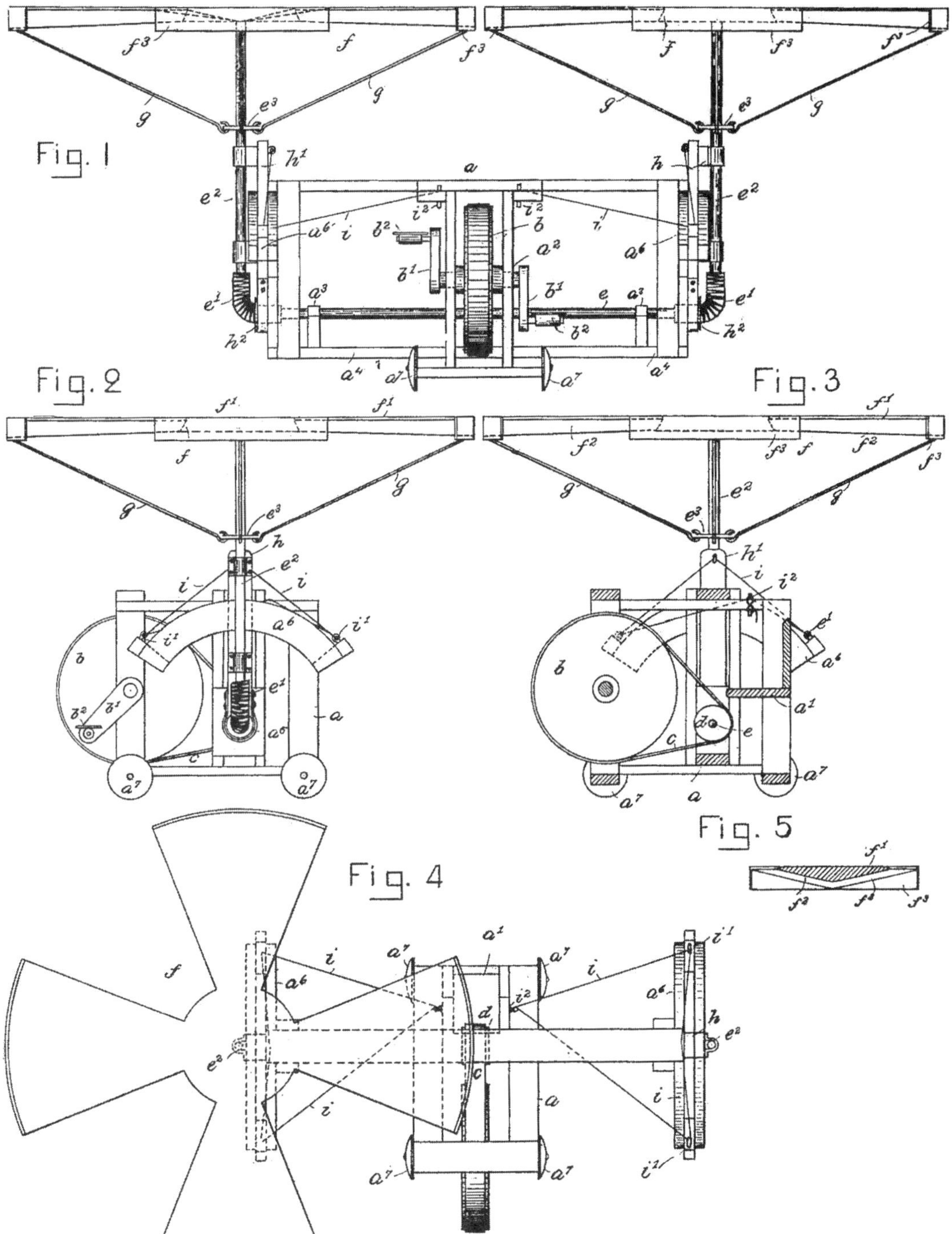

Plate 63: Joseph Foster's Airship—June 23, 1885: *Perhaps one of the earliest attempts at producing a helicopter is this frightening machine devised by Mr. Joseph Foster of Massachusetts. Basically two crank-turned counter rotating scythes, it can only be hoped that the pilot—located behind the foot pedals—had the good sense to keep his head down while he feverishly churned the driveshaft that promises him aviation immortality—or at very least martyrdom.*

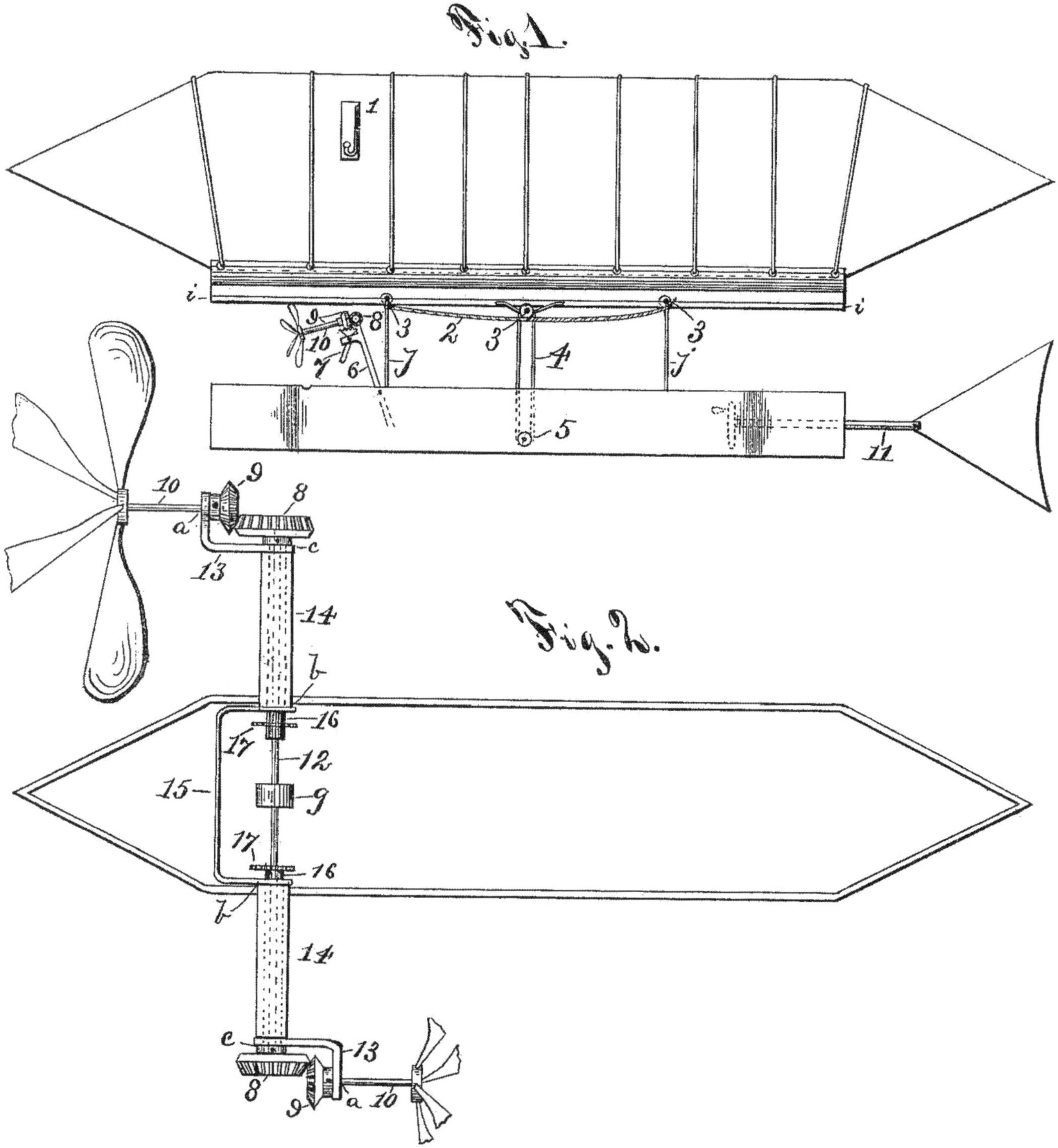

Plate 64: Hurgar Hunnicutt's Aerial Ship—January 26, 1886: *Another combination of propellers being turned by some as yet unnamed but obviously powerful engine, thanks to Arkansas inventor Hurgar (who would name their kid Hurgar?) Hunnicutt, who has at least simplified the thing down to just two multi-directional propellers. The drawing is a bit confusing, however, in trying to determine just how big and what configuration of propellers he proposes; the starboard propeller looks huge and is shaped more like a ship's screw (with some sort of smaller scythe-like blades attached) while the port propeller looks small enough to be a rotor-rooter attachment. It's also interesting that he has made it possible to turn the rudder onto its side, allowing it to serve as either a rudder or an elevator.*

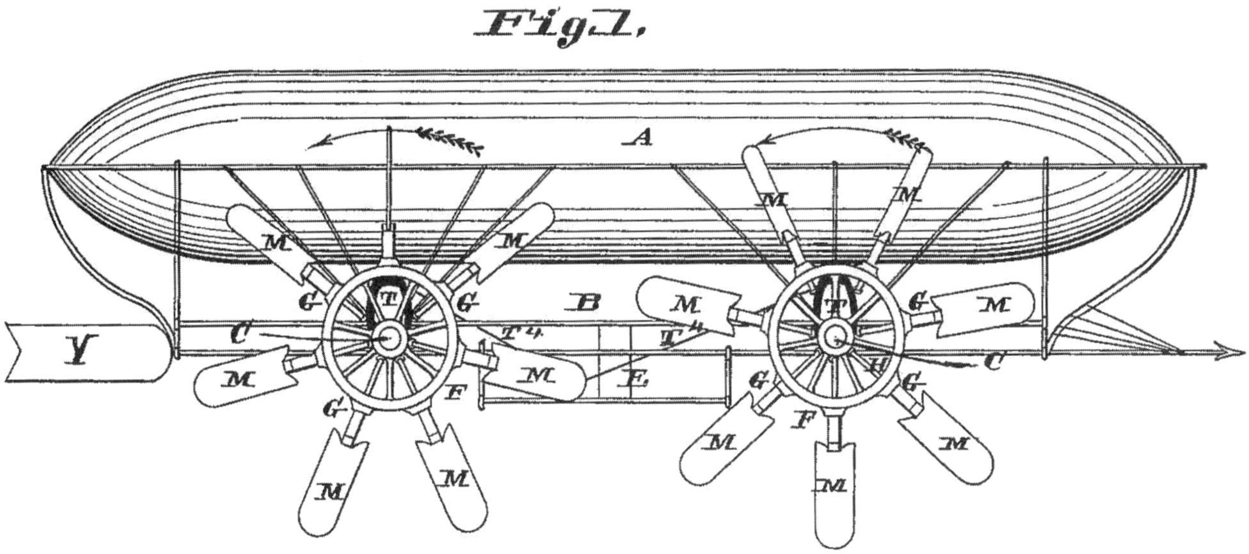

Plate 65: David Stout's Aerial Ship—June 9, 1885: *In yet another attempt to introduce paddle wheels to aviation, St. Louis inventor David Stout presents this very Victorian-looking craft for our consideration. The only real novelty about this design is that he has built a series of cams and springs into each paddle that allows them to swivel into a flat position when at their apex and then swivel to a more diagonal position when at the bottom of their rotation, presumably as a means of reducing wind resistance. This probably would have looked pretty cool in operation but, alas, Mr. Stout's curious-looking craft never made it beyond the concept stage.*

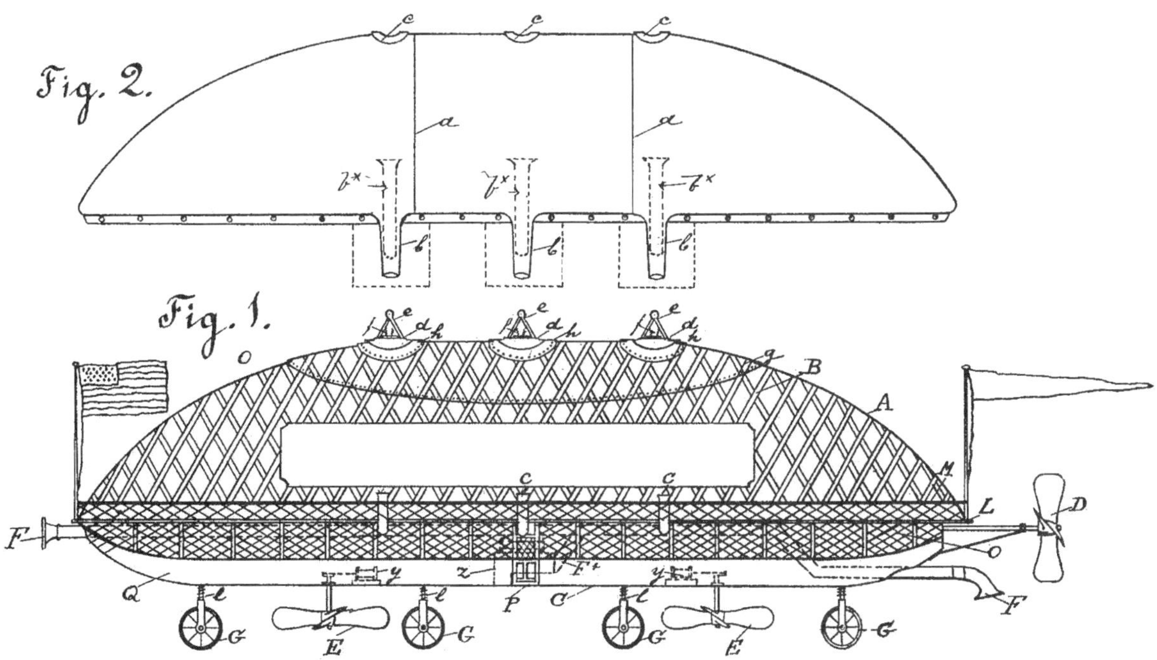

Plate 66: William Patterson's Steam Airship—April 27, 1886: *Reminiscent of a poorly conceived parade float, San Francisco inventor William Patterson presents this steam-powered behemoth as though he were the first to come up with the idea. The only real innovation is that he has whistles attached to the gas relief valves on the top of the envelope (letter E) and he has decided to build almost the entire gondola and frame of the machine out of lightweight iron. On second thought, this may actually be a parade float after all.*

Fig. 1.

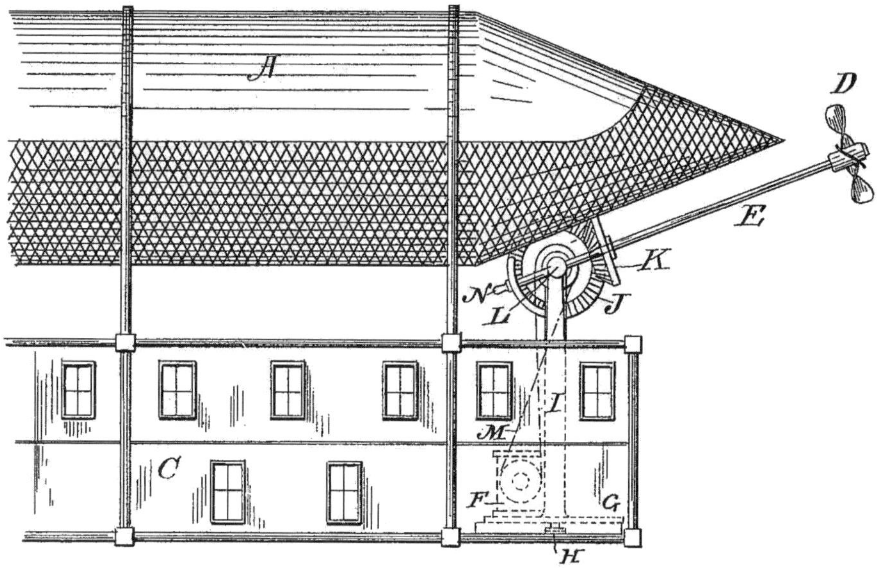

Fig. 3.

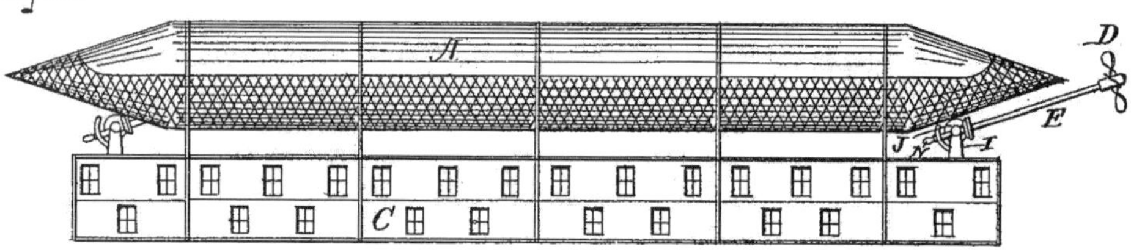

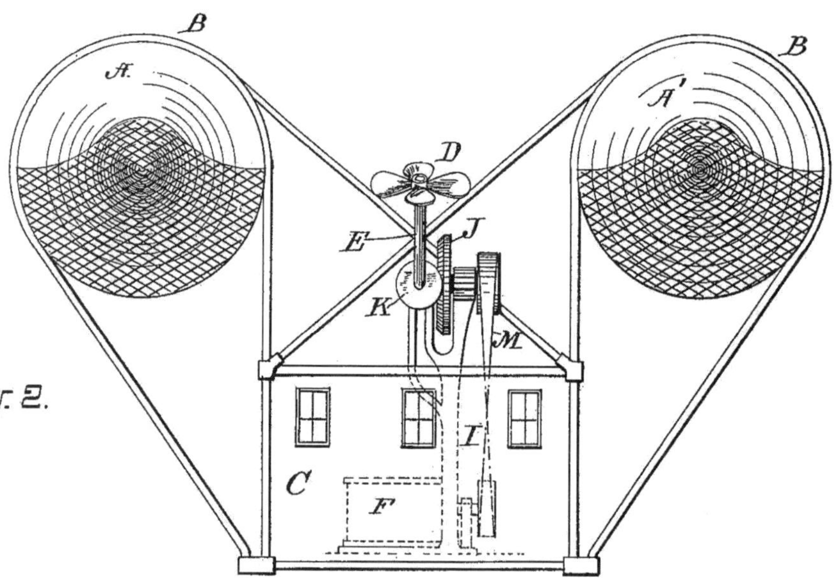

Fig. 2.

Plate 67: Levi Beardsley's Ship for Aerial Navigation—March 2, 1886: *Demonstrating that even a two-story office building is not beyond becoming a subject for aerial locomotion, Dodge City, Kansas' own Levi Beardsley shows us how it's done. Perhaps the most interesting feature is the massive adjustable screw on the end of a very long shaft that's turned by some sort of giant egg-beater. Obviously to get this thing off the ground would require considerably larger balloons and something akin to a train locomotive engine to power the screw, though such appears to be only of passing concern to Dodge City's finest resident airship enthusiast.*

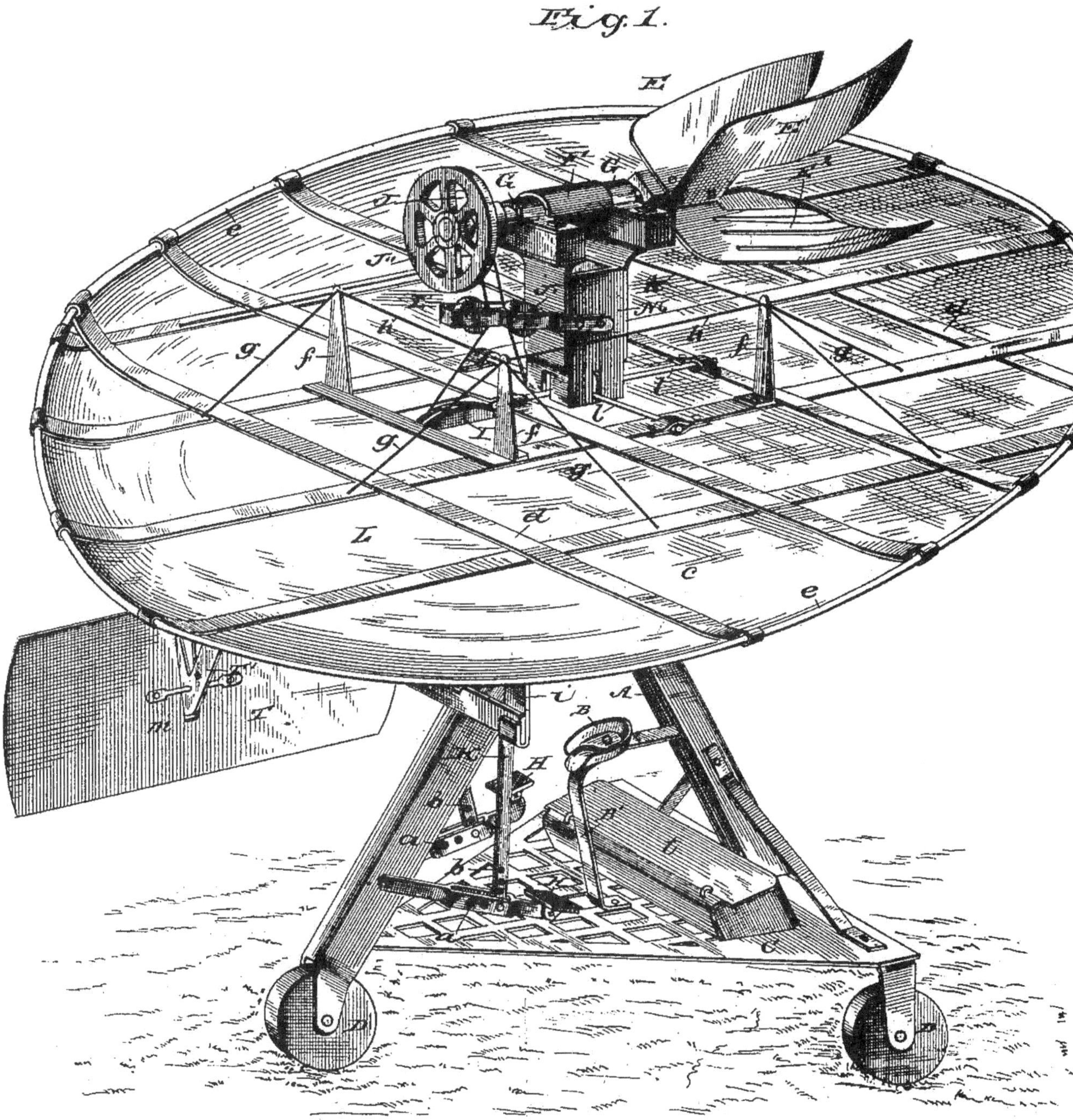

Plate 68: Ringert Jongewaard's Aerial Vessel—March 16, 1886: *Looking a little like a cross between an outdoor grill with a rudder and a satellite dish with feather attachments, this design from the old West (Mr. Jongewaard hailing from the Dakota Territory) is guaranteed to get the neighbors' attention. Working something akin to the way a concave sled would travel on ice, this man-operated machine was designed to "rise upon the wind by presenting the underside of an inclined plane thereto while propelling the machine slowly toward it, or to rise in still air, propelling the same more rapidly in the desired direction." It will also, if carefully tuned to the proper frequency, pick up over two hundred television stations, including ESPN.*

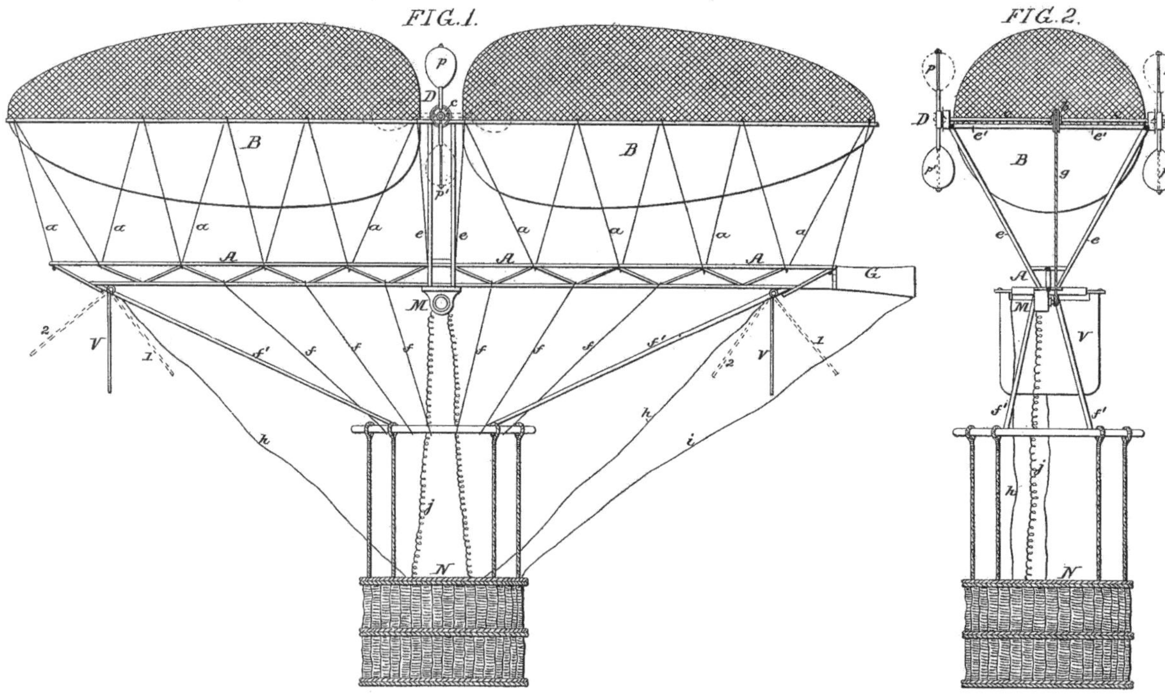

Plate 69: Augustin Hamon's Dirigible Aerostat—October 5, 1886: *This curious design is the first to mention using an electric motor to power the two four-bladed lateral propellers, although considering the state of technology at the time, no motor existed that could have much of an effect on such an unwieldy machine. The dual air bags and overly-extensive undercarriage makes the craft more complex than necessary as well. What's especially unusual about the drawing is the disproportionately large gondola, which gives one the impression that Mr. Hamon's machine is considerably smaller than one might determine at first glance. Also, notice the extremely tiny rudder (figure G) that, if the gondola is of average size, makes it not much larger than a folded road map and, as such, about as useful.*

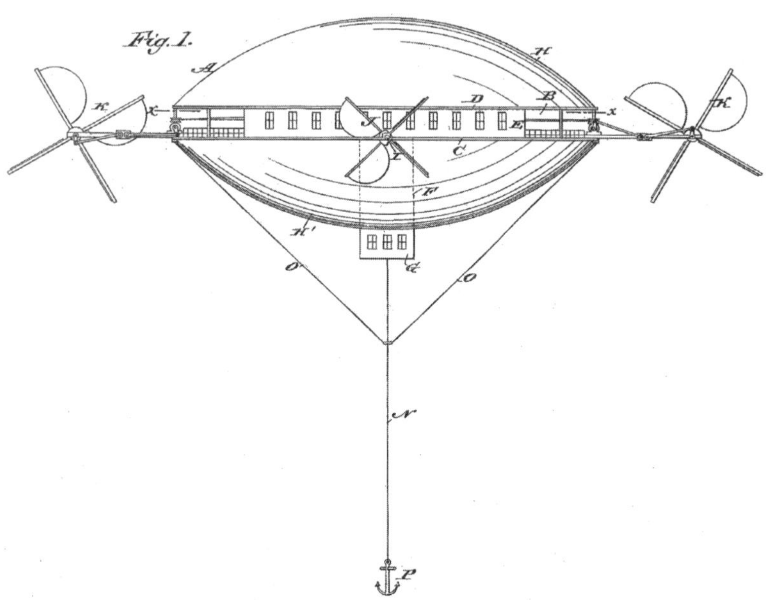

Plate 70: Moses Cole's Aerial Vessel—November 9, 1886: *Looking suspiciously UFO-like, Nicaraguan inventor Moses Cole has simply sandwiched a rather elaborate gondola (it includes staterooms and porticos) between two semi-spherical balloons and placed four massive pivoting propellers on each axis thus hopefully allowing the pilot to steer the ship in any direction desired. The most mysterious aspect is the ship's anchor hanging from the bottom, which he briefly mentions in his description but fails to explain its precise purpose. Perhaps it was used to pluck cattle from the prairies to be experimented upon and then left as early examples of cattle mutilations.*

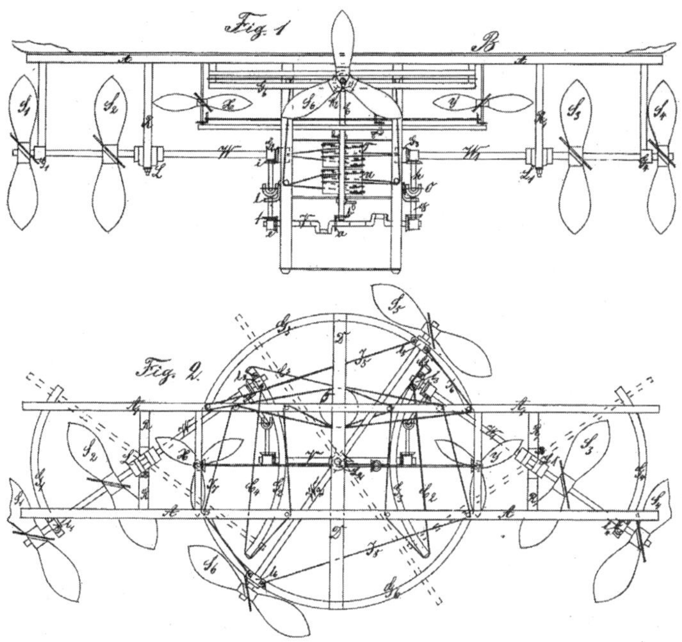

Plate 71: Otto Hartung's Balloon Propulsion System—December 14, 1886: *Working from the premise that if one propeller is good, eight propellers must be even better, German inventor Otto Hartung brings us the ultimate in over-engineering (a trend that was to bedevil the Germans right up through the end of World War II). The only thing missing is space for an engine or motor of some kind, nor does it appear to be open to the idea of seating a pilot (although these might be later attachments not shown in Mr. Hartung's drawings).*

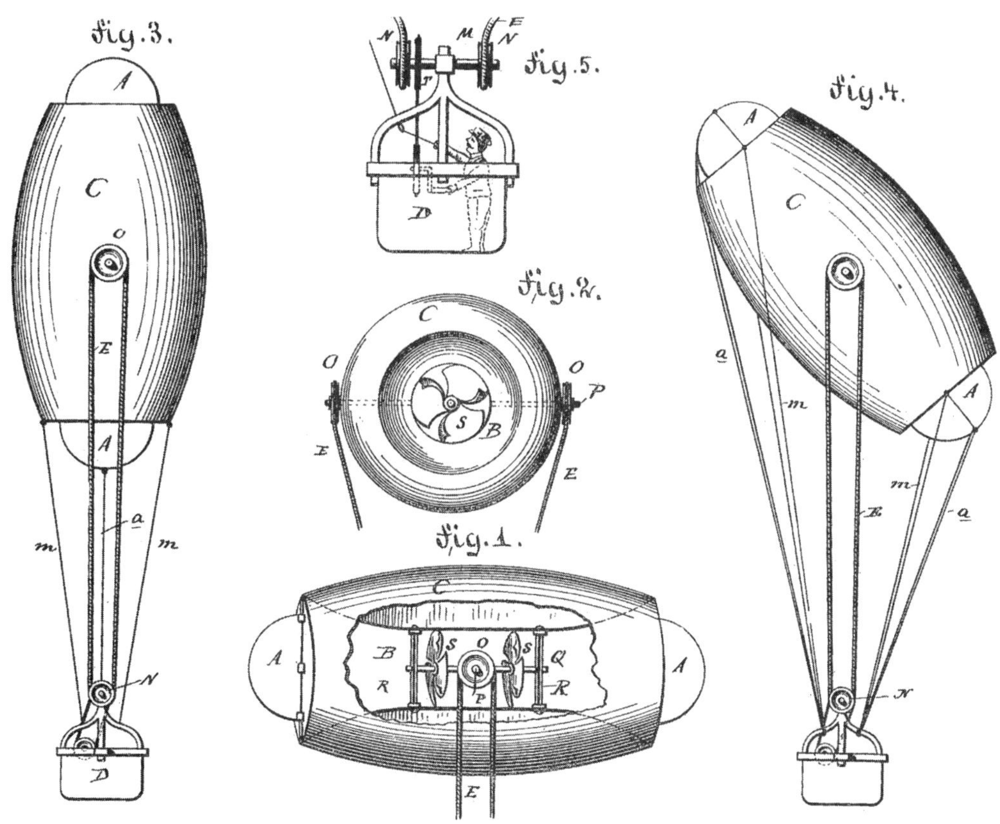

Plate 72: Joseph Neubrand's Aerial Vessel—April 19, 1887: *This novel approach to making a balloon obey the commands of its pilot incorporates twin screws inside the body of the balloon itself, which can then be pointed in the desired direction of travel (not unlike a modern jet engine, one might propose). I'm thinking Mr. Neubrand may want to consider increasing the size of his balloon's hydrogen capacity, however, or only misfortune will accompany him.*

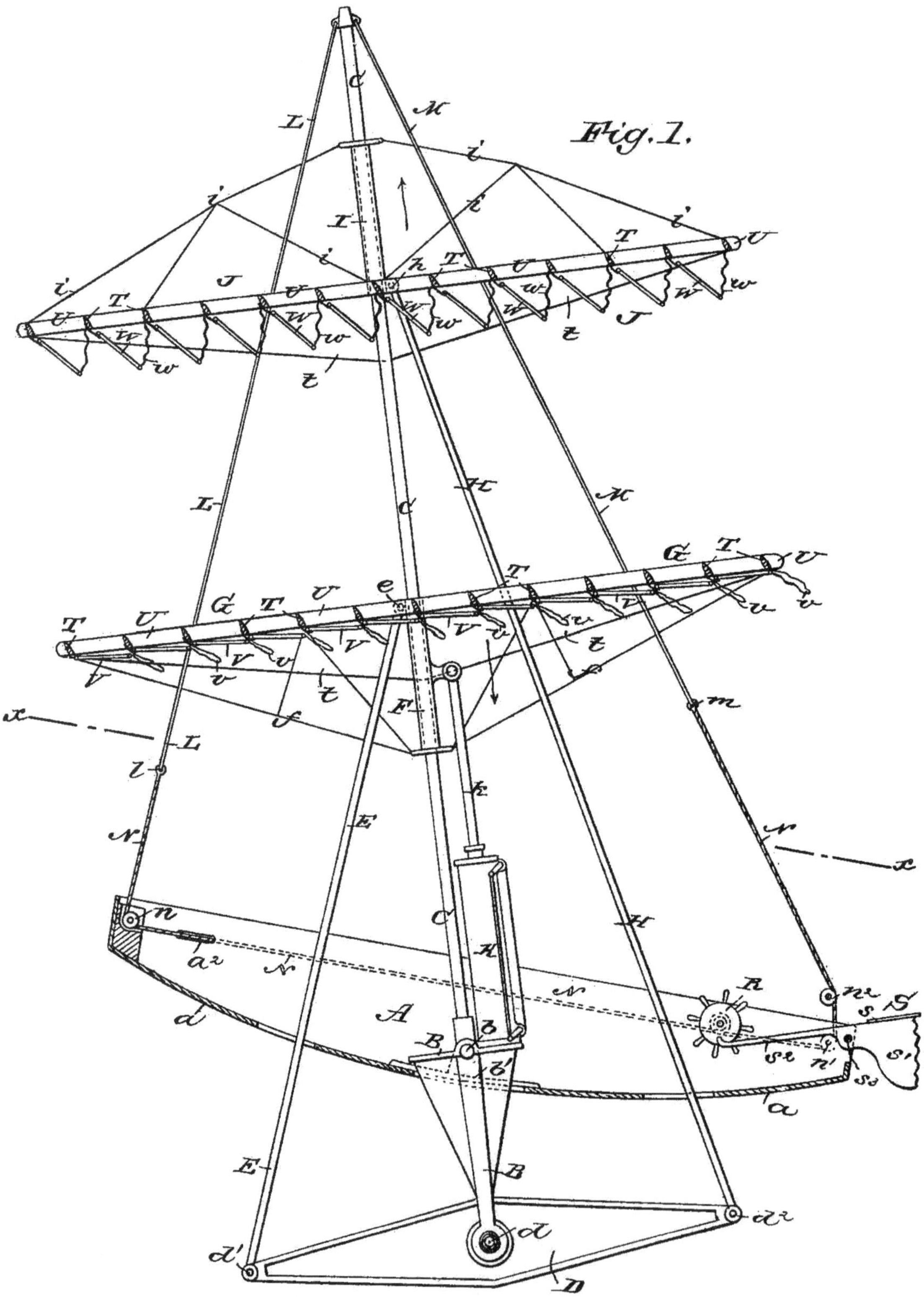

Plate 73: James Wheeler's Airship—April 19, 1887: *This treacherous-looking apparatus is basically a boat with a swiveling mast that flies through the air courtesy of a series of sail-like "aero planes" designed to lift or descend the balloon based on which way the pilot turns the rudder. The machine is absolutely guaranteed to, if not fly, at least provide a satisfyingly spectacular crash.*

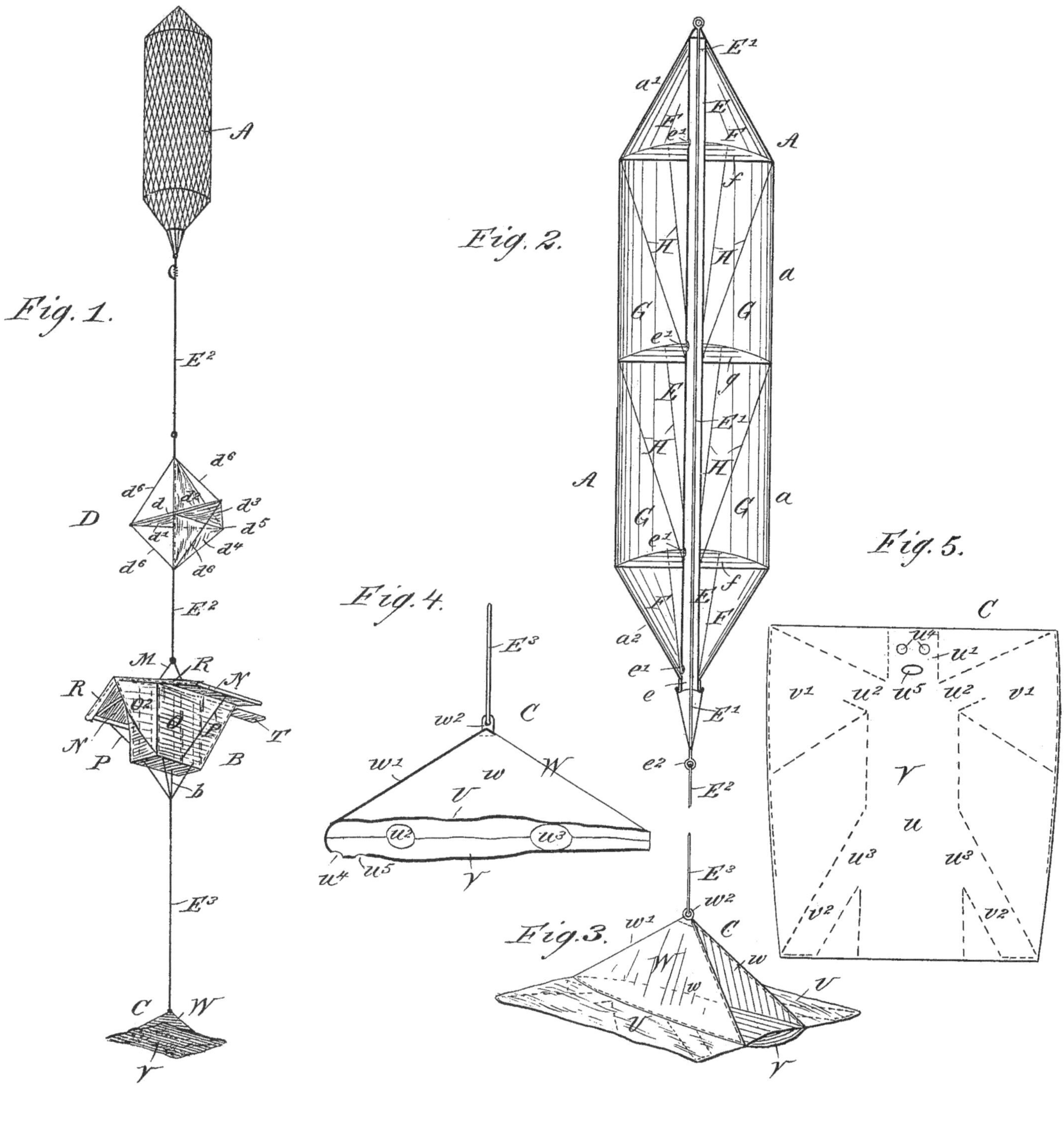

Plate 74: William Beeson's Aerial Vessel—April 26, 1887: *Looking a little like a NOAA weather experiment, William Beeson of the Montana Territory demonstrates that not all great ideas originate in civilized areas of the country, but can even be found in some of the most remote corners of the globe—like Montana. Essentially a pointy vertical balloon with a "flying sail-suit," rudders, and gondola attached, supposedly the pilot only has to shift his weight and the craft will magically glide—thanks to the ingenious sail-suit—in whatever direction the wind happens to push him. It may have been something along this line that crashed near Roswell, New Mexico, in 1947.*

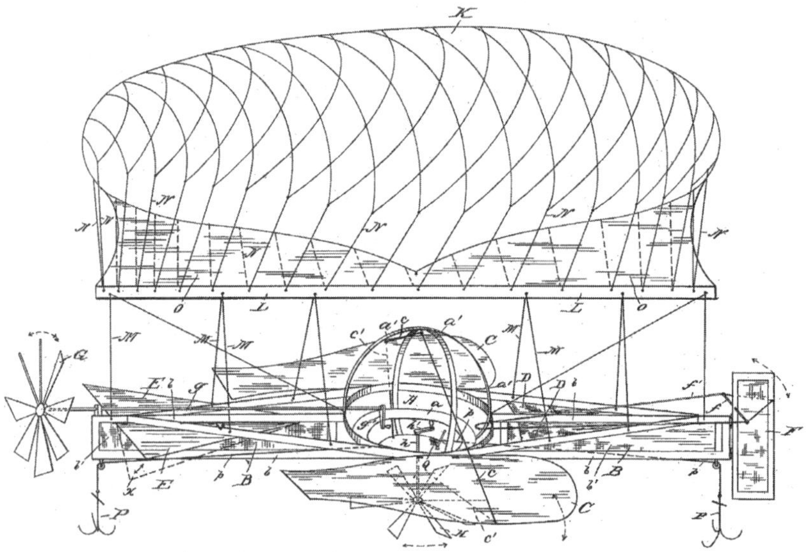

Plate 75: Peter Campbell's Airship—May 10, 1887: *Working from the more-is-better school of dirigible design, Brooklyn designer Peter Campbell presents us with an over-the-top airship configuration that incorporates hydrogen gas, flapping wings, and pusher propellers, all built around a very cool-looking spherical control gondola. Among the many space-age innovations included, and in sharp contrast to earlier designers that had only a single feeble hook anchor on their machines, in a bold stroke of foresight Mr. Campbell provides us with two. Clearly a man of vision.*

FIG. 1.

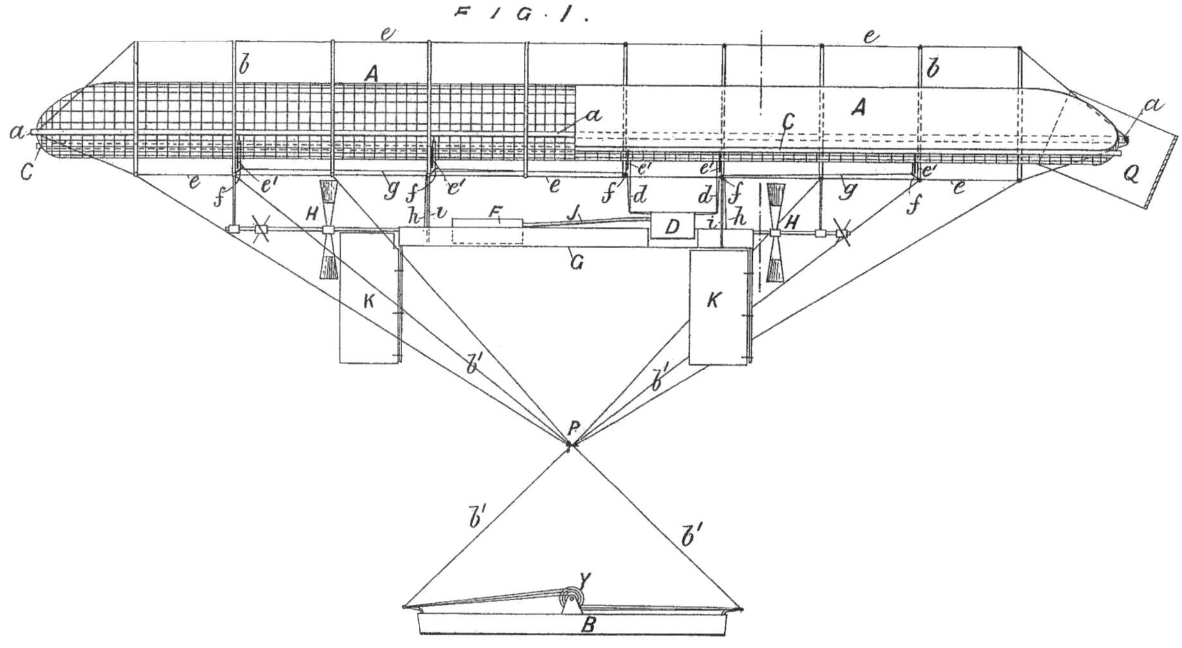

Plate 76: William Hutchinson's Aerial Vessel—October 18, 1887: *This well-thought-out approach to maintaining the shape of a very large envelope (in this case 360 feet by 30 feet in diameter) through the use of triangular bamboo partitions that act like an exoskeleton actually portends advances in rigid construction later seen in von Zeppelin's machines. Like many designs of the era, however, Hutchinson's craft suffers from being needlessly over complex, with both fore-and-aft propellers and oversized rudders (letter K) being placed too far inboard the turning blades to be effective. The device dangling precariously from the bottom (letter B) is a detachable cargo car that, while a novel idea, would probably have proven unworkable without some sort of control surfaces affixed to keep it from spinning uncontrollably during flight.*

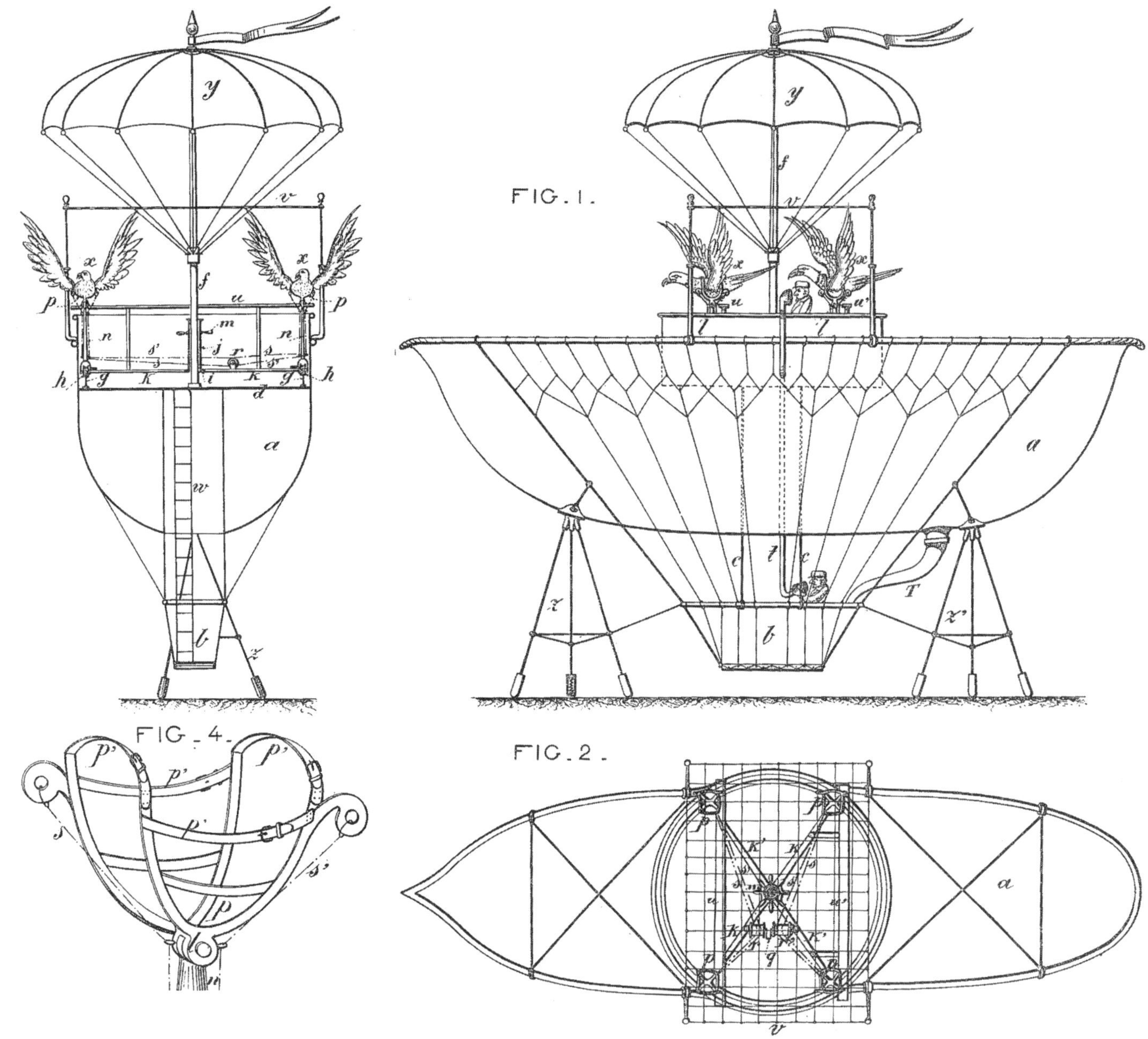

Plate 77: Charles Wulff's "Means and Apparatus for Propelling and Guiding Balloons"—May 17, 1887: *At last, a no-nonsense approach to making a bathtub fly; simply attach the desired number of large birds required to lift it into fixed harnesses (figure 4) and let nature do all the work. (Note the very fine rendering of two sailors/aviators discussing their avian requirements via metal voice tubes in figure 1.) The real mystery behind this design is not in how it flies, but in how Mr. Wulff avoided being shown the door by the patent clerks when he attempted to patent this nonsense. (And you thought people in the nineteenth century didn't have a sense of humor.)*

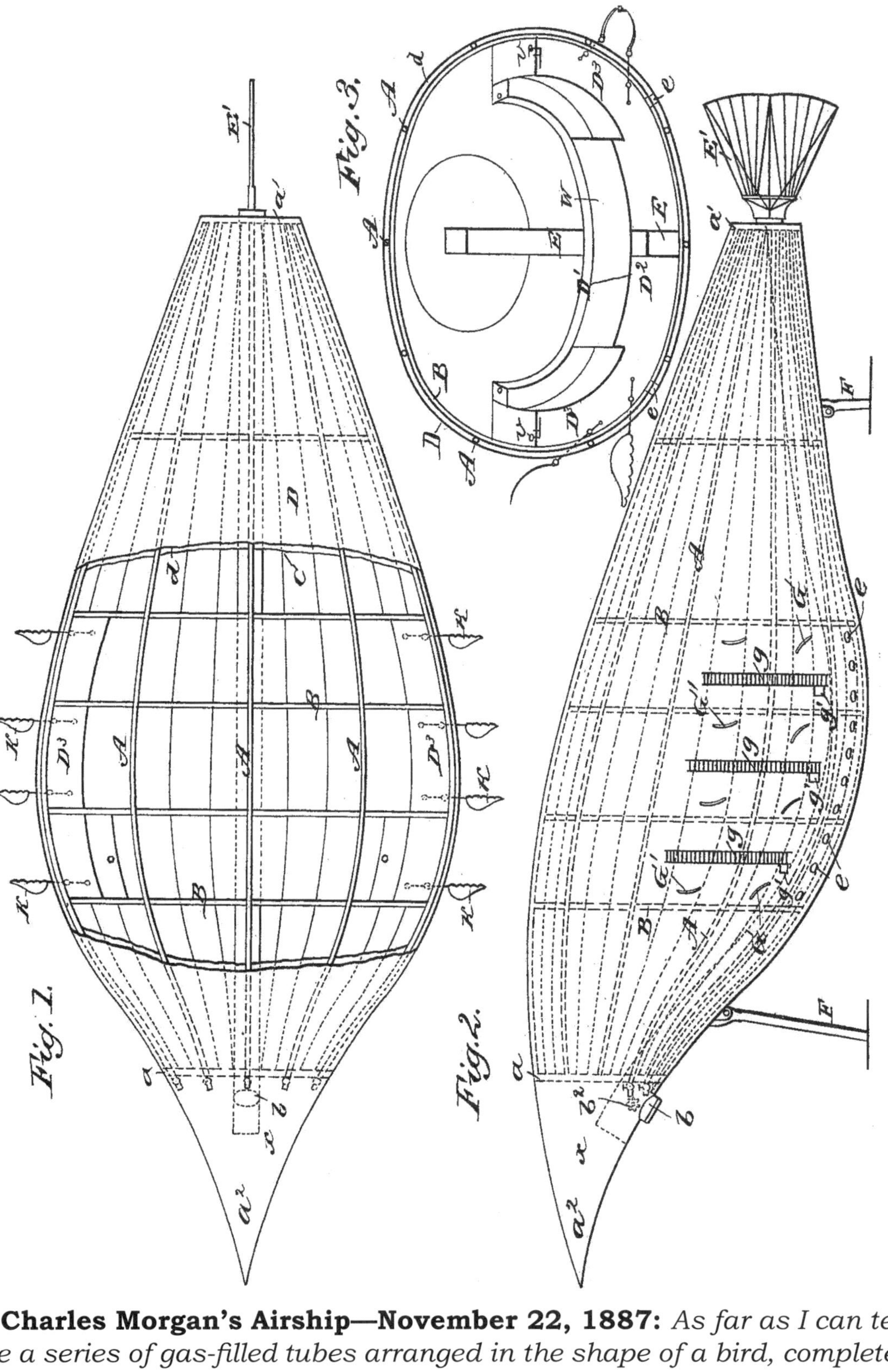

Plate 78: Charles Morgan's Airship—November 22, 1887: *As far as I can tell, this appears to be a series of gas-filled tubes arranged in the shape of a bird, complete with several sets of flapping wings (indicated but not shown) and a vertical tail feather (or perhaps a fish tail.) While its ability to achieve flight appears doubtful, Gunnison, Colorado, inventor Charles Morgan assures us that the craft, besides being light and roomy, is "capable of easy ascension, and which while in passage is readily steered in any direction." I suspect the only direction it would go, however, would be straight down and at high speed.*

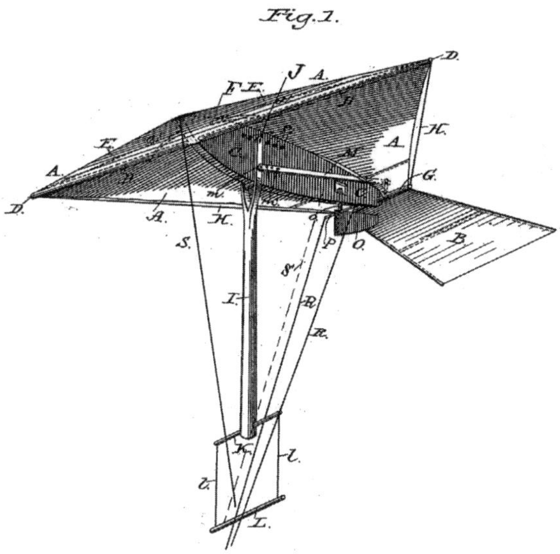

Plate 79: William Beeson's Flying Machine—January 24, 1888: *Montana's finest aeronautical engineer of the nineteenth century is back (see plate 74) and this time he may have finally gotten it right. Though some may consider this nothing more than an oversized kite, that would be a pernicious lie. In reality, this is exactly the right sized kite, and Mr. Beeson has the patent to prove it.*

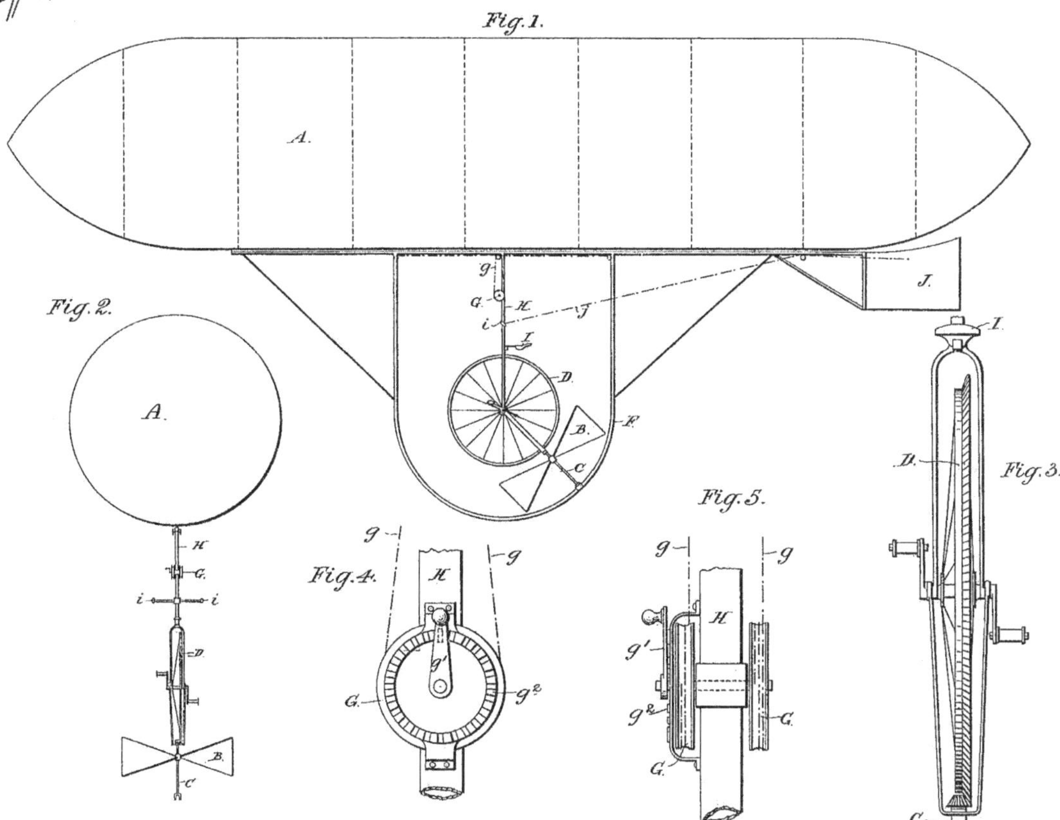

Plate 80: Wald Gustafson's Aerial Vessel—February 21, 1888: *You knew it was only a matter of time until the bicycle-craze of the late nineteenth century finally caught up to airship design, resulting in this pedal powered machine designed to be operated by a single man suspended beneath a gas bag perched upon a unicycle. Lest you imagine it would be difficult to generate the required horsepower needed to spin the adjustable spinning propeller, Mr. Gustafson assures us that his design has simplified things greatly, making pedal power much more desirable to motor power (although he does allow for the fact that larger examples of his design built to carry a hundred people will need to be supplemented by mechanical motors. I wonder if passengers would pedal in shifts, or if all the pedaling would be left to those in steerage?)*

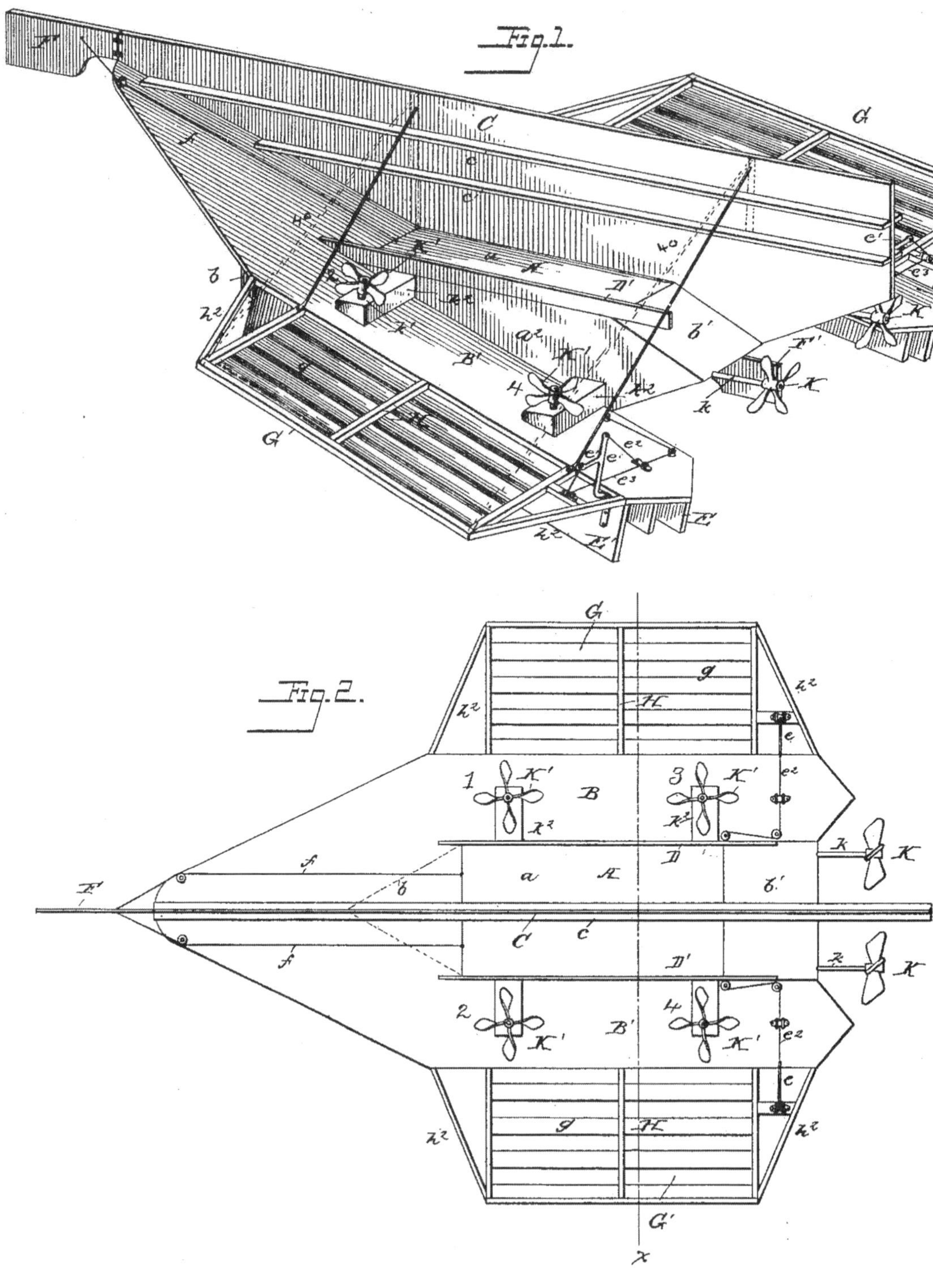

Plate 81: Edward Johnson's Airship—June 5, 1888: *This astonishing vessel, from which Star Wars director George Lucas may have derived at least some of his inspiration for the Empire's Star Destroyers, was the brainchild of Alabaman Edward Johnson, whose engineering acumen was clearly in a league of its own. How, precisely, the machine achieves and maintains lift remains, of course, a complete mystery that defies even the experts, demonstrating that either Mr. Johnston was onto something amazing or, just perhaps, he was not.*

Fig. 1.

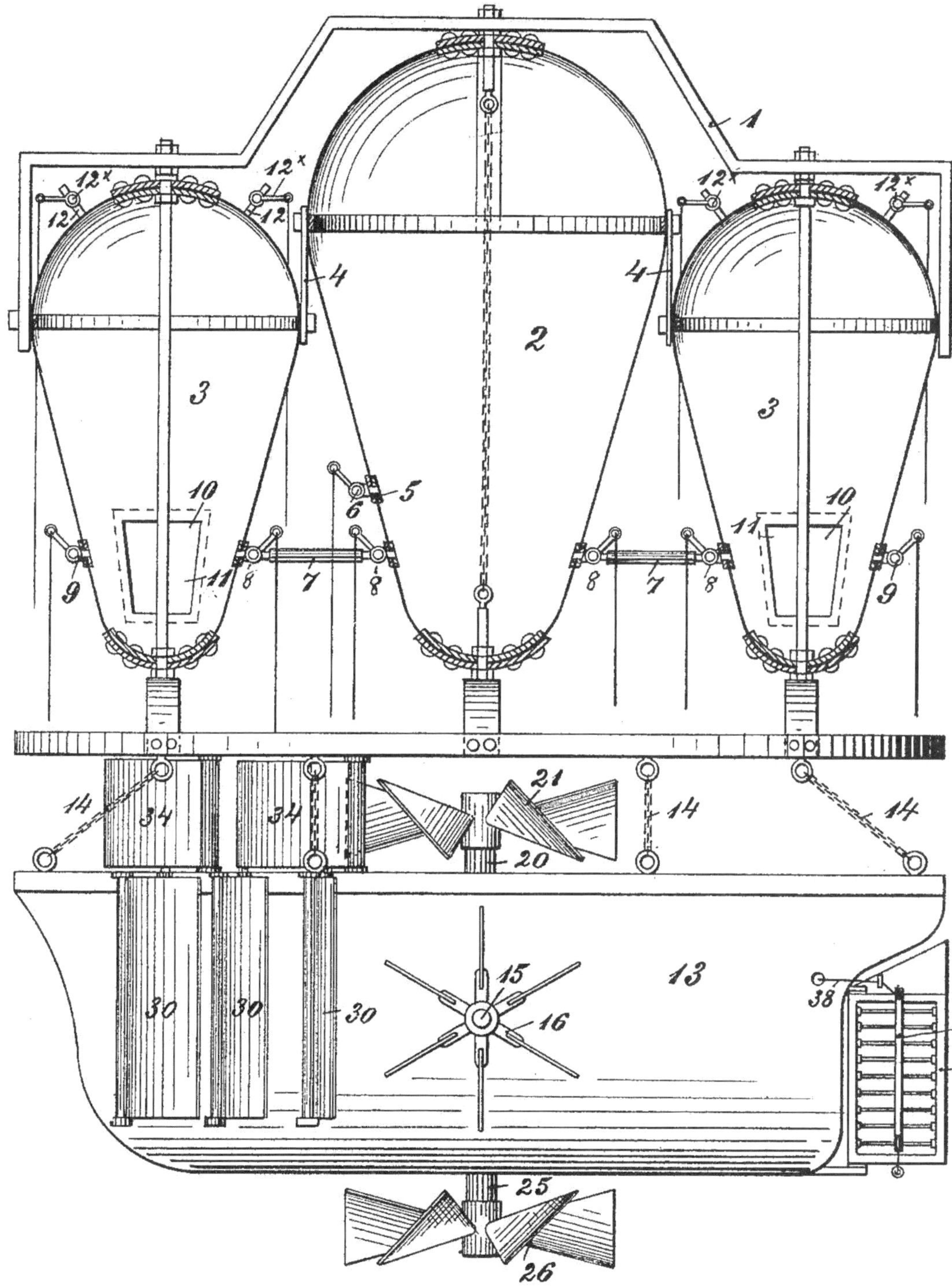

Plate 82: Charles Künzel's Balloon—March 12, 1889: *In perhaps one of the best examples yet of how extensively Jules Verne's magnificent machines influenced inventors of the era, we have Philadelphia inventor Charles Künzel's rendition of the Flying Dutchmen, complete with vertical thrusters, paddle wheels, and pivoted blades for greater maneuverability. One curious feature is the rudder (figures 36 & 37), which is composed of a series of slats and operates much like a window shutter. While the rationale behind this is far from clear, is certainly is original.*

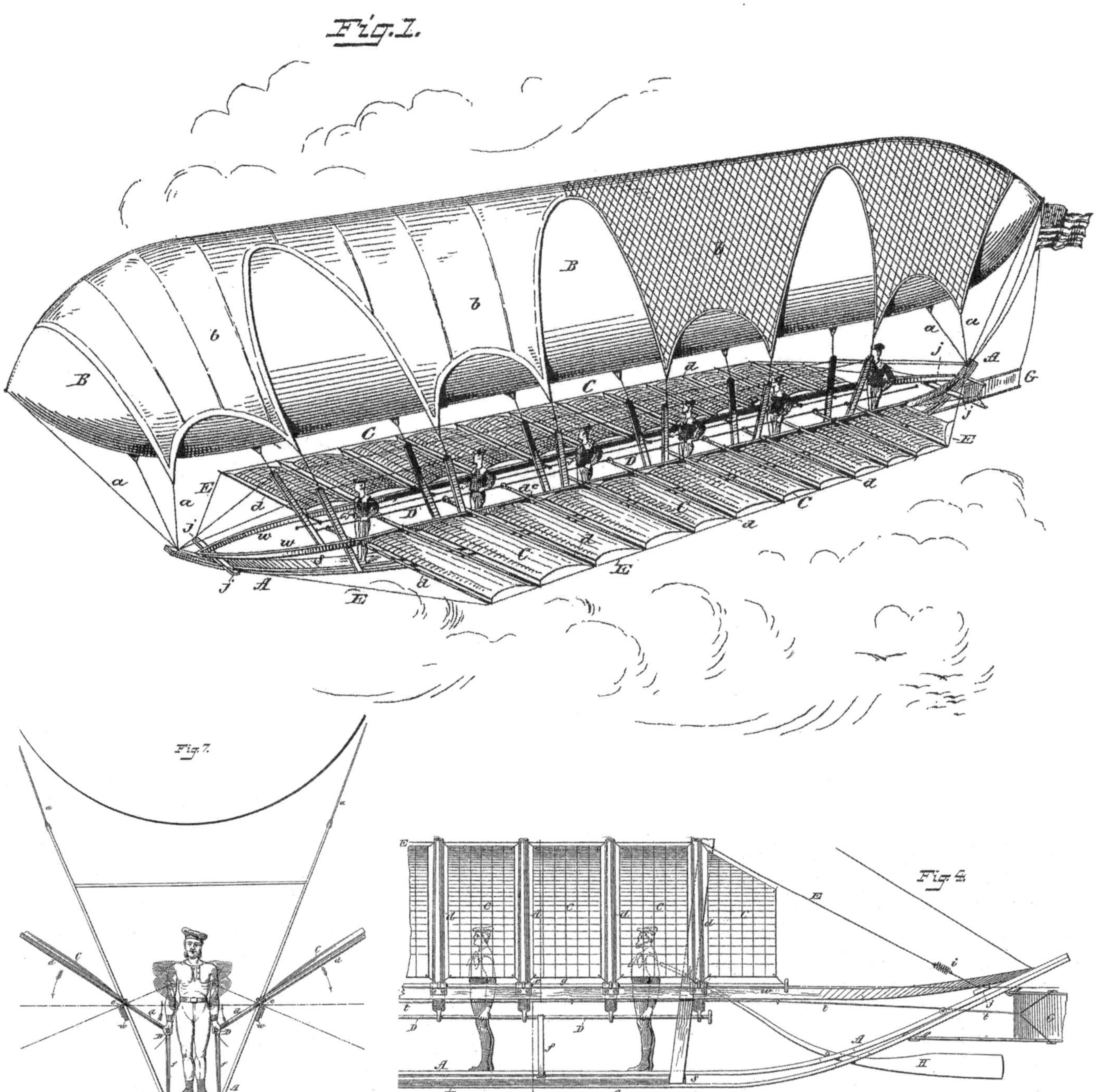

Plate 83: Nicholas Borgfeldt's Flying Machine—October 1, 1889: *Perhaps in an ill-advised attempt to improve Harvard's perennially losing rowing team, Brooklyn inventor Nicholas Borgfeldt's paddle-powered airship has to take the prize for moxie if nothing else. Of course, we can laugh at this now, but if it really did bring the cup back to Harvard, it might have been worth a try.*

Fig. 1.

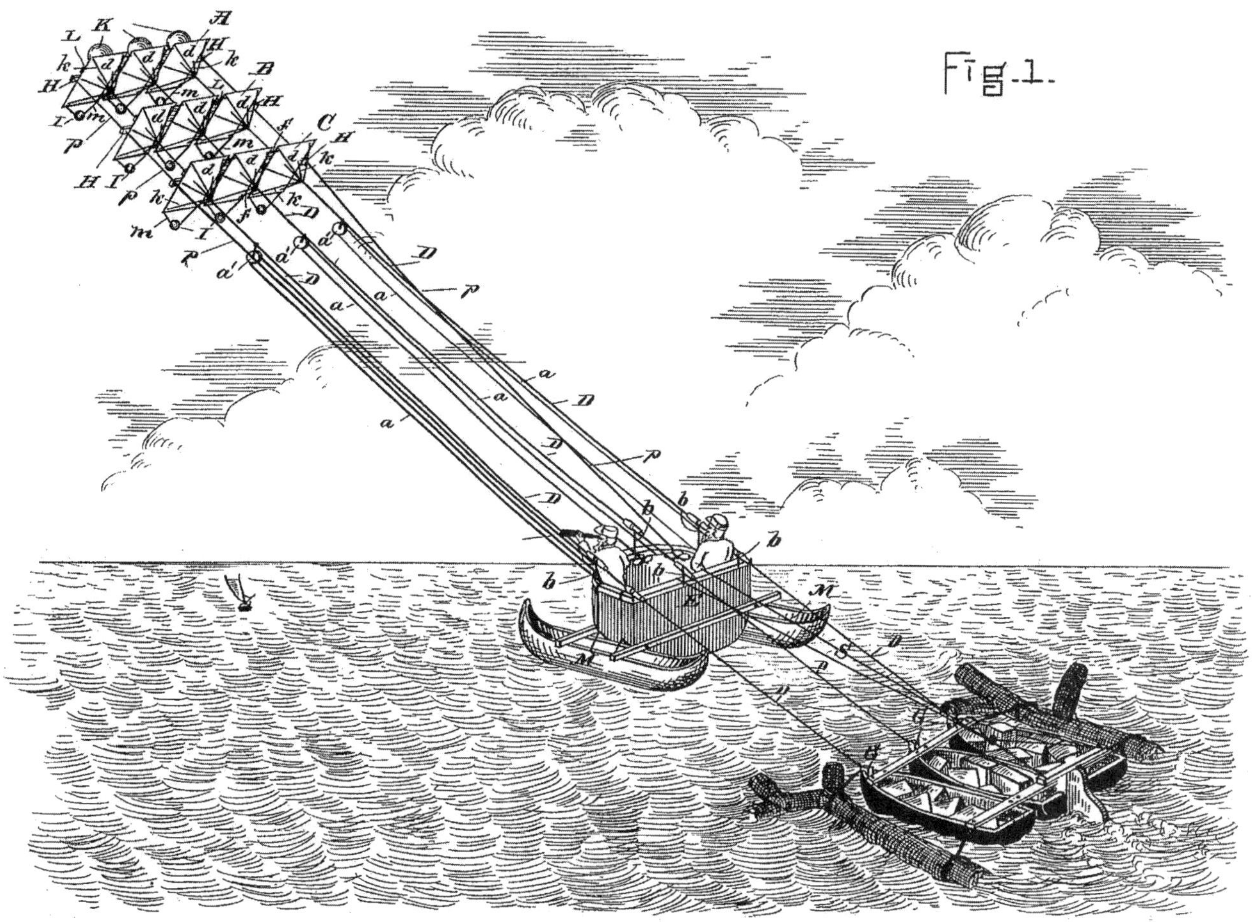

Plate 84: David Thayer's Aerial Apparatus—December 24, 1889: *I know the idea of attaching a boat to a bunch of kites had been proposed before (see plates 10, 26, 43, & 73), but never with such artistic flair. So prolific is Mr. Thayer, in fact, that he attached no fewer than twenty drawings designed to illustrate his utterly unworkable design, each and every one of them a work of art in their own right.*

Fig. 8.

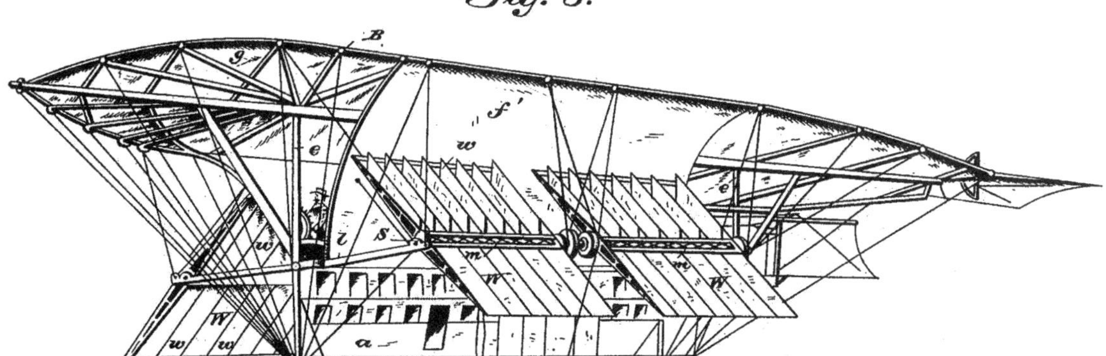

Plate 85: Matthias Baldwin's Aerostat—March 25, 1890: *While the patent drawings continued to grow in complexity and sophistication throughout the waning years of the nineteenth century, the designs they portrayed did not. Here Memphis inventor Matthias Baldwin regales us with the latest attempt to bring flight to the flightless with this array of spinning wings designed, according to Mr. Baldwin, to "produce a lifting and propelling effect similar to that produced by the stroke or movement of a bird's wing in flying."*

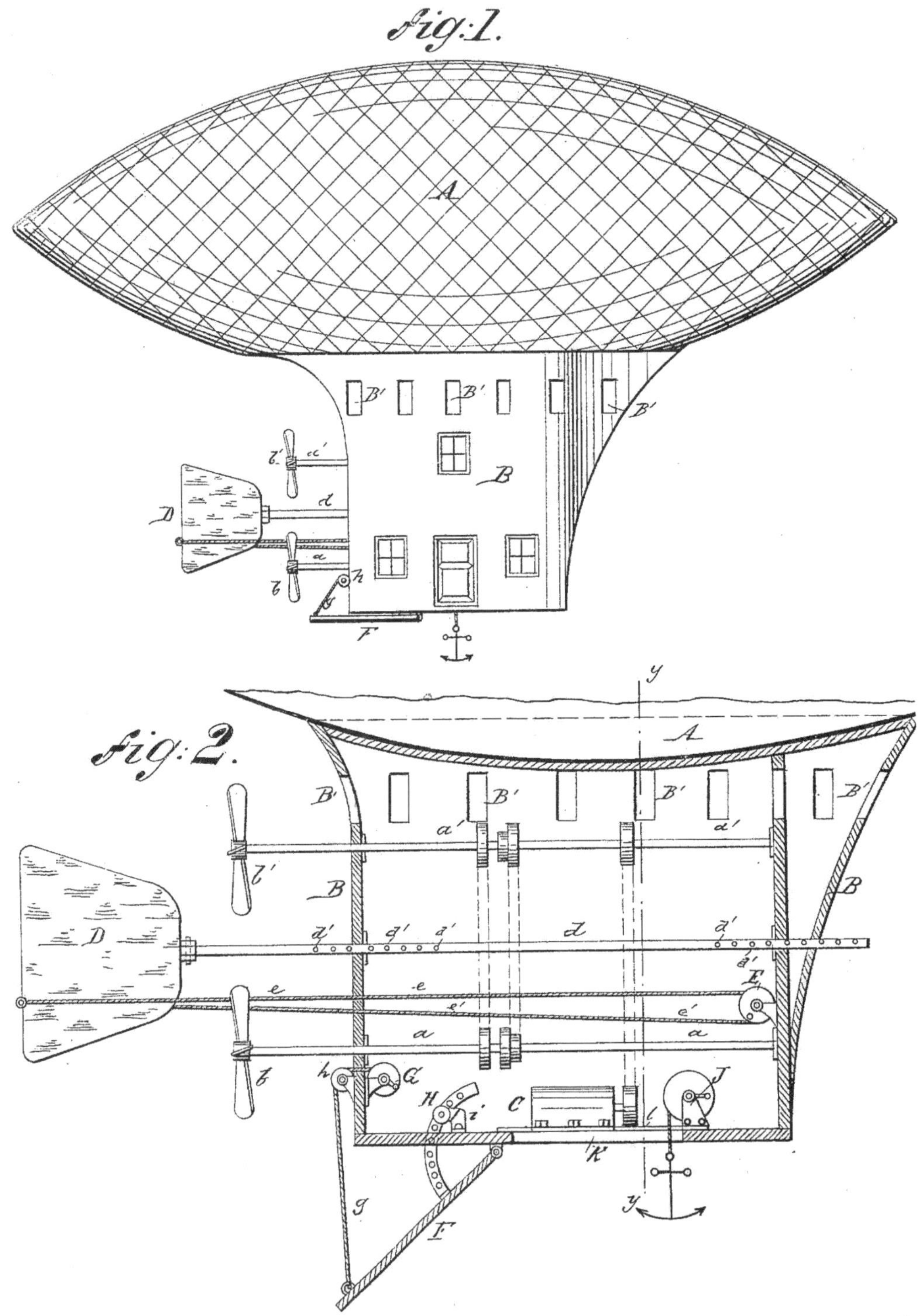

Plate 86: John Pokorny's Airship—April 1, 1890: *Clearly a product of the Gingerbread School of Airship Design, this quaint little doll house is spirited through the sky by a pair of tiny propellers powered by a "gas, electric, or any other motor" considered suitable (such technical details being secondary to the overall ascetics of the design). The strange trap-door looking thing hanging from the bottom of the device (letter F) is a "wind-board" that is lowered by a pulley system. Unfortunately, Mr. Pokorny does not tell us what a "wind-board" is or what its function is, (maybe an early speed brake to slow the vessel's precipitous, headlong velocity or perhaps a means of ejecting unwanted passengers?) leaving such trivialities to the reader's lurid imagination.*

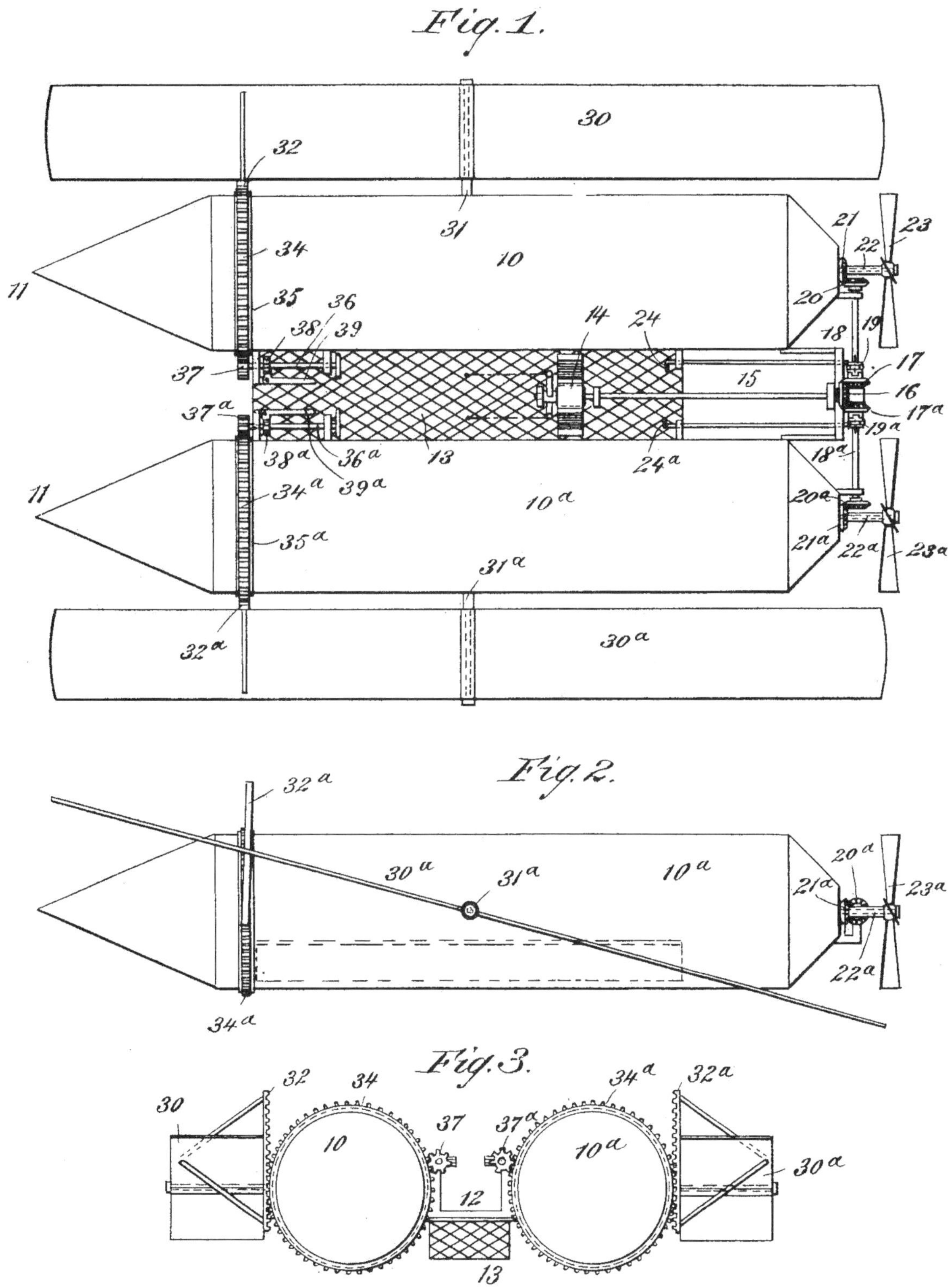

Plate 87: Charles Bechtel's Airship—June 3, 1890: *Mr. Bechtel's faith in the ability of massive movable wings to achieve lift and control is commendable if ill-conceived, for in being completely flat and extremely narrow they can do neither. This is one of the first designs, however, to suggest using an electric motor to drive the twin propellers, demonstrating that designers are becoming increasingly aware of weight considerations and fire safety issues, making this progress of a sort.*

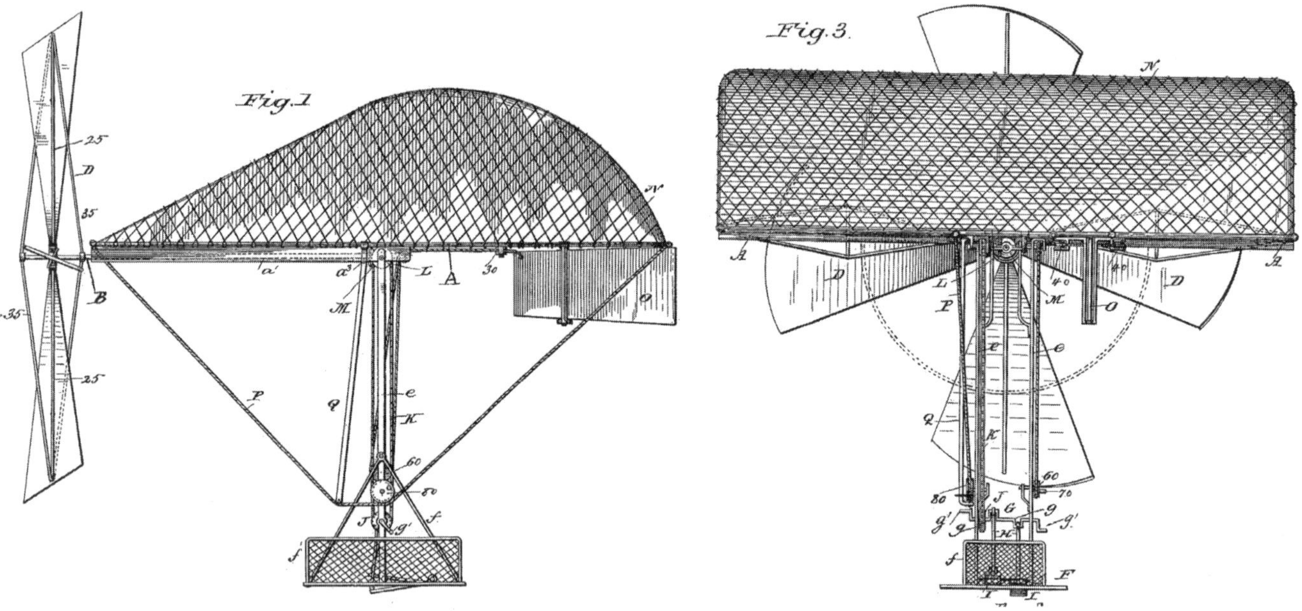

Plate 88: Stewart Cairncross' Aerial Machine—July 22, 1890: *Sporting a propeller large enough to propel a battleship, there is absolutely no doubt that North Dakota inventor Stewart Cairncross has designed absolutely the largest propeller to power an aerial vehicle ever conceived. That he proposes to turn this behemoth with a hand crank is even more astonishing, but in keeping with Victorian-era standards of assumed super-human capabilities.*

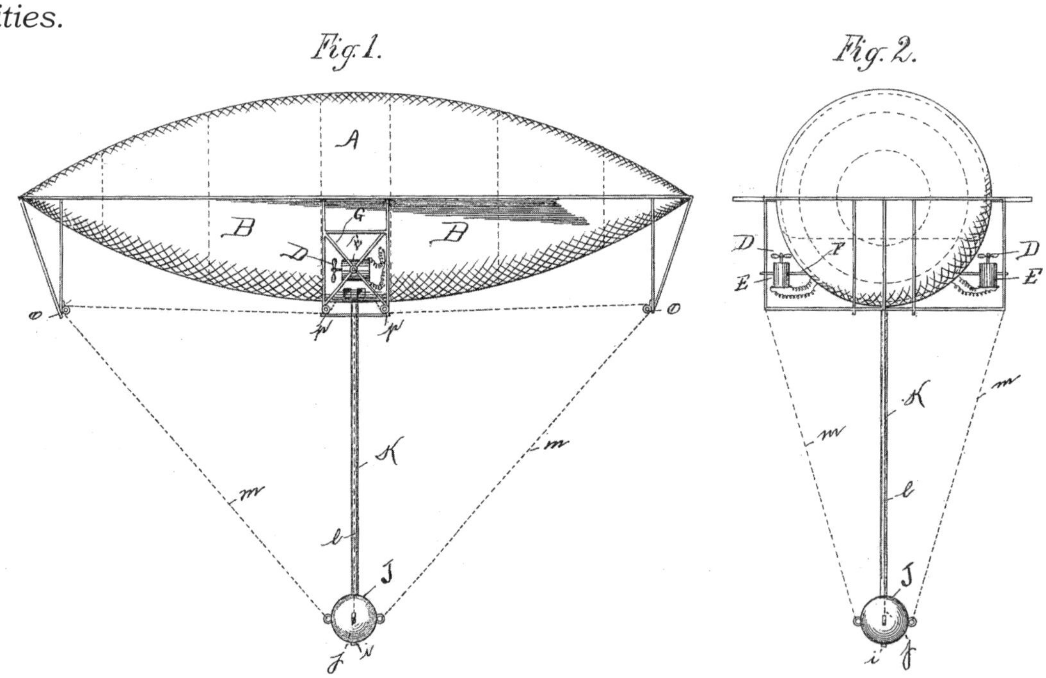

Plate 89: Perham Nahl's Navigable Airship—October 28, 1890: *While Mr. Cairncross (see previous plate) may have designed the largest propeller in history, it is possible that Mr. Nahl's may possess the smallest (letter D). It is fully electric, however, and pivotable, though in being so small, it's doubtful it has enough thrust to cool off an office cubicle. The metal ball (letter J) hanging from the bottom of a long hollow tube (letter K) is filled with sand and supposedly acts as a counterweight for the mass of the balloon. It may also serve as a club with which to knock cattle and other prairie animals senseless during low passes.*

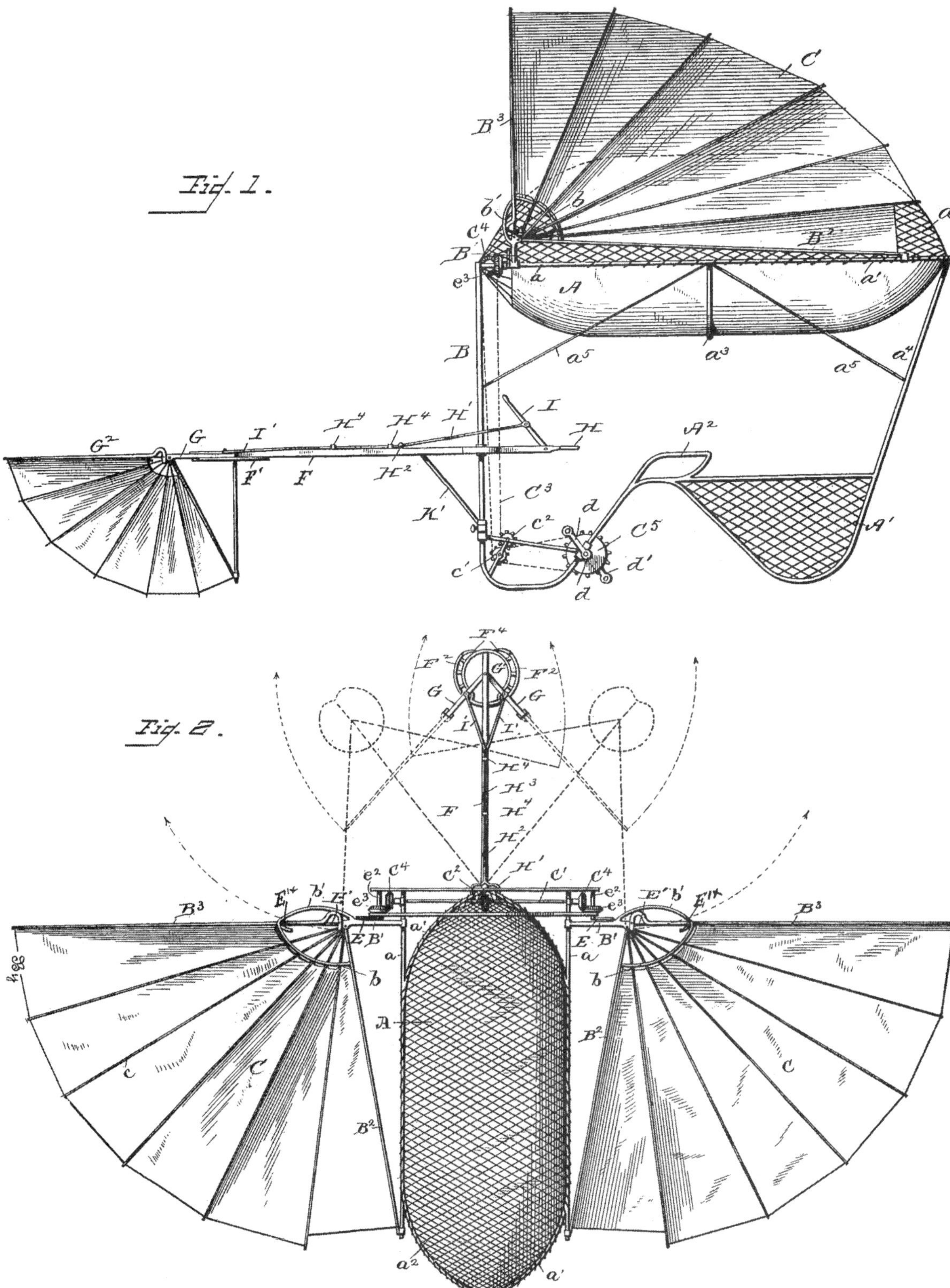

Plate 90: Burr Barnes' Flying Machine—September 29, 1891: *I can see no reason why this curious yet strangely elegant looking machine shouldn't fly; it's got a hydrogen gas cell with bat wings attached to the side, which are capable of being flapped by the eager aeronaut from the comfort of his bicycle seat slung snugly underneath. I mean if this won't work, what will?*

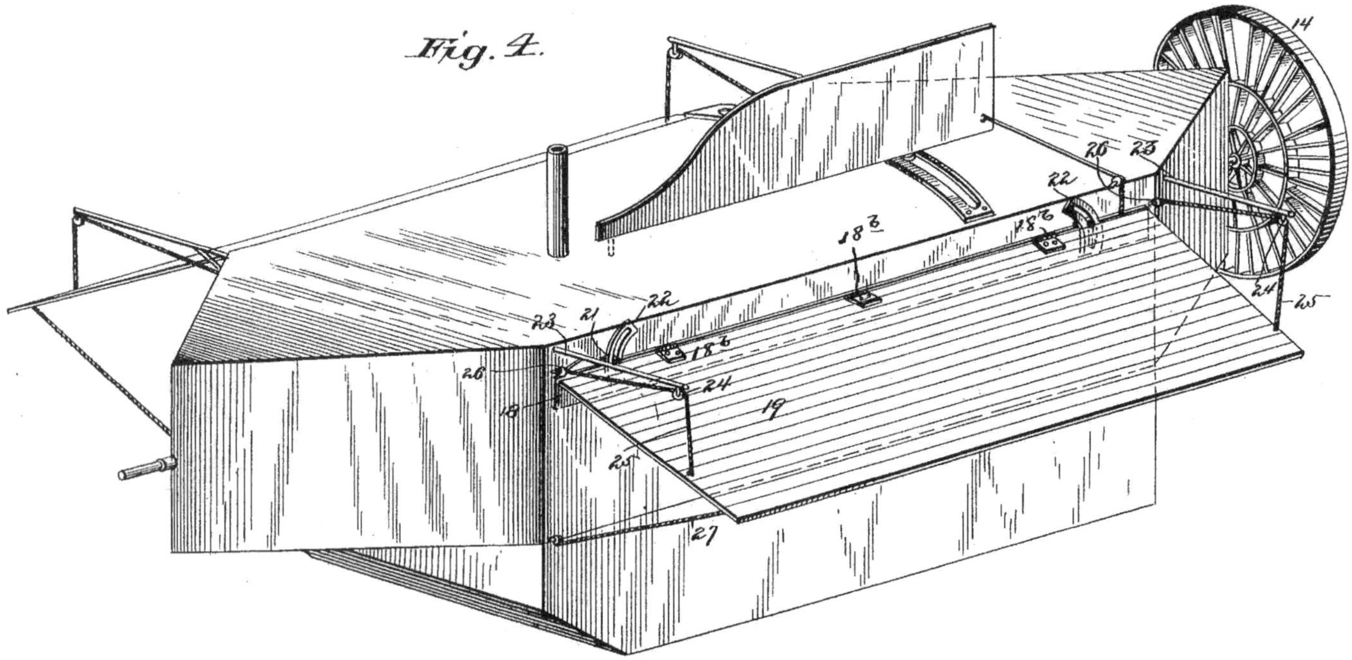

Plate 91: John Arbtin's Airship—November 3, 1891: *While Mr. Cairncross' (see plate 88) massive propeller may the largest, this entry by Des Moines inventor John Arbtin comes in a close second with his multi-bladed monstrosity (there is one on the front as well—not shown, though you can see the axle for it). The most curious part of the design is the vertical pipe just in front of the dorsal fin that is designed to let in gas. It appears that Mr. Arbtin imagines that leaving the interior of his airship open to the atmosphere will fill it with the required amount of lifting agent needed to make it airborne, which, if true, suggests everything we think we know about physics is a bunch of nonsense.*

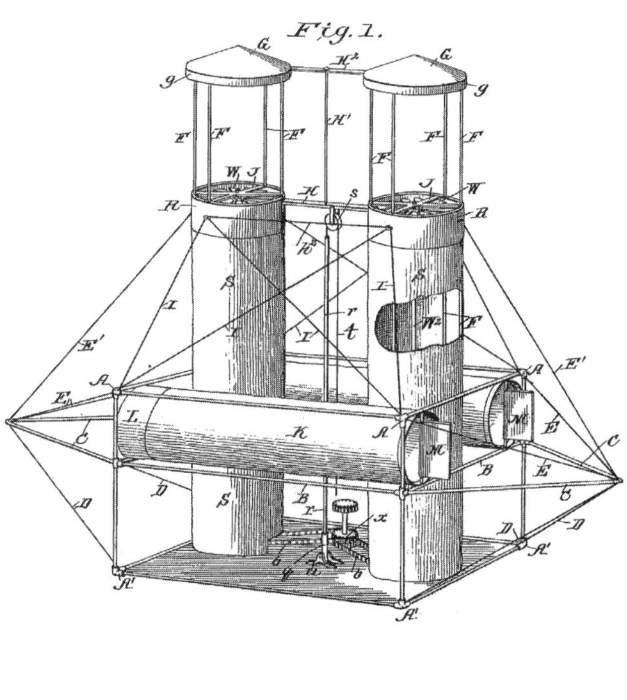

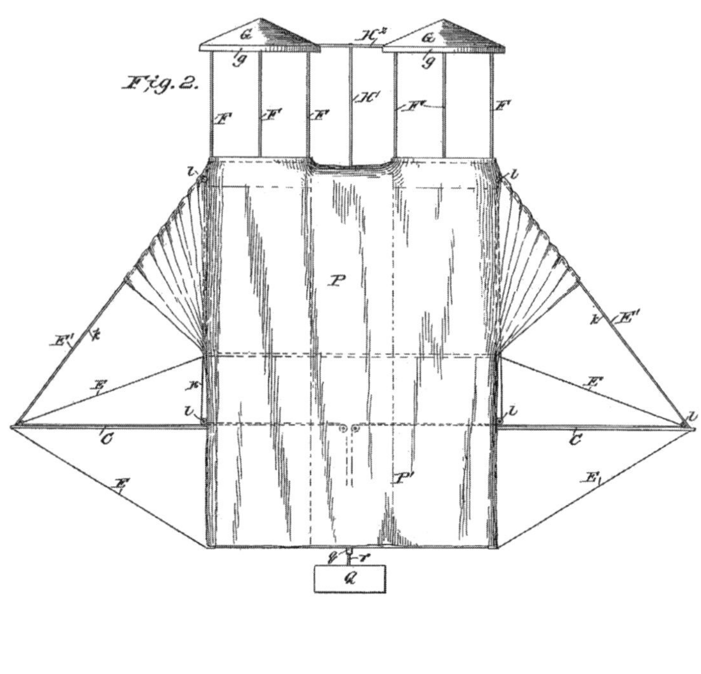

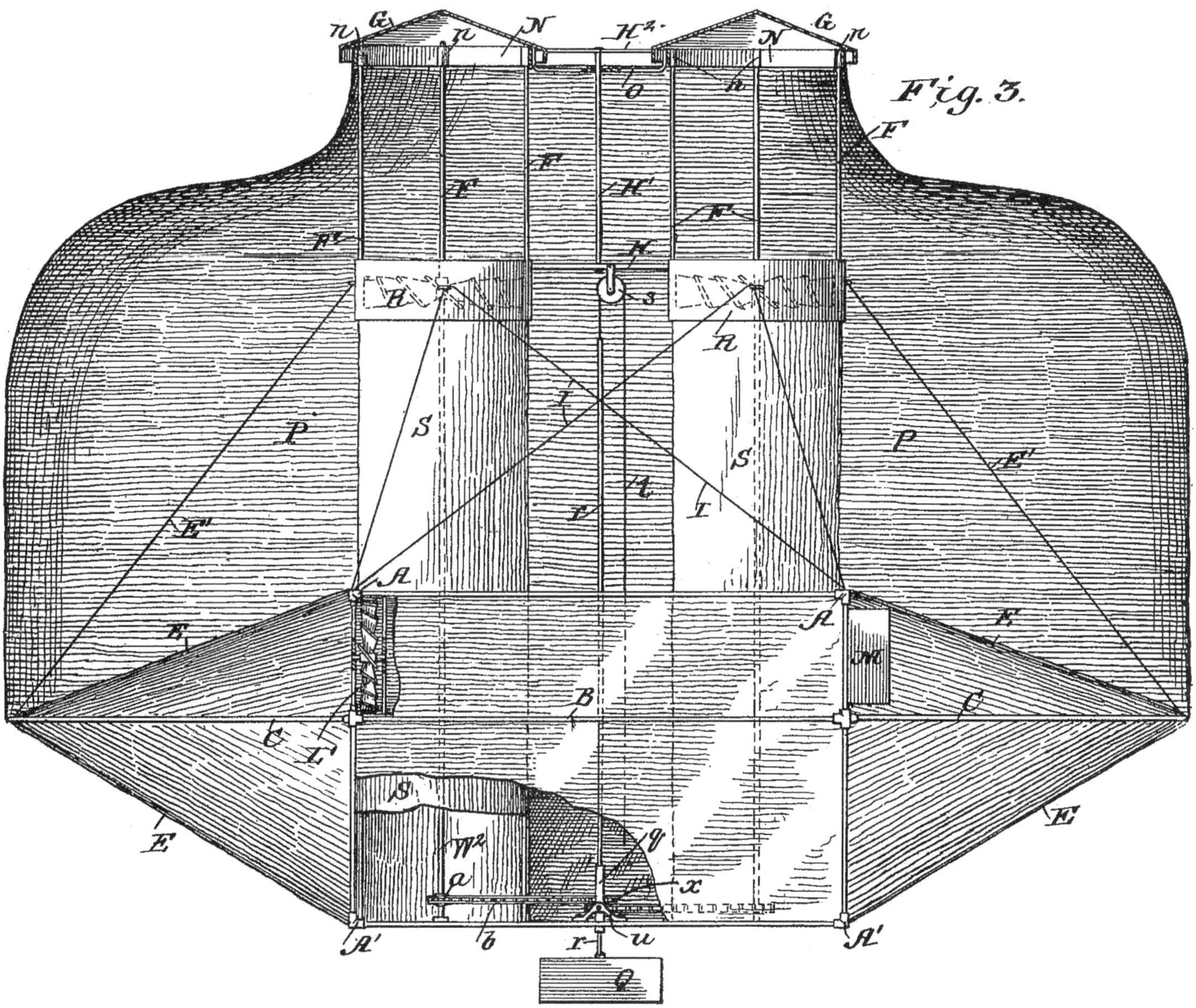

Plate 92: James Walker's Airship—January 12, 1892: *I know it's difficult to figure out what you're looking at in figure 1 (left page, far left), but once it is seen with its parachute shroud attached (also on left page, figure 2, letter P), it becomes much clearer what Mr. Walker was going for. However, in case that doesn't clear things up for you, figure 3 (above) shows the parachute shroud inflated for easy descent, and if it's still not obvious how this machine works after that—well, Mr. Walker has tried his best.*

Fig. 1.

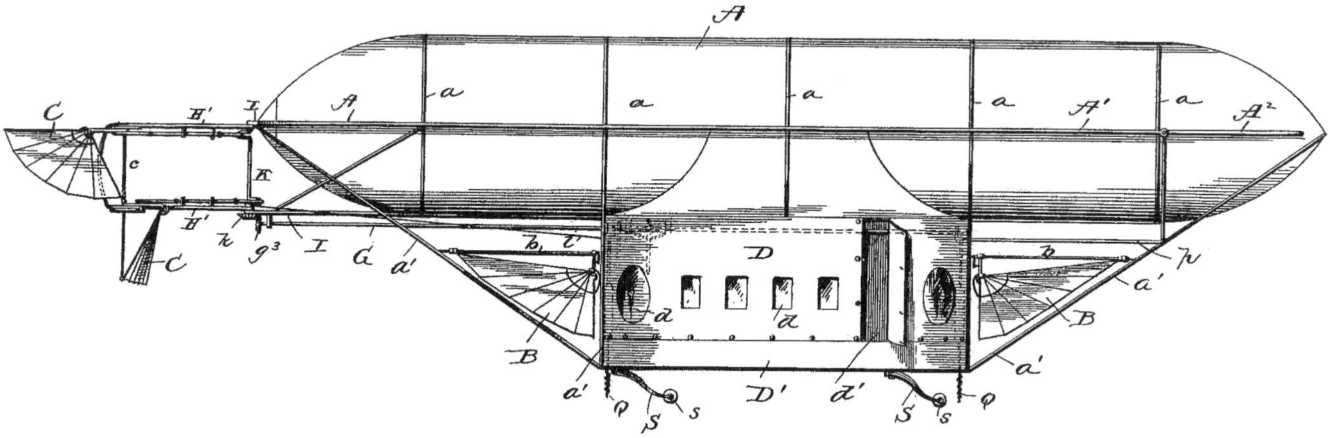

Fig. 2.

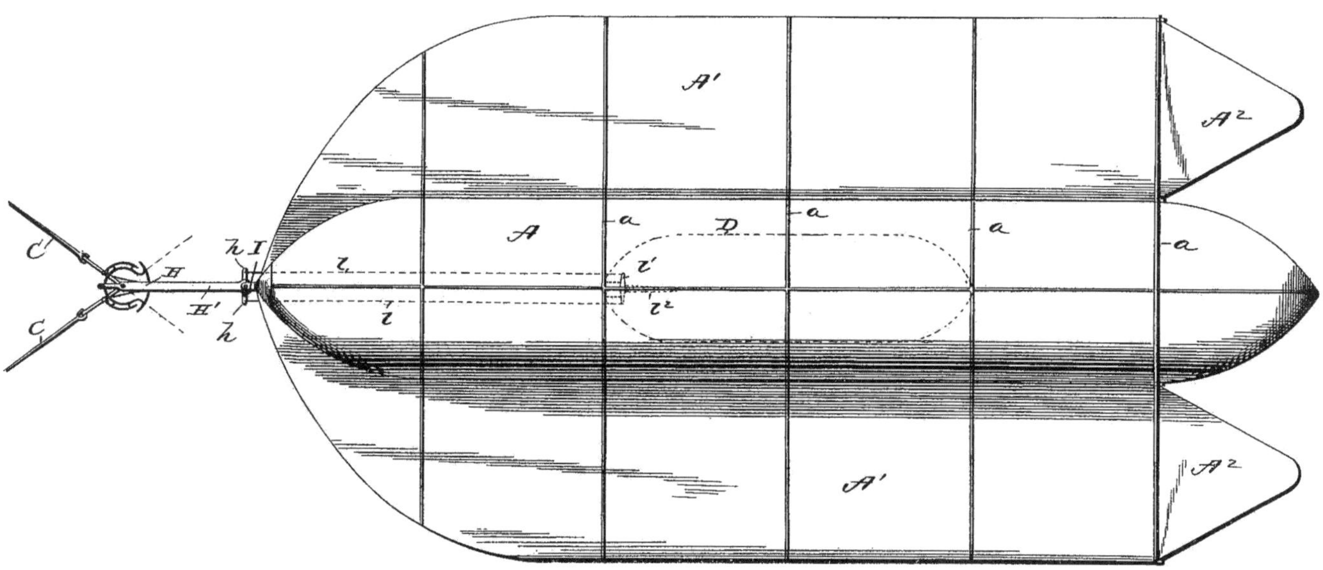

Plate 93: Burr Barnes' Airship—July 12, 1892: *Clearly the beneficiary of a very creative time in his life and still riding the crest of success of his earlier flying machine (see plate 90), Mr. Barnes is back just ten months later with a larger and much improved version of whatever it was he tried to design earlier, only this one can carry up to a dozen passengers in claustrophobic discomfort. Realizing he was on to something with his flapping wings idea, this one works along the same lines, but this time the large flapping wings (letter A) are powered by an engine rather than the super-human efforts of its crew. Notice that though the huge wings have taken on a decidedly more mechanical/geometric shape, his rudders (letter C) still retain the bat-like appearance and feel of his earlier machine.*

Fig. 1.

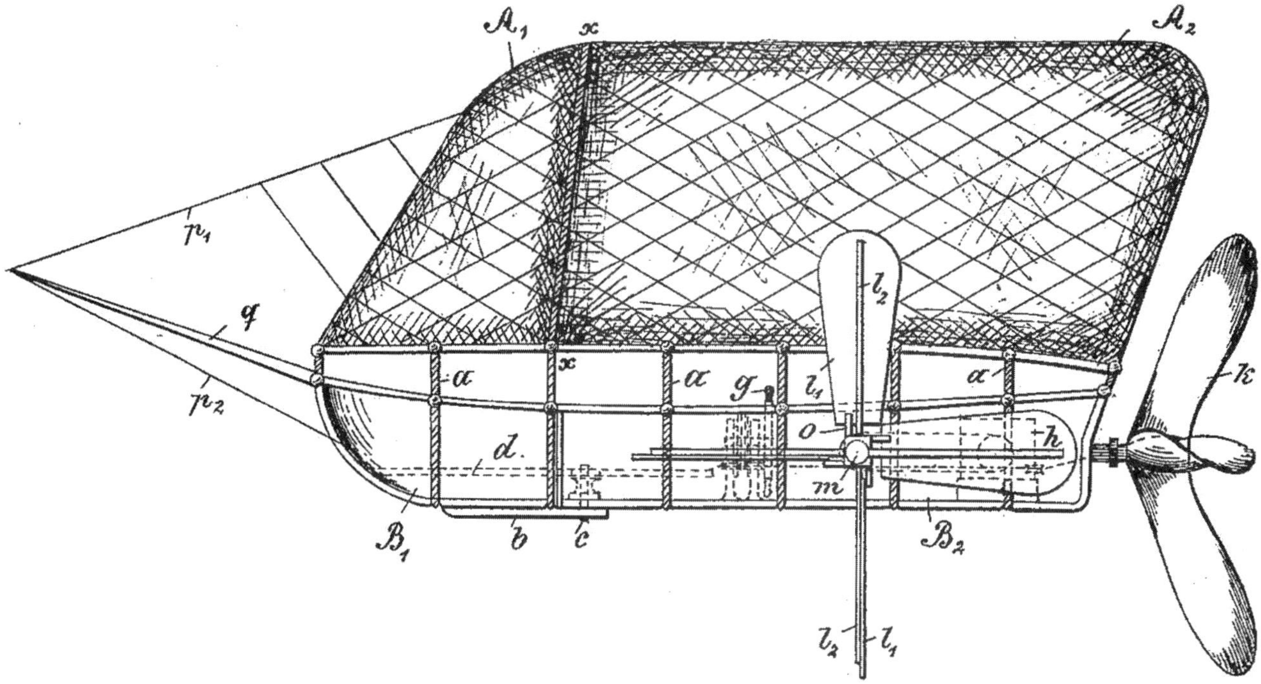

Fig. 2.

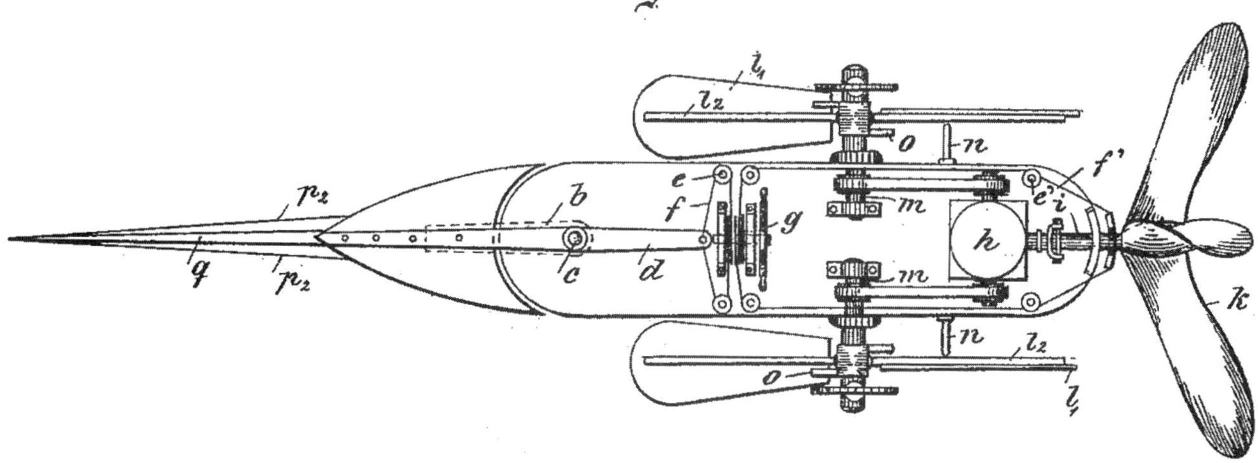

Plate 94: Georg Gabrielii's Balloon Ship—June 27, 1893: *Sleek and pointy, this zippy-looking little craft by Norwegian aeronaut Georg Gabrielii is simplicity at its best. Refreshingly free of all the complexities and protuberances of other designs of the time, it looks like it should easily fly through the air with minimal trouble. Somehow, though, I suspect gravity will not be kind to this contender.*

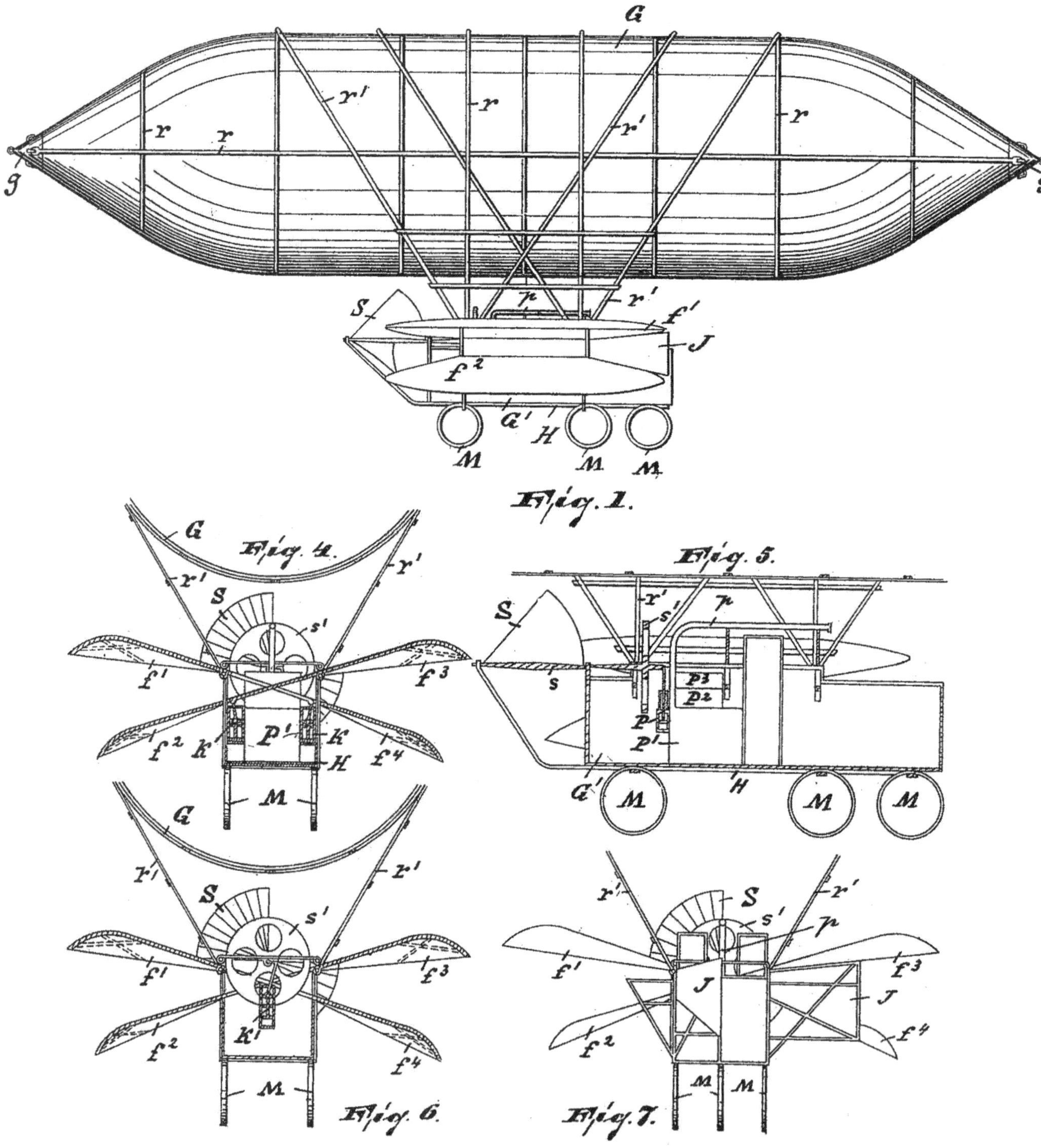

Plate 95: Theodore Steinmann's Airship—July 11, 1893: *In yet one more attempt at using flapping wings to propel an airship forward, Austria-Hungary's own Theodore Steinmann provides the last word in futile airship designs. (Actually, calling this the "last word" is a lie. There were to be many more yet to come.)*

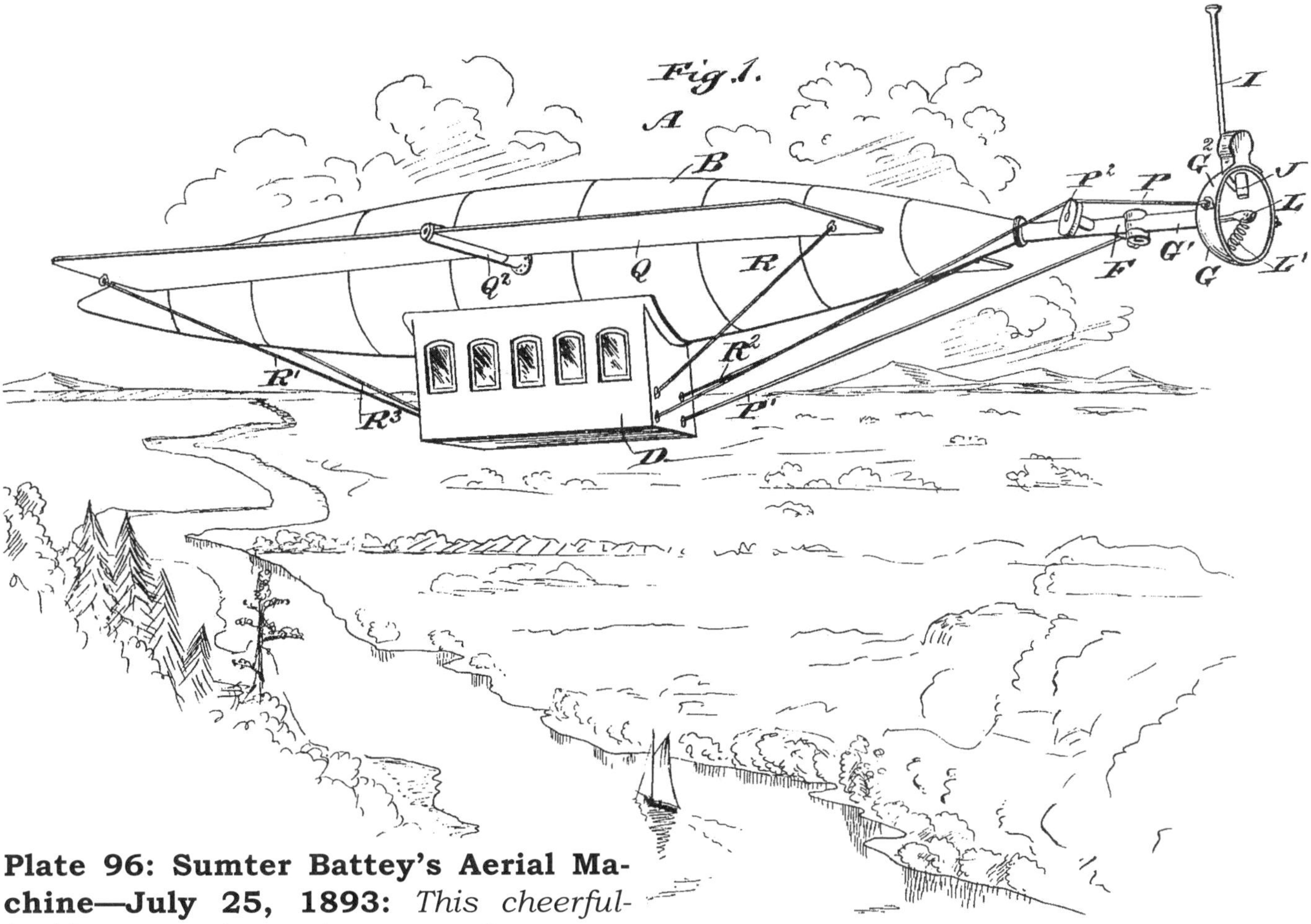

Plate 96: Sumter Battey's Aerial Machine—July 25, 1893: *This cheerful-looking vessel passing serenely over a picturesque countryside is the brainchild of New York City's own Sumter Battey, a man who positively exudes optimism with this fun little craft. The novelty in this design comes from its propulsion system, which incorporates an ingenious system that drops tiny pellets of nitro-glycerin into a multi-directional bowl-shaped dish to be harmlessly detonated by an electrical charge, thereby providing forward thrust (and making it the earliest precursor of the Nuclear Pulse Detonation Engine proposed in the 1960s as a means of getting a heavy-lift rocket to Mars). Mr. Battey doesn't mention what to do should the exploding pellet inadvertently detonate the entire supply of nitro all at once, leaving such indelicate details to be worked out by later crash-scene investigators.*

CLOSE-UP OF ROCKET PROPULSION UNIT

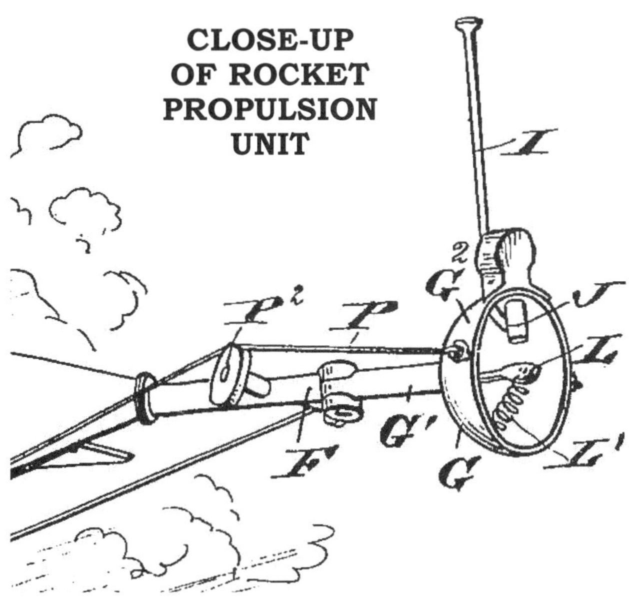

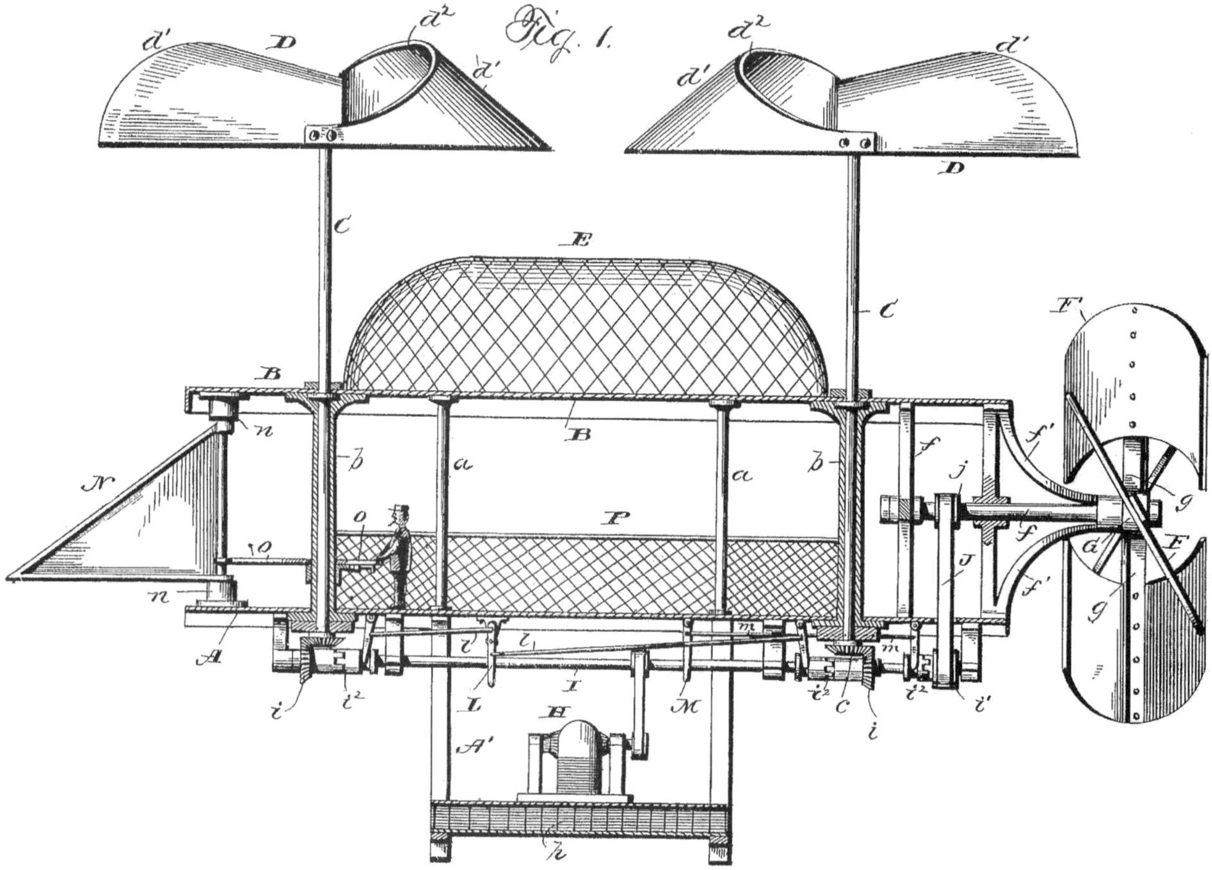

Fig. 2.

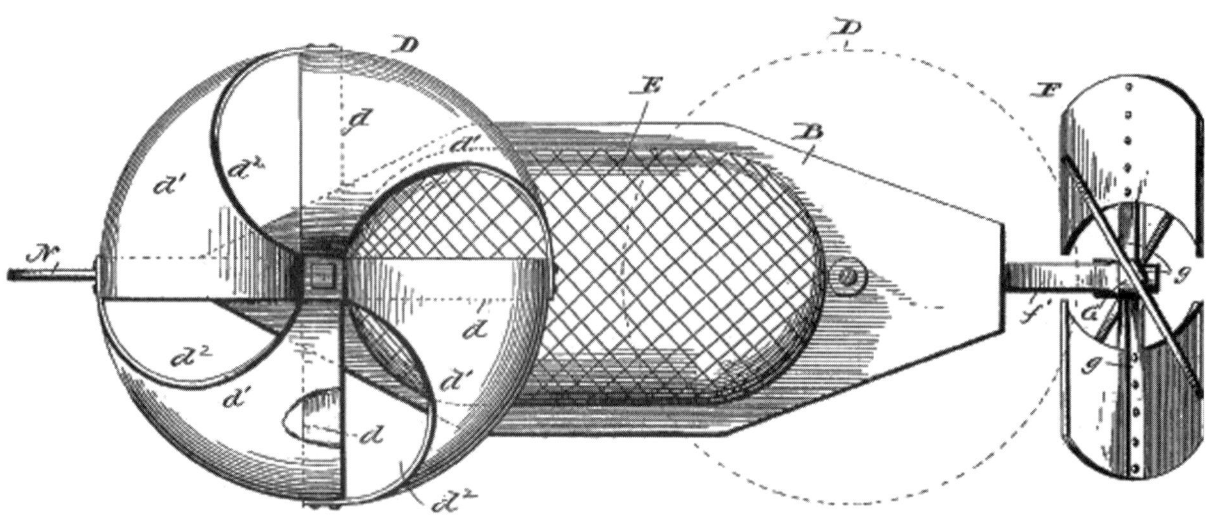

Plate 97: Emory Sowers' Flying Machine—September 5, 1893: *This peculiar-looking hybrid airship/helicopter is probably no more dangerous than it looks. The only real innovation on this craft when compared to others of its ilk is the unusual shape of the upper rotors, that, even though they are unlikely to provide much in terms of thrust, are pretty cool looking. Is it my imagination or does the pilot look bored?*

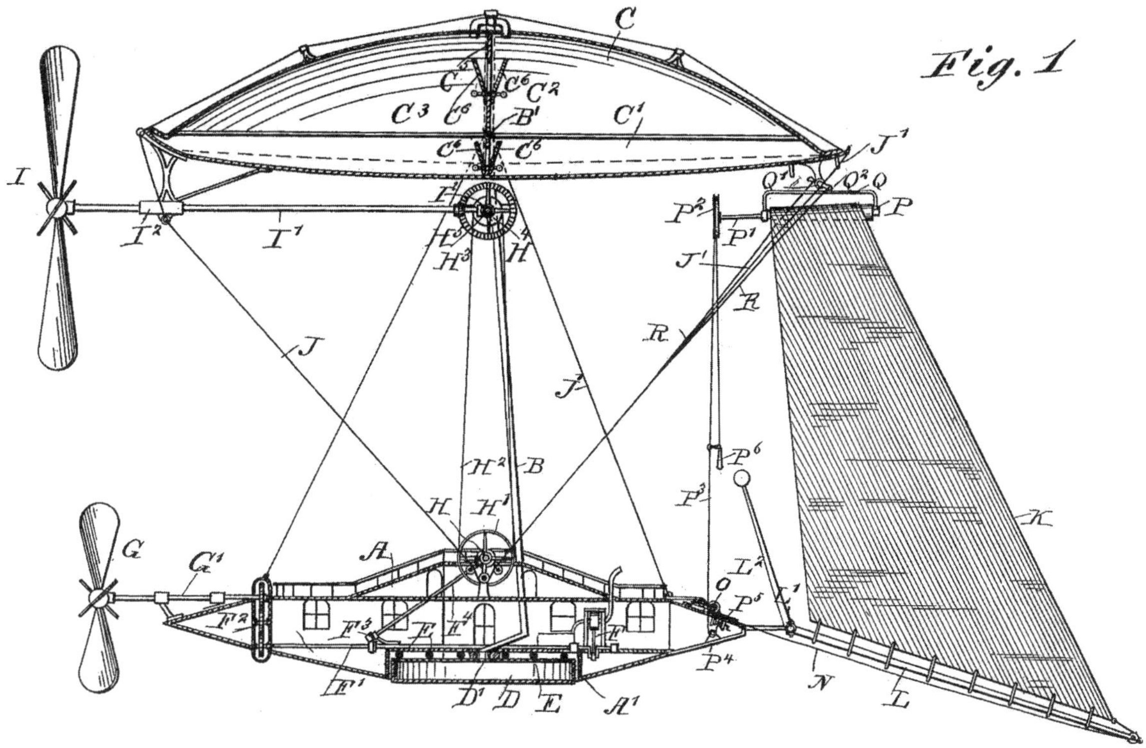

Plate 98: Axel Bergqvist's Aerial Vessel—October 17, 1893: *An unconventional approach to airship design is this machine proposed by Iowa airship enthusiast Axel Bergqvist—a man who knows how to get the most out of a power plant. The big problem with this design is that the gondola and envelope are connected only by a few cables, which would not normally be that much of a problem, but as the second larger propeller is running off a cam coming from the gondola, it would appear that even the slightest twisting motion between the two halves—inevitable in a machine like this—would be highly problematic and likely disastrous. Yet it does demonstrate that even as far back as 1893, Iowans were capable of thinking well outside the box.*

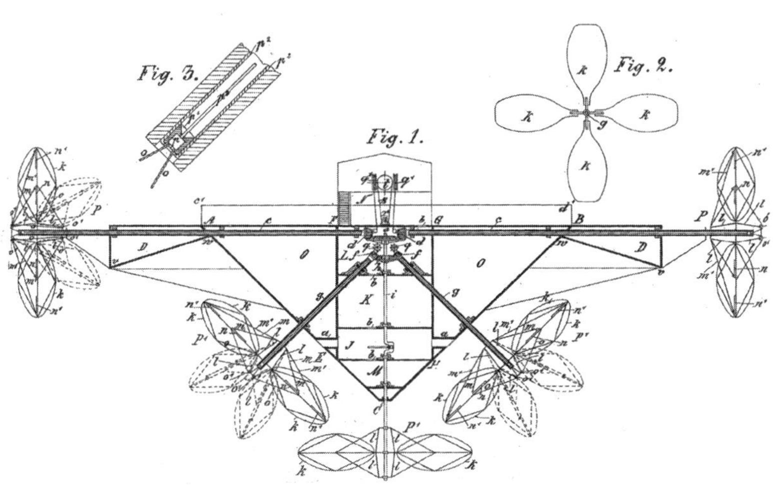

Plate 99: Carl Billwiller's Navigable Airship—October 31, 1893: *Obviously a proponent of the belief that the forces of gravity can only be overcome through the use of lots of multidirectional spinning propellers, Swiss inventor Carl Billwiller has come up with this frightening craft designed to demonstrate precisely that principle. Figure 6 shows the thing perched upon a steel girder ready for take-off, but nowhere does Mr. Billwiller tell us how he proposes getting the craft to land on the device. I imagine such would take a degree of piloting finesse usually reserved for things like close formation flying and aerial refueling.*

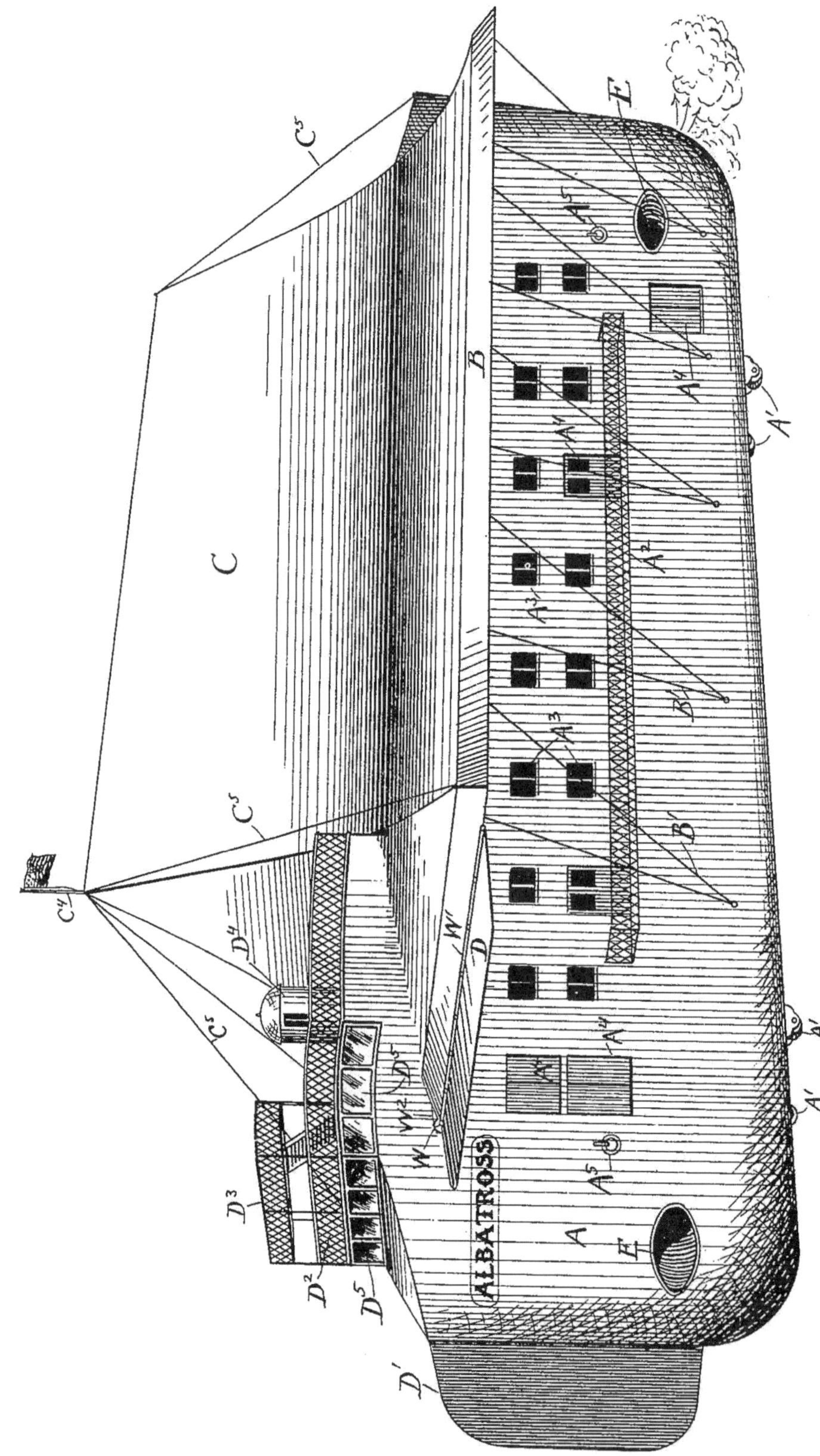

FIG. 7.

Plate 100: Edwin Pynchon's Airship—November 14, 1893: *If Noah had been ordered by God to build an ark in 1893, I suspect this massive—and appropriately named—craft is fairly close to what he would have come up with. No, the little openings in the hull (letters E) are not torpedo tubes (as useful as that might be) but exhaust chambers by which the vessel is propelled forward or backwards by continuous explosions, making it truly a vacationer's paradise.*

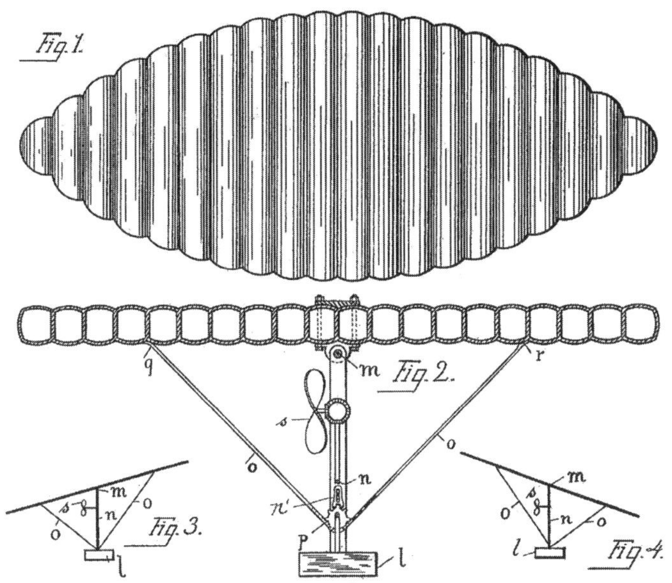

Plate 101: Joachim Sumovski's Balloon—December 12, 1893: *In an early example of Russian engineering prowess, Joachim Sumovski provides us with this curious and even strangely contemporary-looking multi-compartmentalized balloon wing, which, while possibly a useful design for a parachutist, would prove impossible to control with the simple propeller and rudderless assembly shown here. Yet it does demonstrate that the Russian bear was waking from its slumber and looking towards flexing its aeronautical wings. Watch out, twentieth century!*

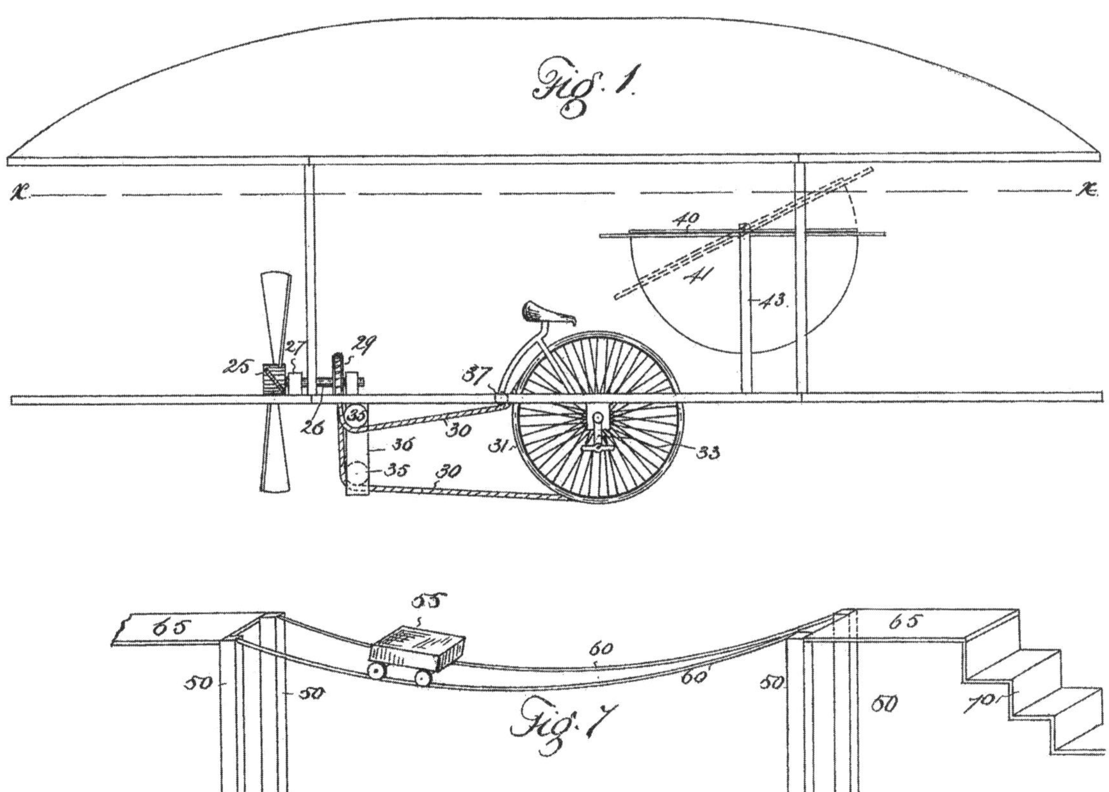

Plate 102: Jerome Blanchard's Aerial Vessel—February 6, 1894: *This rather pedestrian-looking pedal-powered airship design is nothing remarkable, but the way it is launched is unique: it's placed on a precariously suspended tramcar (number 55) that runs down a wobbly track where hopefully it will attain enough speed—and thus lift—by the time it arrives at the second platform to hurtle itself free of the bonds of gravity. I'm not sure but I think Evel Kneivel tried something like this at Caesar's Palace back in the 70s with mixed results.*

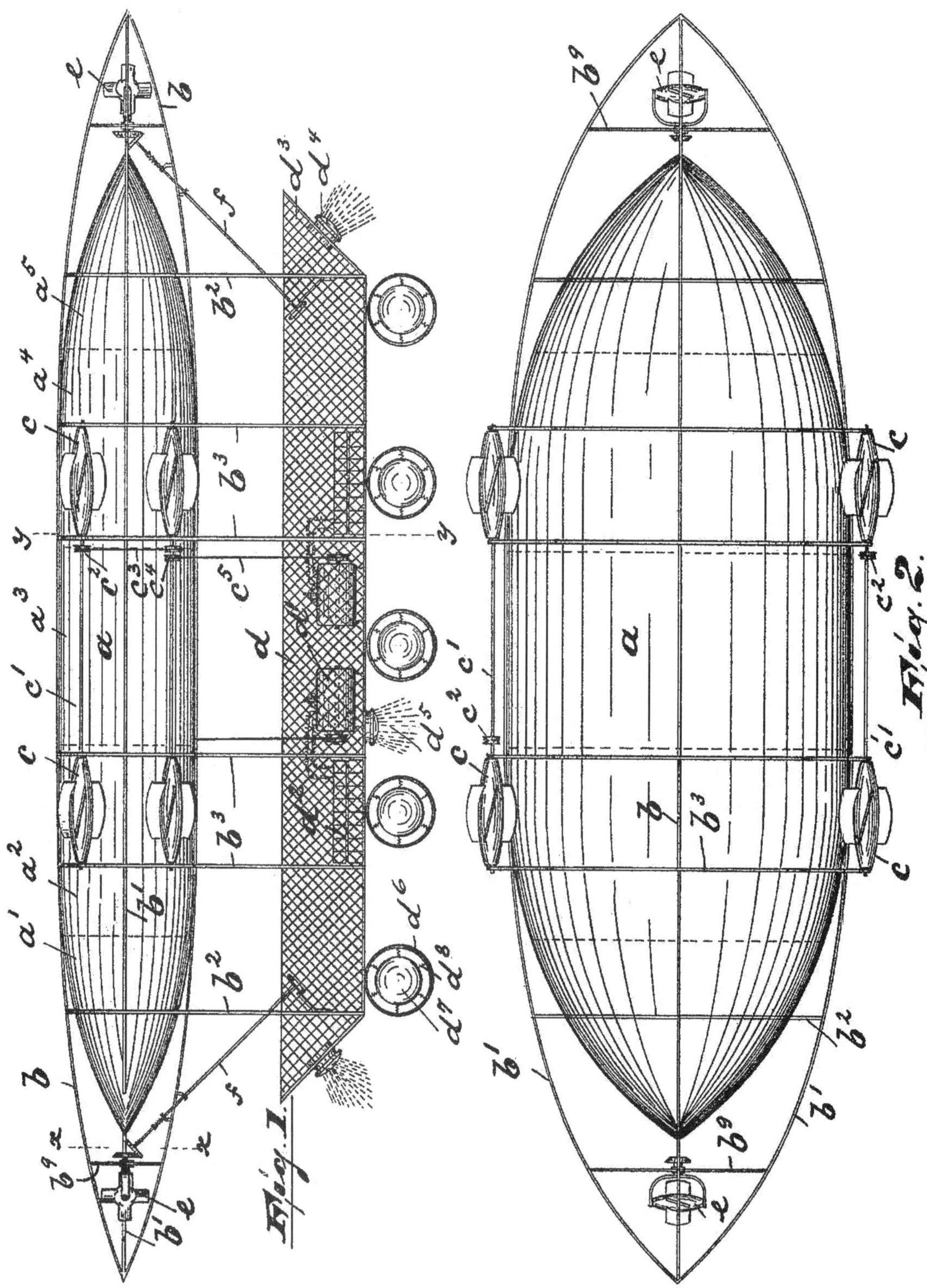

Plate 103: Duane Hurlbut's Airship—September 25, 1894: *This curious-looking craft actually has a few elements going for it: the external skeleton foresees the advent of the fully rigid dirigible design while the much more aerodynamic propeller rotors look positively space age. For some reason, however, Mr. Hurlbut is obsessed with searchlights (item d4), mounting no fewer than three of them to the gondola. Clearly, he was intent on being seen for miles in every direction once he got this behemoth off the ground.*

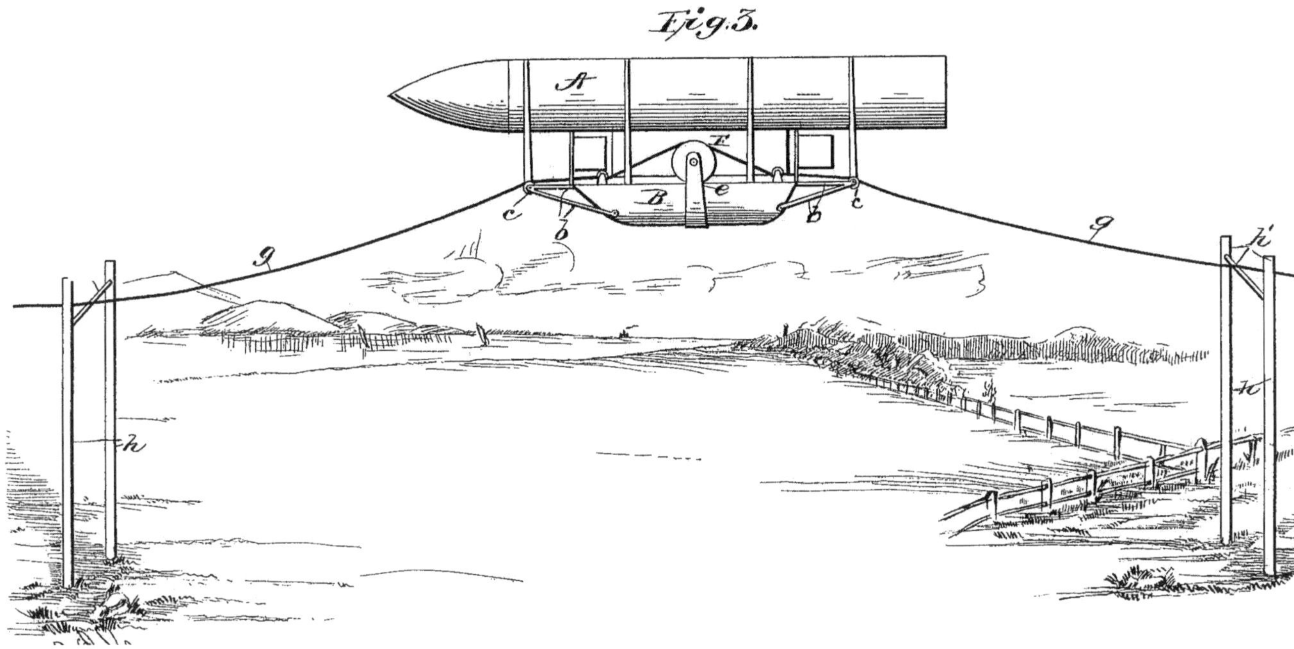

Plate 104: John Lomax's Cable Airship—June 16, 1895: *While more of a floating cable car than a true airship, Arkansas inventor John Lomax's design has merit but would be the more expensive and problematic solution to crossing rivers or open spaces. This is largely due to its very limited carrying capacity (especially when compared to a barge) and the extensive network of girders and cables that would be necessary to string across remote regions of the country to make it work. Might be a nice way to cross the Grand Canyon, perhaps, but a horribly expensive way to see Arkansas from the air.*

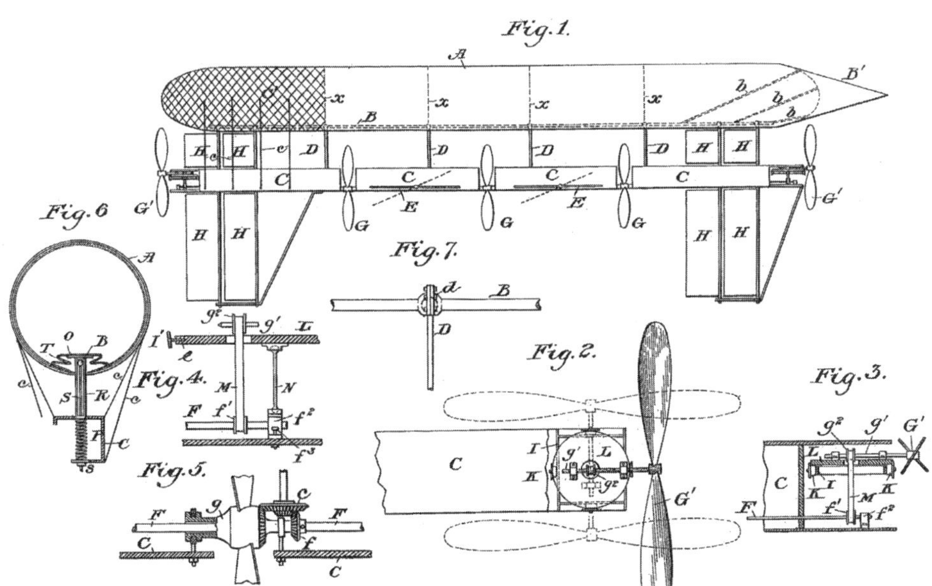

Plate 105: William Hutchinson's Navigable Balloon—October 15, 1895: *Englishman William Hutchinson's design is actually quite sound and may have been workable from an engineering standpoint with the proper power plant. His basic refinements include multiple propellers turned on a single shaft, an aluminum keel (which he calls a "backbone") running the length of the envelope, an aluminum nose cone (letter B), and multidirectional end propellers for lateral control. If the end props worked as planned, the massive rudders (letters H) could probably have been dispensed with or made considerably smaller. All-in-all, an impressive effort.*

FIG. 1

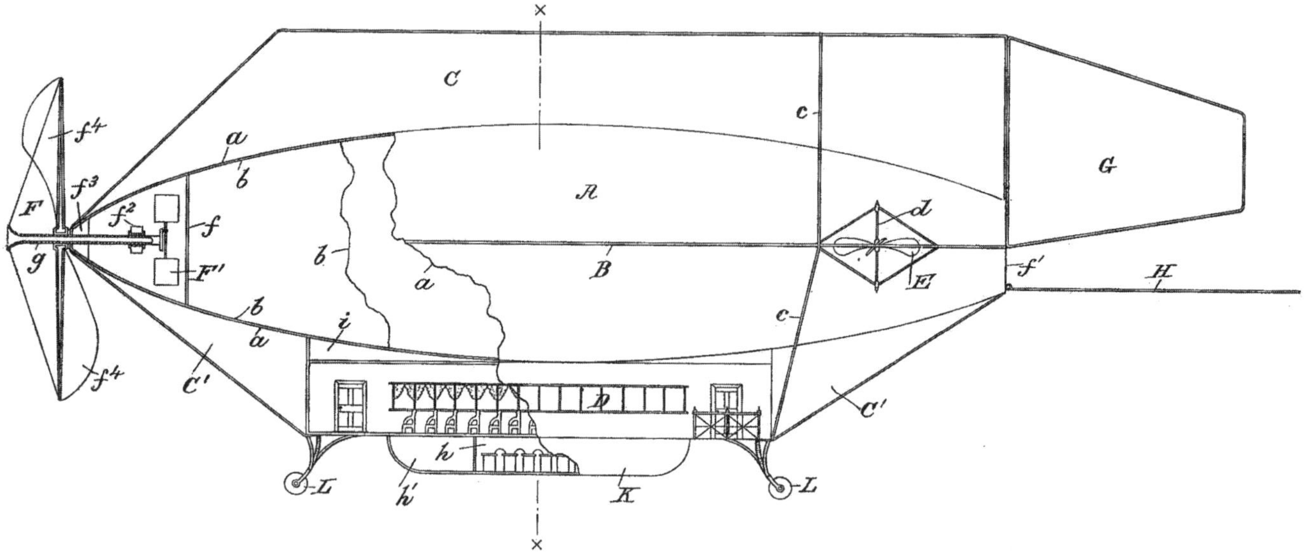

FIG 2

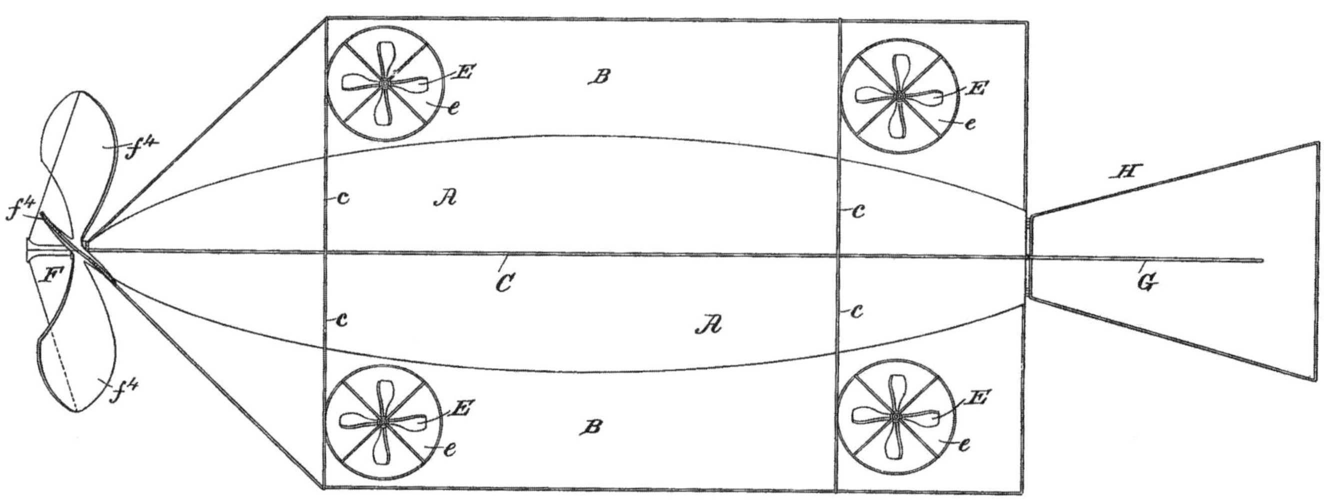

Plate 106: Edward Pennington's Airship—December 31, 1895: *Giving Stewart Cairncross' Aerial Machine (plate 88) a run for his money in terms of who could build and mount the largest propeller possible, Pennington's curious shaped nose propeller is designed to create a vacuum at the front of the craft that will suck the machine forward while simultaneously creating an excess of pressure aft to facilitate the effectiveness of a pair of huge tail rudder (letters G and H). The most mysterious aspect of the machine is Pennington's description of the lifting agent, which he identifies as a fluid that is lighter-than-air, and that permeates every part of the craft. Whatever it was, apparently this buoyant fluid was all the rage in the 1890s, though it seems to have otherwise vanished by the time the twentieth century rolled around, forcing airship captains of the era to rely on that far less romantic standby, hydrogen, to coax their machines skyward.*

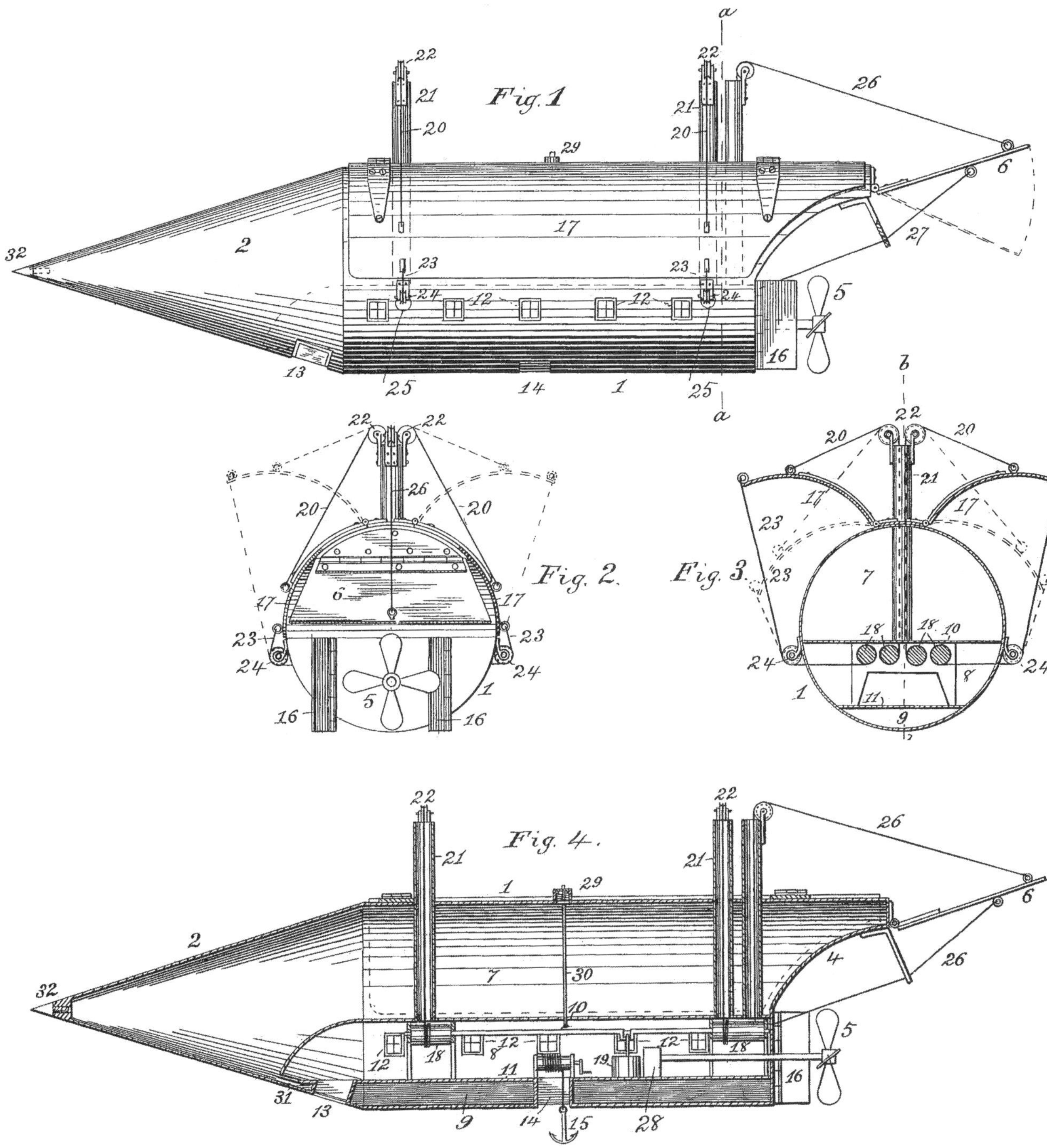

Plate 107: Charles Smith's Airship—August 11, 1896: *This very Jules Verne-looking machine is actually a metal-skinned hydrogen-filled balloon that looks very intent on flying but would, if actually built, only make a rather ornate lawn decoration. It's actually a pity it wouldn't fly, as it would be fun to flit about in such a cool-looking craft.*

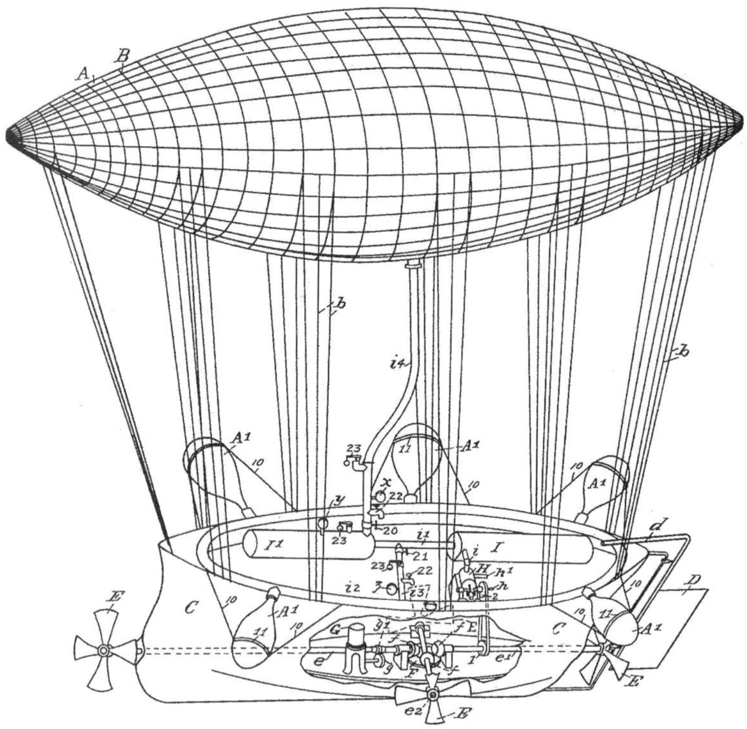

Plate 108: Edward St. Croix's Airship—December 22, 1896: *Looking like something designed by a committee of aeronautical neophytes, the swollen bowling pin-looking devices (items A1) strewn about the perimeter of the gondola are actually additional gas bags, which may be inflated or deflated as required to ascend or descend, thereby dispensing with the need for ballast entirely. Where the unfortunate pilot—not to mention passengers or cargo—might be located within this bewildering menagerie of screws, shafts, cams, and motors can only be guessed at.*

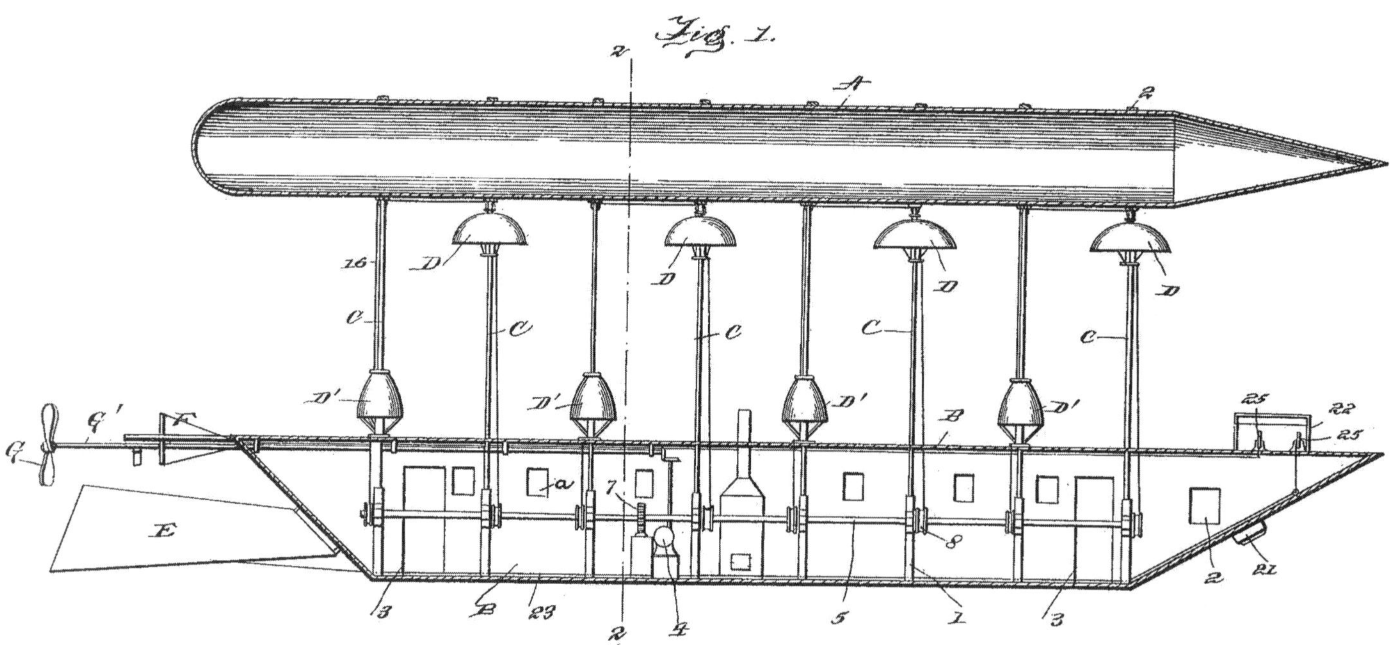

Plate 109: Henry Heintz's Airship—April 20, 1897: *The most novel element of this utterly unworkable design is the Mary Poppins-inspired array of ascending and descending umbrellas (letters D) that run between the oversized gondola and the undersized envelope. These are intended, so we are told, to provide upward thrust through their continual opening and closing, though I suspect this machine would be hard pressed to hoist a small pony skyward.*

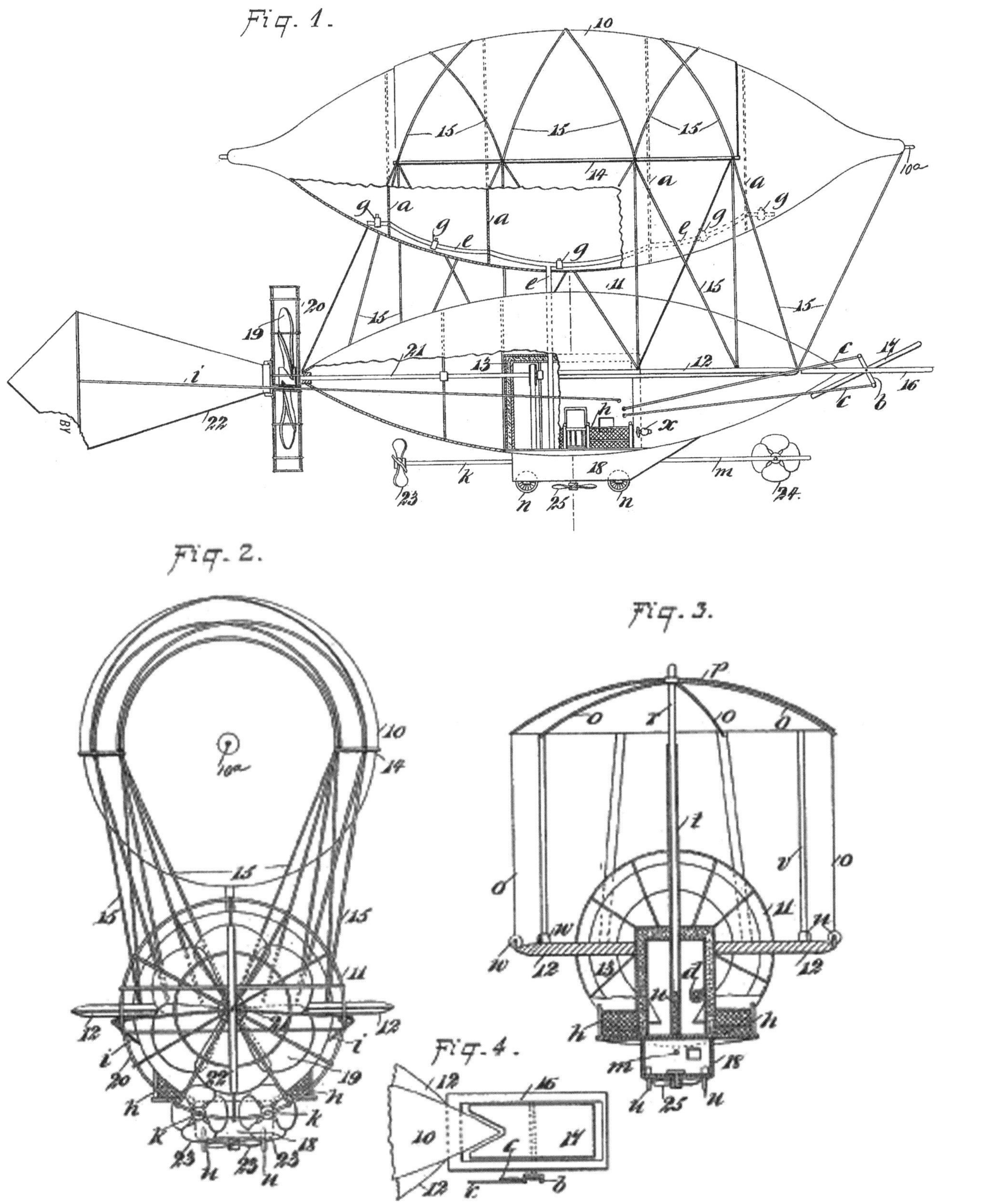

Plate 110: Thomas Crepar's Airship—August 27, 1897: *I'm sure the fact that the envelope on Minnesota inventor Thomas Crepar's no-nonsense approach to lighter-than-air travel looks much like a lemon is purely coincidental and is not a reflection upon the overall practicality of the design nor of its inventor, no matter how well deserved.*

Fig. 1.

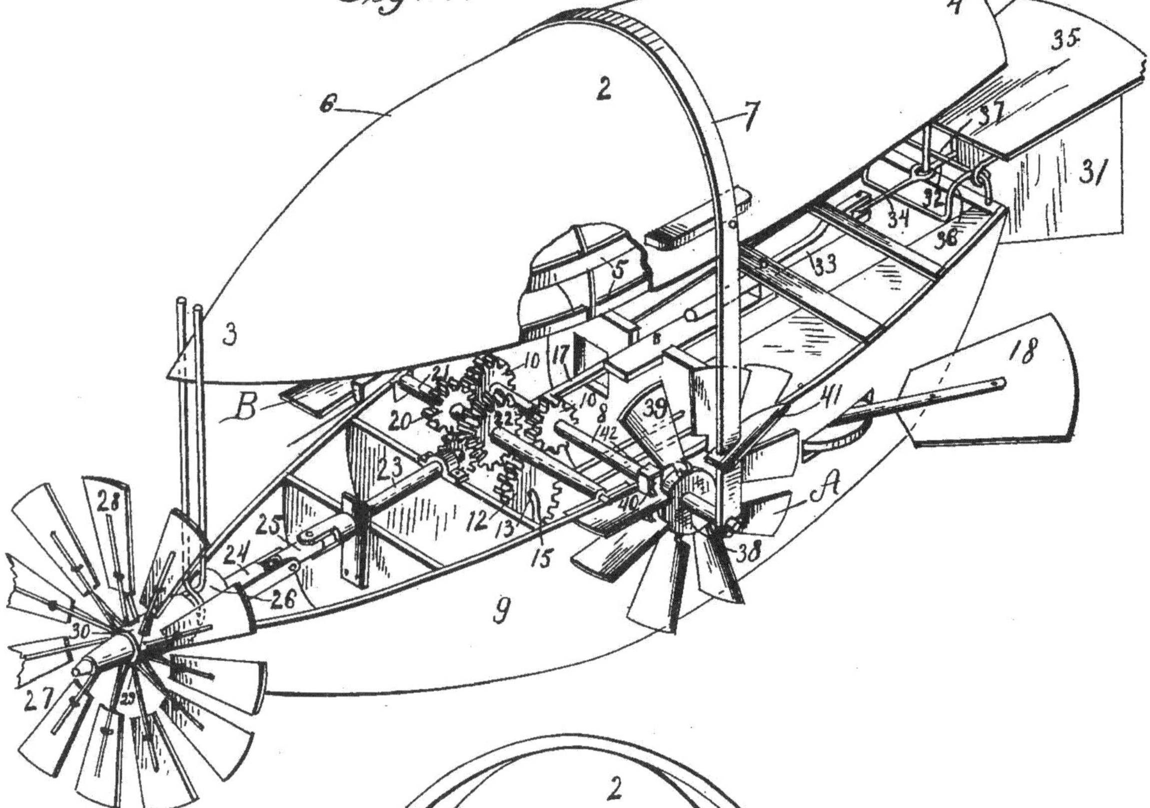

Fig. 4.

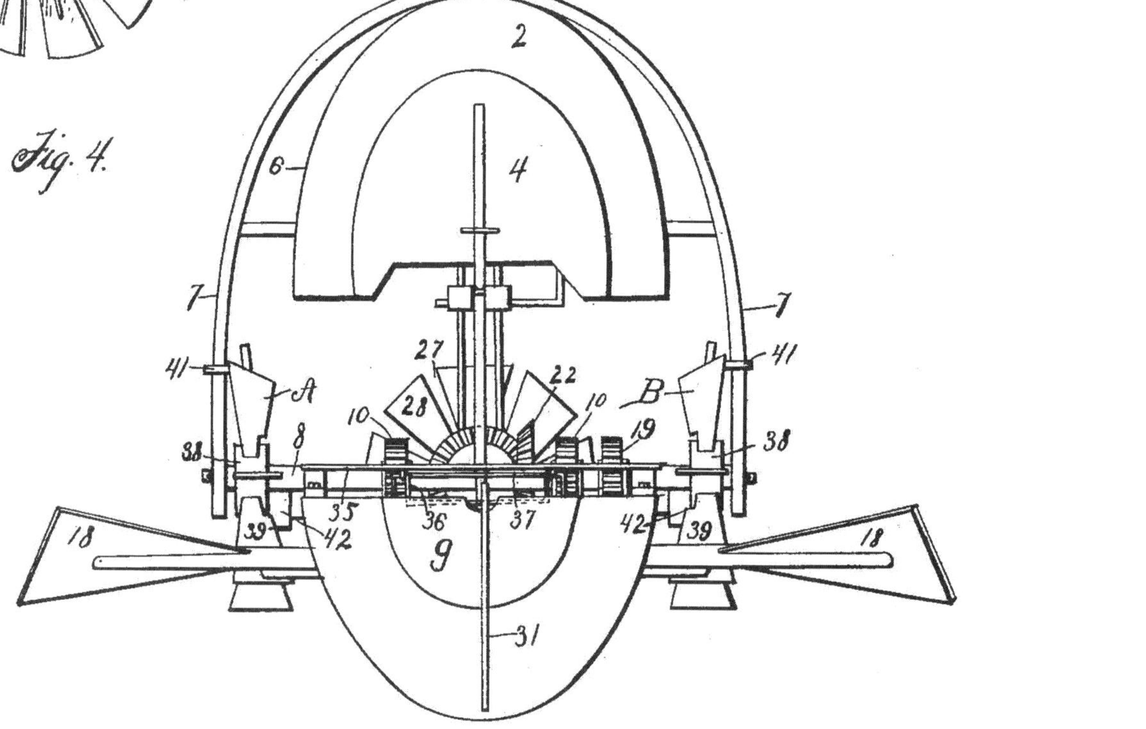

Plate 111: Daniel Reed's Airship—October 12, 1897: *Obviously inspired by Captain Nemo's Nautilus, this submarine turned airship would not only likely prove unflyable, but hazardous to anyone foolish enough to climb into the belly of the beast. It strikes me as astonishing that in 1897 a handful of inventors still hadn't figured out that paddle wheels and oars were useless in the air—a realization the Montgolfier brothers made a century earlier.*

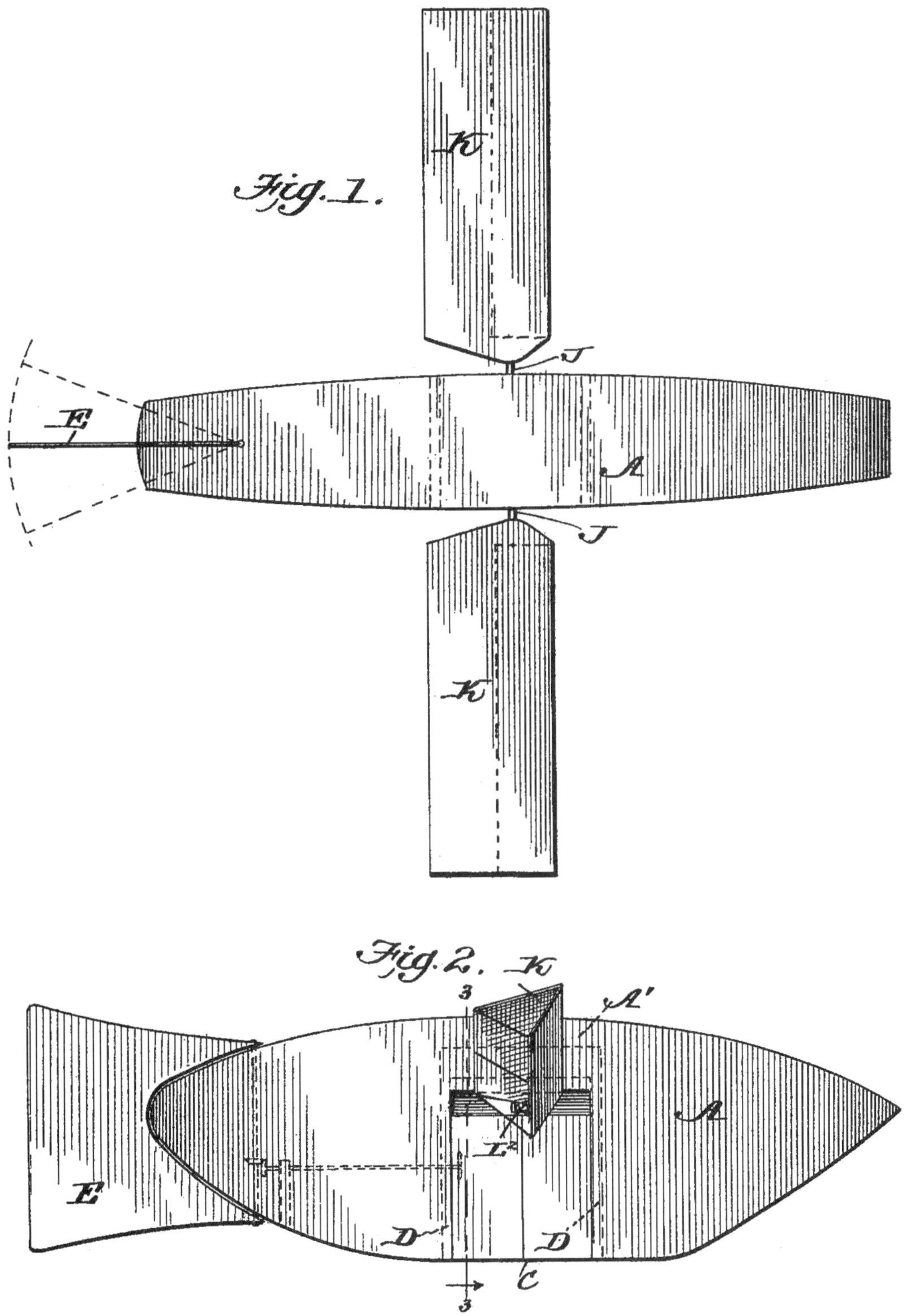

Plate 112: Jacob Graybill's Aerial Machine—October 26, 1897: *In what may have been an early design for one of Flash Gordon's spaceships, New Orleans inventor Jacob Graybill submits this nifty little craft for our consideration. The most mysterious thing about its design is the strange triangular shape of the wings (letter K) affixed to each side of the hull. Designed in such a way as to ensure the least amount of lift possible, it strikes me as curious that at this late date even a neophyte aeronaut wouldn't realize that such a design would render them useless appendages rather than lifting surfaces.*

FIG.1

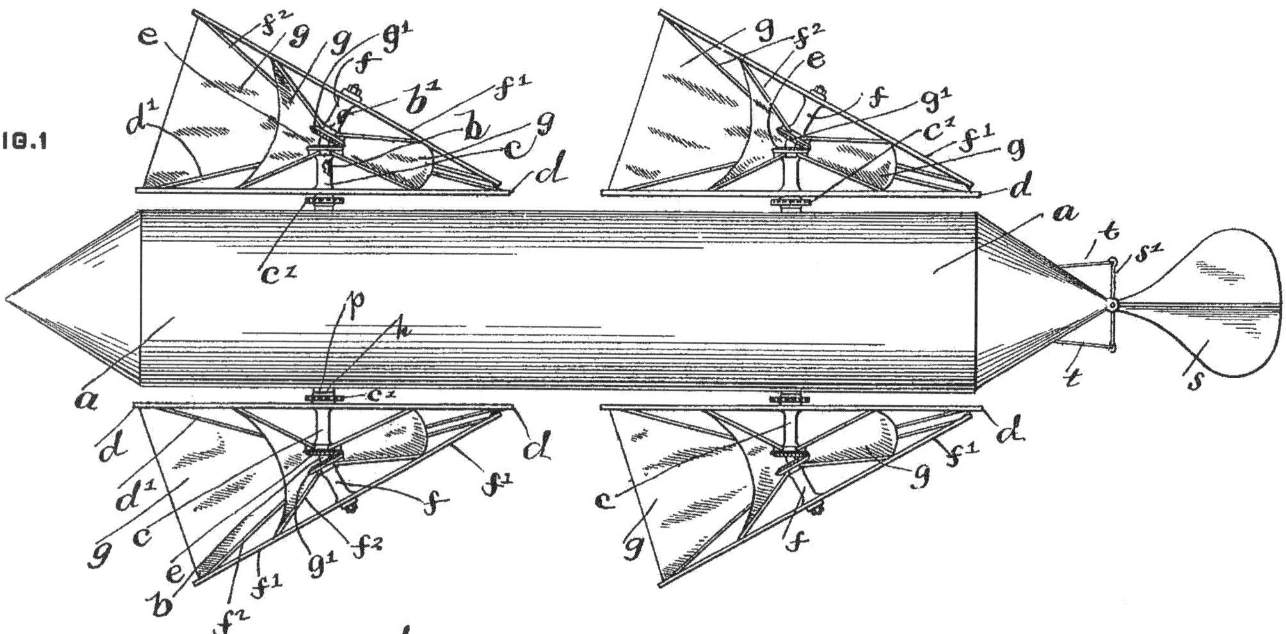

FIG.2

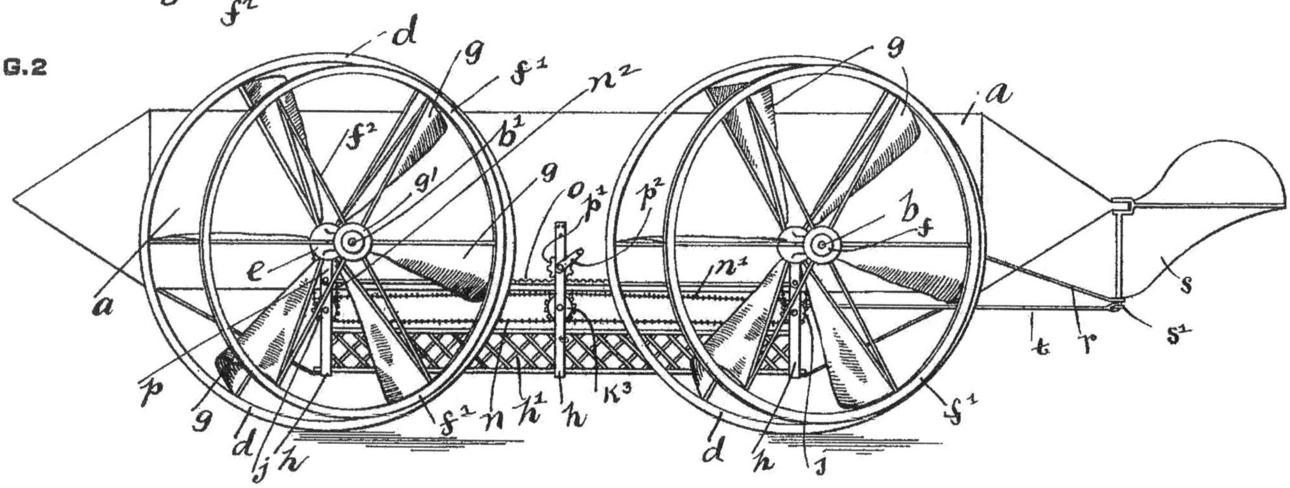

Plate 113: Theodore Leibbrand's Airship—March 22, 1898: *In a novel design not at all unusual for the era, Theodore Leibbrand recommends using giant wing sails that operate somewhat like bellows to propel the craft forward. The dainty rudder/tail assembly is also indicative that Mr. Leibbrand has no problem getting in touch with his feminine side.*

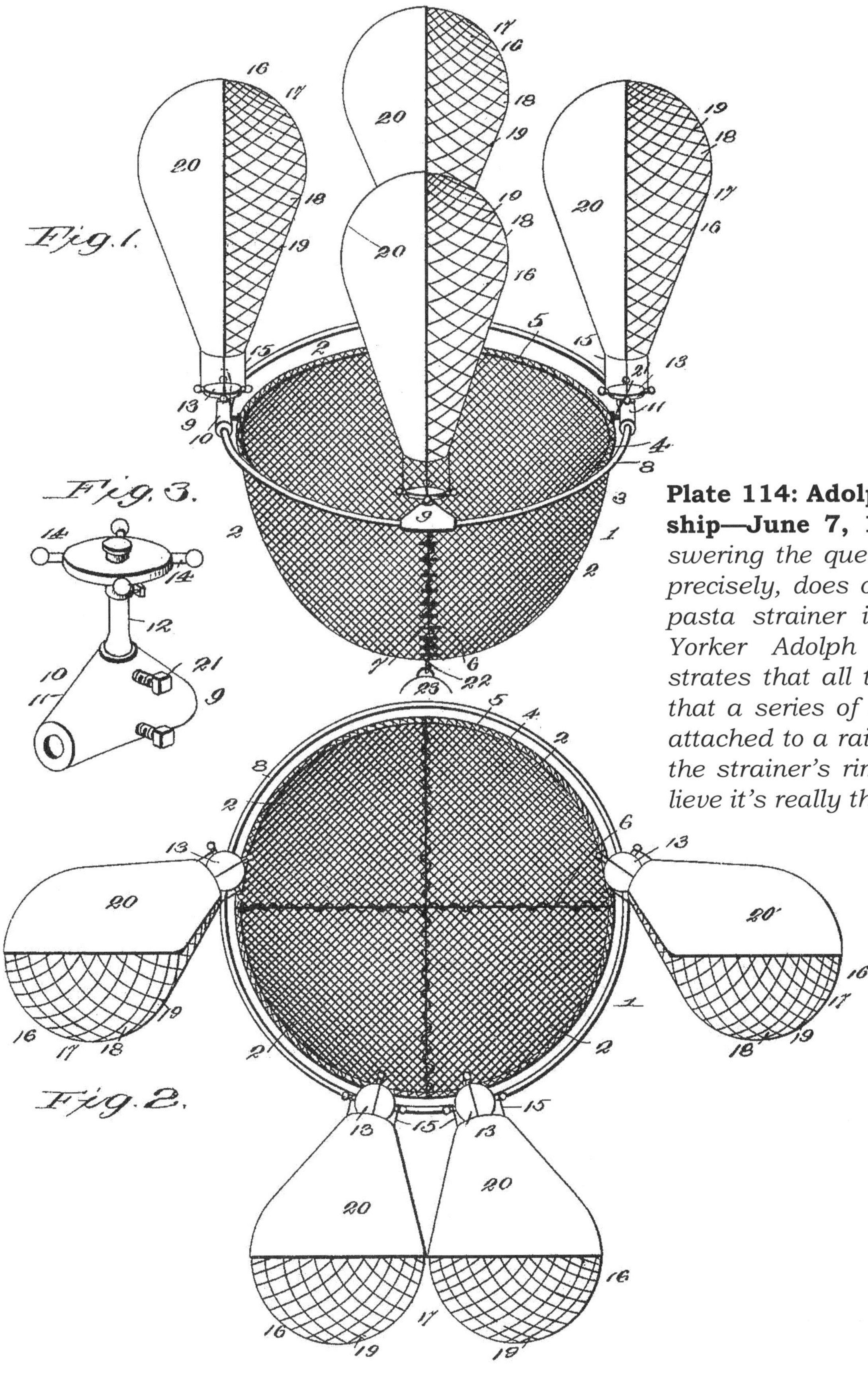

Plate 114: Adolph Koehler's Airship—June 7, 1898: *Finally answering the question of just how, precisely, does one get a massive pasta strainer into the air, New Yorker Adolph Koehler demonstrates that all that is required is that a series of small balloons be attached to a rail that circumvents the strainer's rim. It's hard to believe it's really that simple.*

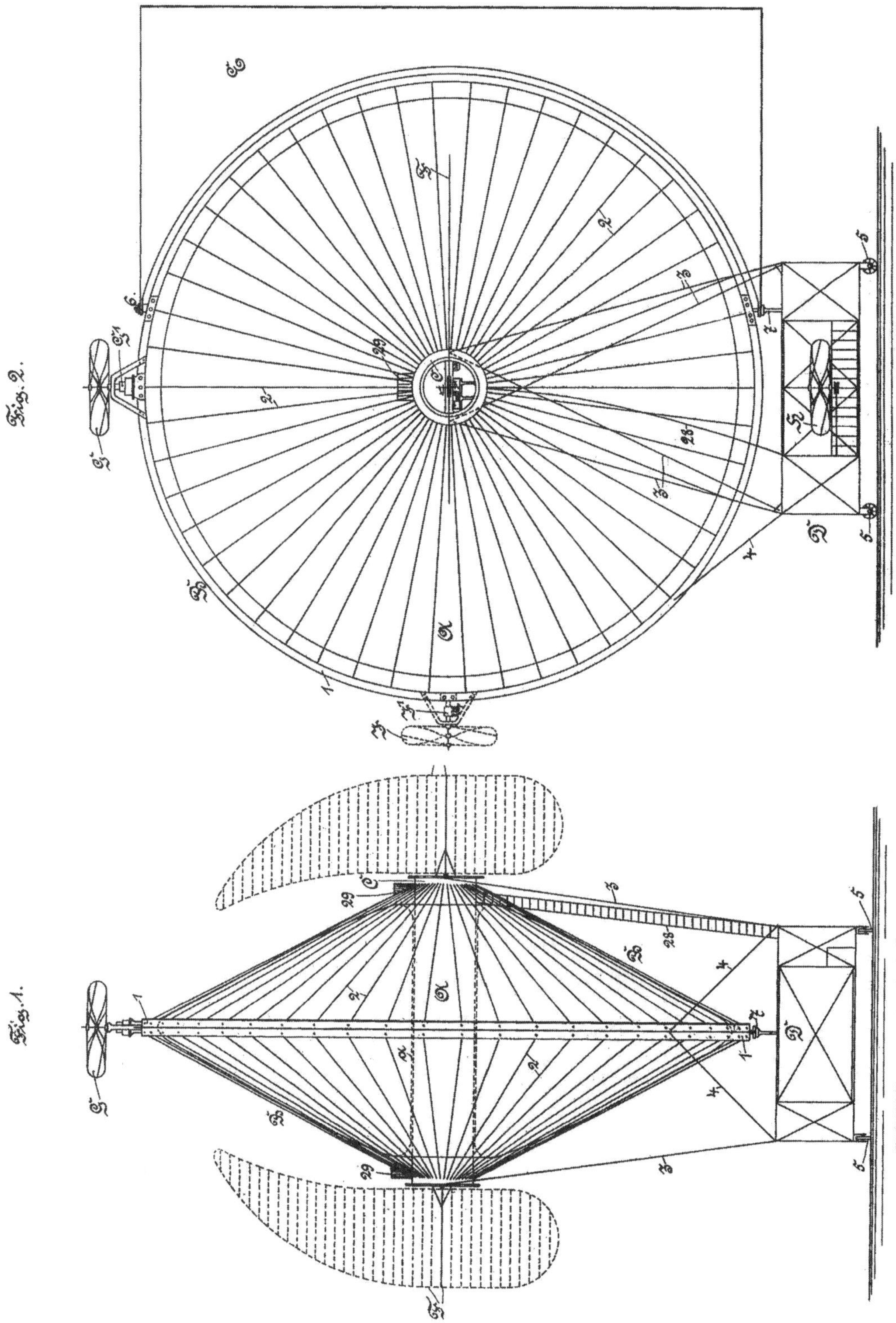

Plate 115: Friedrich Jone's Flying Machine—June 14, 1898: *Just when you thought every possible balloon configuration had been tried, German inventor Friedrich "Fritz" Jone came up with this ferris wheel idea designed to, if not fly, at least impress and frighten. One can only imagine what a sight this would have made as it lurched its way through the peaceful skies over Saxony, filling its viewers with awe and wonder while it searched frantically for a place to crash land.*

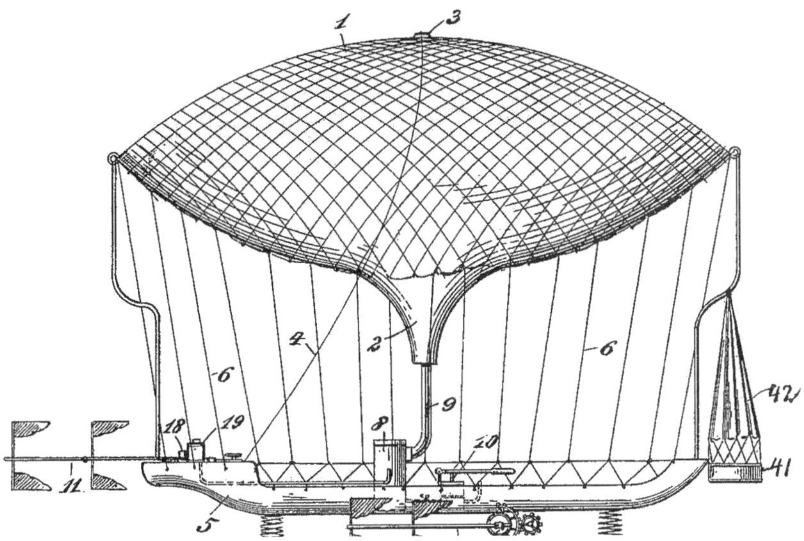

Plate 116: Joel Rice's Airship—July 5, 1898: *This dainty looking craft is nothing more than another variation on the common theme of the era that mandated that all airships contain as many whirling propellers as possible, in this case two fixed and pointing aft and two on a rotating gear that can be pointed in any direction. Mr. Rice's airship is noteworthy for the peculiar wind-vane shaped propellers his design incorporates, along with the massive springs for landing and the whimsical but most practical inclusion of a parachute (letters 41 and 42) just on the remote chance that something went wrong with his brainchild—as difficult as that is to imagine.*

Fig:1.

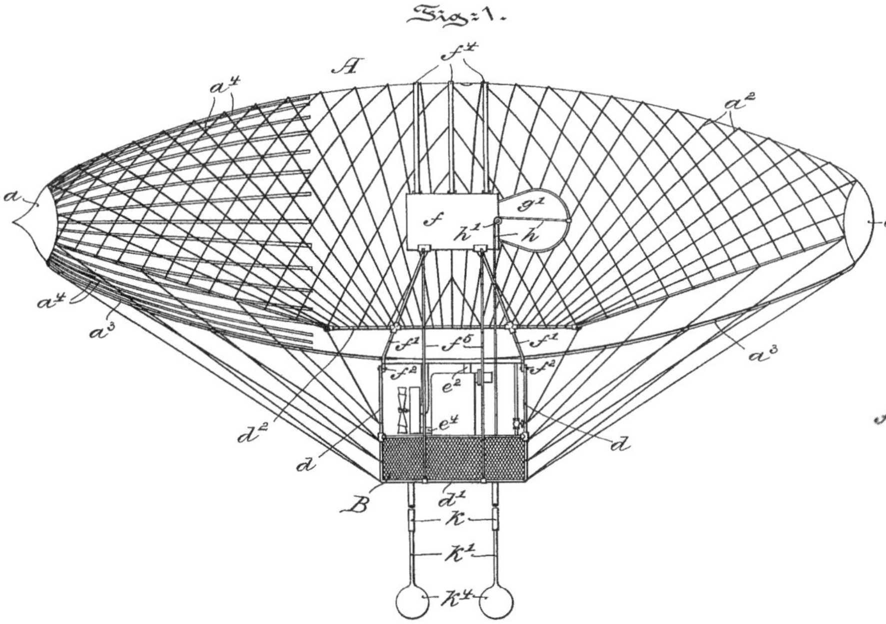

Fig:2.

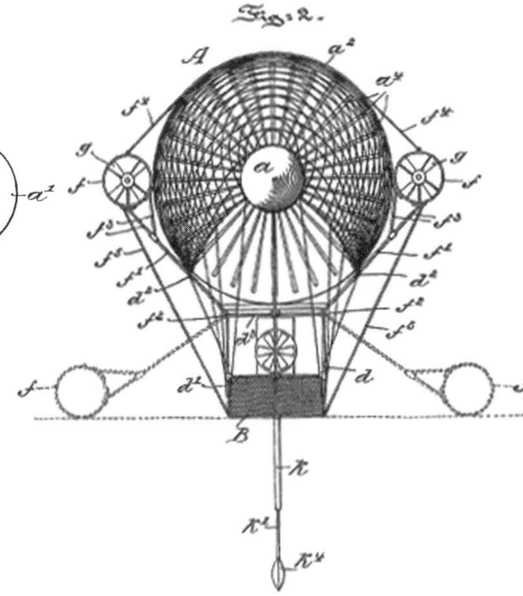

Plate 117: Charles Hite's Airship—July 12, 1898: *Hite's simple craft is, as far as I can determine, a hot-air balloon, sustained in flight by a series of heating tubes that constantly suck cool air out of the envelope, heat it to a high temperature, and then shoot it back up into the envelope, all without the use of an open flame. This craft also has an internal ballonette to maintain the bag's shape and reinforcing ribs along the nose, precisely as modern blimps do today.*

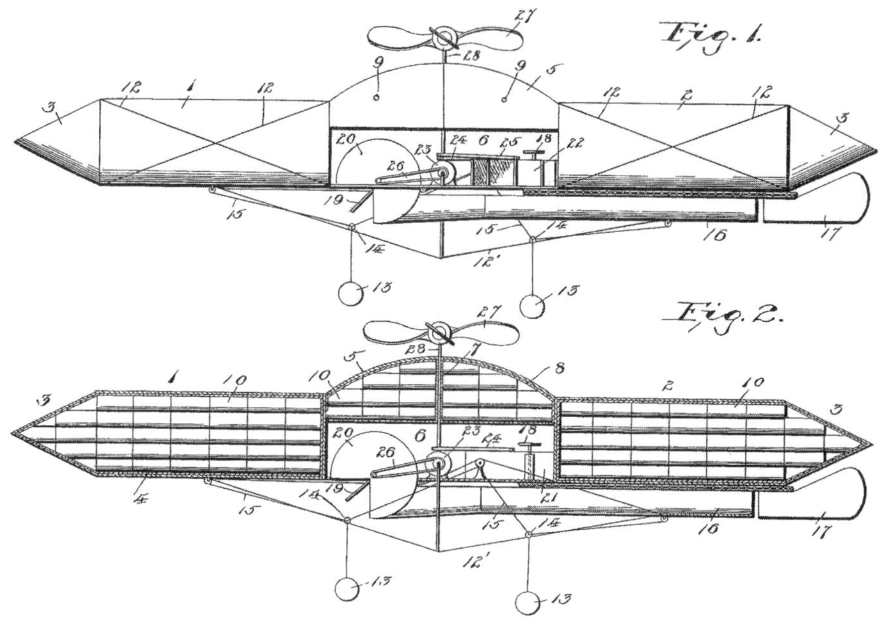

Plate 118: Thomas Brown's Aerial Ship—September 13, 1898: *This imposing looking craft had a couple of novel features built into it. First, the hydrogen was stored in a series of long, narrow cylindrical cells throughout the ship designed to prevent a puncture from sending the craft into a fatal descent. Second, the prop wash is blasted through a conduit along the keel (number 16) against the rudder (number 17), demonstrating at least a primordial understanding that a fixed control surface is only useful if it is being acted upon by an airflow.The most inexplicable element is the massive dorsal propeller on top (number 27) that is presumably designed to assist in the ship's ascent and descent but which would, in all probability, provide little help with either.*

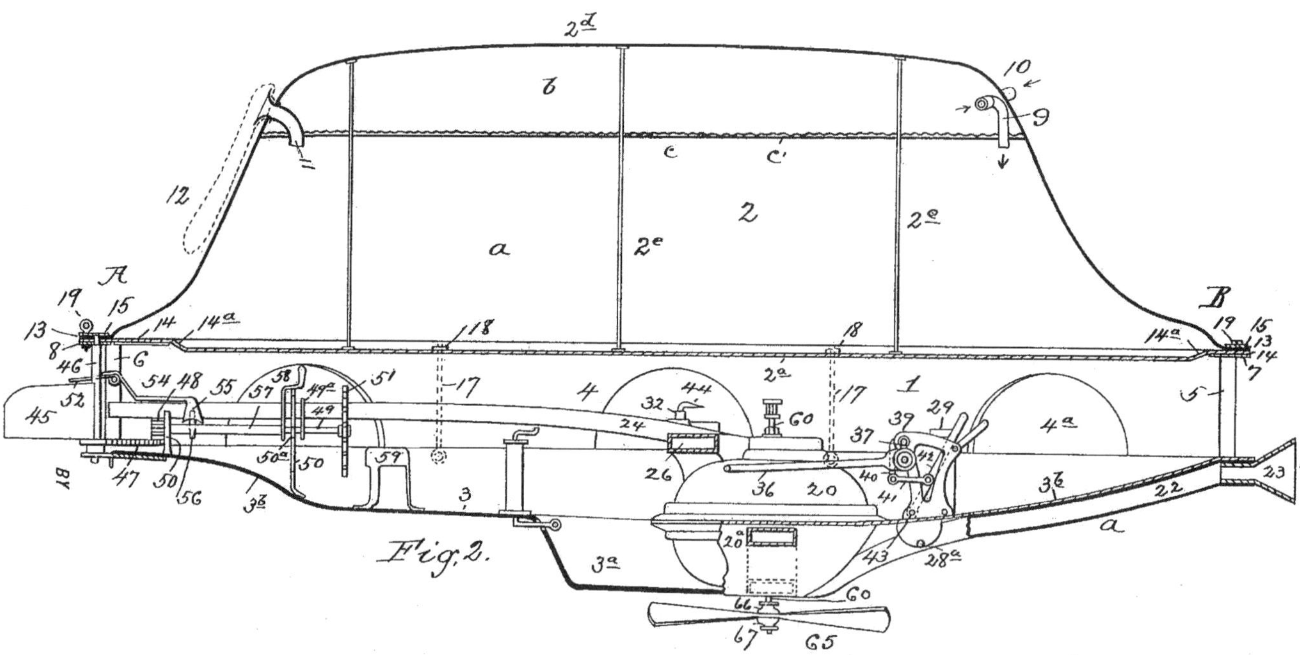

Plate 119: Oscar Smith's Airship—October 18, 1898: *After carefully culling through five pages of techno-babble, I must admit to being no closer to being able to make heads-or-tails of New York inventor Oscar Smith's unusual machine than when I started. All I can discern is that it is some sort of aerial locomotive that uses steam to bully its considerable mass through the air with all the finesse of...well, an aerial locomotive.*

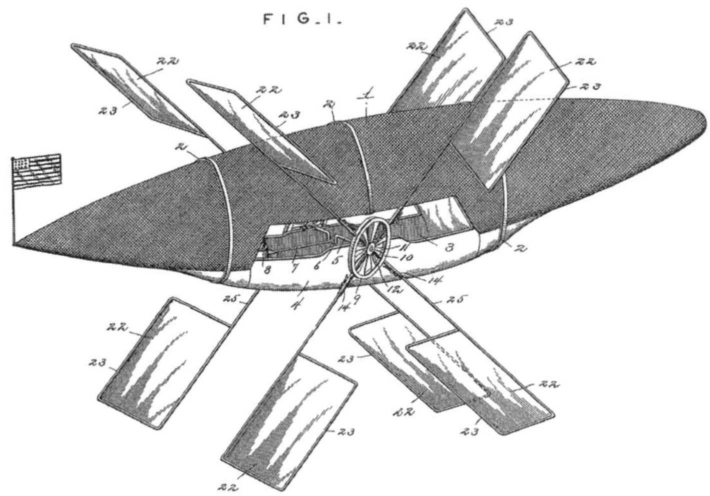

Plate 120: Frederick Merritt's Airship—December 6, 1898: *In a design that would have been considered antiquated in 1850, Oregonian aeronaut Frederick Merritt has decided to go back to the dawn of fanciful airship design to demonstrate that just because an idea may be unworkable, impractical, and done to death doesn't mean it's not good. Well, actually, I guess it does mean that after all. The flag on the nose does a fine job in demonstrating Mr. Merritt's patriotism, however.*

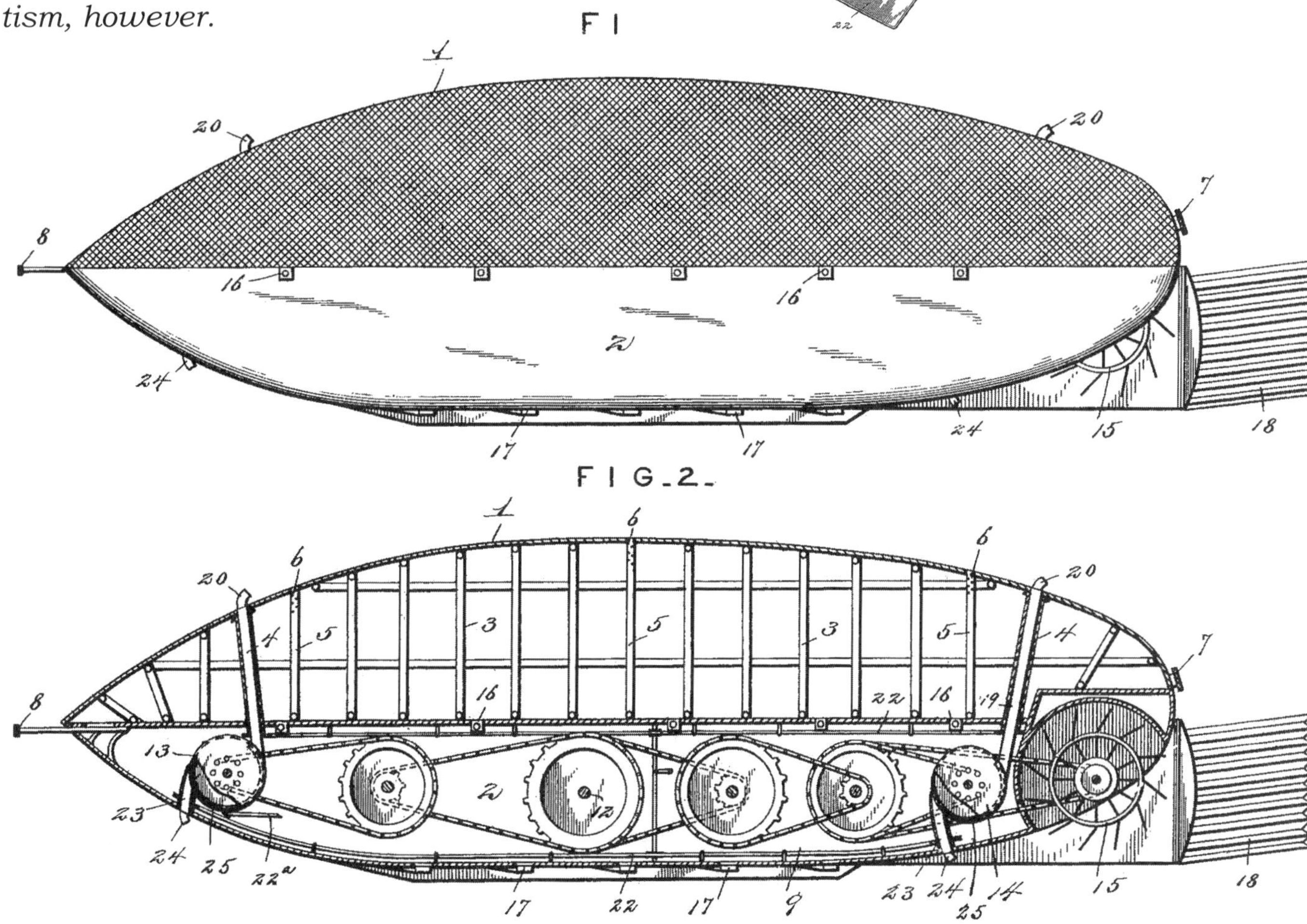

Plate 121: John Pickering's Airship—February 20, 1900: *In the top illustration, Mr. Pickering's airship looks rather ordinary; however, the lower illustration shows us that it is actually one giant mass of spinning gears, all hooked together to churn one massive paddle wheel at the back of the craft. Why Mr. Pickering couldn't have simplified the craft down to a single set of gears is not made clear in the description, but we must trust that his love for over-engineering possesses some sound reasoning beyond merely an unhealthy affection for gears and cams.*

FIG. 1.

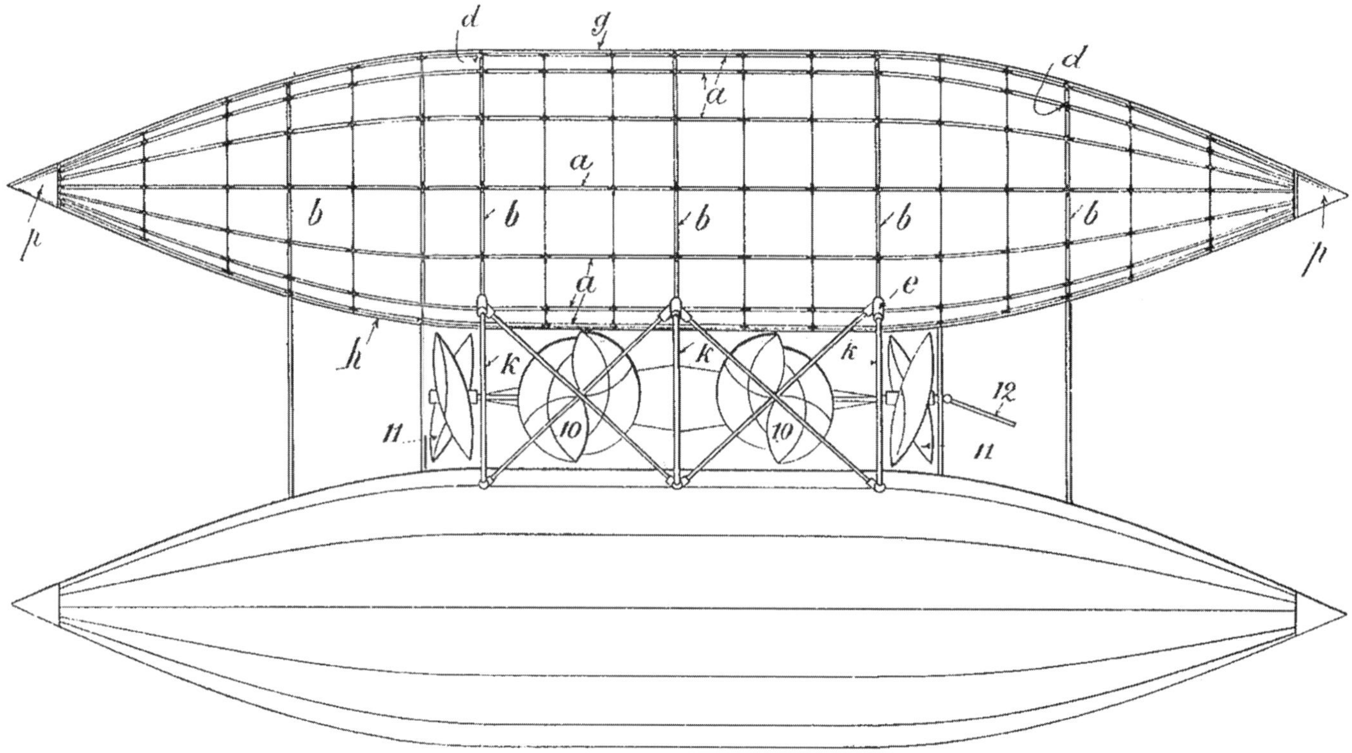

FIG_2_

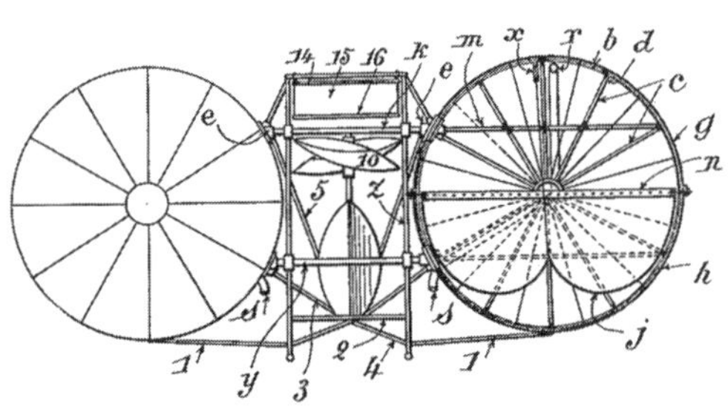

FIG_3_

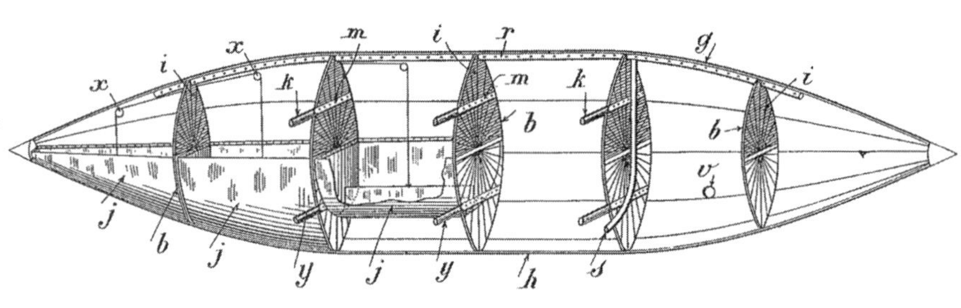

Plate 122: Louis Roze's Airship—May 1, 1900: *What makes this multi-propped monstrosity noteworthy is the addition of diaphragms or partitions within each envelope, which had become a recent innovation in airship design by 1900. Unfortunately, the diaphragms are not to prevent a loss of gas in the event of a puncture, but were open to one another (as well as to the chambers in the opposite envelope via hollow aluminum tubes) as a means of maintaining equilibrium within the envelope and reduce the amount of gas loss. All it would really do is ensure that, in case of a puncture, the entire vessel would slowly bleed off all its hydrogen rather than only a part of it with all the unfortunate consequences that would entail. Another interesting feature is that the many propellers are to be sized "according to the size of the aviator" which is an interesting concept. I wonder if they have to change propellers when there is a change in crew, or if they only hire pilots who are the same identical size?*

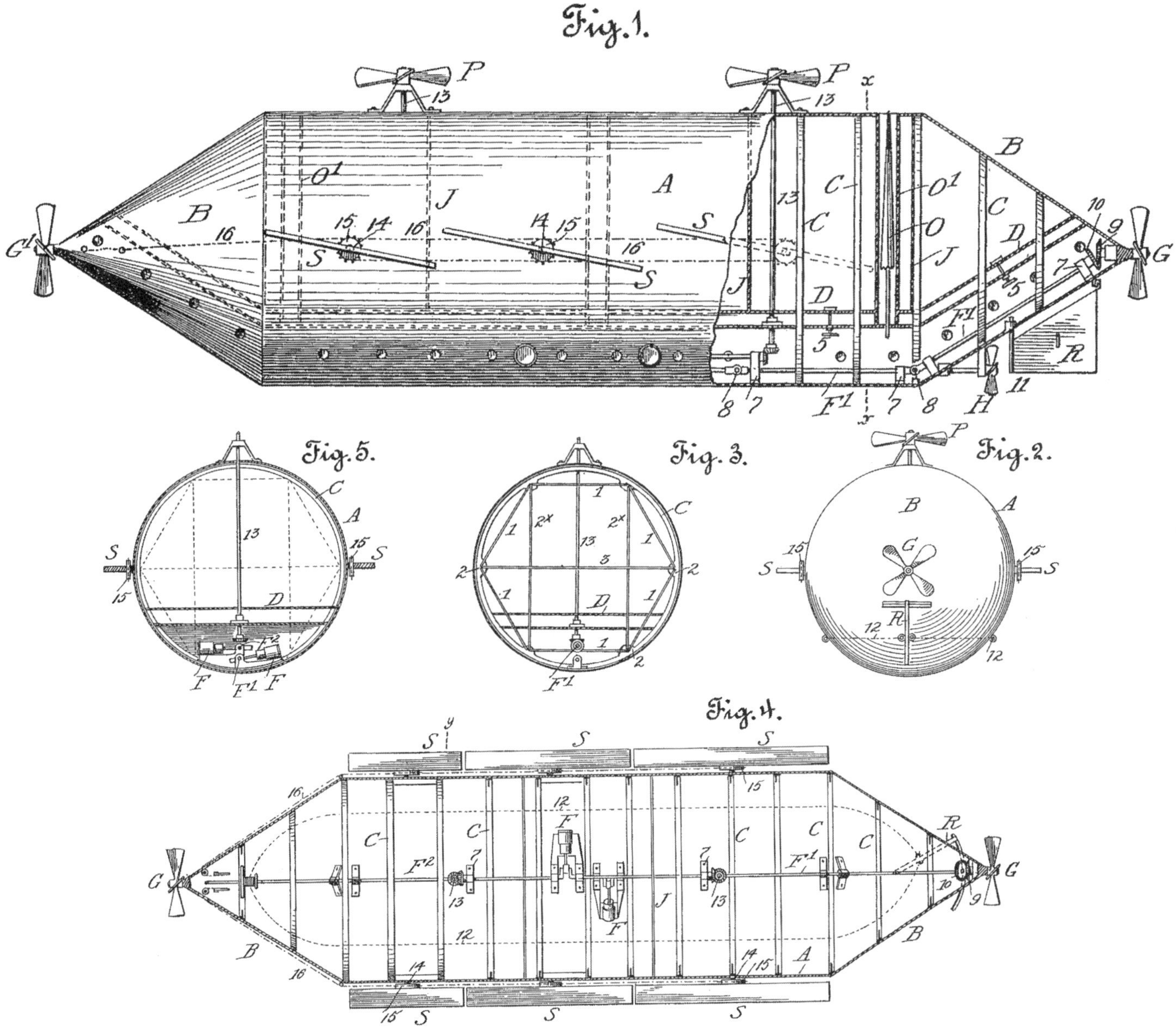

Plate 123: Charles Stanley's Airship—October 9, 1900: *In yet another example of the over-engineering so popular at the time, San Francisco inventor Charles Stanley provides this bewildering array of multi-directional screws and pivoting wing boards in an effort to make LTA travel as difficult as possible. Inventors of the era were still wedded to the idea that a hydrogen-filled envelope required a vertically placed propeller to lift itself into the air, demonstrating a peculiar lack of faith in hydrogen's inherent lifting capacity.*

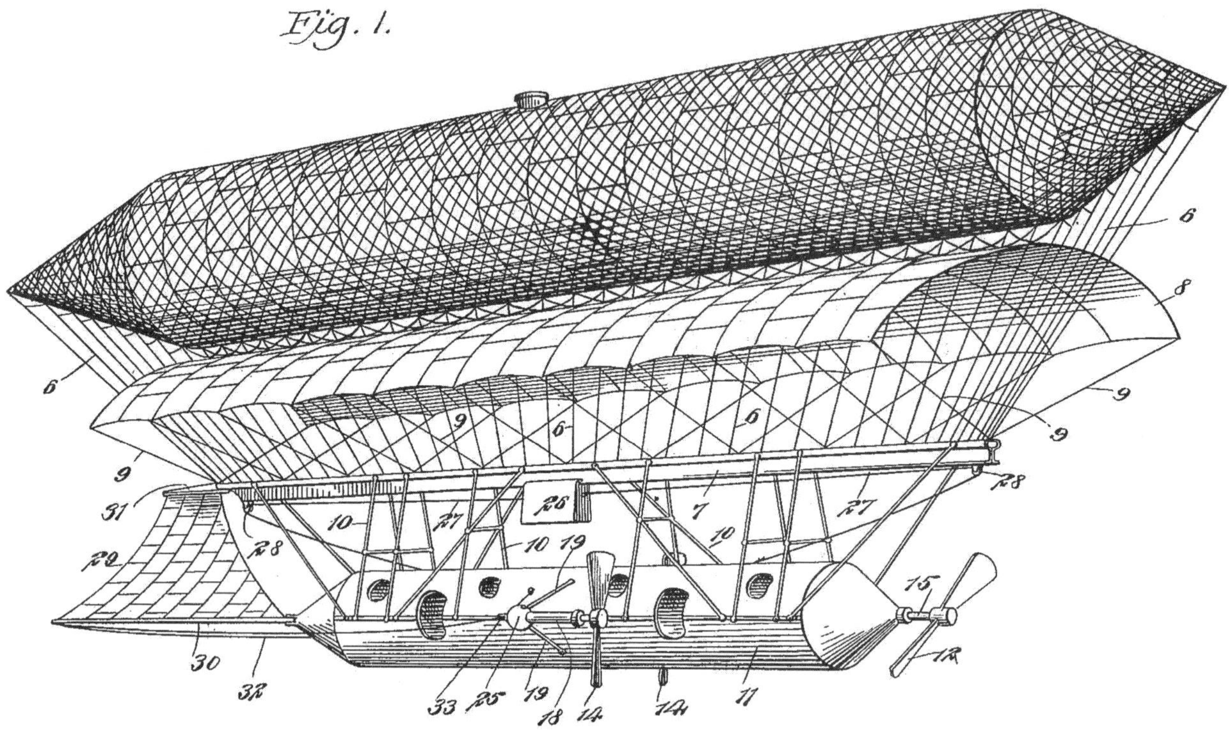

Plate 124: Leo Stevens' Controllable Balloon—November 13, 1900: *This elegant craft has something for everyone: a simple cigar-shaped balloon on top, a massive bow-to-stern parachute/ sail in the middle, and a Buck Rogers-esque propeller-studded gondola at the bottom. I'd imagine keeping each of these elements—all held together by a menagerie of cables and wires—moving in the same direction would be quite a challenge, especially if there was even a hint of a breeze in the offing. Still, you can't help but appreciate the man's imagination.*

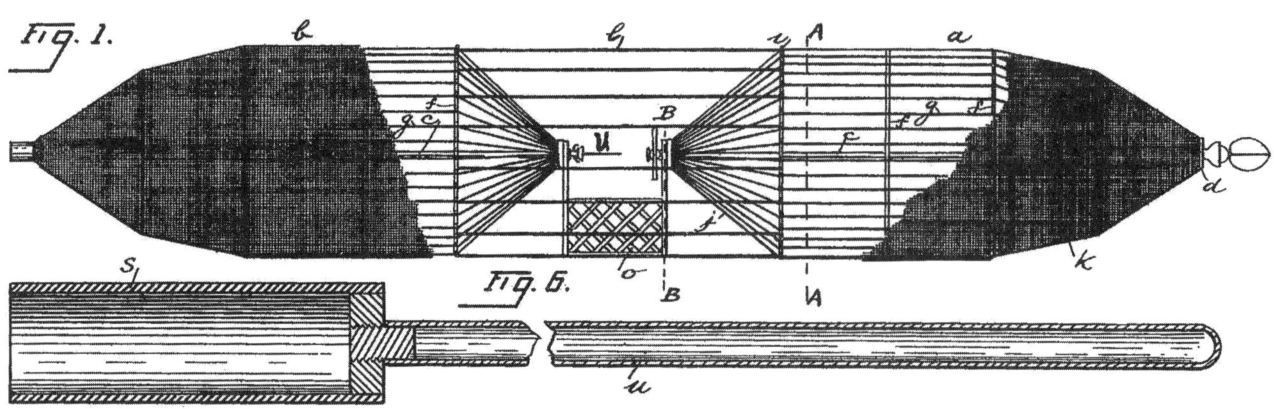

Plate 125: Ole Olsen's Airship—March 26, 1901: *What makes Oakland airship enthusiast Ole Olsen's stately-looking airship different from all the others out there is that it is among the earliest of the rocket-assisted dirigibles to see the light of the twentieth century. Each envelope has a hollow steel tube running its length that fires a "cartridge of any suitable explosive material" designed to blast the craft in the desired direction. While Mr. Olsen is open to the possibility of using hydrogen as his lifting agent of choice, he also suggests that lift could be achieved merely by creating a vacuum within the envelopes, which, he assures us, are strong enough to "resist considerable external pressure." Probably safer than firing a cannon down the middle of a hydrogen airbag, but only marginally.*

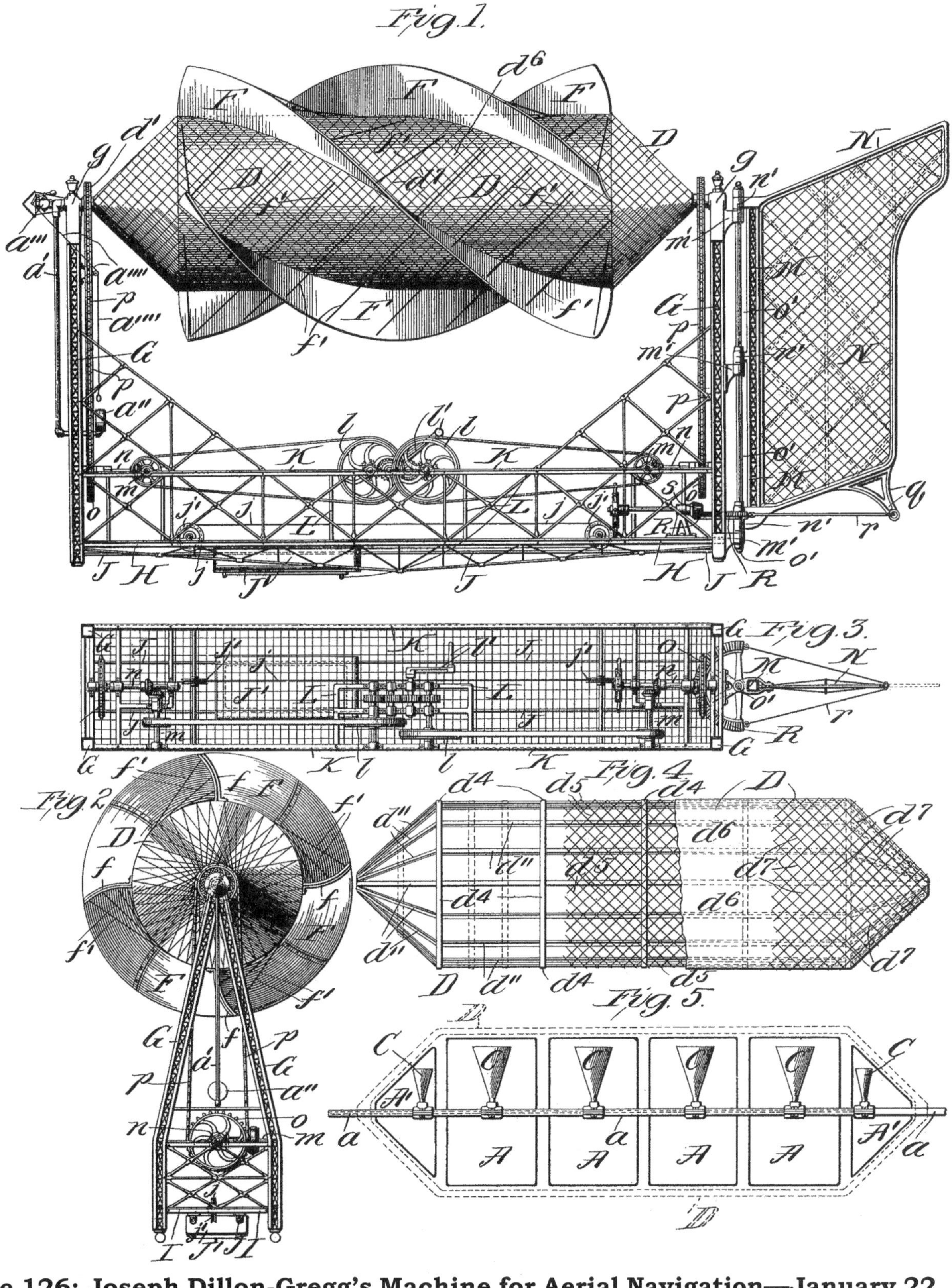

Plate 126: Joseph Dillon-Gregg's Machine for Aerial Navigation—January 22, 1901: *Finally, a common-sense approach to the problem of aerial control: simply turn the entire envelope into one massive spinning, boring screw that can drill its way through the air like a termite through oak, guided by the delicate maneuvering of one of the world's largest rudders. A fine example of the industrial revolution taken to its most impractical extreme.*

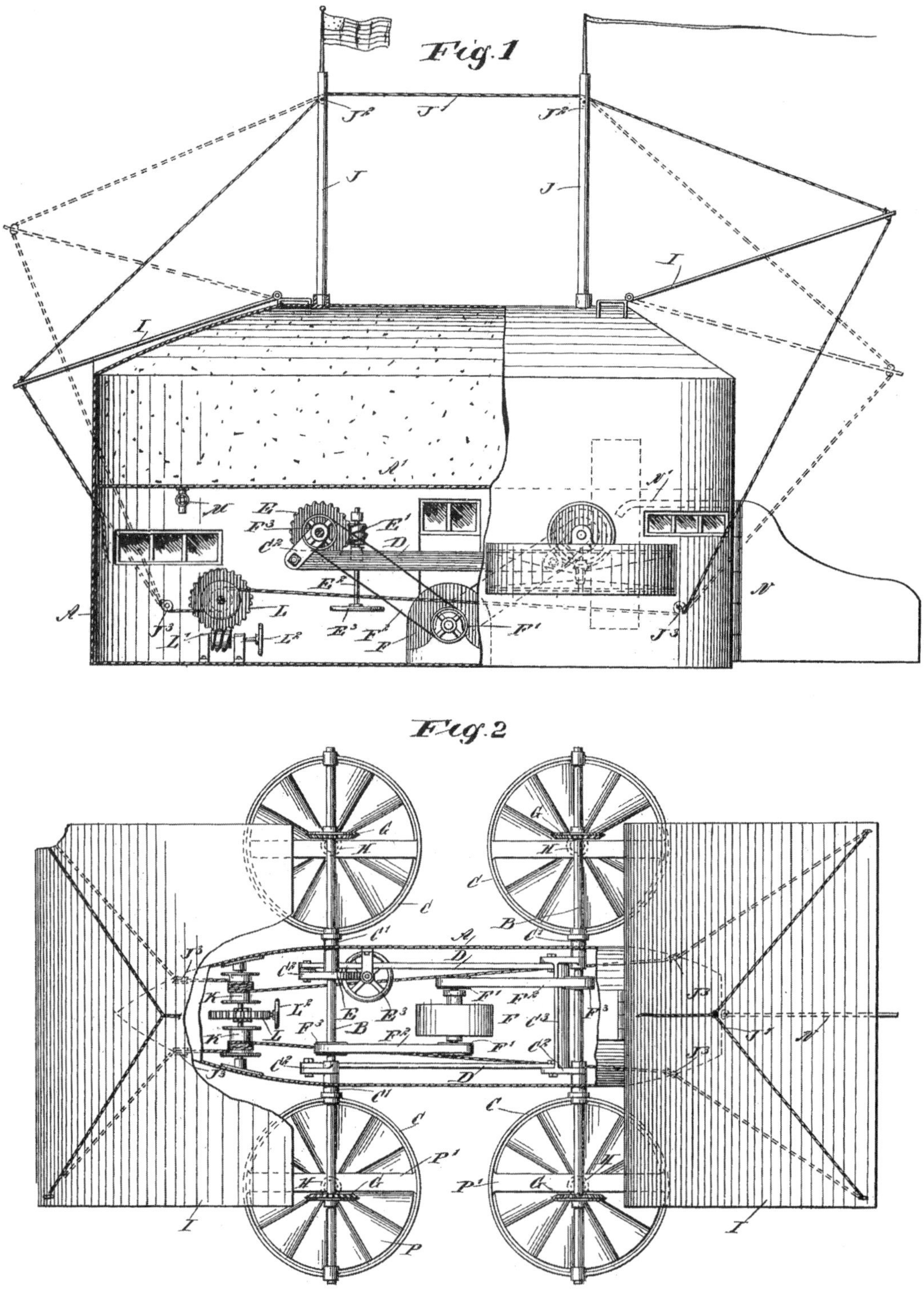

Plate 127: Aristarchus Hubbard's Airship—February 19, 1901: *With a name like Aristarchus Hubbard, fame is almost assured, which this curious contraption very nearly accomplished for the California airship designer. Basically a two-story boat-shaped house in which the upper floor is filled with hydrogen and the lower floor is crammed with machinery, the most curious element is the "aero planes" (sails? Letters I), attached to the masts designed to control the ascent and descent of the craft. While unlikely to be a candidate for flight, the flag on the foremast does add a badly needed patriotic touch.*

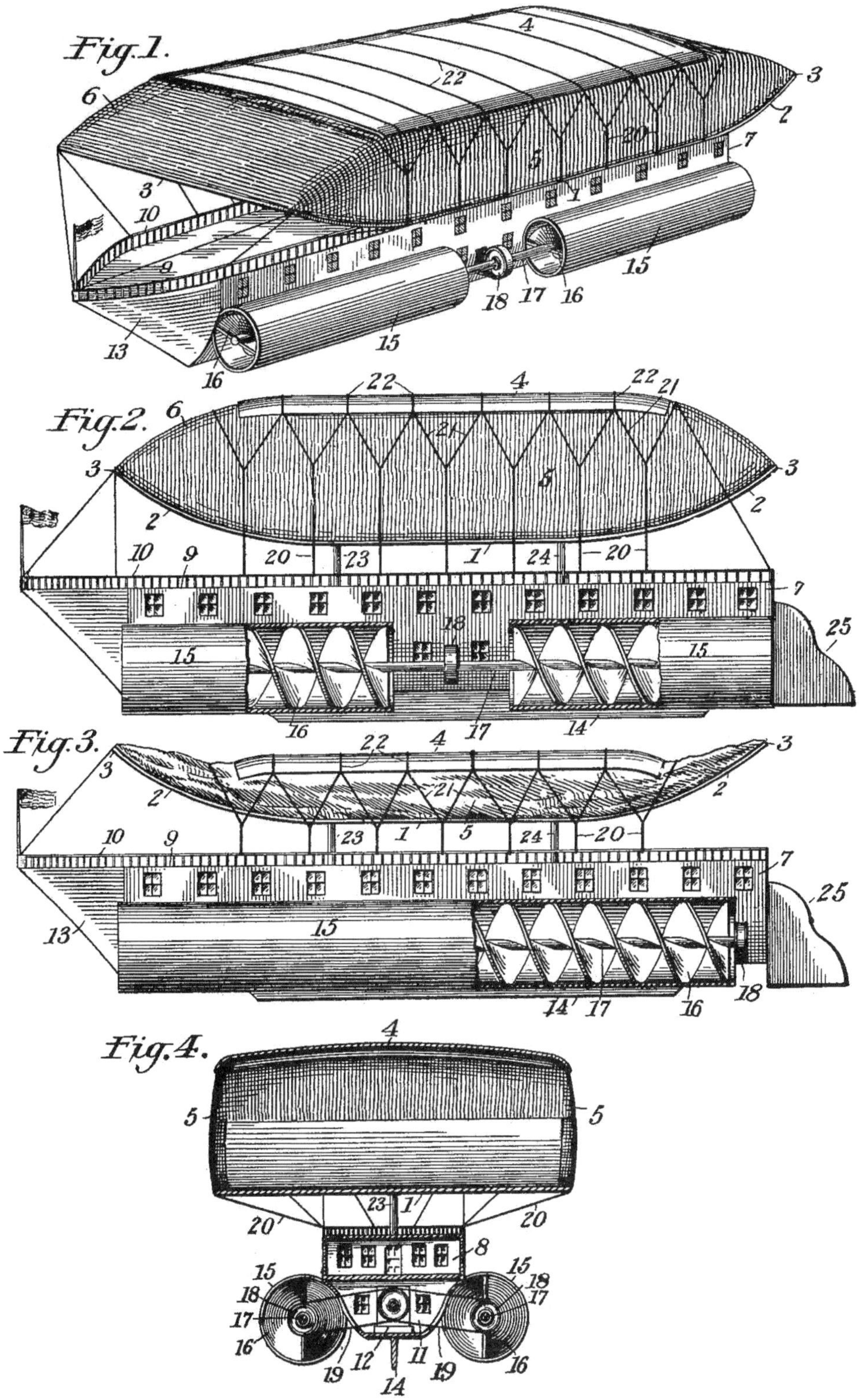

Plate 128: Eugene Farr's Aerial Vessel—July 9, 1901: *In an early effort at once more confounding the forces of gravity, D.C. entrepreneur Eugene Farr's flying ship provides us with a truly unique airscrew-type rotor assembly that foresees the advent of the aqua-jet by a century. Actually, this might have been better tried on an ocean-going vessel before being incorporated into an airship, as it might show promise in a water-borne environment.*

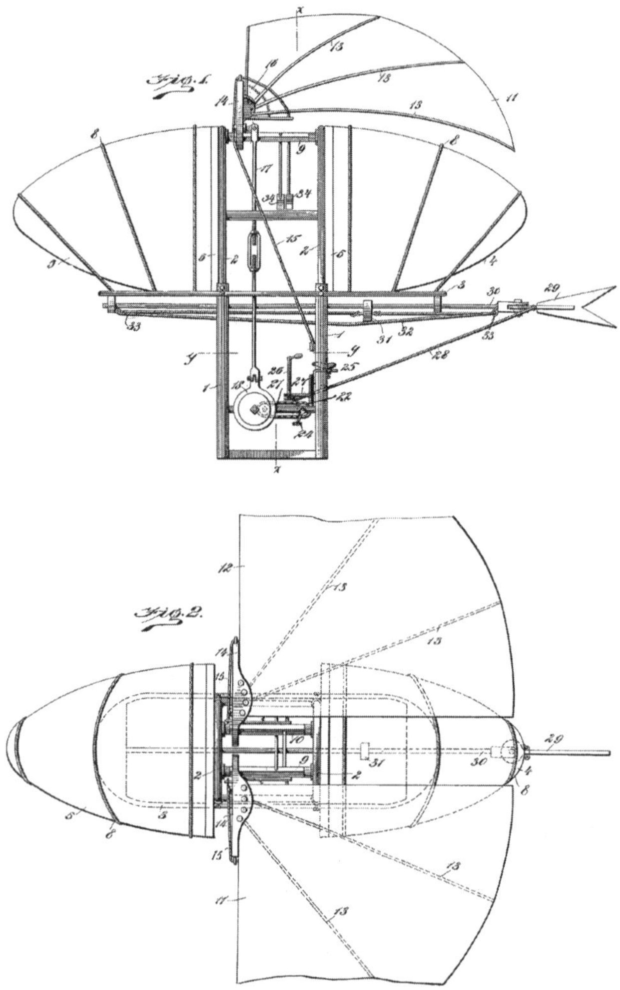

Plate 129: John Spies's Airship—December 3, 1901: *This fine and not at all practical little pedal-powered machine is certain to get the neighbors' attention. Basically a small balloon cut in half, propulsion is provided by a large pair of retractable bat-like wings that are guaranteed to effortlessly propel the craft's lucky pilot through the air with grace and ease, all the while maintaining meticulous control via a tiny fork-tailed rudder of dubious value. It will always remain a mystery as to why Mr. Spies's little machine didn't catch on.*

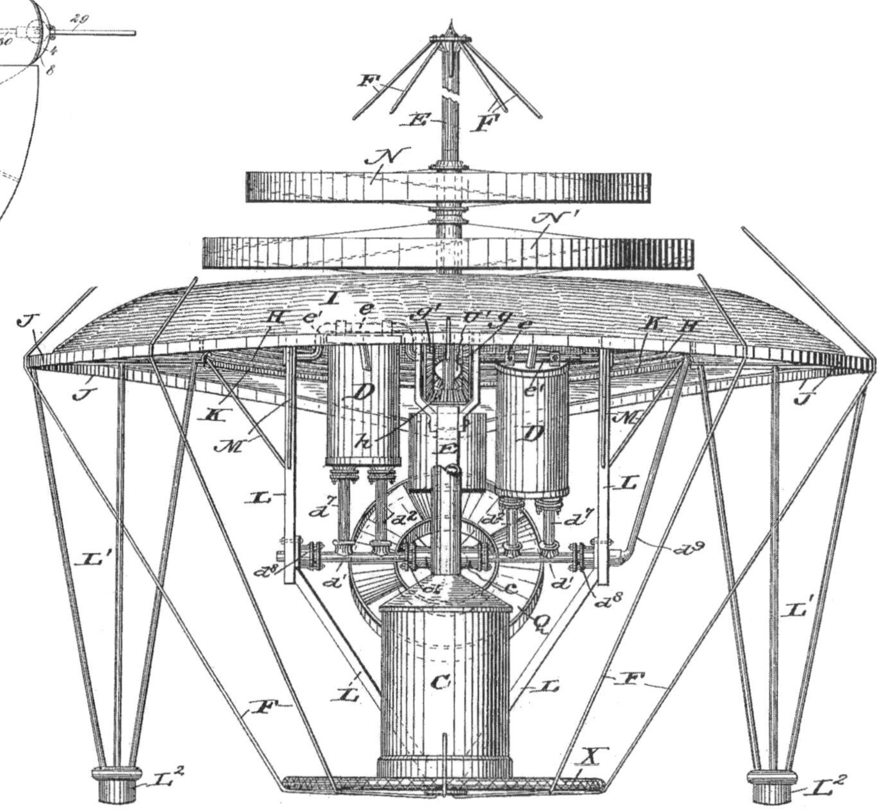

Plate 130: Robert Botts's Airship—December 10, 1901: *Easily one of the most frightening non-military machines ever conceived, this steam-powered airship/helicopter hybrid would be guaranteed to bring, if not fame, at least notoriety, to New Mexico inventor Robert Botts were it ever to be built and started. My only questions are where the unfortunate pilot/crew sits and just how big a crater this machine would leave were it to inexplicably explode.*

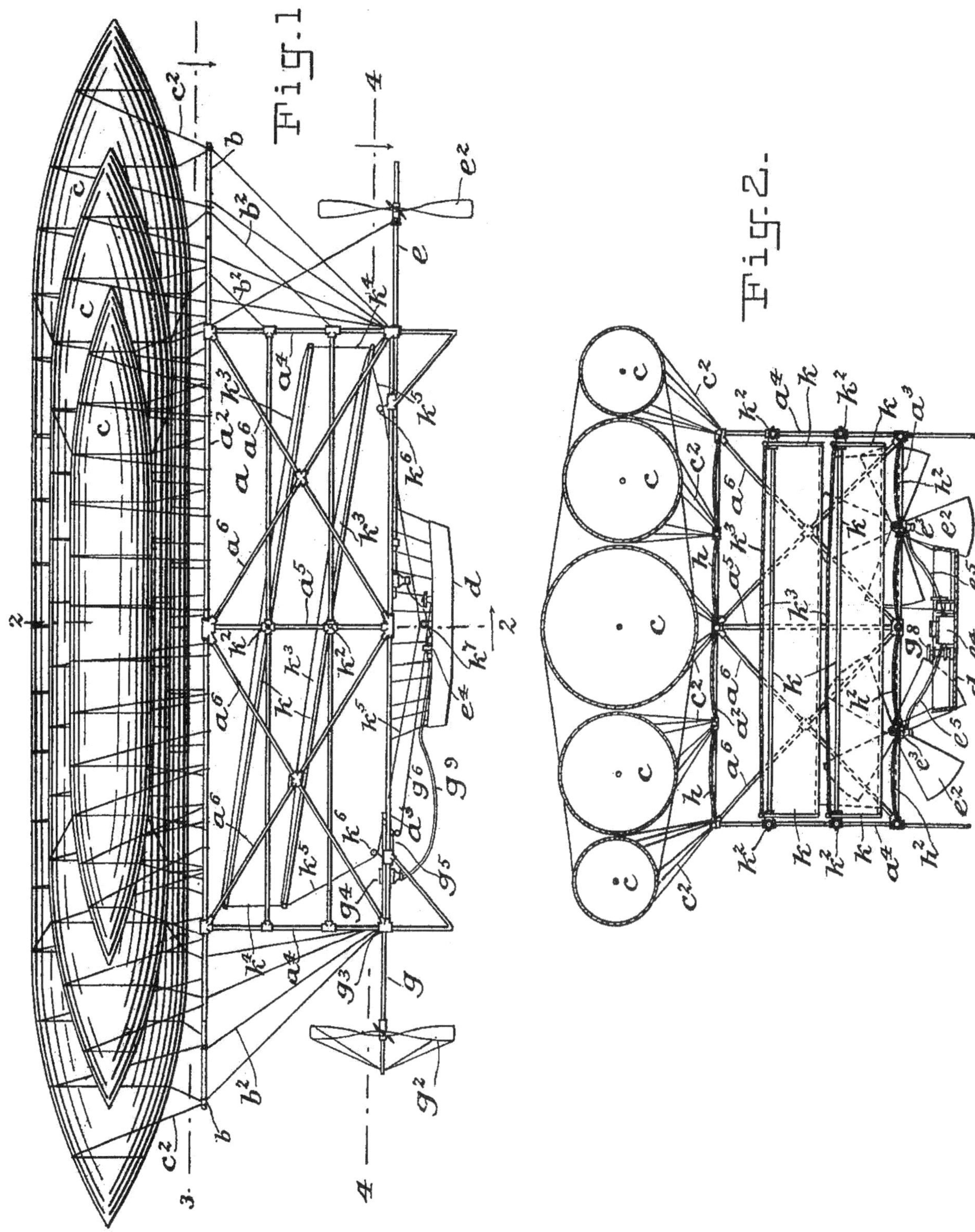

Plate 131: Cassius Richmond's Airship—March 18, 1902: *In another instance of over-engineering gone awry, New Yorker Cassius Montezuma Richmond (his real name) takes bewildering engineering to never before dreamed of heights. Why he chose five balloons when one large one would have made more sense is one thing, but the addition of sails between them and the gondola is a complete mystery. Obviously, Mr. Richmond does not believe that simplicity is highly over-rated, preferring instead to dare fate to tear his house of cards apart.*

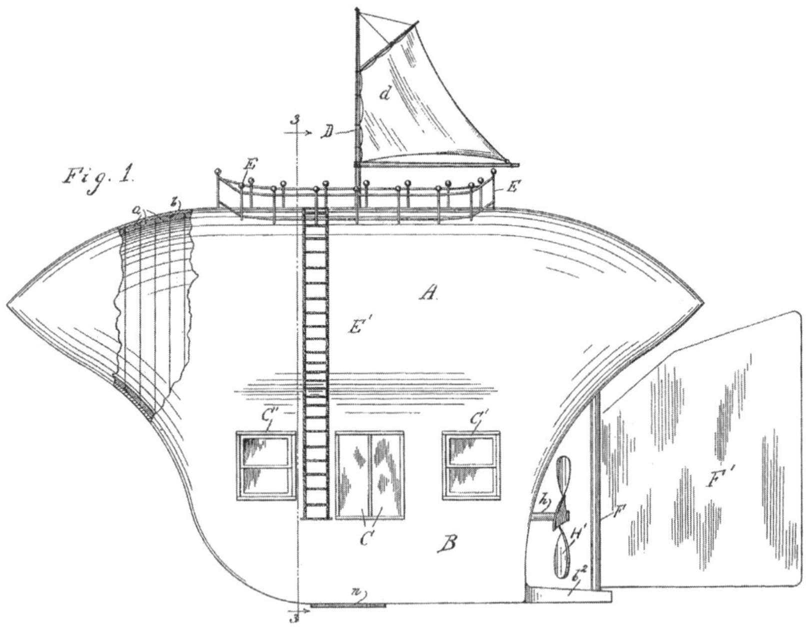

Plate 132: Peter Samorski's Airship—June 3, 1902: *One might imagine that were Mother Goose to design an airship, this might be very close to what she would come up with. Clearly Chicagoan Peter Samorski spent many an hour on the shores of Lake Michigan watching clouds roll by and imagining what life must be like in faraway kingdoms where gnomes ruled and all the trees were giant lollipops.*

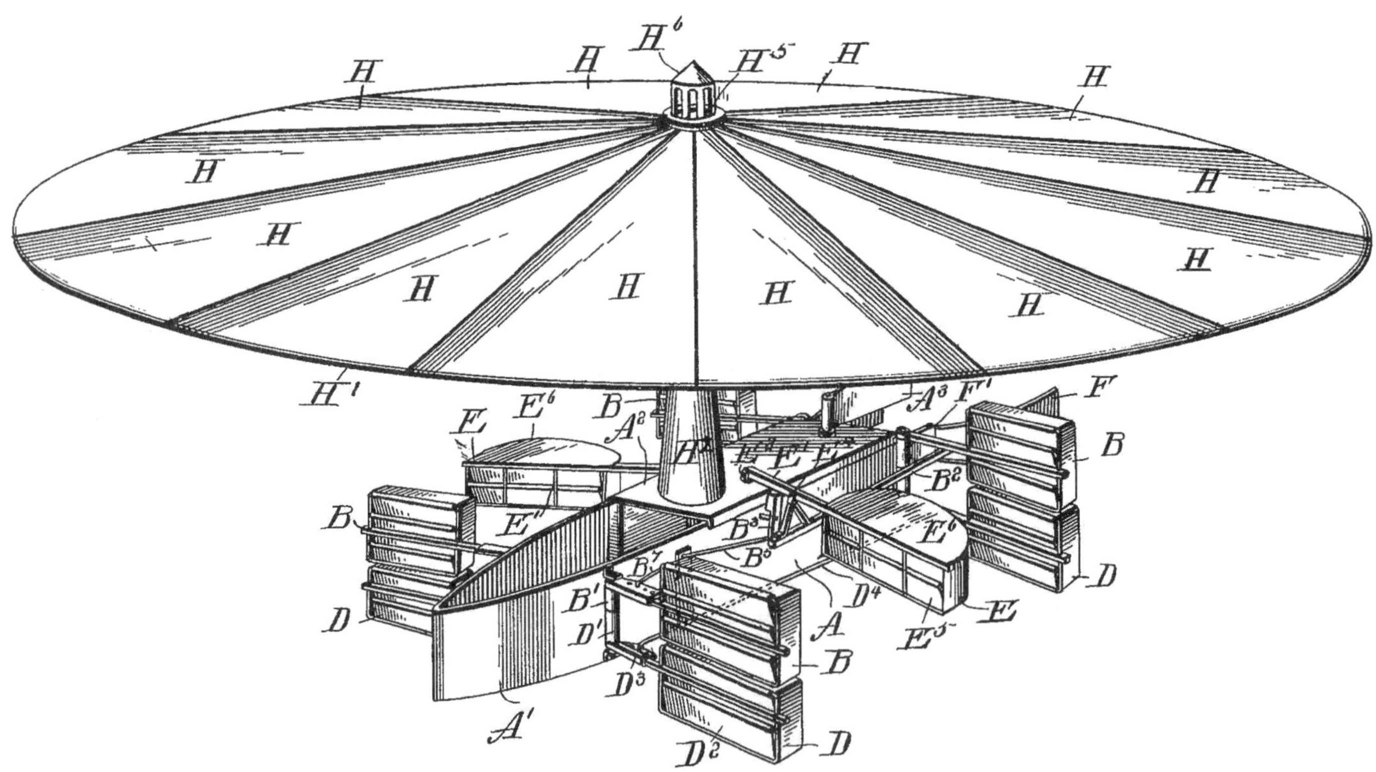

Plate 133: Harvey Van Voorhis' Airship—June 10, 1902: *In a throwback to a clunkier time in aviation history, Harvey Van Voorhis has something for just about everyone: a hot-air-filled disk-shaped balloon for lift, massive flapping wings for propulsion, and even a lighthouse at the top for safe navigation. This is all combined, Harvey tells us, so that the machine can reproduce the driving action of "a bird, a fish, and an insect, such as a bee." Since it's unlikely to produce driving action of any kind, we'll have to take his word for it.*

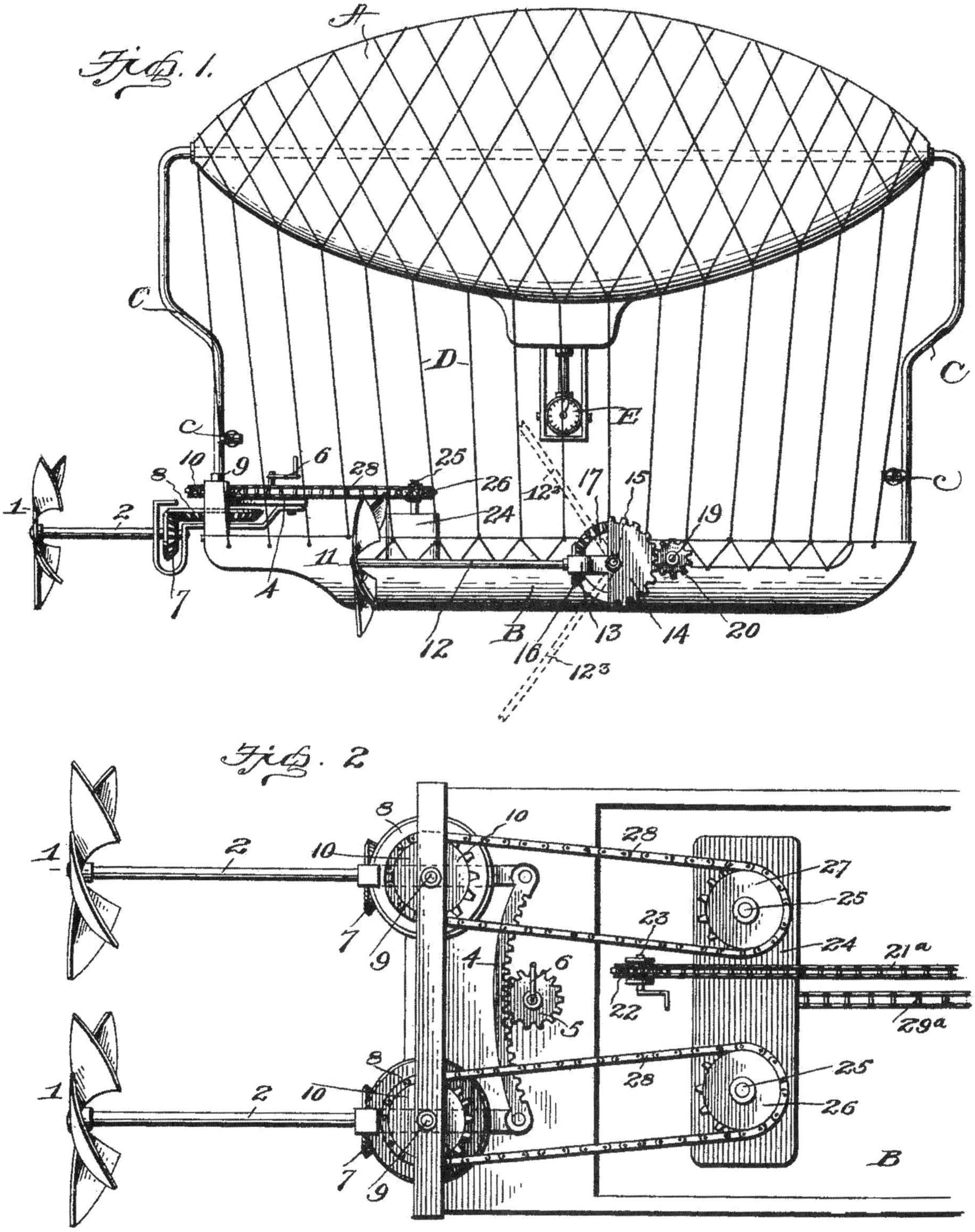

Plate 134: Joel Rice's Flying Machine—July 8, 1902: *In another very Jules Verne-esque attempt at propelling a boat through the air, Hot Springs, Arkansas, inventor Joel Rice simply exudes simplicity with this quaint little craft. It's not entirely clear from the rendering, but the craft possesses side propellers as well as rear ones that can be pointed in various directions, an eminently practical idea formulated in a thoroughly impractical way.*

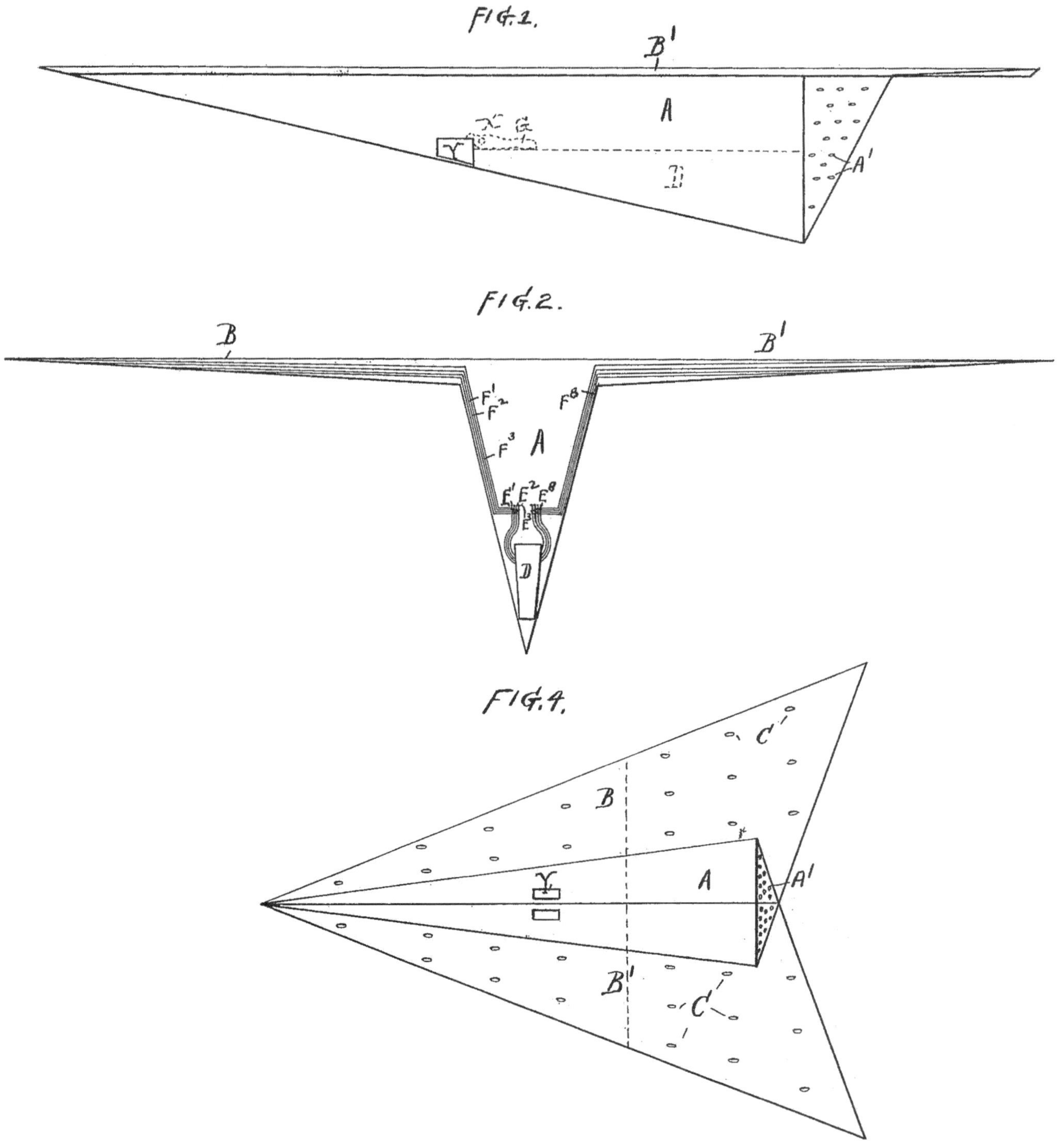

Plate 135: Theodore Gibon's Aeroplane—September 30, 1902: *Despite vicious rumors to the contrary, Mr. Gibon did not base his design upon the legendary paper airplane popularized by high school students during the twentieth century. The truth is that the bane of many a study hall monitor was based loosely on Mr. Gibon's design. A unique aspect of this "aeroplane" is that control is maintained by shifting fluids throughout the body of the craft via a menagerie of hollow aluminum tubes. Certainly a lot simpler solution than ailerons, elevators, rudders, and all that nonsense.*

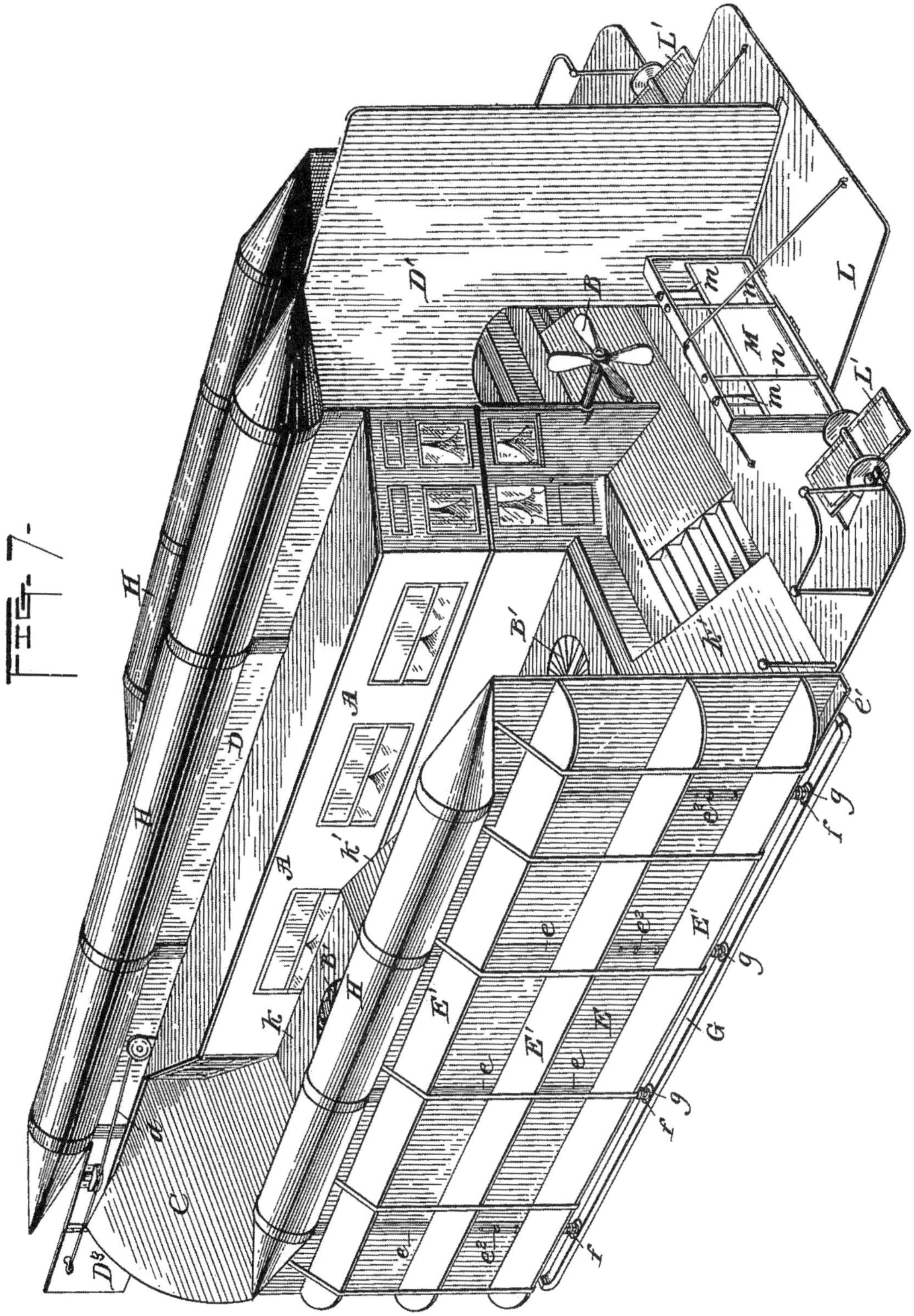

Plate 136: Edward Johnston's Airship—March 10, 1903: *Easily the most magnificent craft ever to roll off a drawing board, Colorado airship aficionado Edward Johnston has managed to turn the humdrum, ordinary airship into a stately mansion of the skies that anyone would be proud to have lumbering overhead—not that there is the slightest chance of it becoming airborne, of course. Even firmly grounded, however, it would make a fine hotel and easily the center of attraction for any small communities' downtown area.*

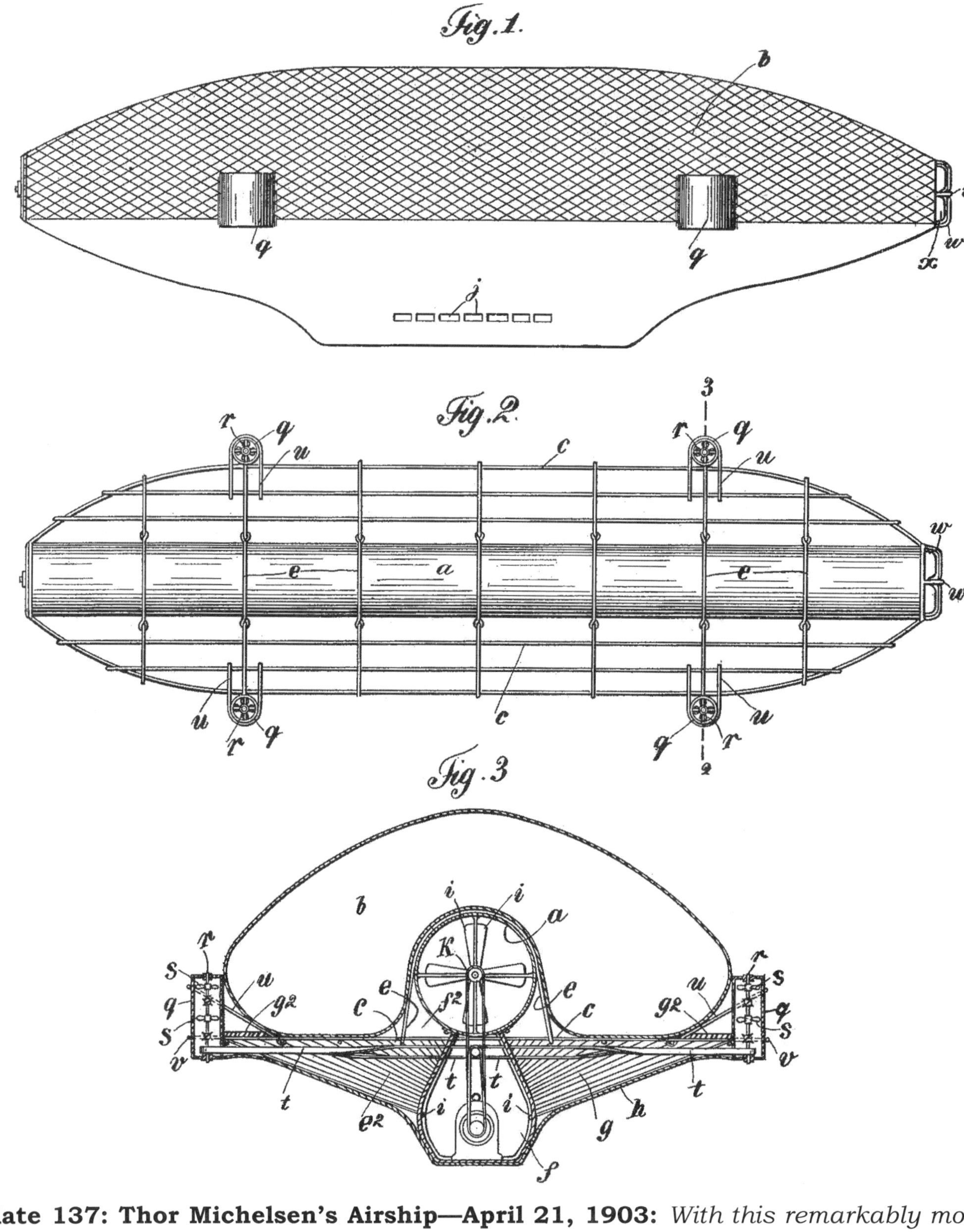

Plate 137: Thor Michelsen's Airship—April 21, 1903: *With this remarkably modern-looking design, Los Angeles inventor Thor Michelsen has fairly successfully managed to greatly simplify things by placing a series of rotating screws inside a steel tube that runs from the bow to the stern of the envelope. An interesting concept that would keep the rotating blades from hitting any nearby object, it's something that twenty-first century airship designers might want to take a look at.*

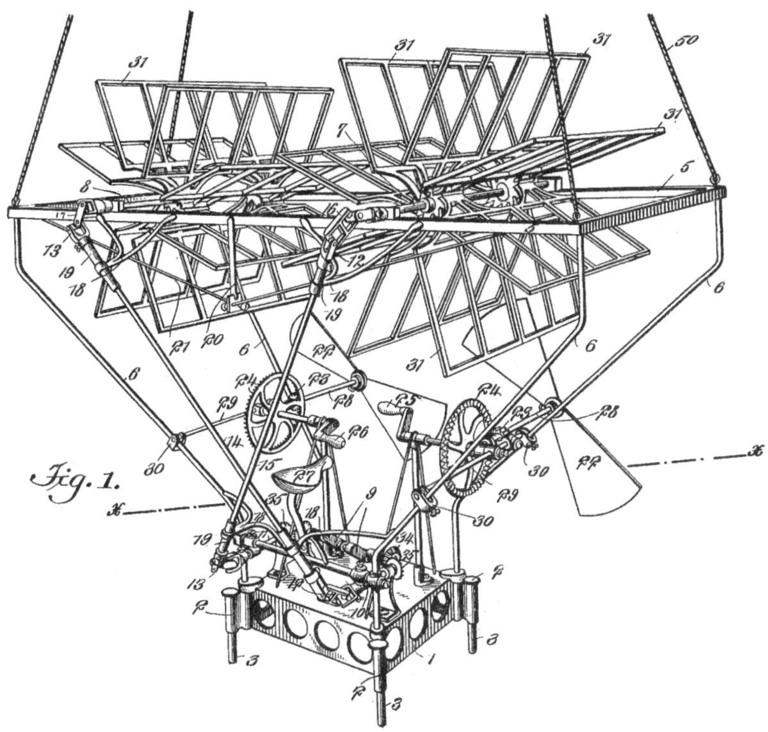

Plate 138: Otto Kaehler's Flying Machine—May 5, 1903: *Clearly challenging the common wisdom that one cannot turn farm machinery into a flying vessel, Michigan's own Otto Kaehler presents this remarkable device for us to ponder. While the chances of getting the craft airborne are clearly nil—pending a suspension of the laws of both gravity and physics—it would have been interesting to have watched Mr. Kaehler get the thing cranked up to full speed—from a safe distance, of course.*

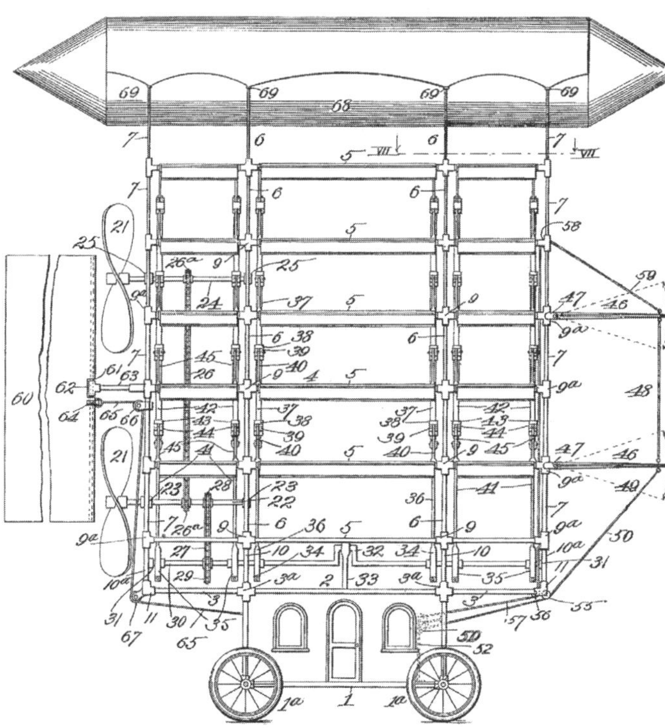

Fig. 3.

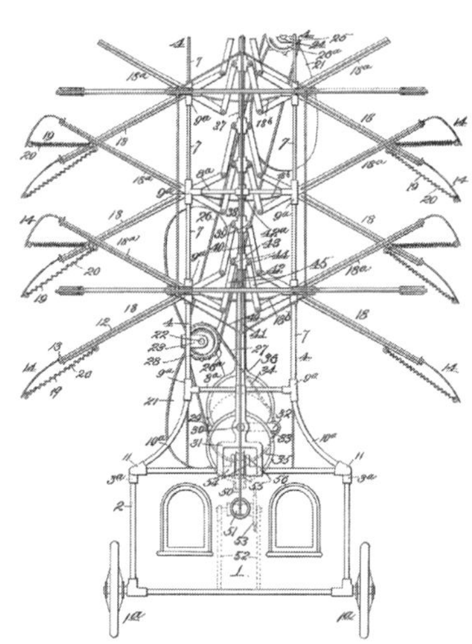

Plate 139: George Shultz's Aerial Vessel—June 2, 1903: *One would think that with the turn of the century, the basics of flight would have been better understood, making contraptions like this the product of a simpler, less scientifically savvy era. Apparently, this assumption is not shared by Kansas City inventor George Shultz, who chose to demonstrate once again how little progress mankind had truly made by introducing this utterly unflyable, multi-winged monstrosity. The arched windows and Conestoga wagon wheels on the carriage-like gondola do give it a nice Victorian touch, however.*

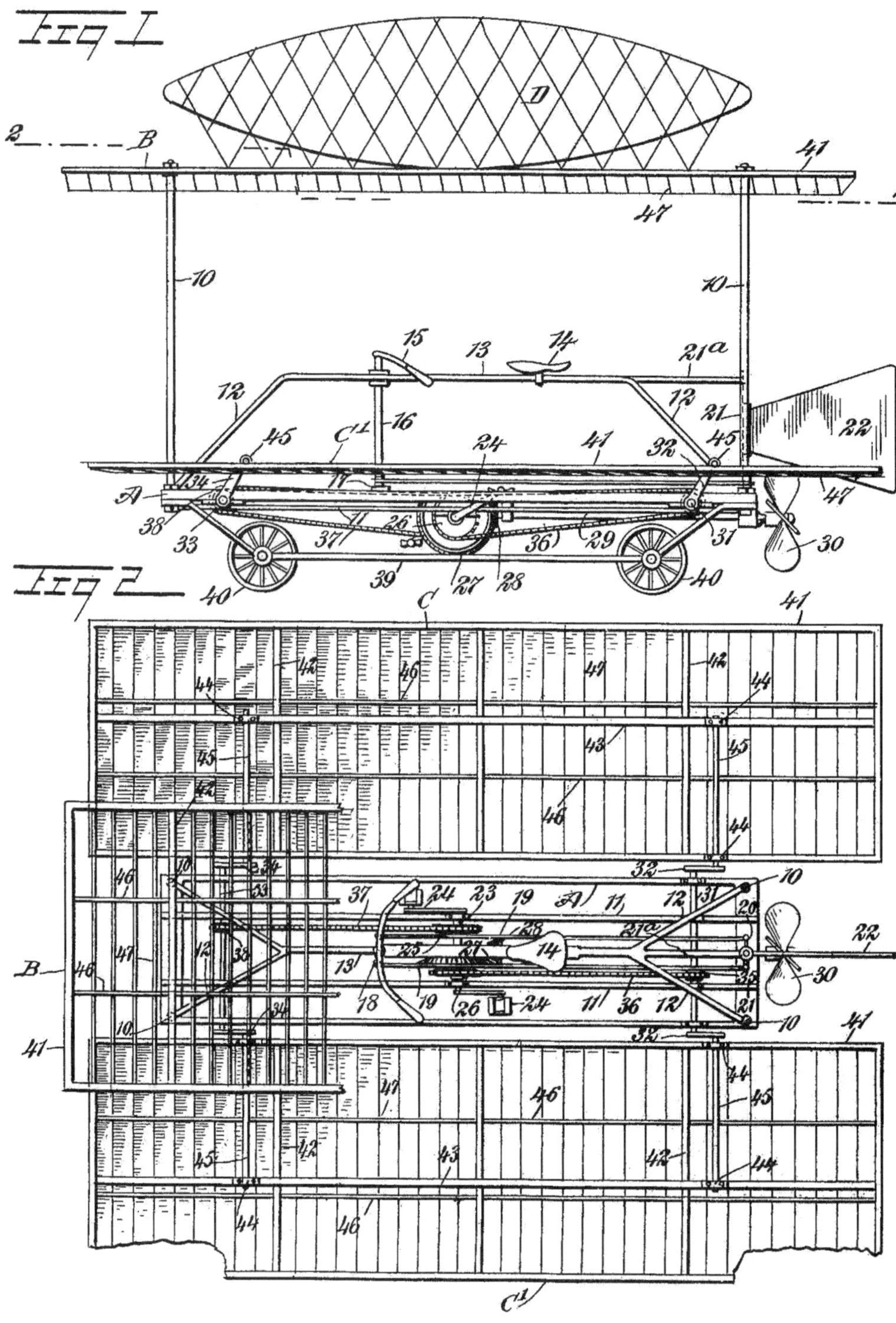

Plate 140: Isaac Morris' Machine for Aerial Navigation—September 1, 1903: *Really nothing more than a bicycle attached to a pair of flimsy wings and a too-small gas bag, Mr. Morris, had he actually built and tested the contraption, would have realized almost immediately that a human-powered machine simply cannot lift itself skyward, even with a balloon attached. Even if he increased the size of the envelope to the point where liftoff was possible, the tiny ship's screw-like propeller and flimsy rudder would have proven useless in propelling the craft forward, making it little more than a source of amusement to anyone fortunate enough to watch it attempt to lift off the ground.*

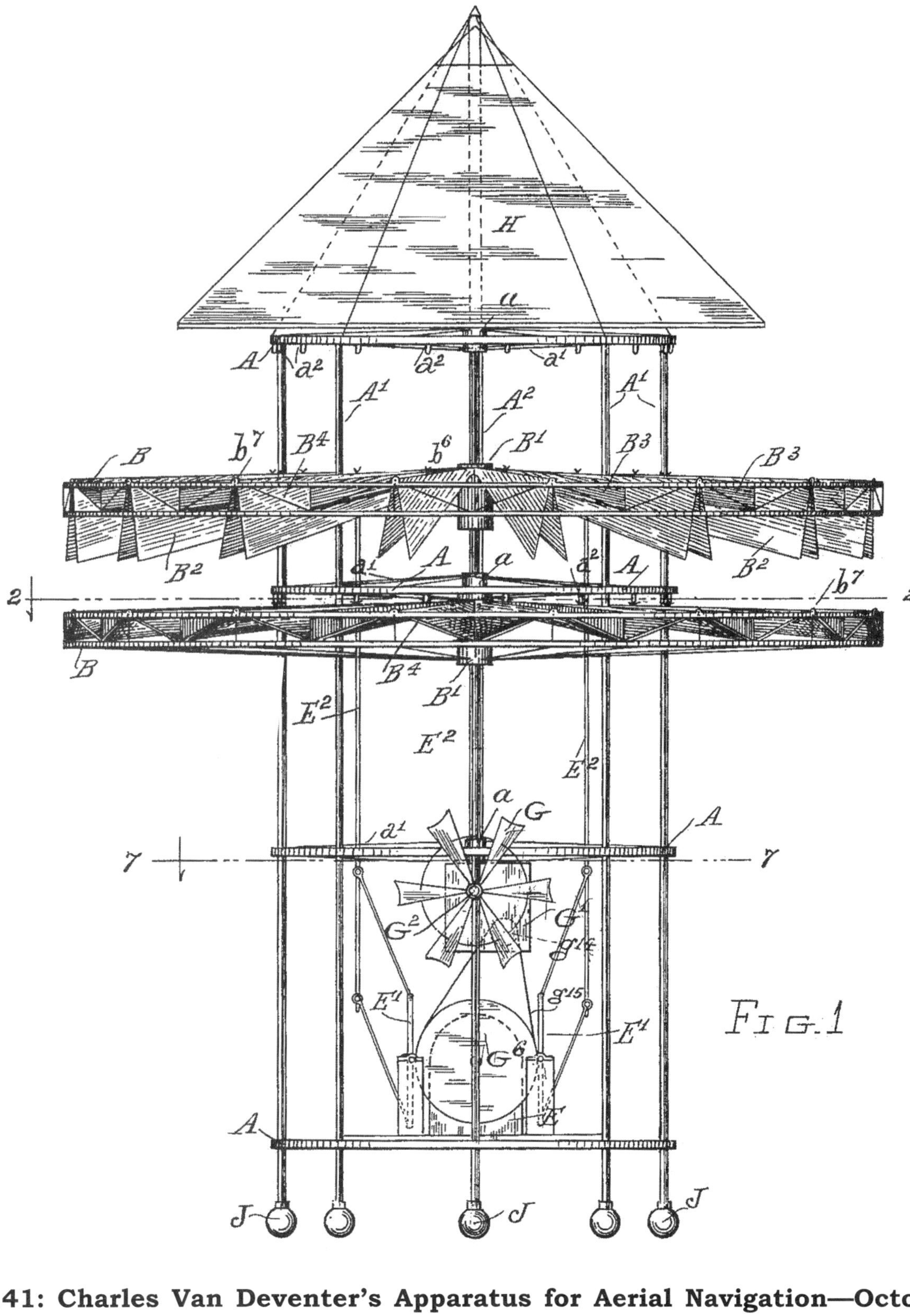

Fig. 1

Plate 141: Charles Van Deventer's Apparatus for Aerial Navigation—October 13, 1903: *Reminiscent of the world's largest candle-powered table windmill, this bizarre contraption is certain to set the science of aeronautics back at least a decade. Little more than one massive vertically mounted turbo-prop, the high center of gravity almost guarantees that the machine will topple onto its side, making it easier for the deadly rotor to make a lasting impression on any spectators unfortunate enough to get too close.*

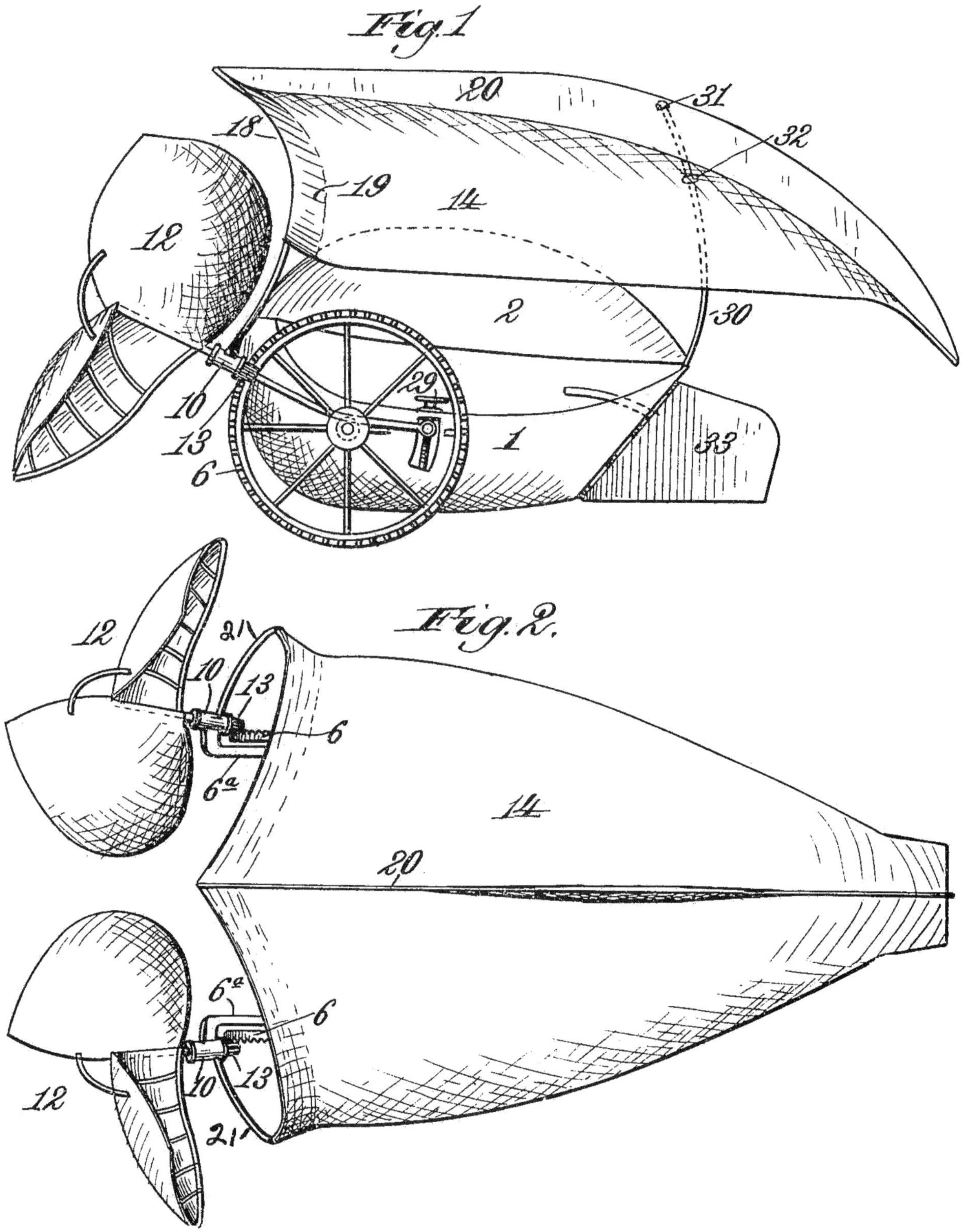

Plate 142: George Thompson's Flying Machine—September 13, 1904: *Looking just a little too much like some sort of high tech girdle, Mr. Thompson's flying machine is guaranteed to do just about anything but fly. This design wins an award for the most unusual propeller blades, however, whose functionality remains a mystery to this day. Definitely not something that might be found in the traditional American garage.*

LIST OF PATENT PLATES

Plate 1: Muzio Muzzi's Flying Machine—October 16, 1844:
Plate 2: Hugh Bell's Balloon "with appendages"—March 26, 1850
Plate 3: Mortimer Nelson's Propeller Driven Balloon—May 21, 1861
Plate 4: W.F. Quinby's Improved Apparatus for Navigating the Air—November 26, 1861
Plate 5: Luther Crowell's Aerial Machine—June 3, 1862
Plate 6: James Connell's Air Balloon—November 17, 1863
Plate 7: A.G. Wright's Improved Balloon—December 27, 1864
Plate 8: Francis Just and August Koellner's Aerial Car—October 10, 1865
Plate 9: John Wootton's Flying Machine—May 22, 1866
Plate 10: Chester Sykes's Marine Balloon—September 11, 1866
Plate 11: J.A. Elston's Aerial Machine—August 13, 1867
Plate 12: Onofrio Abbruzzo's Aerial Car—July 21, 1868
Plate 13: Zaphna Stone's Improved Aerial Navigator—May 12, 1868
Plate 14: W.F. Quinby's Flying Machine—October 5, 1869
Plate 15: Frederick Marriott's Aerial Steam Car – November 23, 1869
Plate 16: Martin Braun's Aerial Ship—February 8, 1870
Plate 17: A.P. Keith's Improved Aerial Car—March 1, 1870
Plate 18: Edward Oakes' Aerial Car—August 25, 1870
Plate 19: Isaac Forbes' Aerial Car—July 16, 1872
Plate 20: Moy and Shill's Apparatus for Aerial Navigation—November 26, 1872
Plate 21: Thomas Green's Apparatus for Navigating the Air—August 12, 1873
Plate 22: Otto Francis' Apparatus for Navigating the Air—February 10, 1874
Plate 23: Daniel Rhone's Aerial Boat—June 1, 1874
Plate 24: Charles McGlashan's "Guiding" Balloon—April 15, 1874
Plate 25: Micajah Dyer's Apparatus for Navigating the Air—September 1, 1874
Plate 26: Peter Fernandez's Aerial and Marine Gyrator—April 6, 1875
Plate 27: Francois Lamboley's Flying Machine—July 19, 1876
Plate 28: Frederick Schroeder's Steerable Airship—July 1, 1875
Plate 29: Melville Murrell's Aerial Navigator—April 16, 1877
Plate 30: Fritz von Ehren's Aerial Vessel—January 15, 1878
Plate 31: Charles Ritchel's Airship—March 12, 1878
Plate 32: William Pearson's Improvement in Balloons—April 23, 1878
Plate 33: John Cameron's Airship—August 28, 1878
Plate 34: W.F. Quinby's Improvement in Aerial Ships—January 16, 1879
Plate 35: Count Antoine Apraxine's Aerial Balloon—March 25, 1879
Plate 36: Henry Bagley's Aerial Machine—April 22, 1879
Plate 37: John Greenough's "Aerobat"—October 14, 1879
Plate 38: Albert Blackman's Vessel for Aerial Navigation—January 6, 1880
Plate 39: Frederick Brearey's Aerial Apparatus—January 23, 1880
Plate 40: Charles Sullivan's Flying Machine—November 30, 1880
Plate 41: John Mackenzie's Aerial Machine—December 21, 1880
Plate 42: Edward Pearse's Machine for Aerial Navigation—January 11, 1881
Plate 43: Edward de Jongh's Aeronautical Apparatus—September 13, 1881

Plate 87: Charles Bechtel's Airship—June 3, 1890
Plate 88: Stewart Cairncross' Aerial Machine—July 22, 1890
Plate 89: Perham Nahl's Navigable Airship—October 28, 1890
Plate 90: Burr Barnes' Flying Machine—September 29, 1891
Plate 91: John Arbtin's Airship—November 3, 1891
Plate 92: James Walker's Airship—January 12, 1892
Plate 93: Burr Barnes' Airship—July 12, 1892
Plate 94: Georg Gabrielii's Balloon Ship—June 27, 1893
Plate 95: Theodore Steinmann's Airship—July 11, 1893
Plate 96: Sumter Battey's Aerial Machine—July 25, 1893
Plate 97: Emory Sowers' Flying Machine—September 5, 1893
Plate 98: Axel Bergqvist's Aerial Vessel—October 17, 1893
Plate 99: Carl Billwiller's Navigable Airship—October 31, 1893
Plate 100: Edwin Pynchon's Airship—November 14, 1893
Plate 101: Joachim Sumovski's Balloon—December 12, 1893
Plate 102: Jerome Blanchard's Aerial Vessel—February 6, 1894
Plate 103: Duane Hurlbut's Airship—September 25, 1894
Plate 104: John Lomax's Cable Airship—June 16, 1895
Plate 105: William Hutchinson's Navigable Balloon—October 15, 1895
Plate 106: Edward Pennington's Airship—December 31, 1895
Plate 107: Charles Smith's Airship—August 11, 1896
Plate 108: Edward St. Croix's Airship—December 22, 1896
Plate 109: Henry Heintz's Airship—April 20, 1897
Plate 110: Thomas Crepar's Airship—August 27, 1897
Plate 111: Daniel Reed's Airship—October 12, 1897
Plate 112: Jacob Graybill's Aerial Machine—October 26, 1897
Plate 113: Theodore Leibbrand's Airship—March 22, 1898
Plate 114: Adolph Koehler's Airship—June 7, 1898
Plate 115: Friedrich Jone's Flying Machine—June 14, 1898
Plate 116: Joel Rice's Airship—July 5, 1898
Plate 117: Charles Hite's Airship—July 12, 1898
Plate 118: Thomas Brown's Aerial Ship—September 13, 1898
Plate 119: Oscar Smith's Airship—October 18, 1898
Plate 120: Frederick Merritt's Airship—December 6, 1898
Plate 121: John Pickering's Airship—February 20, 1900
Plate 122: Louis Roze's Airship—May 1, 1900
Plate 123: Charles Stanley's Airship—October 9, 1900
Plate 124: Leo Stevens' Controllable Balloon—November 13, 1900
Plate 125: Ole Olsen's Airship—March 26, 1901
Plate 126: Joseph Dillon-Gregg's Machine for Aerial Navigation—January 22, 1901
Plate 127: Aristarchus Hubbard's Airship—February 19, 1901
Plate 128: Eugene Farr's Aerial Vessel—July 9, 1901
Plate 129: John Spies's Airship—December 3, 1901
Plate 130: Robert Botts's Airship—December 10, 1901
Plate 131: Cassius Richmond's Airship—March 18, 1902

BIBLIOGRAPHY

BOOKS

Childress, David Hatcher, *Vimana Aircraft of Ancient India & Atlantis*, Adventures Unlimited Press, 1991, ISBN 0-932813-12-7

Crouch, Tom D., *A Dream of Wings: Americans and the Airplane 1875-1905*, W. W. Norton & Company, Inc., 2002, ISBN 0-393-32227-0 pbk.

Danelek, J. Allan, *The Great Airship of 1897: A Provocative Look at the Most Mysterious Aviation Event in History*, Adventures Unlimited Press, 2009, ISBN 1-935487-03-5

DeVries, Col. John A., *Taube: Dove of War, Historical Aviation Album*, 1978, ISBN 0-911852-82-4

Hallion, Richard P., *Taking Flight: Inventing the Aerial Age from Antiquity through the First World War*, Oxford University Press, 2003, ISBN 0-19-516035-5

Kelly, Maurice, *Steam in the Air: The Application of Steam Power in Aviation During the 19th & 20th Centuries*, Pen & Sword Aviation (an imprint of Pen & Sword Books, LTD), 2006, ISBN 1-84415-295-2

Scott, Phil, *The Wrong Stuff*, Hylas Publishing, 2003, ISBN 1-59258-044-0

Spenser, Jay, *The Airplane: How Ideas Gave Us Wings*, HarperCollins Publishers, 2008, ISBN 978-0-06-125919-7

Internet Website Sources
(General Aviation History Sites)

K. O. Eckland's Aerofiles: A Century of American Aviation (www.aerofiles.com); K. O. Eckland

Aerospaceweb.org (www.aerospaceweb.org); Greg Alexander, Aaron Brown, Doug Jackson, Jeff Scott, Molly Swanson, Anton VanderWyst, Matt Walker, Justine Whitman, Joe Yoon and Jayme O'Sullivan

Centennial of Flight (www.centennialofflight.gov)

Century of Flight (www.century-of-flight.freeola.com/Aviation history/)

The Dream of Flight: A Library of Congress Special Presentation Commemorating the Centennial of Flight (www.loc.gov/exhibits/treasures/wb-timeline.html)

The Early Birds of Aviation, Inc. (earlyaviators.com)

The Flying Machines Web Site (www.flyingmachines.org)

Lawrence Hargrave: Australian Aviation Pioneer (www.ctie.monash.edu.au); Professor John Bird & Dr. Russell Naughton (While this site is centered around Lawrence Hargrave's achievments, it contains excellent information on just about every aviation pioneer. Highly recommended.)

Les Modeles Reduits de Machines Volantes au 19eme Siecle (alain.vassel.pagesperso-orange.fr/siecle19.htm)

Old Engine House (Steam Powered Flight page) (oldenginehouse.users.btopenworld.com/flight.htm)

www.twainquotes.com (Special Feature – Frederick Marriott and the AVITOR) (www.twainquotes.com/marriott.html) - A site dedicated to the memorable quotes of Mark Twain (Samuel Clemens) with a page devoted to Frederick Marriott.

Wikipedia (en.wikipedia.org/wiki/); Jimmy Wales and Larry Sanger
(Where would we be without Wikipedia!)

INTERNET WEBSITE SOURCES
(PIONEER-SPECIFIC AVIATION HISTORY SITES – IN ORDER OF APPEARANCE IN THE BOOK)

Saqqara Glider:
http://www.nasca.org.uk/Ancient__Aviation_/ancient__aviation_.html
http://www.ancientmoons.net/ancientaviation.htm
http://www.catchpenny.org/index.html
http://www.catchpenny.org/birdplan.html
http://www.catchpenny.org/birdtest.html
http://www.catchpenny.org/model.html
http://www.catchpenny.org/mercado.html

Pre-Colombian Delta Wing Trinkets:
http://www.nasca.org.uk/Ancient__Aviation_/ancient__aviation_.html

Francesco Lana de Terzi:
www.faculty.fairfield.edu/jmac/sj/scientists/lana.htm

Sir George Cayley:
aviation-history.com/early/mc1852-1.htm
http://aviation-history.com/early/cayley.htm

William Henson:
www.456fis.org/THE_HISTORY_OF_FLIGHT_-_HENSON_AND_STRINGFELLOW.htm
www.historynet.com/william-henson-and-john-stringfellow-pioneer-aviation-strategists.htm/1
www.sciencemuseum.org.uk/images/i004/10216035.aspx

Jean-Marie Le Bris:
www.labarqueailee.org/english/index.php

Octave Chanute:
invention.psychology.msstate.edu/i/Chanute/library/Prog_Illustrations.html

Samuel Langley:
invention.psychology.msstate.edu/library/Magazines/Langley_failure1.html
http://www.aerospaceweb.org/question/history/q0004.shtml

David Schwartz:
croatian-treasure.com/schwarz.html

Gustave Whitehead:
www.weisskopf.de/pageID_9967512.html
gustavewhitehead.org/
www.gatm.com/whitehead/index.html
http://airsports.fai.org/jun98/jun9805.html
http://www.wright-brothers.org/TBR/History/History%20of%20Airplane/Whitehead.htm
http://tchf.wikia.com/wiki/Wright_Brothers_not_the_First_to_Fly
en.Wikipedia.org/wiki/Gustav_Whitehead#Reproductions
www.go-star.com/antiquing/cliffrobertson.htm

Wright Brothers:
tchf.wikia.com/wiki/Wright_Brothers_not_the_First_to_Fly
www.wright-brothers.org

Richard Pearse:
chrisbrady.itgo.com/pearse/pearse.htm
chrisbrady.itgo.com/pearse/shuttleworth.htm
chrisbrady.itgo.com/pearse/flying_replica.htm

Karl Jatho:
The Karl Jatho Project (http://www.karl-jatho.com/)

Preston Watson:
chrisbrady.itgo.com/watson/watson.htm

Burrell Cannon:
www.texasescapes.com/AllThingsHistorical/EzekielAirshipBB1103.htm
www.roadsideamerica.com/story/11610
www.cleburnetimesreview.com/lifestyles/x489000534/John-Watson-Texas-man-invents-flying-machine
www.pittsburgtxmuseum.com/airship.html

William Frost:
http://www.spectator.co.uk/books/142296/welsh-wizard-prang.thtml
http://activity-holidays.visitwales.co.uk/15/yanks-steal-man-fly-welsh/

INTERNET PATENT SOURCES

Google Patents (www.google.com/ptshp?ie=UTF-8&oe=UTF-8&hl=en&tab=wt&q=)
(Best Darned Patent Search Site Around!)

Free Patents Online (www.freepatentsonline.com/)
(You need to start a free account to download PDFs of patents.)

VIDEO RESOURCES

On The Wings of Ezekiel: The Story of the Ezekiel Airship

Credits:	Written & Directed by Jim Long Produced by Carol Long Narrated by Allen Farmer
Publisher:	Lightcatcher Productions
Publishing Date/ Release Date:	2007(?)
Available From:	Lightcatcher Productions (www.lightcatcherprod.com) Amazon.com

SPECIAL NOTE

The authors are always interested in hearing from readers regarding the information presented in this volume. If you have any new information, questions, or comments regarding the content of this book, please contact the authors at: phantomaviators@gmail.com.

Also, keep watching for our Phantom Aviators website where we will be publishing information that did not make it into this book along with and some interesting and entertaining speculations and related fiction.

INDEX

OTHER BOOKS BY J. ALLAN DANELEK FROM ADVENTURES UNLIMITED PRESS

THE GREAT AIRSHIP of 1897

In November of 1896, residents of California watched a mysterious bright light—often described as being suspended beneath a "cigar shaped" craft of considerable size—pass slowly over their cities on several occasions, sparking a media frenzy. A few months later, what appeared to be the same craft was seen in the skies over the sparsely populated prairie states of the Midwest making its way methodically eastward and appearing to literally hundreds—if not thousands—of witnesses. Then, as suddenly as the reports began, they abruptly ended, leaving a mystery that has never been satisfactorily explained by either science or historians to this day. What were these strange lights? Where they evidence of visitor from outer space, making an early appearance or was it all merely a media hoax generated by the yellow journalism of the time in an effort to increase sales? Or, most provocative of all, was it some sort of airship built by parties unknown appearing a full decade before von Zeppelin began building the first of his behemoths in Germany?

This book works from the premise that the mysterious vessel may have been a terrestrial craft built and operated by parties unknown from the privacy of the sparsely populated forestland of central California. While admittedly a controversial idea, the book makes its case in a carefully considered, step-by-step way that demonstrates to the reader not only how it would have been possible to produce such a machine with the materials available in 1896, but why it was done in secret and what possibly happened to the remarkable ship. While such a hypothesis is incapable of being either proven or disproven, the book is guaranteed to leave you wondering if the history of aviation is complete as we know it or if there aren't a few missing chapters still waiting to be written.

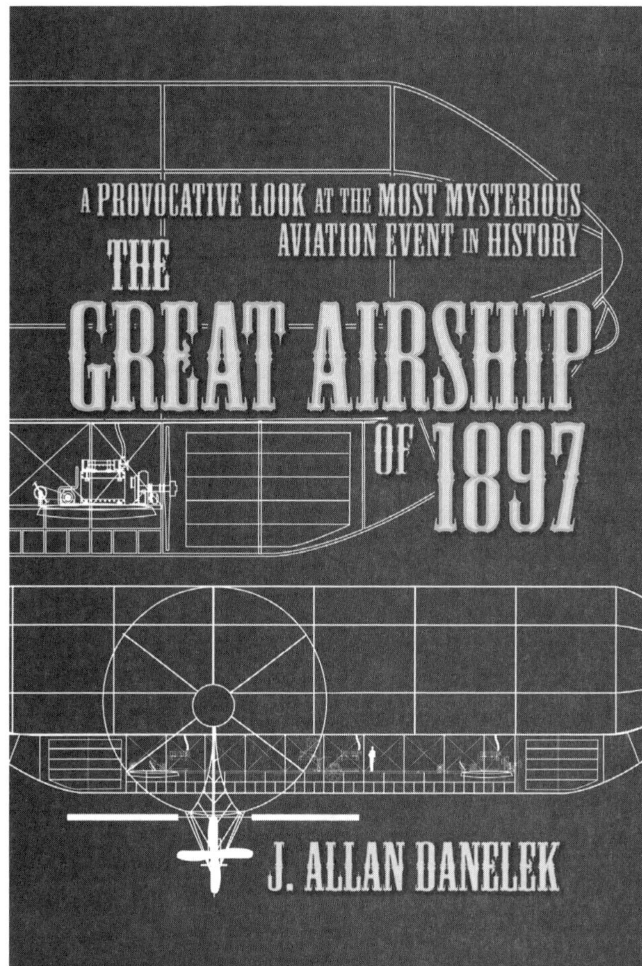

Available now at
www.ourcuriousworld.com
or order directly at
www.adventuresunlimitedpress.com.

OTHER BOOKS BY J. ALLAN DANELEK

Available now at www.llewellyn.com

For those who are open to the idea of reincarnation, the question has often been not whether it's true, but how does it work and what is the real purpose behind it? This book may provide at least some of the answers to this age-old mystery as it attempts to explain the mechanics of how reincarnation brings us closer to a sense of the divine and assists us in our efforts to evolve towards ever greater levels of spiritual consciousness. In it, we will examine the link between the spiritual and physical worlds, how karma works, the role that free will plays in the process, how soul mates assist us in our journey, and even such seldom asked questions such as how is the next physical life chosen, the influence past lives have on our present circumstances, the role of evil in our unfolding spiritual evolution, and what happens between lives. A must read for those who want more than just stories about reincarnation, but a "nuts and bolts" manual that gives real answers to real questions while remaining content to let the reader decide for themselves whether reincarnation remains a valid post-mortem belief system for them.

Available now at www.llewellyn.com

There are literally hundreds of books on the market that talk about 2012 and doomsday in general, but few that challenge the basic premise behind the idea—that the world is headed for destruction—like this one does. *2012: Extinction or Utopia* gives the reader an opportunity to explore not only doomsday beliefs in general but explore why so many millions of people are willing to so readily embrace even the most outlandish claims made by men and woman of often questionable character and motives. In it, we will probe not only the long and dismal history of failed prophecies of the past, but examine the proclamations made by such popular prophets as Nostradamus and Edgar Cayce. It also explores Bible-based end times beliefs in some detail from a theological and practical standpoint, as well as examines how harmful such beliefs can be and how much havoc they continue to produce in many people's lives today. This is the perfect book for a friend or relative who worries about the future and needs a little reassurance that doomsday is not just around the corner despite all the media hype and the prognostications of the many self-styled "prophets of doom" out there.

OTHER BOOKS BY J. ALLAN DANELEK

More than just another litany of UFO stories and page after page of grainy photographs showing "something" flying through the sky, *UFOs: The Great Debate* goes beyond the sensationalism, paranoia, and general flotsam the issue has produced to take a hard look at a phenomenon that has been a major part of our culture for the past sixty years. In it, the reader will find a fresh approach that examines the subject from the perspective of reason rather than emotion, all designed to give the reader an opportunity to decide for themselves whether we are being visited by extraterrestrials or we are the victims of our own imagination. More than that, however, it also examines possible extraterrestrial rationales for coming to our planet, what techniques they might use in studying us, and whether they may not be above doing a little psychological testing on human beings through the mediums of crop circles, cattle mutilations, and other UFO-attributed phenomena, all in an effort to find out just what "makes us tick." A long overdue and very different take on UFOs that will force you to reexamine the issue in ways you've ever considered before.

Available now at www.llewellyn.com

After carefully examining Plato's writings in the light of history and applying what science has learned about human development and recent earth climate and geography changes, *Atlantis: Lessons from a Lost Civilization* makes a compelling case that everything we think we know about Plato's mysterious and exotic lost continent may be wrong. In this book, we will consider the possibility that the great Greek philosopher was referring to more than merely a place, but to something much more grandiose than we have been led to believe. Instead of some mysterious island continent in the Atlantic Ocean, it suggest that Plato was actually referring to a magnificent global civilization not unlike our own today that flourished over 12,000 years ago before destroying itself so thoroughly that no physical trace of it remains today. In it, we will explore not only how this would be possible, but whether our own society is moving down the same path to destruction—or, perhaps, whether we will get it "right" this time and so avoid the catastrophe that claimed the last great global civilization. A thought-provoking book that does much to explain the many flood legends that seem to be a part of so many cultural traditions around the world.

Available now at www.llewellyn.com

OTHER BOOKS BY J. ALLAN DANELEK

Available now at www.llewellyn.com

More than just another collection of anecdotal stories or a basic "how to" ghost hunting book, *The Case for Ghosts* examines the broader issues dealing with the paranormal that few researchers deal with today. For example, is there evidence that the personality survives death, and if it does, how might a person become—or avoid becoming—a ghost? For that matter, what precisely is a ghost and how are they different from other spiritual energies and supernatural entities that exist in the spiritual realm? And if ghosts do exist, how might they materialize and communicate with us? Think of this book as a paranormal "primer" designed to bring the reader up to speed on a broad range of issues surrounding the subject, which they can then use in their own search for answers to what lies beyond. A must read for both the serious ghost hunter and the merely curious who may have had a paranormal experience and want to learn more.

Available now at www.llewellyn.com

The Mystery of Reincarnation is an effort to fuse the practical and theoretical sides of the concept of reincarnation together in a way that is both educational and theoretically plausible, all told in an easy-to-read, non-academic manner designed specifically with the layman in mind. It is written essentially in two parts; the first half deals with the history of reincarnation in the West, the evidences for and against the concept, a single test case, and some personal anecdotes. The second half works on a more theoretical level, articulating the process of rebirth itself and what the soul is attempting to achieve through each incarnation. The book should give anyone who is truly desirous of learning more about the subject a better appreciation for what reincarnation is, how it works, and what it all means without insisting that they embrace any particular belief system. A good starting point for those curious about the subject but who want to understand it better before making up their own mind as to whether reincarnation has anything to say to them personally.

About the Authors

Researcher/author J. (Jeffrey) Allan Danelek has been writing on the paranormal and Fortean subjects such as Bigfoot, UFOs, and things that go bump in the night since 2000, although his interests also extend to military history and, not remarkably, aviation. A frequent guest on many radio shows dealing with the paranormal, Jeff is regularly featured at seminars and conferences around the country, speaking on everything from ghosts and reincarnation to prophecy and UFOs. He is also a novelist and instructor at Colorado Free University and Front Range Community College. He currently resides with his wife, Carol, and two sons in Lakewood, Colorado.

Chuck Davis has been involved in various aspects of aerospace history for the past 40 plus years. As a draftsman, technical illustrator, and technical writer with over 30 years experience, he has assisted in the publication of over 16 aerospace history books, articles, and magazines. He is also interested in an eclectic array of historical and paranormal subjects, alternate history scenarios, and science fiction topics. His particular areas of specialization are unconventional aviation technologies, space commercialization, and wildlife conservation issues. Chuck currently resides in Westminster, Colorado, with his wife, Claudia. They have a son, Mark, who lives in Denver.

To Contact the Authors

If you wish to contact us directly, we encourage you to go to www.ourcuriousworld.com and select the e-mail link or e-mail directly at jeff@ourcuriousworld.com. We always enjoy dialoguing on a number of issues and promise to respond in a timely manner to all serious queries and comments.

The authors can be mailed as well. All letters and packages should be sent to the publisher, who will forward them to the authors (please specify which author) in a timely manner. If you wish for a response, please attach a SASE (or include an e-mail address for an even quicker reply) and send to:

Jeff Danelek or Chuck Davis
c/o Adventures Unlimited Press
P.O. Box 74
Kempton, Illinois 60946-0074

GUARDIANS OF THE HOLY GRAIL
by Mark Amaru Pinkham

This book presents this extremely ancient Holy Grail lineage from Asia and how the Knights Templar were initiated into it. It also reveals how the ancient Asian wisdom regarding the Holy Grail became the foundation for the Holy Grail legends of the west while also serving as the bedrock of the European Secret Societies, which included the Freemasons, Rosicrucians, and the Illuminati. Also: The Fisher Kings; The Middle Eastern mystery schools, such as the Assassins and Yezidhi; The ancient Holy Grail lineage from Sri Lanka and the Templar Knights' initiation into it; The head of John the Baptist and its importance to the Templars; The secret Templar initiation with grotesque Baphomet, the infamous Head of Wisdom; more.

248 PAGES. 6x9 PAPERBACK. ILLUSTRATED. BIBLIOGRAPHY. $16.95. CODE: GOHG

SECRETS OF THE HOLY LANCE
The Spear of Destiny in History & Legend
by Jerry E. Smith

As Jesus Christ hung on the cross a Roman centurion pieced the Savior's side with his spear. A legend has arisen that "whosoever possesses this Holy Lance and understands the powers it serves, holds in his hand the destiny of the world for good or evil." *Secrets of the Holy Lance* traces the Spear from its possession by Constantine, Rome's first Christian Caesar, to Charlemagne's claim that with it he ruled the Holy Roman Empire by Divine Right, and on through two thousand years of kings and emperors, until it came within Hitler's grasp—and beyond! Did it rest for a while in Antarctic ice? Is it now hidden in Europe, awaiting the next person to claim its awesome power? Neither debunking nor worshiping, *Secrets of the Holy Lance* seeks to pierce the veil of myth and mystery around the Spear. Mere belief that it was infused with magic by virtue of its shedding the Savior's blood has made men kings. But what if it's more? What are "the powers it serves"?

312 PAGES. 6x9 PAPERBACK. ILLUSTRATED. BIBLIOGRAPHY. $16.95. CODE: SOHL

SUNS OF GOD
Krishna, Buddha and Christ Unveiled
by Acharya S

From the author of the controversial and best-selling book *The Christ Conspiracy: The Greatest Story Ever Sold* comes this electrifying journey into the origins and meaning of the world's religions and popular gods. Over the past several centuries, the Big Three spiritual leaders have been the Lords Christ, Krishna and Buddha, whose stories and teachings are so remarkably similar as to confound and amaze those who encounter them. As classically educated archaeologist, historian, mythologist and linguist Acharya S thoroughly reveals, these striking parallels exist not because these godmen were "historical" personages who "walked the earth" but because they are personifications of the central focus of the famous and scandalous "mysteries." These mysteries date back thousands of years and are found globally, reflecting an ancient tradition steeped in awe and intrigue. In unveiling the reasons for this highly significant development, the author presents an in-depth analysis that includes fascinating and original research based on evidence both modern and ancient—captivating information kept secret and hidden for ages.

428 PAGES. 6x9 PAPERBACK. ILLUSTRATED. BIBLIOGRAPHY. INDEX. $18.95. CODE: SUNG

THE ORION PROPHECY
Egyptian and Mayan Prophecies on the Cataclysm of 2012
by Patrick Geryl and Gino Ratinckx

In the year 2012 the Earth awaits a super catastrophe: its magnetic field will reverse in one go. Phenomenal earthquakes and tidal waves will completely destroy our civilization. These dire predictions stem from the Mayans and Egyptians—descendants of the legendary Atlantis. The Atlanteans had highly evolved astronomical knowledge and were able to exactly predict the previous world-wide flood in 9792 BC. They built tens of thousands of boats and escaped to South America and Egypt. In the year 2012 Venus, Orion and several others stars will take the same 'code-positions' as in 9792 BC! For thousands of years historical sources have told of a forgotten time capsule of ancient wisdom located in a labyrinth of secret chambers filled with artifacts and documents from the previous flood.

324 PAGES. 6x9 PAPERBACK. ILLUSTRATED. BIBLIOGRAPHY. $16.95. CODE: ORP

INVISIBLE RESIDENTS
The Reality of Underwater UFOS
by Ivan T. Sanderson

Sanderson puts forward the curious theory that "OINTS"—Other Intelligences—live under the Earth's oceans. This underwater, parallel, civilization may be twice as old as Homo sapiens, he proposes, and may have "developed what we call space flight." Sanderson postulates that the OINTS are behind many UFO sightings as well as the mysterious disappearances of aircraft and ships in the Bermuda Triangle. What better place to have an impenetrable base than deep within the oceans of the planet? Sanderson offers here an exhaustive study of USOs (Unidentified Submarine Objects) observed in nearly every part of the world.

298 PAGES. 6x9 PAPERBACK. ILLUSTRATED. $16.95. CODE: INVS

THE A.T. FACTOR

A Scientists Encounter with UFOs: Piece For A Jigsaw Part 3

by Leonard Cramp

British aerospace engineer Cramp began much of the scientific anti-gravity and UFO propulsion analysis back in 1955 with his landmark book *Space, Gravity & the Flying Saucer* (out-of-print and rare). His next books (available from Adventures Unlimited) *UFOs & Anti-Gravity: Piece for a Jig-Saw* and *The Cosmic Matrix: Piece for a Jig-Saw Part 2* began Cramp's in depth look into gravity control, free-energy, and the interlocking web of energy that pervades the universe. In this final book, Cramp brings to a close his detailed and controversial study of UFOs and Anti-Gravity.

324 PAGES. 6X9 PAPERBACK. ILLUSTRATED. BIBLIOGRAPHY. INDEX. $16.95. CODE: ATF

COSMIC MATRIX

Piece for a Jig-Saw, Part Two

by Leonard G. Cramp

Leonard G. Cramp, a British aerospace engineer, wrote his first book *Space Gravity and the Flying Saucer* in 1954. Cosmic Matrix is the long-awaited sequel to his 1966 book *UFOs & Anti-Gravity: Piece for a Jig-Saw.* Cramp has had a long history of examining UFO phenomena and has concluded that UFOs use the highest possible aeronautic science to move in the way they do. Cramp examines anti-gravity effects and theorizes that this super-science used by the craft—described in detail in the book—can lift mankind into a new level of technology, transportation and understanding of the universe. The book takes a close look at gravity control, time travel, and the interlocking web of energy between all planets in our solar system with Leonard's unique technical diagrams. A fantastic voyage into the present and future!

364 PAGES. 6X9 PAPERBACK. ILLUSTRATED. BIBLIOGRAPHY. $16.00. CODE: CMX

UFOS AND ANTI-GRAVITY

Piece For A Jig-Saw

by Leonard G. Cramp

Leonard G. Cramp's 1966 classic book on flying saucer propulsion and suppressed technology is a highly technical look at the UFO phenomena by a trained scientist. Cramp first introduces the idea of 'anti-gravity' and introduces us to the various theories of gravitation. He then examines the technology necessary to build a flying saucer and examines in great detail the technical aspects of such a craft. Cramp's book is a wealth of material and diagrams on flying saucers, anti-gravity, suppressed technology, G-fields and UFOs. Chapters include Crossroads of Aerodymanics, Aerodynamic Saucers, Limitations of Rocketry, Gravitation and the Ether, Gravitational Spaceships, G-Field Lift Effects, The Bi-Field Theory, VTOL and Hovercraft, Analysis of UFO photos, more.

388 PAGES. 6X9 PAPERBACK. ILLUSTRATED. $16.95. CODE: UAG

THE TESLA PAPERS

Nikola Tesla on Free Energy & Wireless Transmission of Power

by Nikola Tesla, edited by David Hatcher Childress

David Hatcher Childress takes us into the incredible world of Nikola Tesla and his amazing inventions. Tesla's rare article "The Problem of Increasing Human Energy with Special Reference to the Harnessing of the Sun's Energy" is included. This lengthy article was originally published in the June 1900 issue of *The Century Illustrated Monthly Magazine* and it was the outline for Tesla's master blueprint for the world. Tesla's fantastic vision of the future, including wireless power, anti-gravity, free energy and highly advanced solar power. Also included are some of the papers, patents and material collected on Tesla at the Colorado Springs Tesla Symposiums, including papers on: •The Secret History of Wireless Transmission •Tesla and the Magnifying Transmitter •Design and Construction of a Half-Wave Tesla Coil •Electrostatics: A Key to Free Energy •Progress in Zero-Point Energy Research •Electromagnetic Energy from Antennas to Atoms •Tesla's Particle Beam Technology •Fundamental Excitatory Modes of the Earth-Ionosphere Cavity

325 PAGES. 8X10 PAPERBACK. ILLUSTRATED. $16.95. CODE: TTP

THE FANTASTIC INVENTIONS OF NIKOLA TESLA

by Nikola Tesla with additional material by David Hatcher Childress

This book is a readable compendium of patents, diagrams, photos and explanations of the many incredible inventions of the originator of the modern era of electrification. In Tesla's own words are such topics as wireless transmission of power, death rays, and radio-controlled airships. In addition, rare material on German bases in Antarctica and South America, and a secret city built at a remote jungle site in South America by one of Tesla's students, Guglielmo Marconi. Marconi's secret group claims to have built flying saucers in the 1940s and to have gone to Mars in the early 1950s! Incredible photos of these Tesla craft are included. The Ancient Atlantean system of broadcasting energy through a grid system of obelisks and pyramids is discussed, and a fascinating concept comes out of one chapter: that Egyptian engineers had to wear protective metal head-shields while in these power plants, hence the Egyptian Pharoah's head covering as well as the Face on Mars! •His plan to transmit free electricity into the atmosphere. •How electrical devices would work using only small antennas. •Why unlimited power could be utilized anywhere on earth. •How radio and radar technology can be used as death-ray weapons in Star Wars.

342 PAGES. 6X9 PAPERBACK. ILLUSTRATED. $16.95. CODE: FINT

SECRETS OF THE MYSTERIOUS VALLEY

by Christopher O'Brien

No other region in North America features the variety and intensity of unusual phenomena found in the world's largest alpine valley, the San Luis Valley of Colorado and New Mexico. Since 1989, Christopher O'Brien has documented thousands of high-strange accounts that report UFOs, ghosts, crypto-creatures, cattle mutilations, skinwalkers and sorcerers, along with portal areas, secret underground bases and covert military activity. This mysterious region at the top of North America has a higher incidence of UFO reports than any other area of the continent and is the publicized birthplace of the "cattle mutilation" mystery. Hundreds of animals have been found strangely slain during waves of anomalous aerial craft sightings. Is the government directly involved? Are there underground bases here? Does the military fly exotic aerial craft in this valley that are radar-invisible below 18,000 feet? These and many other questions are addressed in this all-new, work by one of America's top paranormal investigators. Take a fantastic journey through one of the world's most enigmatic locales!

460 pages. 6x9 Paperback. Illustrated. Bibliography. $19.95. Code: SOMV

THE ENERGY GRID

Harmonic 695, The Pulse of the Universe

by Captain Bruce Cathie.

This is the breakthrough book that explores the incredible potential of the Energy Grid and the Earth's Unified Field all around us. Cathie's first book, *Harmonic 33*, was published in 1968 when he was a commercial pilot in New Zealand. Since then, Captain Bruce Cathie has been the premier investigator into the amazing potential of the infinite energy that surrounds our planet every microsecond. Cathie investigates the Harmonics of Light and how the Energy Grid is created. In this amazing book are chapters on UFO Propulsion, Nikola Tesla, Unified Equations, the Mysterious Aerials, Pythagoras & the Grid, Nuclear Detonation and the Grid, Maps of the Ancients, an Australian Stonehenge examined, more.

255 PAGES. 6x9 TRADEPAPER. ILLUSTRATED. $15.95. CODE: TEG

LEY LINE & EARTH ENERGIES

An Extraordinary Journey into the Earth's Natural Energy System

by David Cowan & Chris Arnold

The mysterious standing stones, burial grounds and stone circles that lace Europe, the British Isles and other areas have intrigued scientists, writers, artists and travellers through the centuries. How do ley lines work? How did our ancestors use Earth energy to map their sacred sites and burial grounds? How do ghosts and poltergeists interact with Earth energy? How can Earth spirals and black spots affect our health? This exploration shows how natural forces affect our behavior, how they can be used to enhance our health and well being. A fascinating and visual book about subtle Earth energies and how they affect us and the world around them.

368 PAGES. 6x9 PAPERBACK. ILLUSTRATED. BIBLIOGRAPHY. $18.95. CODE: LLEE

QUEST FOR ZERO-POINT ENERGY

Engineering Principles for "Free Energy"

by Moray B. King

King expands, with diagrams, on how free energy and anti-gravity are possible. The theories of zero point energy maintain there are tremendous fluctuations of electrical field energy embedded within the fabric of space. King explains the following topics: Tapping the Zero-Point Energy as an Energy Source; Fundamentals of a Zero-Point Energy Technology; Vacuum Energy Vortices; The Super Tube; Charge Clusters: The Basis of Zero-Point Energy Inventions; Vortex Filaments, Torsion Fields and the Zero-Point Energy; Transforming the Planet with a Zero-Point Energy Experiment; Dual Vortex Forms: The Key to a Large Zero-Point Energy Coherence. Packed with diagrams, patents and photos. With power shortages now a daily reality in many parts of the world, this book offers a fresh approach very rarely mentioned in the mainstream media.

224 PAGES. 6x9 PAPERBACK. ILLUSTRATED. $14.95. CODE: QZPE

TAPPING THE ZERO POINT ENERGY

Free Energy & Anti-Gravity in Today's Physics

by Moray B. King

King explains how free energy and anti-gravity are possible. The theories of the zero point energy maintain there are tremendous fluctuations of electrical field energy imbedded within the fabric of space. This book tells how, in the 1930s, inventor T. Henry Moray could produce a fifty kilowatt "free energy" machine; how an electrified plasma vortex creates anti-gravity; how the Pons/Fleischmann "cold fusion" experiment could produce tremendous heat without fusion; and how certain experiments might produce a gravitational anomaly.

180 PAGES. 5x8 PAPERBACK. ILLUSTRATED. $12.95. CODE: TAP

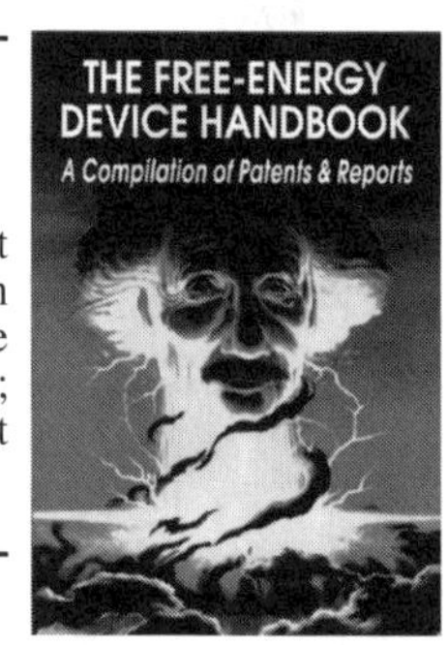

THE FREE-ENERGY DEVICE HANDBOOK

A Compilation of Patents and Reports

by David Hatcher Childress

A large-format compilation of various patents, papers, descriptions and diagrams concerning free-energy devices and systems. *The Free-Energy Device Handbook* is a visual tool for experimenters and researchers into magnetic motors and other "over-unity" devices. With chapters on the Adams Motor, the Hans Coler Generator, cold fusion, superconductors, "N" machines, space-energy generators, Nikola Tesla, T. Townsend Brown, and the latest in free-energy devices.

292 PAGES. 8x10 PAPERBACK. ILLUSTRATED. BIBLIOGRAPHY. $16.95. CODE: FEH

THE ANTI-GRAVITY HANDBOOK
edited by David Hatcher Childress, with Nikola Tesla, T.B. Paulicki, Bruce Cathie, Albert Einstein and others

The new expanded compilation of material on Anti-Gravity, Free Energy, Flying Saucer Propulsion, UFOs, Suppressed Technology, NASA Cover-ups and more. Highly illustrated with patents, technical illustrations and photos. This revised and expanded edition has more material, including photos of Area 51, Nevada, the government's secret testing facility. This classic on weird science is back in a 90s format!

- **How to build a flying saucer.**
- **Arthur C. Clarke on Anti-Gravity.**
- **Crystals and their role in levitation.**
- **Secret government research and development.**

230 PAGES. 7X10 PAPERBACK. ILLUSTRATED. $16.95. CODE: AGH

ANTI–GRAVITY & THE WORLD GRID

Is the earth surrounded by an intricate electromagnetic grid network offering free energy? This compilation of material on ley lines and world power points contains chapters on the geography, mathematics, and light harmonics of the earth grid. Learn the purpose of ley lines and ancient megalithic structures located on the grid. Discover how the grid made the Philadelphia Experiment possible. Explore the Coral Castle and many other mysteries, including acoustic levitation, Tesla Shields and scalar wave weaponry. Browse through the section on anti-gravity patents, and research resources.

274 PAGES. 7X10 PAPERBACK. ILLUSTRATED. $14.95. CODE: AGW

ANTI–GRAVITY & THE UNIFIED FIELD
edited by David Hatcher Childress

Is Einstein's Unified Field Theory the answer to all of our energy problems? Explored in this compilation of material is how gravity, electricity and magnetism manifest from a unified field around us. Why artificial gravity is possible; secrets of UFO propulsion; free energy; Nikola Tesla and anti-gravity airships of the 20s and 30s; flying saucers as superconducting whirls of plasma; anti-mass generators; vortex propulsion; suppressed technology; government cover-ups; gravitational pulse drive; spacecraft & more.

240 PAGES. 7X10 PAPERBACK. ILLUSTRATED. $14.95. CODE: AGU

THE MYSTERY OF THE OLMECS
by David Hatcher Childress

Lost Cities author Childress takes us deep into Mexico and Central America in search of the mysterious Olmecs, North America's early, advanced civilization. The Olmecs, now sometimes called Proto-Mayans, were not acknowledged to have existed as a civilization until an international archeological meeting in Mexico City in 1942. At this time, the megalithic statues, large structures, ceramics and other artifacts were acknowledged to come from this hitherto unknown culture that pre-dated all other cultures of Central America. But who were the Olmecs? Where did they come from? What happened to them? How sophisticated was their culture? How far back in time did it go? Why are many Olmec statues and figurines seemingly of foreign peoples such as Africans, Europeans and Chinese? Is there a link with Atlantis? In this heavily illustrated book, join Childress in search of the lost cites of the Olmecs!

432 Pages. 6x9 Paperback. Illustrated. Bibliography. $20.00. Code: MOLM

PATH OF THE POLE
Cataclysmic Pole Shift Geology
by Charles H. Hapgood

Maps of the Ancient Sea Kings author Hapgood's classic book *Path of the Pole* is back in print! Hapgood researched Antarctica, ancient maps and the geological record to conclude that the Earth's crust has slipped on the inner core many times in the past, changing the position of the pole. *Path of the Pole* discusses the various "pole shifts" in Earth's past, giving evidence for each one, and moves on to possible future pole shifts. Packed with illustrations, this is the sourcebook for many other books on cataclysms and pole shifts.

356 PAGES. 6X9 PAPERBACK. ILLUSTRATED. $16.95. CODE: POP

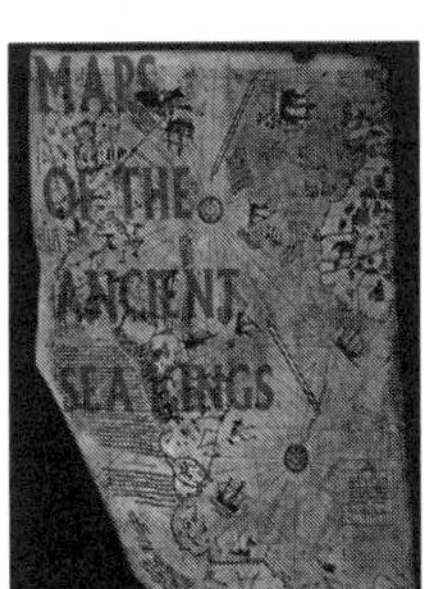

MAPS OF THE ANCIENT SEA KINGS
Evidence of Advanced Civilization in the Ice Age
by Charles H. Hapgood

Charles Hapgood's classic 1966 book on ancient maps produces concrete evidence of an advanced world-wide civilization existing many thousands of years before ancient Egypt. He has found the evidence in the Piri Reis Map that shows Antarctica, the Hadji Ahmed map, the Oronteus Finaeus and other amazing maps. Hapgood concluded that these maps were made from more ancient maps from the various ancient archives around the world, now lost. Not only were these unknown people more advanced in mapmaking than any people prior to the 18th century, it appears they mapped all the continents. The Americas were mapped thousands of years before Columbus. Antarctica was mapped when its coasts were free of ice!

316 PAGES. 7X10 PAPERBACK. ILLUSTRATED. BIBLIOGRAPHY & INDEX. $19.95. CODE: MASK

THE LAND OF OSIRIS

An Introduction to Khemitology

by Stephen S. Mehler

Was there an advanced prehistoric civilization in ancient Egypt who built the great pyramids and carved the Great Sphinx? Did the pyramids serve as energy devices and not as tombs for kings? Mehler has uncovered an indigenous oral tradition that still exists in Egypt, and has been fortunate to have studied with a living master of this tradition, Abd'El Hakim Awyan. Mehler has also been given permission to present these teachings to the Western world, teachings that unfold a whole new understanding of ancient Egypt . Chapters include: Egyptology and Its Paradigms; Asgat Nefer—The Harmony of Water; Khemit and the Myth of Atlantis; The Extraterrestrial Question; more.

272 PAGES. 6X9 PAPERBACK. ILLUSTRATED. COLOR SECTION. BIBLIOGRAPHY. $18.00 CODE: LOOS

REICH OF THE BLACK SUN

Nazi Secret Weapons and the Cold War Allied Legend

by Joseph P. Farrell

Why were the Allies worried about an atom bomb attack by the Germans in 1944? Why did the Soviets threaten to use poison gas against the Germans? Why did Hitler in 1945 insist that holding Prague could win the war for the Third Reich? Why did US General George Patton's Third Army race for the Skoda works at Pilsen in Czechoslovakia instead of Berlin? Why did the US Army not test the uranium atom bomb it dropped on Hiroshima? Why did the Luftwaffe fly a non-stop round trip mission to within twenty miles of New York City in 1944? *Reich of the Black Sun* takes the reader on a scientific-historical journey in order to answer these questions. Arguing that Nazi Germany actually won the race for the atom bomb in late 1944, *Reich of the Black Sun* then goes on to explore the even more secretive research the Nazis were conducting into the occult, alternative physics and new energy sources. The book concludes with a fresh look at the "Nazi Legend" of the UFO mystery by examining the Roswell Majestic-12 documents and the Kecksburg crash in the light of parallels with some of the super-secret black projects being run by the SS. *Reich of the Black Sun* is must-reading for the researcher interested in alternative history, science, or UFOs!

352 PAGES. 6X9 PAPERBACK. ILLUSTRATED. BIBLIOGRAPHY. $16.95. CODE: ROBS

THE GIZA DEATH STAR

The Paleophysics of the Great Pyramid & the Military Complex at Giza

by Joseph P. Farrell

Physicist Joseph Farrell's amazing book on the secrets of Great Pyramid of Giza. *The Giza Death Star* starts where British engineer Christopher Dunn leaves off in his 1998 book, *The Giza Power Plant*. Was the Giza complex part of a military installation over 10,000 years ago? Chapters include: An Archaeology of Mass Destruction, Thoth and Theories; The Machine Hypothesis; Pythagoras, Plato, Planck, and the Pyramid; The Weapon Hypothesis; Encoded Harmonics of the Planck Units in the Great Pyramid; High Fregquency Direct Current "Impulse" Technology; The Grand Gallery and its Crystals: Gravito-acoustic Resonators; The Other Two Large Pyramids; the "Causeways," and the "Temples"; A Phase Conjugate Howitzer; Evidence of the Use of Weapons of Mass Destruction in Ancient Times; more.

290 PAGES. 6X9 PAPERBACK. ILLUSTRATED. $16.95. CODE: GDS

THE GIZA DEATH STAR DEPLOYED

The Physics & Engineering of the Great Pyramid

by Joseph P. Farrell

Farrell expands on his thesis that the Great Pyramid was a chemical maser, designed as a weapon and eventually deployed—with disastrous results to the solar system. Includes: Exploding Planets: The Movie, the Mirror, and the Model; Dating the Catastrophe and the Compound; A Brief History of the Exoteric and Esoteric Investigations of the Great Pyramid; No Machines, Please!; The Stargate Conspiracy; The Scalar Weapons; Message or Machine?; A Tesla Analysis of the Putative Physics and Engineering of the Giza Death Star; Cohering the Zero Point, Vacuum Energy, Flux: Synopsis of Scalar Physics and Paleophysics; Configuring the Scalar Pulse Wave; Inferred Applications in the Great Pyramid; Quantum Numerology, Feedback Loops and Tetrahedral Physics; and more.

290 PAGES. 6X9 PAPERBACK. ILLUSTRATED. BIBLIOGRAPHY. $16.95. CODE: GDSD

THE GIZA DEATH STAR DESTROYED

The Ancient War For Future Science

by Joseph P. Farrell

Recapping his earlier books, Farrell moves on to events of the final days of the Giza Death Star and its awesome power. These final events, eventually leading up to the destruction of this giant machine, are dissected one by one, leading us to the eventual abandonment of the Giza Military Complex—an event that hurled civilization back into the Stone Age. Chapters include: The Mars-Earth Connection; The Lost "Root Races" and the Moral Reasons for the Flood; The Destruction of Krypton: The Electrodynamic Solar System, Exploding Planets and Ancient Wars; Turning the Stream of the Flood: the Origin of Secret Societies and Esoteric Traditions; The Quest to Recover Ancient Mega-Technology; Non-Equilibrium Paleophysics; Monatomic Paleophysics; Frequencies, Vortices and Mass Particles: The Topology of the Aether; "Acoustic" Intensity of Fields; The Pyramid of Crystals; tons more.

292 pages. 6x9 paperback. Illustrated. $16.95. Code: GDES

TECHNOLOGY OF THE GODS
The Incredible Sciences of the Ancients
by David Hatcher Childress

Popular *Lost Cities* author David Hatcher Childress takes us into the amazing world of ancient technology, from computers in antiquity to the "flying machines of the gods." Childress looks at the technology that was allegedly used in Atlantis and the theory that the Great Pyramid of Egypt was originally a gigantic power station. He examines tales of ancient flight and the technology that it involved; how the ancients used electricity; megalithic building techniques; the use of crystal lenses and the fire from the gods; evidence of various high tech weapons in the past, including atomic weapons; ancient metallurgy and heavy machinery; the role of modern inventors such as Nikola Tesla in bringing ancient technology back into modern use; impossible artifacts; and more.

356 PAGES. 6x9 PAPERBACK. ILLUSTRATED. BIBLIOGRAPHY. $16.95. CODE: TGOD

VIMANA AIRCRAFT OF ANCIENT INDIA & ATLANTIS
by David Hatcher Childress, introduction by Ivan T. Sanderson

Did the ancients have the technology of flight? In this incredible volume on ancient India, authentic Indian texts such as the *Ramayana* and the *Mahabharata* are used to prove that ancient aircraft were in use more than four thousand years ago. Included in this book is the entire Fourth Century BC manuscript *Vimaanika Shastra* by the ancient author Maharishi Bharadwaaja, translated into English by the Mysore Sanskrit professor G.R. Josyer. Also included are chapters on Atlantean technology, the incredible Rama Empire of India and the devastating wars that destroyed it. Also an entire chapter on mercury vortex propulsion and mercury gyros, the power source described in the ancient Indian texts. Not to be missed by those interested in ancient civilizations or the UFO enigma.

334 PAGES. 6x9 PAPERBACK. ILLUSTRATED. $15.95. CODE: VAA

LOST CONTINENTS & THE HOLLOW EARTH
I Remember Lemuria and the Shaver Mystery
by David Hatcher Childress & Richard Shaver

Lost Continents & the Hollow Earth is Childress' thorough examination of the early hollow earth stories of Richard Shaver and the fascination that fringe fantasy subjects such as lost continents and the hollow earth have had for the American public. Shaver's rare 1948 book *I Remember Lemuria* is reprinted in its entirety, and the book is packed with illustrations from Ray Palmer's *Amazing Stories* magazine of the 1940s. Palmer and Shaver told of tunnels running through the earth—tunnels inhabited by the Deros and Teros, humanoids from an ancient spacefaring race that had inhabited the earth, eventually going underground, hundreds of thousands of years ago. Childress discusses the famous hollow earth books and delves deep into whatever reality may be behind the stories of tunnels in the earth. Operation High Jump to Antarctica in 1947 and Admiral Byrd's bizarre statements, tunnel systems in South America and Tibet, the underground world of Agartha, the belief of UFOs coming from the South Pole, more.

344 PAGES. 6x9 PAPERBACK. ILLUSTRATED. $16.95. CODE: LCHE

ATLANTIS & THE POWER SYSTEM OF THE GODS
Mercury Vortex Generators & the Power System of Atlantis
by David Hatcher Childress and Bill Clendenon

Atlantis and the Power System of the Gods starts with a reprinting of the rare 1990 book *Mercury: UFO Messenger of the Gods* by Bill Clendenon. Clendenon takes on an unusual voyage into the world of ancient flying vehicles, strange personal UFO sightings, a meeting with a "Man In Black" and then to a centuries-old library in India where he got his ideas for the diagrams of mercury vortex engines. The second part of the book is Childress' fascinating analysis of Nikola Tesla's broadcast system in light of Edgar Cayce's "Terrible Crystal" and the obelisks of ancient Egypt and Ethiopia. Includes: Atlantis and its crystal power towers that broadcast energy; how these incredible power stations may still exist today; inventor Nikola Tesla's nearly identical system of power transmission; Mercury Proton Gyros and mercury vortex propulsion; more. Richly illustrated, and packed with evidence that Atlantis not only existed—it had a world-wide energy system more sophisticated than ours today.

246 PAGES. 6x9 PAPERBACK. ILLUSTRATED. $15.95. CODE: APSG

A HITCHHIKER'S GUIDE TO ARMAGEDDON
by David Hatcher Childress

With wit and humor, popular Lost Cities author David Hatcher Childress takes us around the world and back in his trippy finalé to the Lost Cities series. He's off on an adventure in search of the apocalypse and end times. Childress hits the road from the fortress of Megiddo, the legendary citadel in northern Israel where Armageddon is prophesied to start. Hitchhiking around the world, Childress takes us from one adventure to another, to ancient cities in the deserts and the legends of worlds before our own. Childress muses on the rise and fall of civilizations, and the forces that have shaped mankind over the millennia, including wars, invasions and cataclysms. He discusses the ancient Armageddons of the past, and chronicles recent Middle East developments and their ominous undertones. In the meantime, he becomes a cargo cult god on a remote island off New Guinea, gets dragged into the Kennedy Assassination by one of the "conspirators," investigates a strange power operating out of the Altai Mountains of Mongolia, and discovers how the Knights Templar and their off-shoots have driven the world toward an epic battle centered around Jerusalem and the Middle East.

320 PAGES. 6x9 PAPERBACK. ILLUSTRATED. BIBLIOGRAPHY. INDEX. $16.95. CODE: HGA

ORDER FORM

10% Discount When You Order 3 or More Items!

One Adventure Place
P.O. Box 74
Kempton, Illinois 60946
United States of America
Tel.: 815-253-6390 • Fax: 815-253-6300
Email: auphq@frontiernet.net
http://www.adventuresunlimitedpress.com

ORDERING INSTRUCTIONS

- ✓ Remit by USD$ Check, Money Order or Credit Card
- ✓ Visa, Master Card, Discover & AmEx Accepted
- ✓ Paypal Payments Can Be Made To: info@wexclub.com
- ✓ Prices May Change Without Notice
- ✓ 10% Discount for 3 or more Items

SHIPPING CHARGES

United States

- ✓ Postal Book Rate { $4.00 First Item; 50¢ Each Additional Item
- ✓ POSTAL BOOK RATE Cannot Be Tracked!
- ✓ Priority Mail { $5.00 First Item; $2.00 Each Additional Item
- ✓ UPS { $6.00 First Item; $1.50 Each Additional Item

NOTE: UPS Delivery Available to Mainland USA Only

Canada

- ✓ Postal Air Mail { $10.00 First Item; $2.50 Each Additional Item
- ✓ Personal Checks or Bank Drafts MUST BE US$ and Drawn on a US Bank
- ✓ Canadian Postal Money Orders OK
- ✓ Payment MUST BE US$

All Other Countries

- ✓ Sorry, No Surface Delivery!
- ✓ Postal Air Mail { $16.00 First Item; $6.00 Each Additional Item
- ✓ Checks and Money Orders MUST BE US$ and Drawn on a US Bank or branch.
- ✓ Paypal Payments Can Be Made in US$ To: info@wexclub.com

SPECIAL NOTES

- ✓ RETAILERS: Standard Discounts Available
- ✓ BACKORDERS: We Backorder all Out-of-Stock Items Unless Otherwise Requested
- ✓ PRO FORMA INVOICES: Available on Request

ORDER ONLINE AT: www.adventuresunlimitedpress.com

Please check: ☑

☐ This is my first order ☐ I have ordered before

Name
Address
City
State/Province | Postal Code
Country
Phone day | Evening
Fax | Email

Item Code	Item Description	Qty	Total

Please check: ☑

☐ Postal-Surface
☐ Postal-Air Mail (Priority in USA)
☐ UPS (Mainland USA only)

Subtotal ▸
Less Discount-10% for 3 or more items ▸
Balance ▸
Illinois Residents 6.25% Sales Tax ▸
Previous Credit ▸
Shipping ▸
Total (check/MO in USD$ only) ▸

☐ Visa/MasterCard/Discover/American Express

Card Number
Expiration Date

10% Discount When You Order 3 or More Items!